Introductory Algebra Programmed

Introductory Algebra
Programmed
Second edition

THOMAS J. McHALE
ALLAN A. CHRISTENSON
KEITH J. ROBERTS

ADDISON-WESLEY PUBLISHING COMPANY

Reading, Massachusetts • Menlo Park, California
Don Mills, Ontario • Wokingham, England • Amsterdam
Sydney • Singapore • Tokyo • Mexico City
Bogotá • Santiago • San Juan

Library of Congress Cataloging-in-Publication Data

McHale, Thomas J., 1931-
Introductory algebra, programmed.

Rev. ed. of: Introductory algebra. c1977.
Includes index.
1. Algebra--Programmed instruction. I. Christenson, Allan A. II. Roberts, Keith J. III. McHale, Thomas J., 1931- . Introductory algebra. IV. Title.
QA152.2.M377 1986 512.9 85-22848
ISBN 0-201-15887-6

Reproduced by Addison-Wesley from camera-ready copy supplied by the author.

ISBN 0-201-15887-6
ABCDEFGHIJ-AL-898765

Preface

Introductory Algebra is written for a first course in algebra for college students with little or no knowledge of algebra. It uses a unique approach to deal with the learning difficulties experienced by most students of that type. That is, the content is presented in an interactive format based on a learning task analysis, and the learning is supported by a set of diagnostic and criterion tests. This unique approach, which differs significantly from that of standard texts, has been used successfully with almost 500,000 students in similar courses.

LEARNING TASK ANALYSIS

A learning task analysis was done by the authors to determine how to present the content of the course. That is, each course objective was analyzed to identify the skills, concepts, and procedures needed by the students to master all of the problems covered by that objective. Then based on that analysis, the content of each objective is presented by a carefully planned sequence of worked-out examples proceeding from the simple to the complex. Worked-out examples are given for the full range of problems covered by each objective. In that sense, there are no "gaps" in the instruction from the student's point of view. Because a learning task analysis was used, this book has many more worked-out examples than comparable texts.

INTERACTIVE FORMAT

The format of this book is unique because it forces the students to interact with the content as it is presented. The significant features of the format are described below.

Frames. The content is presented in a step-by-step manner with each step called a frame. Each frame begins with some instruction, including one or more examples, and ends with some problem or problems that the student must do. Since answers for each frame are given to the right of the following frame, the students are given immediate feedback as they proceed from frame to frame. Because they are continually responding and getting feedback, the students can monitor their own learning process.

Assignment Self-Tests. Each chapter is divided into a number of assignments which can ordinarily be covered in one class period. A self-test is provided in the text at the end of each assignment. Students can either use the self-tests as assignment tests or save them for a chapter review after all assignments in the chapter are completed.

Supplementary Problems. Supplementary problems for each assignment are provided at the end of each chapter. They can be assigned as needed by instructors or simply used as further practice by students. Answers for all supplementary problems are given in the back of the book.

DIAGNOSTIC AND CRITERION TESTS

The text is accompanied by the book Tests For Introductory Algebra which contains the various diagnostic and criterion tests described below. Answer keys are provided for all tests. The test book is provided only to instructors. Copies for student use must be made by some copying process.

Diagnostic Pre-Test. This test covers all chapters. It can be used either to simply get a measure of the entry skills of the students or as a basis for prescribing an individualized program.

Assignment Tests. Each chapter is divided into a number of assignments with a total of thirty-eight assignments in the complete text. After the students have completed each assignment and the assignment self-test (in the text), the assignment test (from the test book) can be administered and used as a basis for tutoring or assigning supplementary problems. The assignment tests are simply a diagnostic tool and need not be graded.

Chapter And Multi-Chapter Tests. After the appropriate assignments are completed, either a chapter test or a multi-chapter test can be administered. Ordinarily these tests should be graded. Three parallel forms are provided to facilitate the test administration, including the retesting of students who do not achieve a satisfactory score.

Comprehensive Test. The comprehensive test covers all chapters. Three parallel forms are provided. Since the comprehensive test is a parallel form of the diagnostic pre-test, the difference score can be used as a measure of each student's improvement in the course.

TEACHING MODES

This text can be used in various ways. Some possibilities are described below.

Lecture Class. The text can be used to reinforce lectures and to provide highly structured outside assignments related to the lectures.

Mini-Lecture Class. An instructor can give a brief lecture on difficult points in a completed assignment before the assignment test is administered. An instructor can also give a brief overview of each new assignment before it is begun by students.

No-Lecture Class. The text is ideally suited for a no-lecture class that is either paced or self-paced. Class time can then be used to administer tests and to tutor individual students when tutoring is needed.

Learning Laboratory. The text is also ideally suited for a learning laboratory where students proceed at their own pace. The instructor can manage the instruction by administering tests and tutoring when necessary.

ACKNOWLEDGMENTS

The authors wish to thank Jeffrey M. Pepper of the Addison-Wesley Publishing Company for many suggestions that improved this text. They also thank the following people who reviewed the manuscript: Ron Barnes, University of Houston-Downtown, Kenneth T. Becker, Inver Hills Community College, Robert A. Bilot, Northeast Wisconsin Technical Institute, Donald Buchanan, Jefferson Technical College, Bettyann Daley, University of Delaware, Carol A. McVey, Florence-Darlington Technical College, Bahman Moezzi, Monterey Peninsula College, David E. Stevens, Wentworth Institute of Technology, Jana D. Turner, State Technical Institute of Memphis, and Stephen H. Turner, Jefferson Technical College. They also thank Judy Christenson who typed the rough draft, Arleen D'Amore who typed the camera-ready copy, Peggy McHale who prepared the drawings and made the corrections, and Gail W. Davis who did the final proofreading.

ASSIGNMENTS FOR INTRODUCTORY ALGEBRA

Chapter	Assignment	Pages
Chapter 1:	#1	pp. 1-17
	#2	pp. 18-28
	#3	pp. 29-45
Chapter 2:	#4	pp. 49-61
	#5	pp. 62-71
	#6	pp. 72-86
	#7	pp. 87-99
Chapter 3:	#8	pp. 104-116
	#9	pp. 117-127
	#10	pp. 128-138
	#11	pp. 139-153
Chapter 4:	#12	pp. 157-169
	#13	pp. 170-180
	#14	pp. 180-195
Chapter 5:	#15	pp. 199-213
	#16	pp. 214-223
	#17	pp. 224-234
Chapter 6:	#18	pp. 237-246
	#19	pp. 247-259
	#20	pp. 260-273
Chapter 7:	#21	pp. 276-288
	#22	pp. 289-303
	#23	pp. 304-319
	#24	pp. 320-337
Chapter 8:	#25	pp. 342-358
	#26	pp. 359-371
	#27	pp. 372-383
Chapter 9:	#28	pp. 388-399
	#29	pp. 400-408
	#30	pp. 409-417
Chapter 10:	#31	pp. 420-433
	#32	pp. 434-446
	#33	pp. 447-458
Chapter 11:	#34	pp. 461-472
	#35	pp. 473-484
	#36	pp. 485-496
Chapter 12:	#37	pp. 500-512
	#38	pp. 513-523

Contents

CHAPTER 3: SOLVING EQUATIONS (Pages 104-156)

CHAPTER 4: WORD PROBLEMS (Pages 157-198)

CHAPTER 5: POLYNOMIALS (Pages 199-236)

CHAPTER 8: GRAPHING LINEAR EQUATIONS (Pages 342-387)

CHAPTER 9: SYSTEMS OF EQUATIONS (Pages 388-419)

CHAPTER 10: RADICAL EXPRESSIONS (Pages 420-460)

Integers

1

The purpose of this chapter is to review some topics that you have probably seen before and to introduce you to algebra. We will define integers, discuss the basic operations with integers, and define powers with positive integer exponents. We will discuss the proper order of operations for expressions involving more than one operation. We will also define algebraic expressions and translate English phrases to algebraic expressions.

1-1 INTEGERS AND THE NUMBER LINE

In this section, we will define integers and show them on the number line. We will also define is greater than, is less than, and absolute value.

1. The numbers used for counting are called natural numbers. They are: 1, 2, 3, 4, 5, 6, 7, 8, 9, 10, and so on. The whole numbers include the natural numbers and zero. They are: 0, 1, 2, 3, 4, 5, 6, 7, 8, 9, 10, and so on. Which number is a whole number, but not a natural number? ______	
	0 (or zero)

2. The whole numbers are shown on a number line below. They are equally spaced to the right of 0. The arrow means that other whole numbers continue to the right of 7.

A "-" in front of a number is used to show that it goes to the left of 0. Therefore, numbers like -1, -2, -3, and so on, are placed on the number line as shown below.

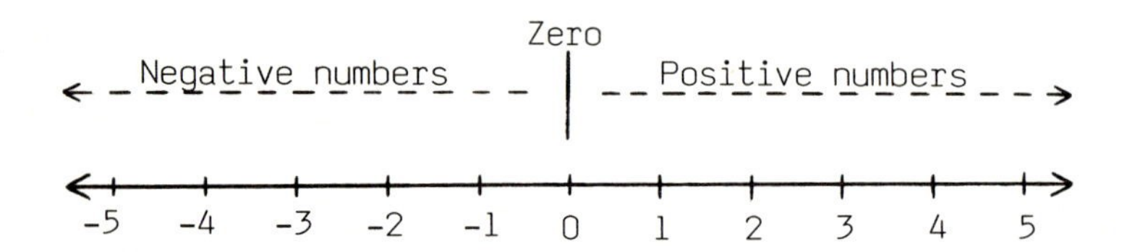

Note: 1. The numbers to the right of 0 are positive numbers.
2. The numbers to the left of 0 are negative numbers.
3. The number 0 is neither positive nor negative.

a) If we put 8 on a number line, it would go to the ____________ (right/left) of 0.

b) If we put -7 on a number line, it would go to the ____________ (right/left) of 0.

Answers: a) right b) left

---

3. Positive and negative numbers are called signed numbers. The "+" sign and "-" sign are used with signed numbers.

For negative numbers, we always use the "-" sign.

"Negative 3" is always written -3.

For positive numbers, we can use the "+" sign or no sign, but we usually use no sign.

"Positive 5" is written 5 instead of +5.

a) Instead of writing +9 for "positive 9", we usually write ______.

b) Would it make sense to write either +0 or -0? ______

Answers: a) 9 b) No. The number 0 is not a signed number.

4. Signed numbers are useful for stating things like gains or losses, temperatures above or below 0°, and distances above or below sea level. For example:

$10 is used for a gain of $10.

-$10 is used for a loss of $10.

15° means 15° above zero.

-15° means 15° below zero.

100 feet means 100 feet above sea level.

-100 feet means 100 feet below sea level.

Write each of these as a signed number.

a) a loss of $25 ______ c) 250 feet below sea level ______

b) 20° above zero ______ d) 40° below zero ______

---

a) -$25

b) 20°

c) -250 feet

d) -40°

5. The numbers shown on the number line below are called integers. Integers include all whole numbers plus negative numbers like -1, -2, -3, and so on.

-5 -4 -3 -2 -1 0 1 2 3 4 5

a) Is 4 both a whole number and an integer? ______

b) Is 0 both a whole number and an integer? ______

c) Is -5 both a whole number and an integer? ______

---

a) Yes

b) Yes

c) No. -5 is an integer, but not a whole number.

6. The two symbols, = and ≠, are used to say that two numbers are either "equal" or "not equal". That is:

= means "is equal to".
≠ means "is not equal to".

Therefore: instead of "5 is equal to 5", we write $5 = 5$.

instead of "4 is not equal to 7", we write ____________.

$4 \neq 7$

7. When two numbers are not equal, we frequently say that one "is greater than" the other or that one "is less than" the other. The two symbols, $>$ and $<$, are used for "is greater than" and "is less than". That is:

$>$ means "is greater than".
$<$ means "is less than".

Therefore: instead of "9 is greater than 4", we write $9 > 4$.

instead of "2 is less than 6", we write ____________.

---

8. To keep the symbols $>$ and $<$ straight, remember that the symbol always points to the smaller number. For example:

In $5 > 3$, the $>$ points to the 3.

In $4 < 7$, the $<$ points to the 4.

Write either $>$ or $<$ in each blank.

a) 6 ____ 9   b) 10 ____ 4   c) 1 ____ 5

$2 < 6$

---

9. The number line below contains integers.

-5 -4 -3 -2 -1 0 1 2 3 4 5

The definition of "is greater than" for integers is:

A first integer is greater than a second integer if the first is located to the right of the second on the number line.

Therefore: since 4 is to the right of -1, $4 > -1$.

since 0 is to the right of -3, $0 > -3$.

since -2 is to the right of -5, ____________ .

a) $6 < 9$

b) $10 > 4$

c) $1 < 5$

$-2 > -5$

10. The definition of "is less than" for integers is:

> A first integer is less than a second integer if the first is located to the left of the second on the number line.

Therefore: since -2 is to the left of 3, $-2 < 3$.

since -1 is to the left of 0, $-1 < 0$.

since -5 is to the left of -3, __________.

$-5 < -3$

11. Write either $>$ or $<$ in each blank.

a) 7 ____ -4 c) -8 ____ 9

b) -6 ____ 0 d) 0 ____ -7

---

a) $7 > -4$ c) $-8 < 9$

b) $-6 < 0$ d) $0 > -7$

12. Write either $>$ or $<$ in each blank.

a) -6 ____ -2 c) -7 ____ -8

b) -4 ____ -9 d) -9 ____ -3

---

a) $-6 < -2$

b) $-4 > -9$

c) $-7 > -8$

d) $-9 < -3$

13. A number line with integers is shown below.

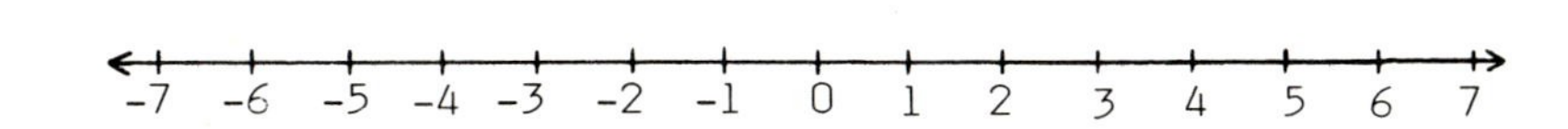

The absolute value of an integer is its distance from 0 on the number line.

Since 4 is 4 units from 0, the absolute value of 4 is 4.

Since -7 is 7 units from 0, the absolute value of -7 is 7.

The symbol $| \ |$ is used for absolute value. That is:

$|4|$ means "the absolute value of 4".

$|-7|$ means "the absolute value of -7".

Therefore: $|4| = 4$ $|-7| = 7$ $|-2| =$ ______

2

14. Since distance is a measurement that is never negative, the absolute value of a number is never negative. Therefore, pairs of integers that are the same distance from 0 on the number line have the same absolute value.

Both 3 and -3 have the same absolute value.

$|3| = 3$ and $|-3| = 3$

Both 9 and -9 have the same absolute value.

$|9| = 9$ and $|-9| = 9$

Name the pair of integers whose absolute value is 6.

______ and ______

---

15. Since 0 is 0 units from 0 on the number line, the absolute value of 0 is 0.

Answer to 14: 6 and -6

Complete: a) $|-10| =$ ______ c) $|0| =$ ______

b) $|27| =$ ______ d) $|-45| =$ ______

Answer to 15: a) 10 c) 0

b) 27 d) 45

1-2 ADDITION OF INTEGERS

In this section, we will use the number line to show the rules for the addition of integers.

---

16. In an addition, the numbers added are called the terms; the answer is called the sum. For example:

In $3 + 2 = 5$: 3 and 2 are the terms.

5 is the sum.

Continued on following page.

16. Continued

We did 3 + 2 = 5 on the number line below. Notice that the arrows for positive terms go to the right. The arrow for 3 goes 3-units to the right; the arrow for 2 goes 2-units to the right.

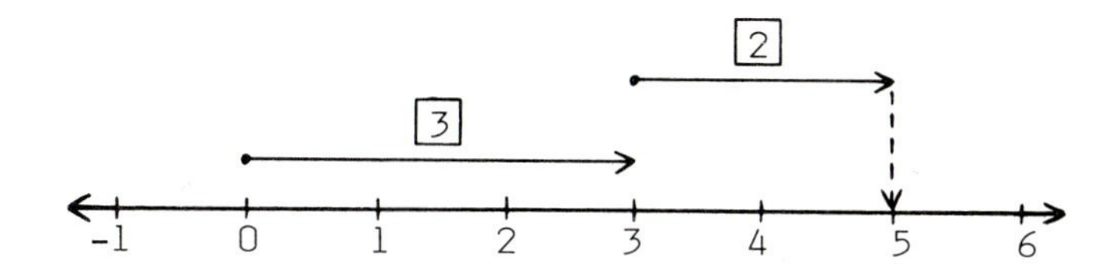

From the example, you can see the rule for adding two positive terms.

To add two positive terms, add their absolute values. The sign of the sum is positive.

Use the rule for these:

a) 8 + 6 = ______ b) 40 + 30 = ______

---

17. We did -2 + (-4) = -6 on the number line below. Notice that the arrows for negative terms go to the left. The arrow for -2 goes 2-units to the left; the arrow for -4 goes 4-units to the left.

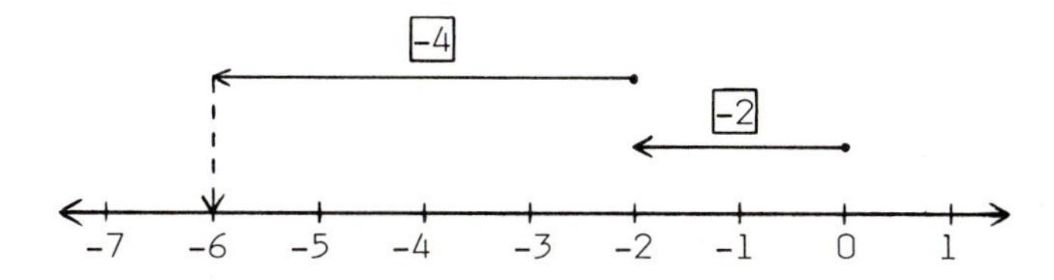

From the example, you can see the rule for adding two negative terms.

To add two negative terms, add their absolute values. The sign of the sum is negative.

Use the rule for these:

a) -6 + (-10) = ______ b) -50 + (-40) = ______

a) 14 b) 70

a) -16 b) -90

18. We did $5 + (-2) = 3$ and $-6 + 4 = -2$ on the number lines below.

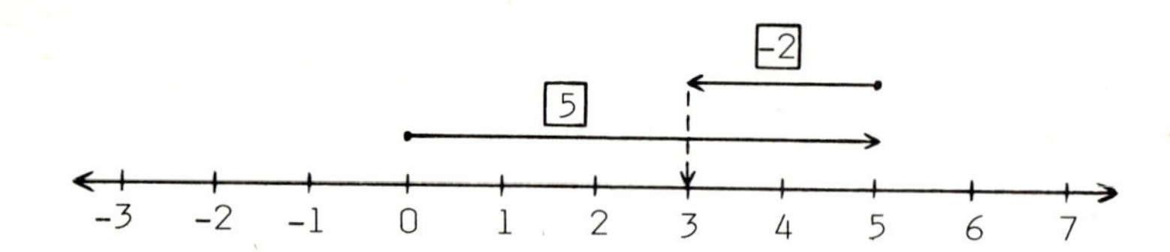

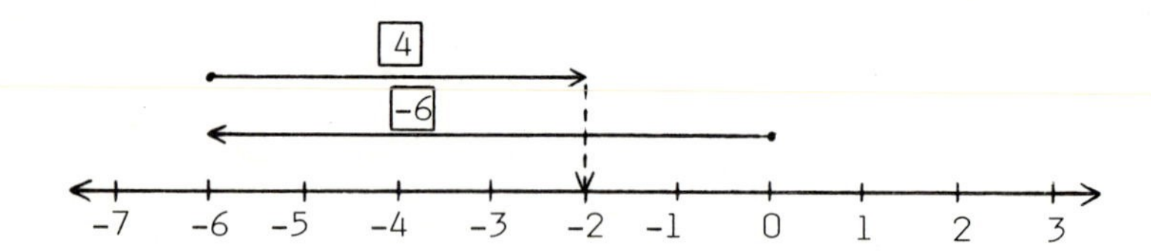

From the examples, you can see the rule for adding a positive term and a negative term.

> <u>To add a positive term and a negative term:</u>
>
> 1. <u>Subtract their absolute values to get the absolute value of the sum</u>.
> 2. <u>Give the sum the same sign as the term with the larger absolute value</u>.

Let's apply the rule to the two examples above.

$5 + (-2) = 3$	Since $5 - 2 = 3$, the absolute value of the sum is 3. Since 5 has the larger absolute value, the sum is positive.
$-6 + 4 = -2$	Since $6 - 4 = 2$, the absolute value of the sum is 2. Since -6 has the larger absolute value, the sum is negative.

Use the rule for these:

a) $4 + (-9) =$ ______ c) $-10 + 3 =$ ______

b) $-3 + 7 =$ ______ d) $12 + (-6) =$ ______

---

a) -5 c) -7

b) 4 d) 6

19. Complete: a) $-40 + 10 =$ ______ c) $20 + (-70) =$ ______

b) $80 + (-55) =$ ______ d) $-30 + 90 =$ ______

---

20. To add three or more terms, we can add "two at a time" from left to right as we have done below.

$$\underbrace{3 + (-5)} + 6 + (-8) =$$
$$\downarrow$$
$$\underbrace{-2 \quad + 6} + (-8) =$$
$$\downarrow$$
$$4 \quad + (-8) = -4$$

By adding "two at a time", find each sum.

a) $-6 + 2 + (-1) =$ b) $10 + (-4) + (-9) + 6 =$

Answers to 19: a) -30 b) 25 c) -50 d) 60

---

21. By adding "two at a time", find each sum.

a) $-5 + (-4) + 10 =$ b) $20 + (-10) + (-30) + 5 =$

Answers to 20: a) -5 b) 3

Answers to 21: a) 1 b) -15

1-3 PROPERTIES OF ADDITION

In this section we will discuss three properties of addition - the identity property, the commutative property, and the associative property.

---

22. When 0 is one term in an addition, the sum is identical to the other term. For example:

$$8 + 0 = 8 \qquad 0 + (-5) = -5$$

The property above is called the identity property of addition. The number 0 is called the identity element for addition.

The identity property is stated for any number a below.

$$a + 0 = a \quad \text{and} \quad 0 + a = a$$

Using the identity property, complete these:

a) $15 + 0 =$ ______ b) $0 + (-12) =$ ______

Answers: a) 15 b) -12

23. If we interchange the two terms in an addition, we get the same sum. For example:

Both $-6 + (-2)$ and $-2 + (-6)$ equal -8.

Both $10 + (-15)$ and $-15 + 10$ equal -5.

The property above is called the commutative property of addition. The commutative property is stated for any numbers a and b below.

$$a + b = b + a$$

Using the commutative property, complete these:

a) $-8 + 7 = 7 + (\ \ )$ b) $12 + (-5) = (\ \ ) + 12$

Answers: a) $7 + (-8)$ b) $(-5) + 12$

24. In the expression below, 5 + 7 is enclosed in parentheses ( ). When parentheses contain an addition, they are grouping symbols. The operation within the grouping is always performed first. For example:

$$3 + \underbrace{(5 + 7)}$$
$$3 + 12 = 15$$

When the grouping contains a negative number in parentheses, we use brackets [ ] as the grouping symbols to avoid a series of parentheses. For example:

$$\underbrace{[7 + (-9)]} + (-6)$$
$$(-2) + (-6) = -8$$

Complete each addition by performing the operation within the grouping first.

a) $\underbrace{(-1 + 8)} + (-4)$

$(\quad) + (-4) = \_\_\_\_$

b) $-6 + \underbrace{[4 + (-7)]}$

$-6 + (\quad) = \_\_\_\_$

Answers: a) 7 + (-4) = 3   b) -6 + (-3) = -9

25. To add three or more terms, we add "two at a time". As the groupings below show, we can add the terms in any order and get the same sum.

$\underbrace{(5 + 3)} + 6$

$8 + 6 = 14$

$5 + \underbrace{(3 + 6)}$

$5 + 9 = 14$

The property above is called the associative property of addition. The associative property is stated for any numbers a, b, and c below.

$$(a + b) + c = a + (b + c)$$

Complete these additions which also illustrate the associative property.

a) $\underbrace{(-6 + 4)} + (-7)$

$\_\_\_\_ + (-7) = \_\_\_\_$

b) $-6 + \underbrace{[4 + (-7)]}$

$-6 + (\quad) = \_\_\_\_$

Answers: a) -2 + (-7) = -9   b) -6 + (-3) = -9

26. The important point about the commutative and associative properties is this: <u>the terms can be written in any order or added in any order without changing the sum</u>. Therefore, when adding three or more integers, we can add the positive terms and the negative terms and then add the results. For example:

Original addition: $8 + (-5) + (-7) + 3 + 4 + (-6)$

Rearranged addition: $\underbrace{(8 + 3 + 4)}_{15} + \underbrace{[(-5) + (-7) + (-6)]}_{(-18)} = -3$

Sometimes we can add the positive terms and negative terms mentally without rewriting them. Let's do the addition below mentally.

$-5 + 10 + 4 + (-9) + (-1) + 3$

a) The sum of the positive terms is ______.

b) The sum of the negative terms is ______.

c) The total sum is ______.

Answers: a) 17 b) -15 c) 2

27. Use the mental shortcut for these:

a) $3 + (-5) + 6 =$ ______

b) $-10 + 7 + (-8) + 3 =$ ______

c) $-4 + 1 + (-6) + 3 + (-5) =$ ______

Answers: a) 4, from $9 + (-5)$ b) -8, from $10 + (-18)$ c) -11, from $4 + (-15)$

28. When it is difficult to do the addition mentally, we can rewrite the numbers in columns. For example:

$27 + (-11) + 34 + (-48) + (-37) + 76 + (-15)$

Add the positives	Add the negatives	Add the sums
27	-11	137
34	-48	-111
76	-37	26
137	-15	
	-111	

Continued on following page.

28. Continued

Use the method just explained to do this one:

-35 + 68 + (-79) + (-92) + 41 + 27

Add the positives	Add the negatives	Add the sums

-70, from:

$$\begin{array}{r} 136 \\ \underline{-206} \\ -70 \end{array}$$

1-4 SUBTRACTION OF INTEGERS

In this section, we will begin by defining opposites. Then we will show how any subtraction of integers can be done by rewriting the subtraction as an equivalent addition.

---

29. Two numbers with the same absolute value but different signs are called opposites. They are called opposites because they are on opposite sides of 0 on the number line.

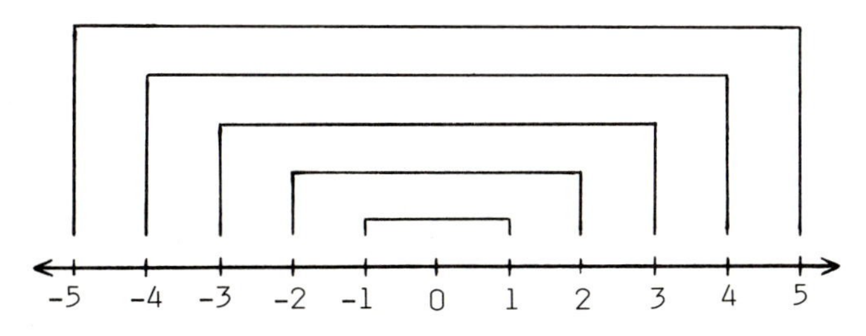

From the number line above, you can see that:

1 and -1 are opposites

3 and -3 are opposites

5 and __ are opposites

30. Since 6 and -6 are opposites, we say: The opposite of 8 is -8. The opposite of -8 is 8. Name the opposite of each number. a) -10 ______ b) 17 ______ c) -42 ______	-5
31. When two opposites are added, their sum is 0. That is: $4 + (-4) = 0$ $-25 + 25 =$ ______	a) 10 b) -17 c) 42
32. Since $5 - 2 = 3$ and $5 + (-2) = 3$, the subtraction $5 - 2$ and the addition $5 + (-2)$ are equal. That is: $5 - 2 = 5 + (-2)$ Notice that subtracting 2 from 5 is the same as adding the opposite of 2 to 5. The definition above can be extended to any number $\underline{a}$ and $\underline{b}$. That is: $\boxed{a - b = a + (-b)}$ Notice that subtracting $\underline{b}$ from $\underline{a}$ is the same as adding the opposite of $\underline{b}$ to $\underline{a}$. Using the definition above, we can convert any subtraction to an equivalent addition by ADDING THE OPPOSITE OF THE SECOND TERM TO THE FIRST TERM. For example: Change - to + Opposite of 7 $4 - 7 = 4 + (-7)$ Complete each conversion from subtraction to addition. a) $10 - 5 = 10 + (\quad)$ b) $2 - 6 = 2 + (\quad)$	0
	a) $10 + (-5)$ b) $2 + (-6)$

33. The same definition applies when the second term is negative. That is, we can convert the subtraction to an equivalent addition by ADDING THE OPPOSITE OF THE SECOND TERM TO THE FIRST TERM. For example:

Change - to +
Opposite of -3

$$6 - (-3) = 6 + 3$$

Complete each conversion from subtraction to addition.

a) $2 - (-5) = 2 +$ ______ b) $-8 - (-4) = -8 +$ ______

---

a) $2 + 5$

b) $-8 + 4$

34. We performed each subtraction below by converting to addition and then finding the difference.

$$9 - 4 = 9 + (-4) = 5$$

$$3 - 7 = 3 + (-7) = -4$$

$$-2 - 5 = -2 + (-5) = -7$$

Following the examples, convert to addition and then find the difference.

	ADDITION		DIFFERENCE
a) $4 - 10 =$	______	=	______
b) $-6 - 1 =$	______	=	______

---

a) $4 + (-10) = -6$

b) $-6 + (-1) = -7$

35. We performed each subtraction below by converting to addition and then finding the difference.

$$5 - (-2) = 5 + 2 = 7$$

$$-1 - (-6) = -1 + 6 = 5$$

$$-9 - (-3) = -9 + 3 = -6$$

Following the examples, convert to addition and then find the difference.

	ADDITION		DIFFERENCE
a) $5 - (-10) =$	______	=	______
b) $-7 - (-9) =$	______	=	______
c) $-10 - (-5) =$	______	=	______

36. Complete each subtraction by adding the opposite of the second term.

a) $8 - 9 =$ ________ c) $10 - (-2) =$ ________

b) $-1 - 2 =$ ________ d) $-12 - (-7) =$ ________

Answers: a) $5 + 10 = 15$ b) $-7 + 9 = 2$ c) $-10 + 5 = -5$

---

37. Use the same method for these.

a) $40 - 50 =$ ________ c) $20 - (-30) =$ ________

b) $-5 - 20 =$ ________ d) $-10 - (-80) =$ ________

Answers: a) -1 c) 12 b) -3 d) -5

---

38. Complete:

a) $10 - 15 =$ ________ c) $-17 - 13 =$ ________

b) $25 - (-25) =$ ________ d) $-41 - (-41) =$ ________

Answers: a) -10 c) 50 b) -25 d) 70

---

39. In each example below, an integer is subtracted from 0.

$$0 - 7 = 0 + (-7) = -7$$

$$0 - (-3) = 0 + 3 = 3$$

In each example below, 0 is subtracted from an integer. The answer is identical to the integer.

$$4 - 0 = 4 \qquad -8 - 0 = -8$$

Complete: a) $0 - 10 =$ ________ c) $0 - (-20) =$ ________

b) $10 - 0 =$ ________ d) $-20 - 0 =$ ________

Answers: a) -5 c) -30 b) 50 d) 0

---

40. Earlier we showed the commutative property of addition. That is, the terms can be interchanged without changing the sum. That is:

$$-8 + 5 = 5 + (-8), \text{ since both equal } -3$$

However, if we interchange the terms in a subtraction, we do not get the same difference. For example:

$$9 - 4 = 5, \text{ but } 4 - 9 = -5$$

$$-2 - (-6) = 4, \text{ but } -6 - (-2) = -4$$

Continued on following page.

Answers: a) -10 c) 20 b) 10 d) -20

40. Continued

Which of the following statements are true? ____________

a) 8 + (-3) = -3 + 8

b) 7 - 8 = 8 - 7

c) 10 - (-5) = -5 - 10

d) -4 + (-1) = -1 + (-4)

Only (a) and (d)

SELF-TEST 1 (pages 1-17)

Write either > or < in each blank.

1. -3 ______ 9
2. 0 ______ -6
3. -10 ______ -4
4. -5 ______ -15

Find each absolute value.

5. |12| = ______
6. |-1| = ______

Write the opposite of each number.

7. 13 ______
8. -7 ______

Do each addition.

9. 40 + (-10) = ______
10. -6 + (-5) = ______
11. -9 + 0 = ______
12. -4 + 7 + (-6) = ______
13. 8 + (-5) + 6 + (-9) = ______
14. -12 + (-1) + 10 + (-3) + 9 = ______

Do each subtraction.

15. 4 - 7 = ______
16. -6 - 6 = ______
17. 8 - (-2) = ______
18. 0 - (-1) = ______
19. -10 - (-4) = ______
20. -2 - (-2) = ______

ANSWERS:

1. <
2. >
3. <
4. >
5. 12
6. 1
7. -13
8. 7
9. 30
10. -11
11. -9
12. -3
13. 0
14. 3
15. -3
16. -12
17. 10
18. 1
19. -6
20. 0

1-5 MULTIPLICATION OF INTEGERS

In this section, we will discuss the rules for the multiplication of integers.

---

41. We can use the word "times", the symbols "x" or "·", or parentheses to indicate a multiplication. For example, all six expressions below mean: multiply 3 and 7.

3 times 7	3(7)
3 x 7	(3)7
3 · 7	(3)(7)

In a multiplication, the numbers multiplied are called factors; the answer is called the product. For example:

In 3(7) = 21: 3 and 7 are the factors.
21 is the product.

Any multiplication stands for an addition of identical terms. For example:

4(2) = four 2's = 2 + 2 + 2 + 2

3(-8) = three -8's = -8 + (-8) + (-8)

Write each multiplication as an addition.

a) 2(10) = ___________ b) 5(-4) = ___________

Answers to 41: a) 10 + 10 b) -4 + (-4) + (-4) + (-4) + (-4)

42. When both factors are positive, the product is positive. For example:

3(6) = 6 + 6 + 6 = 18

Using the above rule, complete these:

a) 5(30) = ___________ b) 9(100) = ___________

Answers to 42: a) 150 b) 900

43. When one factor is positive and one is negative, the product is negative. For example:

$$4(-7) = -7 + (-7) + (-7) + (-7) = -28$$

The above rule applies whether the one negative factor is written first or second. Complete these:

a) $9(-10) =$ ______ b) $(-5)(20) =$ ______

---

a) -90 b) -100

44. When both factors are negative, the product is positive. To see that rule, look at the pattern below.

This number decreases by 1's. This number increases by 5's.

$$\begin{aligned} 3(-5) &= -15 \\ 2(-5) &= -10 \\ 1(-5) &= -5 \\ 0(-5) &= 0 \\ (-1)(-5) &= 5 \\ (-2)(-5) &= 10 \\ (-3)(-5) &= 15 \end{aligned}$$

Using the above rule, complete these:

a) $(-2)(-7) =$ ______ b) $(-5)(-100) =$ ______

---

a) 14 b) 500

45. The rules for multiplying integers are summarized below.

1) When the two factors have the same sign, the product is positive.

2) When the two factors have different signs, the product is negative.

Using the above rules, complete these:

a) $8(-7) =$ ______ c) $(-2)(40) =$ ______

b) $(-6)(-9) =$ ______ d) $(-10)(-7) =$ ______

---

a) -56 c) -80

b) 54 d) 70

46. When one factor is 0, the product is 0. For example:

$$7(0) = 0 \qquad 0(-9) = 0$$

Complete: a) $4(-8) =$ ______ c) $(-9)(-5) =$ ______

b) $0(5) =$ ______ d) $(-6)(0) =$ ______

47. When a multiplication contains three factors, we can multiply "two at a time" as we did at the left below. Complete the other multiplication.

$2(-4)(5)$	$(-4)(-5)(3)$
↓	↓
$(-8)(5) = -40$	$(\quad)(3) = \_\_\_\_$

Answers: a) -32 b) 0 c) 45 d) 0

---

48. We multiplied "two at a time" at the left below. Complete the other multiplication.

$(-4)(3)(-1)(2)$	$(-1)(5)(-3)(-2)$
↓	↓
$(-12)\ (-1)(2)$	$(-5)\ (-3)(-2)$
↓	↓
$(12)\ (2) = 24$	$(\quad)\ (-2) = \_\_\_\_$

Answer: $(20)(3) = 60$

---

49. Using the "two at a time" method, complete these:

a) $(-8)(-5)(-2) = \_\_\_\_$

b) $2(-1)(-3)(-4) = \_\_\_\_$

Answer: $(15)(-2) = -30$

---

50. When one of the factors is 0, the product is always 0. For example:

$7(-9)(0)$	$(-4)(0)(8)(-10)$
↓	↓
$(-63)(0) = 0$	$(0)\ (8)(-10)$
	↓
	$(0)\ (-10) = \_\_\_\_$

Answers: a) -80 b) -24

Answer: 0

1-6 PROPERTIES OF MULTIPLICATION

In this section, we will discuss four properties of multiplication - the identity property, the opposite property, the commutative property, and the associative property.

---

51. When "1" is one factor in a multiplication, the product is identical to the other factor. For example:

$1(5) = 5$ $(-3)(1) = -3$

The property above is called the identity property of multiplication. The number "1" is called the identity element for multiplication.

The identity property is stated for any number a below.

$a \cdot 1 = a$ and $1 \cdot a = a$

Using the identity property, complete these:

a) $1(-17) =$ ______ b) $20(1) =$ ______

---

52. When -1 is one factor in a multiplication, the product is the opposite of the other factor. For example:

$4(-1) = -4$ $(-1)(-7) = 7$

The property above is called the opposite property of multiplication. The opposite property is stated for any number a below.

$a \cdot -1 = -a$ and $-1 \cdot a = -a$

Using the opposite property, complete these:

a) $(-10)(-1) =$ ______ b) $(-1)(5) =$ ______

Answers to 51: a) -17 b) 20

---

53. Complete: a) $1(-1) =$ ______ c) $(-1)(0) =$ ______

b) $0(1) =$ ______ d) $(-1)(-1) =$ ______

Answers to 52: a) 10 b) -5

Answers to 53: a) -1 c) 0 b) 0 d) 1

54. If we interchange the two factors in a multiplication, we get the same product. For example:

Both $3(-4)$ and $(-4)(3)$ equal -12.

Both $(-7)(-8)$ and $(-8)(-7)$ equal 56.

The property above is called the <u>commutative</u> property of multiplication. The commutative property is stated for any numbers <u>a</u> and <u>b</u> below.

$$a \cdot b = b \cdot a$$

Using the commutative property, complete these:

a) $(-5)(6) = (\quad)(\quad)$ b) $(-9)(-4) = (\quad)(\quad)$

---

55. For $4(-2)(5)$, we get the same product whether we multiply 4 and -2 first or -2 and 5 first. That is:

$\underbrace{4(-2)}(5)$ → $(-8)(5) = -40$

$4\underbrace{(-2)(5)}$ → $4(-10) = -40$

a) $(6)(-5)$

b) $(-4)(-9)$

The property above is called the <u>associative</u> property of multiplication. The associative property is stated for any numbers <u>a</u>, <u>b</u>, and <u>c</u> below.

$$(a \cdot b) \cdot c = a \cdot (b \cdot c)$$

Using the associative property, complete these:

a) $(-4 \cdot 5) \cdot 6 = -4 \cdot (5 \cdot \quad)$ b) $3 \cdot (-5 \cdot -9) = (3 \cdot \quad) \cdot -9$

---

56. The important point about the commutative and associative properties of multiplication is this: <u>the factors can be written in any order or multiplied in any order without changing the product</u>.

a) $-4 \cdot (5 \cdot 6)$

b) $(3 \cdot -5) \cdot -9$

Using the above fact, complete these:

a) $(-2)(4)(8) = 4(\quad)(-2)$ b) $7(-1)(-9) = (\quad)(-9)(7)$

---

a) $4(8)(-2)$

b) $(-1)(-9)(7)$

1-7 DIVISION OF INTEGERS

In this section, we will discuss the rules for the division of integers.

---

57. In algebra, a division is usually written as a fraction. For example:

$24 \div 6$ is written $\frac{24}{6}$ $-10 \div 5$ is written ______

Answer: $\frac{-10}{5}$

58. In $\frac{24}{6} = 4$: 24 is called the numerator.
6 is called the denominator.
4 is called the quotient.

In $\frac{14}{-2} = -7$:
a) the denominator is ________.
b) the quotient is ________.
c) the numerator is ________.

Answers: a) -2 b) -7 c) 14

59. The rules for the division of integers are similar to the rules for the multiplication of integers. They are:

1) IF BOTH NUMBERS ARE POSITIVE, THE QUOTIENT IS POSITIVE.

$\frac{35}{5} = 7$ $\frac{48}{4} = 12$

2) IF BOTH NUMBERS ARE NEGATIVE, THE QUOTIENT IS POSITIVE.

$\frac{-30}{-6} = 5$ $\frac{-90}{-10} = 9$

3) IF ONE NUMBER IS POSITIVE AND THE OTHER IS NEGATIVE, THE QUOTIENT IS NEGATIVE.

$\frac{-32}{8} = -4$ $\frac{14}{-2} = -7$

Using the rules above, complete these:

a) $\frac{-10}{5} =$ ______ b) $\frac{-56}{-7} =$ ______ c) $\frac{28}{-4} =$ ______

60. When a division is done correctly, the product of the denominator and the quotient equals the numerator. For example:

$$\frac{12}{3} = 4 \text{ , since } 3(4) = 12$$

The fact above can be used to justify the rules for dividing integers. That is:

$$\frac{-20}{4} = -5 \text{ , since } 4(-5) = -20 \qquad \frac{-18}{-6} = 3 \text{ , since } (-6)(3) = \_\_\_\_\_$$

a) -2

b) 8

c) -7

---

61. Four properties of division are listed below.

1) When a non-zero number is divided by itself, the quotient is +1.

$$\frac{9}{9} = 1 \qquad \frac{-12}{-12} = 1$$

2) When a number is divided by +1, the quotient is <u>identical</u> to the number.

$$\frac{6}{1} = 6 \qquad \frac{-8}{1} = -8$$

3) When a number is divided by -1, the quotient is the <u>opposite</u> of the number.

$$\frac{5}{-1} = -5 \qquad \frac{-10}{-1} = 10$$

4) When 0 is divided by a non-zero number, the quotient is 0.

$$\frac{0}{4} = 0 \qquad \frac{0}{-7} = 0$$

Using the above properties, complete these.

a) $\frac{-15}{-15} = \_\_\_\_\_$ b) $\frac{40}{-1} = \_\_\_\_\_$ c) $\frac{-59}{1} = \_\_\_\_\_$ d) $\frac{0}{-13} = \_\_\_\_\_$

-18

---

62. Complete these:

a) $\frac{-1}{1} = \_\_\_\_\_$ b) $\frac{-1}{-1} = \_\_\_\_\_$ c) $\frac{0}{-1} = \_\_\_\_\_$ d) $\frac{1}{-1} = \_\_\_\_\_$

a) 1 c) -59

b) -40 d) 0

---

a) -1 c) 0

b) 1 d) -1

1-8 POSITIVE EXPONENTS

In this section, we will discuss exponential expressions with positive integer exponents.

---

63. Expressions like 5^3 and 4^7 are in exponential form. The number on the lower left is called the base; the number on the upper right is called the exponent. That is:

In 5^3 : the base is 5.
the exponent is 3.

In 4^7 : a) the base is ______.
b) the exponent is ______.

---

64. An exponential expression is a short way of writing a multiplication of identical factors. The exponent tells us how many times to use the base as a factor. For example:

$5^3 = (5)(5)(5)$	(The exponent tells us to multiply three 5's.)
$2^4 = (2)(2)(2)(2)$	(The exponent tells us to multiply four 2's.)

Write each expression as a multiplication of identical factors.

a) $7^3 =$ ______________ b) $8^5 =$ ______________

Answers to 63: a) 4 b) 7

---

65. Any multiplication of identical factors can be written in exponential form.

$(4)(4)(4) = 4^3$	(Since there are three 4's, the exponent is 3.)
$(6)(6)(6)(6) = 6^4$	(Since there are four 6's, the exponent is 4.)

Write each multiplication in exponential form.

a) $(7)(7) =$ ______ b) $(10)(10)(10)(10)(10)(10) =$ ______

Answers to 64: a) $(7)(7)(7)$ b) $(8)(8)(8)(8)(8)$

---

66. Any exponential expression is called a power of the base. For example:

6^2, 6^3, and 6^5 are called "powers of 6".

2^3, 2^4, and 2^7 are called "powers of 2".

Answers to 65: a) 7^2 b) 10^6

Continued on following page.

66. Continued

The following language is used to distinguish powers of the same base.

4^2 is called "4 to the second power".

4^3 is called "4 to the third power".

4^5 is called "4 to the fifth power".

Similarly: a) 2^4 is called "2 to the ________ power".

b) 10^7 is called "10 to the ________ power".

---

67. The words squared and cubed are frequently used for the second and third power of any base. That is:

5^2 is called "5 to the second power" or "5 squared".

9^3 is called "9 to the third power" or "9 cubed".

Write the exponential expression for each of these:

a) 3 squared = ________ b) 7 cubed = ________

Answers to 66: a) fourth b) seventh

---

68. We can find the value of a power by performing the multiplication. That is:

$$4^3 = (4)(4)(4) = 64$$

a) $3^4 = (3)(3)(3)(3) =$ ________ b) $2^5 = (2)(2)(2)(2)(2) =$ ________

Answers to 67: a) 3^2 b) 7^3

---

69. The base in an exponential expression can be a negative number. Some examples are shown. Notice that the value at the right is positive when the exponent is even and negative when the exponent is odd.

$$(-3)^2 = (-3)(-3) = 9$$

$$(-5)^3 = (-5)(-5)(-5) = -125$$

$$(-4)^4 = (-4)(-4)(-4)(-4) = 256$$

$$(-2)^5 = (-2)(-2)(-2)(-2)(-2) = -32$$

Following the examples, complete these:

a) $(-7)^2 =$ ________ b) $(-4)^3 =$ ________ c) $(-2)^4 =$ ________

Answers to 68: a) 81 b) 32

70. When the base is negative, it is <u>always</u> written in parentheses. Therefore, don't confuse $(-5)^2$ with -5^2.

$$(-5)^2 \text{ means } (-5)(-5) = 25$$

$$-5^2 \text{ means } -(5)(5) = -25$$

Following the pattern above, complete these:

a) $(-2)^4 =$ ________ b) $-2^4 =$ ________

Answers: a) 49 b) -64 c) 16

---

71. Any power whose exponent is "1" is equal to its base. That is:

$2^1 = 2$ $5^1 = 5$

The definition above fits the pattern for powers.

$2^3 = (2)(2)(2)$ $5^3 = (5)(5)(5)$

$2^2 = (2)(2)$ $5^2 = (5)(5)$

$2^1 = (2)$ or 2 $5^1 = (5)$ or 5

Using the definition, complete these:

a) $8^1 =$ ________ b) $10^1 =$ ________ c) $25^1 =$ ________

Answers: a) 16 b) -16

---

72. Any positive power of "1" equals "1". Any positive power of 0 equals 0. Any positive power of -1 equals either "1" or -1. For example:

$1^2 = (1)(1) = 1$ $1^5 = (1)(1)(1)(1)(1) = 1$

$0^3 = (0)(0)(0) = 0$ $0^4 = (0)(0)(0)(0) = 0$

$(-1)^2 = (-1)(-1) = 1$ $(-1)^3 = (-1)(-1)(-1) = -1$

Complete these:

a) $1^7 =$ ______ c) $(-1)^4 =$ ______

b) $0^{10} =$ ________ d) $(-1)^5 =$ ______

Answers: a) 8 b) 10 c) 25

---

73. Complete these:

a) $(-3)^2 =$ ______ c) $-1^4 =$ ______

b) $-3^2 =$ ______ d) $(-1)^4 =$ ______

Answers: a) 1 b) 0 c) 1 d) -1

Answers: a) 9 b) -9 c) -1 d) 1

SELF-TEST 2 (pages 18-28)

Do each multiplication.

1. $4(-7) =$ ______	3. $(-1)(5) =$ ______	5. $(-1)(-3)(-4) =$ ______
2. $(-6)(-9) =$ ______	4. $(-12)(0) =$ ______	6. $(-5)(1)(-6)(2) =$ ______

Do each division.

7. $\frac{-24}{6} =$ ______	9. $\frac{-63}{-9} =$ ______	11. $\frac{0}{-3} =$ ______
8. $\frac{42}{-7} =$ ______	10. $\frac{-12}{-12} =$ ______	12. $\frac{49}{-1} =$ ______

Find the value of each power.

13. $3^4 =$ ______	15. $8^1 =$ ______	17. $(-4)^2 =$ ______
14. $2^6 =$ ______	16. $(-1)^5 =$ ______	18. $-4^2 =$ ______

ANSWERS:

1. -28	4. 0	7. -4	10. 1	13. 81	16. -1
2. 54	5. -12	8. -6	11. 0	14. 64	17. 16
3. -5	6. 60	9. 7	12. -49	15. 8	18. -16

1-9 ORDER OF OPERATIONS

In this section, we will discuss the proper order of operations for evaluating expressions that contain more than one operation.

---

74. We want to find the value of $3 + 2(4)$. Should we add first or multiply first?

Adding first, we get: $3 + 2(4) = 5(4) = 20$ (Incorrect)

Multiplying first, we get: $3 + 2(4) = 3 + 8 = 11$ (Correct)

To evaluate expressions like those above, we use the following order of operations.

Order of Operations

1. If the expression does not contain grouping symbols like parentheses ( ) or brackets [ ], use the following steps:

   a) Evaluate any power.

   b) Do all multiplications and divisions from left to right.

   c) Do all additions and subtractions from left to right.

2. If the expression contains grouping symbols like parentheses ( ) or brackets [ ], do the operations within the grouping symbols first before evaluating the whole expression.

3. If the expression contains a fraction bar, do the operations above and below the fraction bar before dividing.

Many examples of the proper order of operations are given in the following frames.

75. To evaluate the expressions below, we converted each subtraction to addition and then performed the addition.

$$4 + 8 - (-5) = 4 + 8 + 5 = 17$$

$$9 - (-2) - 7 = 9 + 2 + (-7) = 4$$

Using the same method, evaluate these:

a) $-1 - 8 + 10 =$ ____________

b) $6 - (-3) + 5 - 20 =$ ____________

---

76. One or both terms in an addition or subtraction can be a power. For example:

a) 1

b) -6

In $-6 + 2^3$, the second term is a power.

In $(-5)^2 - 3^3$, both terms are powers.

To evaluate the expressions, we convert the powers to ordinary numbers first. We get:

$-6 + 2^3$	$(-5)^2 - 3^3$
$-6 + 8 = 2$	$25 - 27$
	$25 + (-27) = -2$

Using the same method, evaluate these:

a) $4^2 - 10$

b) $(-1)^5 + 6^2$

---

77. One or both terms in an addition or subtraction can be a multiplication. For example:

a) 6 b) 35

In $4(-2) + 10$, the terms are $4(-2)$ and 10.

In $3(8) - (-1)(5)$, the terms are $3(8)$ and $(-1)(5)$.

Continued on following page.

77. Continued

To evaluate the expressions, <u>we perform the multiplications first</u>.

$\underbrace{4(-2)} + 10$ → $-8 + 10 = 2$

$\underbrace{3(8)} - \underbrace{(-1)(5)}$ → $24 - (-5)$ → $24 + 5 = 29$

Using the same method, evaluate these:

a) $3(-10) - (-9)$ b) $(-1)(6) + (-7)(-2)$

---

a) -21 b) 8

78. To evaluate the expressions below, we converted the power to an ordinary number before multiplying.

$2(5)^2 - 30$

$2(25) - 30$

$50 - 30 = 20$

$50 + 2(-3)^3$

$50 + 2(-27)$

$50 + (-54) = -4$

Using the same method, evaluate these:

a) $30 - 2(4)^2$ b) $5(-2)^3 + 40$

---

a) -2 b) 0

79. One term in an addition or subtraction can be a grouping. For example:

In $(7 - 11) + 10$, the terms are $(7 - 11)$ and 10.

In $6 - (-7 + 5)$, the terms are 6 and $(-7 + 5)$.

To evaluate the expressions, <u>we perform the operations within the grouping first</u>.

$$\underbrace{(7 - 11)}_{\downarrow} + 10$$
$$-4 + 10 = 6$$

$$6 - \underbrace{(-7 + 5)}_{\downarrow}$$
$$6 - (-2)$$
$$6 + 2 = 8$$

Using the same method, evaluate these:

a) $-4 - (10 - 2)$

b) $[3 + (-9)] + 20$

---

a) -12 b) 14

80. To evaluate the expressions below, we simplified each term first.

$$\underbrace{7(-5)}_{\downarrow} + \underbrace{(6 - 10)}_{\downarrow}$$
$$-35 + (-4) = -39$$

$$\underbrace{[-1 - (-3)]}_{\downarrow} - \underbrace{2(4)^2}_{\downarrow}$$
$$(-1 + 3) - 2(16)$$
$$2 - 32 = -30$$

Using the same method, evaluate these:

a) $10(-1)^4 - (12 - 8)$

b) $[-2 - (-1)] + 5(-3)$

---

a) 6 b) -16

81. One of the factors in a multiplication can be a grouping. For example:

In $(5 - 2)6$, the factors are $(5 - 2)$ and 6.

In $4[-1 + (-7)]$, the factors are 4 and $[-1 + (-7)]$.

To evaluate the expressions, we perform the operations within the grouping first.

$(5 - 2)6$ → $(3)\ 6 = 18$

$4[-1 + (-7)]$ → $4\ (-8) = -32$

Using the same method, evaluate these:

a) $[6 + (-9)]7$

b) $2(4 - 10)$

---

a) -21 b) -12

82. In the expression below, the terms are $5(3 + 4)$ and 10. To evaluate it, we simplified $5(3 + 4)$ first.

$$5(3 + 4) - 10$$

$$5(7) - 10$$

$$35 - 10 = 25$$

Using the same method, evaluate these:

a) $30 - 2(4 + 5)$

b) $4[-2 + (-3)] + 3^2$

---

a) 12 b) -11

83. To evaluate the expression below, we simplified each term first.

$$5(-2) + 3(-2 + 4)$$
$$-10 + 3(2)$$
$$-10 + 6 = -4$$

Using the same method, evaluate these:

a) $6(5 + 2) - 3(2)^2$

b) $5(-8) - 2[-5 - (-15)]$

Answers: a) 30 b) -60

84. One or both terms in an addition or subtraction can be a division. For example:

In $\frac{-12}{4} + 7$, the terms are $\frac{-12}{4}$ and 7.

In $\frac{10}{2} - \frac{20}{-5}$, the terms are $\frac{10}{2}$ and $\frac{20}{-5}$.

To evaluate the expressions, <u>we perform the divisions first</u>.

$$\frac{-12}{4} + 7$$
$$\downarrow$$
$$-3 + 7 = 4$$

$$\frac{10}{2} - \frac{20}{-5}$$
$$\downarrow \quad \downarrow$$
$$5 - (-4)$$
$$5 + 4 = 9$$

Using the same method, evaluate these:

a) $\frac{6}{-1} + \frac{12}{2}$

b) $2 - \frac{-9}{3}$

Answers: a) 0 b) 5

85. When the numerator or denominator of a fraction is a multiplication, <u>we multiply before dividing</u>. For example:

$$\frac{3(-4)}{2} = \frac{-12}{2} = -6$$

Evaluate these:

a) $\frac{24}{2(-3)}$ b) $\frac{2(-20)}{4(-2)}$

Answers: a) -4   b) 5

86. When the numerator or denominator of a fraction is an addition or subtraction, <u>we add or subtract before dividing</u>. For example:

$$\frac{-11+7}{2} = \frac{-4}{2} = -2 \qquad \frac{21}{5-2} = \frac{21}{3} = 7$$

Evaluate these:

a) $\frac{6}{-5+2}$ c) $\frac{2-20}{2-8}$

b) $\frac{-5+(-2)}{-7}$ d) $\frac{18-(-7)}{-8+3}$

Answers: a) -2   c) 3   b) 1   d) -5

87. When an expression contains a fraction bar, <u>we do all operations above and below the fraction bar before dividing</u>. For example:

$$\frac{4(3+5)-2}{10-4} = \frac{4(8)-2}{6} = \frac{32-2}{6} = \frac{30}{6} = 5$$

Using the same method, evaluate this one:

$$\frac{5(-1-9)}{3(4)-2}$$

Answer: -5

88. Evaluate:

a) $2^4 - \left(\frac{14}{2} - 5\right)$ b) $7\left[\frac{-9+(-1)}{2}\right]$

Answers: a) 14   b) -35

1-10 ALGEBRAIC EXPRESSIONS

In this section, we will define algebraic expressions and use evaluations to show their meaning.

---

89. An algebraic expression is a collection of numbers, letters, operation symbols, and grouping symbols. Some examples are:

$x + 7$ $\quad$ $3a - 2b$ $\quad$ $5y^2 + 3y - 1$

$\frac{xy}{10}$ $\quad$ $\frac{2(d+1)}{5}$ $\quad$ $\frac{x+2}{y+3}$

Each letter in an algebraic expression is a variable. A variable is a letter used to represent a number or set of numbers.

To evaluate an algebraic expression, we must substitute some number for each variable. Let's evaluate $x + 7$ when x is 10, -3, and -10.

When $x = 10$, $x + 7 = 10 + 7 = 17$

When $x = -3$, $x + 7 = -3 + 7 = 4$

When $x = -10$, $x + 7 = -10 + 7 =$ ______

---

90. Expressions like $3x$ or $5y$ indicate a multiplication of a number and a variable. That is:

-3

$3x$ means "3 times x".

$5y$ means "5 times y".

Let's evaluate $3x$ when $x = 4$ and $5y$ when $y = -1$.

When $x = 4$, $3x = 3(4) = 12$

When $y = -1$, $5y = 5(-1) =$ ______

---

91. Let's evaluate $2d + 5$ when $d = 10$ and $d = -4$.

-5

When $d = 10$, $2d + 5 = 2(10) + 5 = 20 + 5 = 25$

When $d = -4$, $2d + 5 = 2(-4) + 5 = -8 + 5 =$ ______

-3

92. Evaluate each expression below when $x = 6$.

a) $\frac{2x}{3} = \frac{2(6)}{3} =$ ______

b) $\frac{4x - 3}{7} =$ ______

c) $\frac{2(x + 4)}{5} =$ ______

---

93. Evaluate each expression below when $y = -2$.

a) $10 - (y + 5) = 10 - (-2 + 5) =$ ______

b) $9 + (y + 2) =$ ______

c) $3(2y - 1) =$ ______

d) $4y - (y - 6) =$ ______

Answers to 92: a) 4 b) 3 c) 4

---

94. To evaluate an expression containing two variables, we must substitute for both variables.

Let's evaluate each expression below when $p = 4$ and $q = -1$.

$$p - q = 4 - (-1) = 4 + 1 = 5$$

a) $4p + 3q =$ ______

b) $p - (4q + 5) =$ ______

Answers to 93: a) 7 b) 9 c) -15 d) 0

---

95. Evaluate each expression below when $x = 9$ and $y = 3$.

a) $\frac{x + 1}{y - 1} = \frac{9 + 1}{3 - 1} =$ ______

b) $\frac{2(x + 3)}{5y - 3} =$ ______

Answers to 94: a) 13 b) 3

---

96. Expressions like <u>xy</u> or <u>abc</u> also indicate a multiplication. That is:

xy means "<u>x</u> times <u>y</u>".

abc means "<u>a</u> times <u>b</u> times <u>c</u>".

Continued on following page.

Answers to 95: a) 5 b) 2

96. Continued

Let's substitute to evaluate each expression on the preceding page.

When $x = 2$ and $y = -5$, $xy = 2(-5) = -10$

When $a = 4$, $b = 3$, and $c = 2$, $abc = (4)(3)(2) =$ ______

---

97. The expression $2xy$ also indicates a multiplication. That is:

$2xy$ means "2 times x times y".

Let's evaluate each expression below when $x = 3$ and $y = -4$.

a) $2xy = 2(3)(-4) =$ ______

b) $-5xy =$ ______

24

---

98. Evaluate each expression below when $c = -8$ and $d = 5$.

a) $\frac{cd}{10} = \frac{(-8)(5)}{10} =$ ______

b) $2cd - 15 =$ ______

a) -24

b) 60

---

99. The expression $3x^2$ indicates a multiplication. That is:

$3x^2$ means "3 times x^2".

Let's evaluate the expression when $x = 5$, $x = -2$, and $x = -5$.

When $x = 5$, $3x^2 = 3(5)^2 = 3(25) = 75$

a) When $x = -2$, $3x^2 = 3(-2)^2 =$ ______

b) When $x = -5$, $3x^2 =$ ______

a) -4

b) -95

---

100. Let's evaluate $-4t^2$ when $t = 3$ and $t = -1$.

a) $-4t^2 = -4(3)^2 =$ ______ b) $-4t^2 = -4(-1)^2 =$ ______

a) 12

b) 75

---

101. To evaluate the expressions below, let $x = -2$ and $y = 4$.

a) $x^3 + 5x =$ ______

b) $3y^2 - 2y + 10 =$ ______

a) -36, from $-4(9)$

b) -4, from $-4(1)$

---

a) -18, from $-8 + (-10)$

b) 50, from $48 - 8 + 10$

1-11 TRANSLATING PHRASES TO ALGEBRAIC EXPRESSIONS

In this section, we will show how English phrases can be translated to algebraic expressions.

---

102. The phrase "five plus some number" can be translated to the algebraic expression $5 + n$ by substituting mathematical symbols for the words.

five	plus	some number
↓	↓	↓
5	+	n

The following phrases also translate to the expression $5 + n$.

"the sum of five and some number"

"five increased by some number"

The following phrases translate to the expression $n + 5$.

"some number plus five"

"the sum of a number and five"

"some number increased by five"

"five more than a number"

Translate each phrase to an algebraic expression.

a) the sum of a number and ten ____________

b) eight increased by some number ____________

c) a number plus fifty ____________

d) three more than a number ____________

---

a) $n + 10$

b) $8 + n$

c) $n + 50$

d) $n + 3$

103. The phrase "a number decreased by ten" can be translated to the algebraic expression $n - 10$.

a number	decreased by	ten
↓	↓	↓
n	-	10

The following phrases also translate to the expression $n - 10$.

"a number less ten"

"a number minus ten"

"some number diminished by ten"

"some number reduced by ten"

"ten subtracted from some number"

"ten less than a number"

The following phrases translate to the expression $10 - n$.

"ten decreased by some number"

"ten less a number"

"ten minus some number"

"ten diminished by a number"

"ten reduced by a number"

"some number subtracted from ten"

Translate each phrase to an algebraic expression.

a) a number reduced by one ________

b) twenty less some number ________

c) twelve decreased by a number ________

d) five subtracted from some number ________

---

a) $n - 1$

b) $20 - n$

c) $12 - n$

d) $n - 5$

104. The phrase "seven times a number" can be translated to the algebraic expression 7 x n or 7n.

seven	times	a number	
↓	↓	↓	
7	x	n	(or 7n)

Note: In algebra, we usually do not use the multiplication symbol "x". Therefore, we write 7n instead of 7 x n.

The following phrases also translate to 7n.

"seven multiplied by a number"

"the product of seven and a number"

Translate each phrase to an algebraic expression.

a) three times a number ____________

b) the product of nine and some number ____________

c) twenty multiplied by a number ____________

---

a) 3n

b) 9n

c) 20n

105. When multiplying a number and a variable in algebra, we always write the number first. That is:

We write 5n instead of n5 .

Therefore, each phrase below translates to 5n.

"a number times five"

"some number multiplied by five"

"the product of a number and five"

Translate each phrase to an algebraic expression.

a) a number times two ____________

b) some number multiplied by twelve ____________

c) the product of a number and forty ____________

---

a) 2n

b) 12n

c) 40n

106. The phrases "twice a number" and "double a number" translate to $2n$; the phrase "triple a number" translates to $3n$. Using those facts, we can translate "six plus twice a number" to $6 + 2n$ and "two minus triple a number" to $2 - 3n$.

six	plus	twice a number
↓	↓	↓
6	+	2n

two	minus	triple a number
↓	↓	↓
2	-	3n

Translate each phrase to an algebraic expression.

a) ten decreased by double a number ____________

b) the sum of triple a number and five ____________

c) seven more than six times a number ____________

d) twenty reduced by four times a number ____________

---

107. The phrase "the square of a number" translates to n^2; the phrase "the cube of a number" translates to n^3. Using those facts, translate each phrase below to an algebraic expression.

a) triple the square of a number ____________

b) five times the cube of a number ____________

Answers to 106: a) $10 - 2n$ b) $3n + 5$ c) $6n + 7$ d) $20 - 4n$

---

108. Two translations are shown below.

"triple a number, plus two" $3n + 2$

"the square of a number, minus one" $n^2 - 1$

Translate each of these to an algebraic expression.

a) double a number, minus ten ____________

b) four times the cube of a number, plus seven ____________

Answers to 107: a) $3n^2$ b) $5n^3$

Answers to 108: a) $2n - 10$ b) $4n^3 + 7$

109. The phrase "a number divided by three" can be translated to $n \div 3$ or $\frac{n}{3}$.

a number	divided by	three	
↓	↓	↓	
n	÷	3	$\left(\text{or } \frac{n}{3}\right)$

Note: In algebra, we usually do not use the division symbol "÷".
We write $\frac{n}{3}$ instead of $n \div 3$.

The following phrase also translates to $\frac{n}{3}$.

"the quotient of a number and three"

Translate each phrase to an algebraic expression.

a) ten divided by a number ______

b) the quotient of a number and five ______

c) the quotient of five and a number ______

---

110. Translate each phrase to an algebraic expression.

a) seven divided by double a number ______

b) the quotient of the square of a number and twelve ______

Answers to 109: a) $\frac{10}{n}$ b) $\frac{n}{5}$ c) $\frac{5}{n}$

---

111. When translating to an algebraic expression, we can use any letter for "a number" or "some number". For example, two plus a number can be written $2 + x$, $2 + y$, $2 + n$, $2 + a$, and so on. Use the letter x to translate these to algebraic expressions.

a) the square of a number ______

b) five times the cube of a number ______

c) twice a number, increased by 7 ______

d) forty decreased by triple a number ______

Answers to 110: a) $\frac{7}{2n}$ b) $\frac{n^2}{12}$

---

112. Use the letter y to translate these to algebraic expressions.

a) a number divided by two ______

b) the quotient of twenty and three times a number ______

c) triple the square of a number, minus one ______

Answers to 111: a) x^2 b) $5x^3$ c) $2x + 7$ d) $40 - 3x$

113. Each phrase below involves two letters. Translate each phrase to an algebraic expression.

a) p more than q ______

b) b subtracted from a ______

c) x divided by y ______

d) the product of c and d ______

Answers (frame 112): a) $\frac{y}{2}$ b) $\frac{20}{3y}$ c) $3y^2 - 1$

---

114. We used the letters x and y for two numbers to make the translations below.

the sum of two numbers	$x + y$
the difference of two numbers	$x - y$
the product of two numbers	xy
the quotient of two numbers	$\frac{x}{y}$

Using the letters x and y, translate these to algebraic expressions.

a) the sum of two numbers minus five ______

b) six more than the difference of two numbers ______

c) the product of two numbers, plus one ______

d) triple the product of two numbers ______

Answers (frame 113): a) $q + p$ b) $a - b$ c) $\frac{x}{y}$ d) cd

---

115. Use the letters a and b to translate these.

a) the sum of the squares of two numbers ______

b) the difference of the cubes of two numbers ______

c) the product of the squares of two numbers ______

d) the quotient of the cubes of two numbers ______

Answers (frame 114): a) $x + y - 5$ b) $x - y + 6$ c) $xy + 1$ d) $3xy$

---

116. Sometimes we need parentheses to translate to algebraic expressions. Two examples are shown.

"ten minus the sum of two numbers"	$10 - (x + y)$
"the product of two, and the difference of a number and five"	$2(x - 5)$

Continued on following page.

Answers (frame 115): a) $a^2 + b^2$ b) $a^3 - b^3$ c) a^2b^2 d) $\frac{a^3}{b^3}$

116. Continued

Using $\underline{x}$ for one number and $\underline{x}$ and $\underline{y}$ for two numbers, translate these.

a) the product of three, and the sum of the square of a number and one ______

b) fifty minus the difference of two numbers ______

c) the product of ten, and the sum of two numbers ______

a) $3(x^2 + 1)$ b) $50 - (x - y)$ c) $10(x + y)$

SELF-TEST 3 (pages 29-45)

Evaluate each expression.

1. $-5 - (-9) - 1 =$ ______
2. $4(-7) + 6^2 =$ ______
3. $\frac{-20 - (-4)}{3(-6) + 10} =$ ______
4. $5(-1)^3 - (7 - 10) =$ ______
5. $10(-9 + 3) =$ ______
6. $2^5 - \left[\frac{3(1 - 5)}{2}\right] =$ ______

Evaluate each expression when $x = 2$.

7. $5(3x - 4) =$ ______
8. $\frac{x^3 + 10}{3} =$ ______
9. $x^2 - 5x + 6 =$ ______
10. $\frac{x - 10}{x - 4} =$ ______

Evaluate each expression when $x = -3$ and $y = -1$.

11. $xy - 2y =$ ______
12. $\frac{5(x - 1)}{y + 3} =$ ______

Translate each phrase to an algebraic expression. Use $\underline{x}$ for one variable and $\underline{x}$ and $\underline{y}$ for two variables.

13. the product of a number and 7 ______
14. 5 times the difference of a number and 3 ______
15. 8 times the square of a number, plus 4 ______
16. 10 divided by triple a number ______
17. the product of two numbers, less 10 ______
18. the difference of the cubes of two numbers ______

ANSWERS:

1. 3	4. -2	7. 10	10. 4	13. $7x$	16. $\frac{10}{3x}$
2. 8	5. -60	8. 6	11. 5	14. $5(x - 3)$	17. $xy - 10$
3. 2	6. 38	9. 0	12. -10	15. $8x^2 + 4$	18. $x^3 - y^3$

SUPPLEMENTARY PROBLEMS - CHAPTER 1

Note: Answers for all supplementary problems are in the back of the text.

Assignment 1

Write either $>$ or $<$ in each blank.

1. 4 ______ -3 2. -5 ______ -7 3. -2 ______ 0 4. 0 ______ 3

5. -6 ______ 8 6. -9 ______ -4 7. 16 ______ 31 8. -12 ______ -15

Find each absolute value.

9. $|6| =$ ______ 10. $|-1| =$ ______ 11. $|0| =$ ______ 12. $|-18| =$ ______

Do each addition.

13. -2 + (-4) 14. 4 + (-7) 15. -9 + 5 16. 8 + (-1)

17. -6 + (-1) 18. -8 + 0 19. 7 + (-7) 20. 10 + (-5)

21. 0 + (-30) 22. -10 + 15 23. 20 + (-50) 24. 6 + (-7) + 5

25. -9 + 8 + (-3) 26. 2 + (-3) + 9 + (-5) 27. -8 + 9 + (-7) + (-6) + 9

Write the opposite of each number.

28. 9 ______ 29. -6 ______ 30. 5 ______ 31. -1 ______

Do each subtraction.

32. 5 - 13 33. 6 - (-4) 34. -1 - 7 35. -7 - (-8)

36. 4 - (-4) 37. -9 - (-9) 38. 0 - 3 39. -1 - 1

40. 11 - 22 41. 15 - (-17) 42. -6 - 0 43. -40 - (-30)

Assignment 2

Do each multiplication.

1. 8(-7) 2. (-5)(4) 3. 0(-3) 4. (-1)(-10)

5. 1(-12) 6. (-7)(0) 7. (-6)(-4) 8. (-10)(5)

9. (-7)(-1)(4) 10. (-3)(-5)(-2) 11. (-9)(0)(-1)(-2) 12. (1)(-6)(3)(-7)

Do each division.

13. $\frac{42}{-7}$ 14. $\frac{-20}{4}$ 15. $\frac{0}{-6}$ 16. $\frac{15}{-1}$ 17. $\frac{-9}{1}$

18. $\frac{-32}{-4}$ 19. $\frac{-64}{2}$ 20. $\frac{-40}{-10}$ 21. $\frac{-5}{-1}$ 22. $\frac{0}{7}$

Find the value of each power.

23. 8^2 24. 2^3 25. 4^4 26. 1^8 27. 9^1

28. 0^5 29. $(-9)^2$ 30. -9^2 31. $(-1)^7$ 32. -1^7

Assignment 3

Find the value of each expression.

1. $7 - (-4) - 9$
2. $2(4)^2 - 10$
3. $(-2)^5 + 3^3$
4. $(-3)(6) - 2(-1)$
5. $4(-1)^4 - (10 - 6)$
6. $[-5 - (-3)] + 4(-2)$
7. $10(-1) + 2(-3 + 4)$
8. $4(3^2 - 11)$
9. $7(1 - 6) + 5^2$
10. $\frac{6 - 7}{7 - 6}$
11. $\frac{24}{-8} - \frac{-8}{2}$
12. $\frac{35}{5 - 2(-1)}$
13. $\frac{17 - (-3)}{-8 + 3(1)}$
14. $\frac{10 - 30}{6 - 4(1 - 2)}$
15. $10\left[\frac{-2 - (-11)}{3}\right]$

Evaluate each expression when $x = 2$ and $y = 3$.

16. $2x^3$ 17. $x - y$ 18. $4x - 5y$ 19. $10(3x - 1)$

20. $5x - (y + 4)$ 21. $x^2 - 6x$ 22. $\frac{x + 4}{y - 5}$ 23. $\frac{4xy}{3}$

Evaluate each expression when $x = -1$ and $y = -4$.

24. $-2xy$ 25. $3x - (y - 5)$ 26. $x^2 - 4x + 5$ 27. $\frac{3(y - 1)}{x - 4}$

Translate each phrase to an algebraic expression. Use $\underline{x}$ as the variable.

28. the difference of 20 and a number
29. the product of 6 and the square of a number
30. 8 more than triple a number
31. double a number, minus 3
32. 6 times the cube of a number, plus 9
33. the quotient of 5 and three times a number
34. the square of a number, minus double that number
35. the product of 7, and the sum of a number and 2

Translate each phrase to an algebraic expression. Use $\underline{x}$ and $\underline{y}$ as the variables.

36. four times the product of two numbers
37. the quotient of two numbers, plus 2
38. the difference of the squares of two numbers
39. the product of the cubes of two numbers
40. 20 minus the sum of two numbers
41. the product of 5, and the difference of two numbers

Rational Numbers

2

In this chapter, we will discuss the basic operations with signed fractions and decimal numbers. We will define rational numbers and powers with zero and negative exponents. We will introduce three basic laws of exponents, and discuss scientific notation as an application of exponents.

2-1 MULTIPLICATION OF FRACTIONS

In this section, we will discuss the procedure for multiplications involving signed fractions.

1. To multiply fractions, we multiply their numerators and denominators. That is:

$$\left(\frac{2}{3}\right)\left(\frac{5}{7}\right) = \frac{(2)(5)}{(3)(7)} = \frac{10}{21} \qquad \left(\frac{1}{2}\right)\left(\frac{1}{4}\right) = \frac{(1)(1)}{(2)(4)} = \underline{\qquad}$$

$\frac{1}{8}$

2. Any whole number is equal to a fraction whose numerator is that number and whose denominator is "1".

$$1 = \frac{1}{1} \qquad 3 = \frac{3}{1} \qquad 15 = \frac{15}{1}$$

Therefore, any multiplication of a whole number and a fraction can be converted into a multiplication of two fractions.

$$3\left(\frac{4}{7}\right) = \frac{3}{1}\left(\frac{4}{7}\right) = \frac{3(4)}{1(7)} = \frac{12}{7} \qquad \frac{7}{8}(1) = \frac{7}{8}\left(\frac{1}{1}\right) = \frac{7(1)}{8(1)} = \frac{7}{8}$$

Continued on following page.

2. Continued

However, we ordinarily use the shorter method below. That is, we get the numerator of the product by multiplying the whole number and the original numerator and keep the same denominator.

$3\left(\frac{4}{7}\right) = \frac{3(4)}{7} = \frac{12}{7}$ $\qquad \frac{7}{8}(1) = \frac{7(1)}{8} = \frac{7}{8}$

Use the shorter method for these:

a) $4\left(\frac{2}{9}\right) =$ ______ $\qquad$ b) $\frac{4}{3}(1) =$ ______

---

3. The rules for multiplication of integers also apply to multiplications of signed fractions.

When one factor is positive and the other is negative, the product is negative. For example:

$\frac{1}{3}\left(-\frac{4}{5}\right) = -\frac{4}{15}$ $\qquad (-7)\left(\frac{1}{5}\right) = -\frac{7}{5}$

When both factors have the same sign, the product is positive. For example:

$\left(\frac{2}{3}\right)\left(\frac{1}{5}\right) = \frac{2}{15}$ $\qquad \left(-\frac{3}{7}\right)(-4) = \frac{12}{7}$

Using the above rules, complete these:

a) $\left(-\frac{5}{2}\right)\left(\frac{3}{7}\right) =$ ______ $\qquad$ c) $\left(-\frac{3}{2}\right)\left(-\frac{5}{4}\right) =$ ______

b) $(-2)\left(-\frac{1}{3}\right) =$ ______ $\qquad$ d) $\left(-\frac{3}{5}\right)(2) =$ ______

Answers: a) $\frac{8}{9}$ $\qquad$ b) $\frac{4}{3}$

---

4. Three properties of multiplication are listed below.

1) When a fraction is multiplied by "1", the product is identical to the fraction.

$1\left(\frac{2}{5}\right) = \frac{2}{5}$ $\qquad \left(-\frac{7}{3}\right)(1) = -\frac{7}{3}$

2) When a fraction is multiplied by -1, the product is the opposite of the fraction.

$(-1)\left(\frac{7}{4}\right) = -\frac{7}{4}$ $\qquad \left(-\frac{1}{9}\right)(-1) = \frac{1}{9}$

3) When a fraction is multiplied by 0, the product is 0.

$0\left(\frac{3}{8}\right) = 0$ $\qquad \left(-\frac{13}{8}\right)(0) = 0$

Continued on following page.

Answers: a) $-\frac{15}{14}$ $\qquad$ b) $\frac{2}{3}$ $\qquad$ c) $\frac{15}{8}$ $\qquad$ d) $-\frac{6}{5}$

4. Continued

Using the properties above, complete these:

a) $(-1)\left(\frac{1}{2}\right) =$ ______ c) $0\left(-\frac{9}{7}\right) =$ ______

b) $\left(-\frac{5}{4}\right)(1) =$ ______ d) $\left(-\frac{3}{5}\right)(-1) =$ ______

a) $-\frac{1}{2}$ b) $-\frac{5}{4}$ c) 0 d) $\frac{3}{5}$

2-2 EQUIVALENT FRACTIONS

In this section, we will define equivalent fractions. Then we will discuss the procedures for raising signed fractions to higher terms and reducing signed fractions to lowest terms.

5. The numerator and denominator of a fraction are called its terms. Fractions that are equal even though they have different terms are called equivalent fractions.

$\frac{1}{2}$ and $\frac{3}{6}$ are equivalent fractions.

When two fractions are equivalent, we say that the one with larger terms is in higher terms. That is:

$\frac{3}{6}$ is in higher terms than $\frac{1}{2}$, because
3 is larger than "1" and 6 is larger than 2.

In each pair of equivalent fractions, which one is in higher terms?

a) $\frac{4}{3}$ and $\frac{8}{6}$ ____ b) $\frac{10}{20}$ and $\frac{2}{4}$ ____ c) $-\frac{3}{2}$ and $-\frac{6}{4}$ ____

a) $\frac{8}{6}$ b) $\frac{10}{20}$ c) $-\frac{6}{4}$

6. Any fraction whose numerator and denominator are the same is equal to "1".

$$1 = \frac{2}{2} = \frac{3}{3} = \frac{4}{4} = \frac{5}{5} = \frac{7}{7} = \frac{10}{10}$$

To raise a fraction to higher terms, we can multiply it by any fraction that equals "1".

By substituting $\frac{2}{2}$ for the "1" we raised $\frac{3}{4}$ to higher terms.

$$\frac{3}{4} = \frac{3}{4}(1) = \frac{3}{4}\left(\frac{2}{2}\right) = \frac{6}{8}$$

By substituting $\frac{5}{5}$ for the "1", raise $-\frac{1}{2}$ to higher terms.

$$-\frac{1}{2} = -\frac{1}{2}(1) = -\frac{1}{2}(\quad)$$

$-\frac{1}{2}\left(\frac{5}{5}\right) = -\frac{5}{10}$

7. In the problem below, we are asked to raise $\frac{2}{3}$ to an equivalent fraction whose denominator is 12. Since the 3 has to be multiplied by 4 to get the new denominator 12, we multiply $\frac{2}{3}$ by $\frac{4}{4}$ to get the equivalent fraction $\frac{8}{12}$.

$$\frac{2}{3} = \frac{\square}{12} \qquad \frac{2}{3} = \frac{2}{3}\left(\frac{4}{4}\right) = \frac{8}{12}$$

A more direct method can be used for the same problem. The two steps are:

1. Divide 12 by 3 and get 4.
2. Multiply 2 by 4 and get 8.

$$\frac{2}{3} = \frac{\boxed{8}}{12}$$

Using the more direct method, do these:

a) $\frac{3}{10} = \frac{\square}{20}$ b) $-\frac{5}{4} = -\frac{\square}{24}$ c) $-\frac{1}{2} = -\frac{\square}{100}$

---

8. To convert a whole number to a fraction, we can also multiply it by a fraction that equals "1".

Answers to 7:	a) $\frac{6}{20}$ b) $-\frac{30}{24}$ c) $-\frac{50}{100}$

By substituting $\frac{2}{2}$ for the "1". we converted 3 to a fraction.

$$3 = 3(1) = 3\left(\frac{2}{2}\right) = \frac{6}{2}$$

By substituting $\frac{4}{4}$ for the "1", convert -6 to a fraction.

$$-6 = -6(1) = -6(\quad) = \_\_\_\_$$

---

9. In the problems below, we are asked to convert 4 to an equivalent fraction whose denominator is 3 and -2 to an equivalent fraction whose denominator is 10. To do so, we multiplied 4 by $\frac{3}{3}$ and -2 by $\frac{10}{10}$.

Answer to 8: $-6\left(\frac{4}{4}\right) = -\frac{24}{4}$

$$4 = \frac{\square}{3} \qquad 4 = 4\left(\frac{3}{3}\right) = \frac{\boxed{12}}{3}$$

$$-2 = \frac{\square}{10} \qquad -2 = -2\left(\frac{10}{10}\right) = -\frac{\boxed{20}}{10}$$

The same problem can be done directly <u>by multiplying the absolute value of the integer by the desired denominator</u>.

$$4 = \frac{\boxed{12}}{3} \qquad \text{(Since } 4 \times 3 = 12\text{)}$$

$$-2 = -\frac{\boxed{20}}{10} \qquad \text{(Since } 2 \times 10 = 20\text{)}$$

Using the direct method, do these:

a) $3 = \frac{\square}{5}$ b) $-1 = -\frac{\square}{16}$ c) $-5 = -\frac{\square}{6}$

10. To factor a number, we write it as a multiplication. For example:

15 can be factored into 3(5) .

42 can be factored into 7(6) .

To reduce a fraction to lower terms, we can factor the original terms to get a fraction that equals "1". That is:

$$\frac{6}{10} = \frac{3(2)}{5(2)} = \frac{3}{5}\left(\frac{2}{2}\right) = \frac{3}{5}(1) = \frac{3}{5}$$

Using the same method, complete each reduction.

a) $\frac{6}{9} = \frac{2(3)}{3(3)} = \frac{2}{3}\left(\frac{3}{3}\right) = \frac{2}{3}(\quad) = $ ______

b) $\frac{24}{20} = \frac{6(4)}{5(4)} = \frac{6}{5}\left(\frac{4}{4}\right) = \frac{6}{5}(\quad) = $ ______

Answers: a) $\frac{15}{5}$ b) $-\frac{16}{16}$ c) $-\frac{30}{6}$

---

11. A fraction can be reduced to lower terms only if both terms contain a common factor.

$\frac{10}{14}$ can be reduced because of the common factor 2 .

$\frac{6}{11}$ cannot be reduced because there is no common factor.

When a fraction cannot be reduced, we say that it is in lowest terms. Which fractions below are in lowest terms? ______

a) $\frac{1}{3}$ b) $\frac{14}{16}$ c) $\frac{8}{12}$ d) $\frac{5}{8}$

Answers: a) $\frac{2}{3}(1) = \frac{2}{3}$ b) $\frac{6}{5}(1) = \frac{6}{5}$

---

12. The terms of $\frac{8}{24}$ have three common factors: 2, 4, and 8. We used each to reduce the fraction to lower terms below. Only $\frac{1}{3}$ is in lowest terms; $\frac{4}{12}$ and $\frac{2}{6}$ can be reduced further.

$$\frac{8}{24} = \left(\frac{4}{12}\right)\left(\frac{2}{2}\right) = \left(\frac{4}{12}\right)(1) = \frac{4}{12}$$

$$\frac{8}{24} = \left(\frac{2}{6}\right)\left(\frac{4}{4}\right) = \left(\frac{2}{6}\right)(1) = \frac{2}{6}$$

$$\frac{8}{24} = \left(\frac{1}{3}\right)\left(\frac{8}{8}\right) = \left(\frac{1}{3}\right)(1) = \frac{1}{3}$$

When the terms of a fraction have more than one common factor:

1) We can reduce to lower terms by factoring out any of them.

2) We can reduce to lowest terms in one step only by factoring out the largest one.

Answer: Only (a) and (d)

Continued on following page.

12. Continued

Identify the largest common factor for the terms of each fraction.

a) $\frac{12}{18}$ ____ b) $\frac{36}{24}$ ____ c) $\frac{44}{66}$ ____

Answer: a) 6 b) 12 c) 22

13. Since it is sometimes difficult to identify the largest common factor, we sometimes need more than one step to reduce a fraction to lowest terms. For example, we used 8 as the common factor to reduce the fraction below.

$$\frac{32}{48} = \left(\frac{4}{6}\right)\left(\frac{8}{8}\right) = \frac{4}{6}(1) = \frac{4}{6}$$

Since $\frac{4}{6}$ is not in lowest terms, we must reduce further. We get:

$$\frac{32}{48} = \frac{4}{6} = \_\_\_\_$$

Answer: $\frac{2}{3}$

14. Reduce each fraction to lowest terms.

a) $\frac{4}{8} =$ ____ b) $\frac{25}{10} =$ ____ c) $\frac{48}{18} =$ ____

Answer: a) $\frac{1}{2}$ b) $\frac{5}{2}$ c) $\frac{8}{3}$

15. Negative fractions can also be reduced to lowest terms. For example:

$$-\frac{8}{12} = -\frac{2}{3} \qquad -\frac{20}{16} = -\frac{5}{4}$$

Reduce each fraction to lowest terms.

a) $-\frac{16}{12} =$ ____ b) $-\frac{14}{21} =$ ____ c) $-\frac{27}{81} =$ ____

Answer: a) $-\frac{4}{3}$ b) $-\frac{2}{3}$ c) $-\frac{1}{3}$

16. A fraction that equals an integer can be converted to an integer by reducing to lowest terms. For example:

$$\frac{24}{8} = \left(\frac{3}{1}\right)\left(\frac{8}{8}\right) = \left(\frac{3}{1}\right)(1) = \frac{3}{1} \text{ or } 3$$

However, we usually make such conversions by simply performing a division. That is:

$\frac{24}{8} = 24 \div 8 = 3$ a) $\frac{30}{6} =$ ____ b) $-\frac{14}{2} =$ ____

Answer: a) 5 b) -7

2-3 CANCELLING IN MULTIPLICATIONS

In this section, we will discuss multiplications with products that need to be reduced to lowest terms. We will show how the "cancelling" process can be used to simplify multiplications of that type.

---

17. When multiplying fractions, <u>the product should always be reduced to lowest terms</u>. For example:

$$\left(\frac{1}{2}\right)\left(\frac{4}{5}\right) = \frac{4}{10} = \frac{2}{5} \qquad\qquad 8\left(\frac{3}{4}\right) = \frac{24}{4} = 6$$

Complete. Reduce each product to lowest terms:

a) $\left(\frac{4}{3}\right)\left(-\frac{6}{7}\right) =$ ______ b) $\left(-\frac{1}{5}\right)(-5) =$ ______

---

18. We had to reduce the product at the left below to lowest terms. To avoid that reduction, we can use a process called <u>cancelling</u> in which we take out common factors <u>before</u> multiplying. We did so at the right below.

$$\left(\frac{4}{5}\right)\left(\frac{3}{8}\right) = \frac{12}{40} = \frac{3}{10} \qquad\qquad \left(\frac{\overset{1}{\cancel{4}}}{5}\right)\left(\frac{3}{\underset{2}{\cancel{8}}}\right) = \frac{3}{10}$$

Notice these points about the cancelling process.

1) We divided the "4" and the "8" by 4.

2) We got the terms of the product by the multiplications (1)(3) and (5)(2).

Use cancelling to complete these:

a) $\left(\frac{3}{4}\right)\left(\frac{5}{6}\right) =$ ______ b) $\left(-\frac{7}{6}\right)\left(\frac{4}{5}\right) =$ ______

Answers to 17: a) $-\frac{8}{7}$ b) 1

---

19. Two cancellings were possible below. We divided both 9 and 6 by 3. We divided both 5 and 10 by 5. Use cancelling to complete the other multiplications.

$$\left(\frac{\overset{3}{\cancel{9}}}{\underset{2}{\cancel{10}}}\right)\left(\frac{\overset{1}{\cancel{5}}}{\underset{2}{\cancel{6}}}\right) = \frac{3}{4} \qquad\qquad \left(-\frac{9}{14}\right)\left(-\frac{7}{12}\right) = \text{\_\_\_\_\_\_}$$

Answers to 18:

a) $\left(\frac{\overset{1}{\cancel{3}}}{4}\right)\left(\frac{5}{\underset{2}{\cancel{6}}}\right) = \frac{5}{8}$

b) $\left(-\frac{7}{\underset{3}{\cancel{6}}}\right)\left(\frac{\overset{2}{\cancel{4}}}{5}\right) = -\frac{14}{15}$

---

20. We got the numerator of the product below by multiplying "1" and "1". Complete the other multiplication.

$$\left(\frac{\overset{1}{\cancel{4}}}{\underset{3}{\cancel{21}}}\right)\left(\frac{\overset{1}{\cancel{7}}}{\underset{2}{\cancel{8}}}\right) = \frac{1}{6} \qquad\qquad \left(\frac{3}{10}\right)\left(-\frac{5}{6}\right) = \text{\_\_\_\_\_\_}$$

Answer to 19:

$\left(-\frac{\overset{3}{\cancel{9}}}{\underset{2}{\cancel{14}}}\right)\left(-\frac{\overset{1}{\cancel{7}}}{\underset{4}{\cancel{12}}}\right) = \frac{3}{8}$

21. We got the denominator of the product below by multiplying "1" and "1". Complete the other multiplication.

$$\left(\frac{\cancel{9}^{3}}{\cancel{5}_{1}}\right)\left(\frac{\cancel{5}^{1}}{\cancel{2}_{1}}\right) = \frac{3}{1} = 3 \qquad \left(-\frac{11}{15}\right)\left(-\frac{15}{11}\right) = \_\_\_\_$$

Answer: $\left(\frac{\cancel{3}^{1}}{\cancel{10}_{2}}\right)\left(-\frac{\cancel{5}^{1}}{\cancel{6}_{2}}\right) = -\frac{1}{4}$

---

22. We can also cancel when multiplying a whole number and a fraction. For example:

$$\cancel{3}^{1}\left(\frac{5}{\cancel{9}_{3}}\right) = \frac{5}{3} \qquad \cancel{10}^{2}\left(\frac{2}{\cancel{5}_{1}}\right) = \frac{4}{1} = 4$$

Use cancelling to complete these:

a) $8\left(-\frac{5}{6}\right) = \_\_\_\_$ b) $\left(-\frac{4}{7}\right)(-7) = \_\_\_\_$

Answer: $\left(-\frac{\cancel{11}^{1}}{\cancel{15}_{1}}\right)\left(-\frac{\cancel{15}^{1}}{\cancel{11}_{1}}\right) = \frac{1}{1} = 1$

---

23. Cancelling is not possible with all multiplications. Some examples are shown.

$$\left(\frac{3}{7}\right)\left(\frac{4}{5}\right) = \frac{12}{35} \qquad 24\left(-\frac{2}{5}\right) = -\frac{48}{5} \qquad \left(-\frac{8}{9}\right)(-13) = \frac{104}{9}$$

Note: Even if you do cancel, always check to see that the product is in lowest terms.

Answers:

a) $\cancel{8}^{4}\left(-\frac{5}{\cancel{6}_{3}}\right) = -\frac{20}{3}$

b) $\left(-\frac{4}{\cancel{7}_{1}}\right)(-\cancel{7}^{1}) = \frac{4}{1} = 4$

2-4 RECIPROCALS AND DIVISION

In this section, we will define reciprocals (also called multiplicative inverses). We will show how any division of integers can be converted to an equivalent multiplication. We will also show that division by 0 is an impossible operation.

---

24. Two numbers whose product is "1" are called reciprocals of each other. For example:

Since $\left(\frac{1}{3}\right)(3) = 1$: The reciprocal of $\frac{1}{3}$ is 3. The reciprocal of 3 is $\frac{1}{3}$.

Since $\left(-\frac{1}{5}\right)(-5) = 1$: The reciprocal of $-\frac{1}{5}$ is -5. The reciprocal of -5 is $-\frac{1}{5}$.

Write the reciprocals of these:

a) $\frac{1}{7}$ ____ b) $-\frac{1}{6}$ ____ c) 10 ____ d) -4 ____

25. A pair of fractions can also be reciprocals. For example:

Since $\left(\frac{5}{3}\right)\left(\frac{3}{5}\right) = 1$: The reciprocal of $\frac{5}{3}$ is $\frac{3}{5}$.

The reciprocal of $\frac{3}{5}$ is $\frac{5}{3}$.

Since $\left(-\frac{2}{7}\right)\left(-\frac{7}{2}\right) = 1$: The reciprocal of $-\frac{2}{7}$ is $-\frac{7}{2}$.

The reciprocal of $-\frac{7}{2}$ is $-\frac{2}{7}$.

Write the reciprocals of these:

a) $\frac{3}{4}$ ______ b) $-\frac{9}{8}$ ______ c) $\frac{6}{11}$ ______ d) $-\frac{25}{33}$ ______

Answers: a) 7 b) -6 c) $\frac{1}{10}$ d) $-\frac{1}{4}$

---

26. There are two numbers that are their own reciprocals.

a) Since (1)(1) = 1, the reciprocal of "1" is ______.

b) Since (-1)(-1) = 1, the reciprocal of -1 is ______.

Answers: a) $\frac{4}{3}$ b) $-\frac{8}{9}$ c) $\frac{11}{6}$ d) $-\frac{33}{25}$

---

27. To see that 0 has no reciprocal, answer these:

a) Whenever one factor is 0, the product is always ______.

b) Can we get "1" as a product when one factor is 0? ______

c) Therefore, does 0 have a reciprocal? ______

Answers: a) 1 b) -1

---

28. Any division can be converted to an equivalent multiplication by multiplying the numerator by the reciprocal of the denominator. For example:

the reciprocal of 3

$$\frac{12}{3} = 12\left(\frac{1}{3}\right)$$

the reciprocal of 8

$$\frac{7}{8} = 7\left(\frac{1}{8}\right)$$

Divisions involving signed numbers can be converted to multiplications in the same way. For example:

the reciprocal of -5

$$\frac{10}{-5} = 10\left(-\frac{1}{5}\right)$$

the reciprocal of -4

$$\frac{-3}{-4} = -3\left(-\frac{1}{4}\right)$$

Continued on following page.

Answers: a) 0 b) No c) No

28. Continued

Convert each division to an equivalent multiplication.

a) $\frac{-16}{8} =$ ______ b) $\frac{20}{-2} =$ ______ c) $\frac{-35}{-7} =$ ______

Answers: a) $-16\left(\frac{1}{8}\right)$ b) $20\left(-\frac{1}{2}\right)$ c) $-35\left(-\frac{1}{7}\right)$

29. When a division by 0 is converted to multiplication, one of the factors is "the reciprocal of 0". For example:

$$\frac{6}{0} = 6 \text{ (the reciprocal of 0)}$$

$$\frac{-4}{0} = -4 \text{ (the reciprocal of 0)}$$

$$\frac{0}{0} = 0 \text{ (the reciprocal of 0)}$$

Since 0 has no reciprocal, DIVISION BY 0 IS IMPOSSIBLE. Perform each possible division below.

a) $\frac{0}{1} =$ ______ b) $\frac{1}{0} =$ ______ c) $\frac{0}{-3} =$ ______ d) $\frac{-3}{0} =$ ______

Answers: a) 0 b) impossible c) 0 d) impossible

30. Any fraction with a negative numerator alone or a negative denominator alone is equal to a negative fraction. For example:

$$\frac{-8}{2} \text{ and } \frac{8}{-2} \text{ equal } -\frac{8}{2}, \text{ because each equals } -4$$

The same fact applies to fractions that do not equal whole numbers. That is:

$$\frac{-3}{4} \text{ and } \frac{3}{-4} \text{ equal } -\frac{3}{4}$$

Write each fraction in two equivalent forms.

a) $\frac{-15}{3} =$ ______ and ______ b) $-\frac{1}{5} =$ ______ and ______

Answers: a) $\frac{15}{-3}$ and $-\frac{15}{3}$ b) $\frac{-1}{5}$ and $\frac{1}{-5}$

2-5 DIVISION OF FRACTIONS

In this section, we will discuss the procedure for divisions involving signed fractions.

---

31. Any division involving a fraction is written as a complex fraction in algebra. For example:

$$\frac{2}{3} \div \frac{4}{7} \text{ is written } \frac{\frac{2}{3}}{\frac{4}{7}}$$

$$\frac{5}{8} \div 6 \text{ is written } \frac{\frac{5}{8}}{6}$$

$10 \div \frac{1}{2}$ is written ______

32. In a complex fraction, the longer fraction line is called the major fraction line. The numerator or denominator or both are fractions. For example:

In $\dfrac{\frac{1}{4}}{\frac{3}{8}}$: the numerator is $\frac{1}{4}$; the denominator is $\frac{3}{8}$.

In $\dfrac{\frac{3}{10}}{9}$: the numerator is $\frac{3}{10}$; the denominator is 9 .

In $\dfrac{2}{\frac{5}{6}}$: a) the numerator is ______; b) the denominator is ______

Answer: $\dfrac{10}{\frac{1}{2}}$

---

33. To perform a division in complex-fraction form, we multiply the numerator by the reciprocal of the denominator. For example:

the reciprocal of $\frac{4}{9}$

$$\dfrac{\frac{3}{7}}{\frac{4}{9}} = \frac{3}{7}\left(\frac{9}{4}\right) = \frac{27}{28}$$

the reciprocal of $\frac{5}{3}$

$$\dfrac{6}{\frac{5}{3}} = 6\left(\frac{3}{5}\right) = \frac{18}{5}$$

the reciprocal of 10

$$\dfrac{\frac{1}{4}}{10} = \frac{1}{4}\left(\quad\right) = \_\_\_\_$$

Answer: a) 2 b) $\frac{5}{6}$

---

34. The rules for dividing integers also apply to divisions involving signed fractions.

When one term is positive and the other negative, the quotient is negative.

$$\dfrac{-\frac{2}{3}}{\frac{7}{11}} = \left(-\frac{2}{3}\right)\left(\frac{11}{7}\right) = -\frac{22}{21}$$

When both terms are positive or both terms are negative, the quotient is positive.

$$\dfrac{-\frac{1}{2}}{-3} = \left(-\frac{1}{2}\right)\left(-\frac{1}{3}\right) = \frac{1}{6}$$

Complete each division:

a) $\dfrac{2}{-\frac{3}{7}} =$ ________ b) $\dfrac{-\frac{1}{5}}{-\frac{3}{2}} =$ ________

Answer: $\frac{1}{4}\left(\frac{1}{10}\right) = \frac{1}{40}$

35. When dividing fractions, <u>we always reduce the quotient to lowest terms</u>. For example:

$$\frac{\frac{9}{8}}{\frac{3}{4}} = \left(\frac{\overset{3}{\cancel{9}}}{\underset{2}{\cancel{8}}}\right)\left(\frac{\overset{1}{\cancel{4}}}{\underset{1}{\cancel{3}}}\right) = \frac{3}{2}$$

Complete. Reduce each quotient to lowest terms.

a) $\frac{-9}{\frac{6}{5}} =$ ______ b) $\frac{-\frac{2}{3}}{-8} =$ ______

a) $-\frac{14}{3}$ b) $\frac{2}{15}$

---

36. Three properties of division are extended to divisions involving fractions below.

1) When a fraction is divided by itself, the quotient is "1".

$$\frac{\frac{3}{5}}{\frac{3}{5}} = \left(\frac{3}{5}\right)\left(\frac{5}{3}\right) = 1 \qquad \frac{-\frac{7}{2}}{-\frac{7}{2}} = \left(-\frac{7}{2}\right)\left(-\frac{2}{7}\right) = 1$$

2) When a fraction is divided by "1", the quotient is <u>identical</u> to the fraction.

$$\frac{\frac{9}{8}}{1} = \frac{9}{8}(1) = \frac{9}{8} \qquad \frac{-\frac{1}{4}}{1} = \left(-\frac{1}{4}\right)(1) = -\frac{1}{4}$$

3) When a fraction is divided by -1, the quotient is the <u>opposite</u> of the fraction.

$$\frac{\frac{4}{5}}{-1} = \left(\frac{4}{5}\right)(-1) = -\frac{4}{5} \qquad \frac{-\frac{8}{3}}{-1} = \left(-\frac{8}{3}\right)(-1) = \frac{8}{3}$$

Using the properties above, complete these:

a) $\frac{-\frac{5}{8}}{-\frac{5}{8}} =$ ______ b) $\frac{-\frac{7}{16}}{1} =$ ______ c) $\frac{-\frac{1}{2}}{-1} =$ ______

a) $-\frac{15}{2}$ b) $\frac{1}{12}$

---

37. Two more properties of division are extended to divisions involving fractions below.

1) When 0 is divided by a fraction, the quotient is 0.

$$\frac{0}{\frac{3}{4}} = 0\left(\frac{4}{3}\right) = 0 \qquad \frac{0}{-\frac{6}{5}} = 0\left(-\frac{5}{6}\right) = 0$$

2) Dividing a fraction by 0 is <u>IMPOSSIBLE</u> because division by 0 is impossible.

The divisions $\frac{\frac{5}{2}}{0}$ and $\frac{-\frac{6}{7}}{0}$ are not possible.

Continued on following page.

a) 1

b) $-\frac{7}{16}$

c) $\frac{1}{2}$

37. Continued

Using the properties, complete these:

a) $\dfrac{0}{\frac{1}{4}} =$ ____ b) $\dfrac{\frac{1}{4}}{0} =$ ____ c) $\dfrac{0}{-\frac{7}{2}} =$ ____ d) $\dfrac{-\frac{7}{2}}{0} =$ ____

a) 0 b) impossible c) 0 d) impossible

SELF-TEST 4 (pages 49-61)

Multiply. Report each product in lowest terms.

1. $\left(\frac{5}{8}\right)\left(\frac{1}{7}\right) =$ ____

2. $\left(-\frac{2}{3}\right)\left(\frac{3}{8}\right) =$ ____

3. $\left(-\frac{6}{5}\right)\left(-\frac{15}{16}\right) =$ ____

4. $(10)\left(-\frac{3}{4}\right) =$ ____

Complete:

5. $\frac{4}{5} = \frac{\square}{15}$

6. $-\frac{1}{8} = -\frac{\square}{40}$

Reduce each fraction to lowest terms.

7. $-\frac{27}{45} =$ ____

8. $\frac{60}{12} =$ ____

Divide. Report each quotient in lowest terms.

9. $\dfrac{\frac{1}{2}}{\frac{3}{4}} =$ ____

10. $\dfrac{-\frac{5}{2}}{-\frac{1}{6}} =$ ____

11. $\dfrac{2}{-\frac{8}{5}} =$ ____

12. $\dfrac{-\frac{2}{3}}{12} =$ ____

ANSWERS:

1. $\frac{5}{56}$
2. $-\frac{1}{4}$
3. $\frac{9}{8}$
4. $-\frac{15}{2}$
5. $\frac{\boxed{12}}{15}$
6. $-\frac{\boxed{5}}{40}$
7. $-\frac{3}{5}$
8. 5
9. $\frac{2}{3}$
10. 15
11. $-\frac{5}{4}$
12. $-\frac{1}{18}$

2-6 ADDITION OF FRACTIONS

In this section, we will discuss the procedure for additions involving signed fractions.

---

38. To add fractions with like or common denominators, we add their numerators and keep the same denominator. For example:

$$\frac{4}{7} + \frac{1}{7} = \frac{4+1}{7} = \frac{5}{7}$$

When adding signed fractions, we put all negative signs in their numerators. For example:

$$\frac{7}{9} + \left(-\frac{2}{9}\right) = \frac{7+(-2)}{9} = \frac{5}{9}$$

$$-\frac{2}{5} + \left(-\frac{4}{5}\right) = \frac{-2+(-4)}{5} = \frac{-6}{5} = -\frac{6}{5}$$

Complete these additions:

a) $-\frac{1}{3} + \frac{2}{3} =$ ______ b) $-\frac{5}{7} + \left(-\frac{5}{7}\right) =$ ______

Answers to 38: a) $\frac{1}{3}$ b) $-\frac{10}{7}$

39. When adding fractions, we always reduce the sum to lowest terms. For example:

$$\frac{1}{10} + \frac{3}{10} = \frac{4}{10} = \frac{2}{5} \qquad \frac{2}{3} + \frac{7}{3} = \frac{9}{3} = 3$$

Complete. Reduce each sum to lowest terms.

a) $\frac{3}{4} + \frac{3}{4} =$ ______ b) $\frac{8}{9} + \frac{1}{9} =$ ______

Answers to 39: a) $\frac{3}{2}$ b) 1

40. In each addition below, we reduced the sum to lowest terms.

$$\frac{3}{8} + \left(-\frac{7}{8}\right) = \frac{-4}{8} = -\frac{1}{2}$$

$$-\frac{5}{2} + \left(-\frac{3}{2}\right) = \frac{-8}{2} = -4$$

Complete. Reduce each sum to lowest terms.

a) $-\frac{11}{16} + \left(-\frac{5}{16}\right) =$ ______ b) $-\frac{5}{6} + \frac{1}{6} =$ ______

Answers to 40: a) -1 b) $-\frac{2}{3}$

41. Two fractions whose sum is 0 are called opposites. Therefore:

Since $\frac{3}{4} + \left(-\frac{3}{4}\right) = 0$: The opposite of $\frac{3}{4}$ is $-\frac{3}{4}$.

The opposite of $-\frac{3}{4}$ is $\frac{3}{4}$.

Write the opposite of each fraction.

a) $\frac{3}{8}$ ______ b) $-\frac{7}{4}$ ______ c) $-\frac{1}{2}$ ______ d) $\frac{11}{16}$ ______

42. To add a fraction and a whole number, we convert the whole number to a fraction with the same denominator. For example:

$$3 + \frac{1}{2} = \frac{6}{2} + \frac{1}{2} = \frac{7}{2}$$

$$\frac{5}{8} + (-1) = \frac{5}{8} + \left(-\frac{8}{8}\right) = -\frac{3}{8}$$

Using the same steps, complete these:

a) $1 + \frac{5}{6} =$ ______ b) $-\frac{4}{3} + (-2) =$ ______

Answers: a) $-\frac{3}{8}$ b) $\frac{7}{4}$ c) $\frac{1}{2}$ d) $-\frac{11}{16}$

---

43. We cannot add the fractions below until we get a common denominator.

$$\frac{5}{6} + \frac{1}{8}$$

A common denominator must be a multiple of both original denominators. The multiples of a number can be obtained by counting by that number. For example:

Multiples of 6: 6, 12, 18, 24, 30, 36, 42, 48, 54...

Multiples of 8: 8, 16, 24, 32, 40, 48, 56, 64, 72...

Since both 24 and 48 are multiples of both 6 and 8, we can use either for the addition.

Using 24, we get: $\frac{5}{6} + \frac{1}{8} = \frac{20}{24} + \frac{3}{24} = \frac{23}{24}$

Using 48, we get: $\frac{5}{6} + \frac{1}{8} = \frac{40}{48} + \frac{6}{48} = \frac{46}{48} = \frac{23}{24}$

Though either 24 or 48 can be used for the addition above, it is simpler to use 24 because we get smaller numbers and avoid reducing to lowest terms. 24 is the lowest common multiple of 6 and 8. The lowest common multiple of both denominators is called the lowest common denominator (or LCD).

For $\frac{5}{6} + \frac{1}{8}$, the lowest common denominator is ______

Answers: a) $\frac{11}{6}$ b) $-\frac{10}{3}$

24

44. When adding fractions, we try to use the lowest common denominator (LCD) because using it is simpler. When the larger denominator is a multiple of the smaller, the larger denominator is the LCD.

For $\frac{3}{4} + \frac{1}{8}$, the LCD is 8.

$$\frac{3}{4} + \frac{1}{8} = \frac{6}{8} + \frac{1}{8} = \frac{7}{8}$$

For $-\frac{9}{10} + \frac{2}{5}$, the LCD is 10.

$$-\frac{9}{10} + \frac{2}{5} = -\frac{9}{10} + \frac{4}{10} = -\frac{5}{10} = -\frac{1}{2}$$

Complete. Reduce each sum to lowest terms.

a) $\frac{5}{6} + \left(-\frac{1}{3}\right) =$ ______

b) $-\frac{4}{5} + \left(-\frac{8}{15}\right) =$ ______

Answers: a) $\frac{1}{2}$, from $\frac{3}{6}$ b) $-\frac{4}{3}$, from $-\frac{20}{15}$

45. In the additions below, neither denominator is the lowest common denominator.

For $\frac{1}{5} + \frac{2}{3}$, the LCD is 15.

$$\frac{1}{5} + \frac{2}{3} = \frac{3}{15} + \frac{10}{15} = \frac{13}{15}$$

For $\frac{5}{6} + \left(-\frac{9}{10}\right)$, the LCD is 30.

$$\frac{5}{6} + \left(-\frac{9}{10}\right) = \frac{25}{30} + \left(-\frac{27}{30}\right) = -\frac{2}{30} = -\frac{1}{15}$$

Find the LCD for each addition.

a) $\frac{1}{2} + \frac{1}{3}$ ______

b) $-\frac{5}{6} + \frac{3}{4}$ ______

c) $\frac{6}{7} + \left(-\frac{1}{3}\right)$ ______

d) $-\frac{5}{8} + \left(-\frac{7}{12}\right)$ ______

Answers: a) 6 b) 12 c) 21 d) 24

46. Use the LCD for these:

a) $\frac{3}{4} + \left(-\frac{1}{10}\right) =$ ______

b) $-\frac{2}{5} + \left(-\frac{1}{2}\right) =$ ______

47. Complete. Reduce to lowest terms.

a) $-\frac{1}{4} + \frac{7}{12} =$ ______

b) $-\frac{3}{5} + \left(-\frac{5}{6}\right) =$ ______

Answers: a) $\frac{13}{20}$ b) $-\frac{9}{10}$

Answers: a) $\frac{1}{3}$ b) $-\frac{43}{30}$

2-7 SUBTRACTION OF FRACTIONS

In this section, we will discuss the procedure for subtractions involving signed fractions.

48. To subtract fractions with common denominators, we subtract their numerators and keep the same denominator. The difference is always reduced to lowest terms. For example:

$$\frac{11}{4} - \frac{5}{4} = \frac{11-5}{4} = \frac{6}{4} = \frac{3}{2}$$

$$\frac{9}{8} - \frac{1}{8} = \frac{9-1}{8} = \frac{8}{8} = 1$$

Complete. Reduce to lowest terms.

a) $\frac{9}{10} - \frac{7}{10} =$ ______ b) $\frac{9}{2} - \frac{3}{2} =$ ______

49. When a fraction is subtracted from itself, the difference is 0. That is:

$\frac{11}{4} - \frac{11}{4} = \frac{0}{4} = 0$ $\frac{1}{16} - \frac{1}{16} =$ ______

Answers: a) $\frac{1}{5}$ b) 3

50. To subtract fractions with "unlike" denominators, we must get common denominators first. For example:

$$\frac{13}{12} - \frac{1}{4} = \frac{13}{12} - \frac{3}{12} = \frac{10}{12} = \frac{5}{6}$$

$$\frac{1}{2} - \frac{2}{5} = \frac{5}{10} - \frac{4}{10} = \frac{1}{10}$$

Complete each subtraction.

a) $\frac{7}{3} - \frac{5}{6} =$ ______ b) $\frac{11}{6} - \frac{3}{8} =$ ______

Answer: 0

51. In each subtraction below, we converted the whole number to a fraction with the common denominator.

$1 - \frac{6}{7} = \frac{7}{7} - \frac{6}{7} = \frac{1}{7}$ $\frac{5}{2} - 2 = \frac{5}{2} - \frac{4}{2} = \frac{1}{2}$

Answers: a) $\frac{3}{2}$ b) $\frac{35}{24}$

Continued on following page.

51. Continued

Using the same method, complete these:

a) $2 - \frac{7}{8} =$ ________ b) $\frac{7}{4} - 1 =$ ________

Answers to 51: a) $\frac{9}{8}$ b) $\frac{3}{4}$

52. Any subtraction of fractions can be converted to an equivalent addition by ADDING THE OPPOSITE OF THE SECOND TERM. That is:

Change - to +; Opposite of $\frac{4}{5}$

$$\frac{2}{5} - \frac{4}{5} = \frac{2}{5} + \left(-\frac{4}{5}\right)$$

Change - to +; Opposite of $-\frac{1}{9}$

$$-\frac{7}{9} - \left(-\frac{1}{9}\right) = -\frac{7}{9} + \frac{1}{9}$$

Convert each subtraction to an equivalent addition.

a) $-\frac{5}{4} - \frac{7}{4} =$ ________ b) $\frac{7}{8} - \left(-\frac{3}{8}\right) =$ ________

Answers to 52: a) $-\frac{5}{4} + \left(-\frac{7}{4}\right)$ b) $\frac{7}{8} + \frac{3}{8}$

53. To perform each subtraction below, we converted to addition and then found the difference.

$$\frac{3}{8} - \frac{7}{8} = \frac{3}{8} + \left(-\frac{7}{8}\right) = -\frac{4}{8} = -\frac{1}{2}$$

$$-\frac{5}{4} - \left(-\frac{1}{4}\right) = -\frac{5}{4} + \frac{1}{4} = -\frac{4}{4} = -1$$

Using the same method, complete these:

a) $-\frac{1}{5} - \frac{9}{5} =$ ________

b) $\frac{11}{16} - \left(-\frac{3}{16}\right) =$ ________

Answers to 53: a) -2 b) $\frac{7}{8}$

54. In each subtraction below, we had to get common denominators before finding the difference.

$$\frac{1}{4} - \frac{3}{2} = \frac{1}{4} + \left(-\frac{3}{2}\right) = \frac{1}{4} + \left(-\frac{6}{4}\right) = -\frac{5}{4}$$

$$-\frac{1}{6} - \left(-\frac{3}{8}\right) = -\frac{1}{6} + \frac{3}{8} = -\frac{4}{24} + \frac{9}{24} = \frac{5}{24}$$

Using the same method, complete these:

a) $\frac{5}{12} - \left(-\frac{1}{4}\right) =$ ________

b) $-\frac{1}{3} - \frac{1}{2} =$ ________

Answers to 54: a) $\frac{2}{3}$ b) $-\frac{5}{6}$

55. In each subtraction below, we converted the whole number to a fraction.

$$1 - \frac{7}{4} = 1 + \left(-\frac{7}{4}\right) = \frac{4}{4} + \left(-\frac{7}{4}\right) = -\frac{3}{4}$$

$$-\frac{2}{3} - (-2) = -\frac{2}{3} + 2 = -\frac{2}{3} + \frac{6}{3} = \frac{4}{3}$$

Using the same method, complete these:

a) $-\frac{4}{5} - 1 =$ ________________

b) $-\frac{5}{2} - (-3) =$ ________________

Answers: a) $-\frac{9}{5}$ b) $\frac{1}{2}$

2-8 POSITIVE EXPONENTS AND FRACTIONS

In this section, we will discuss powers with positive integer exponents in which the base is a fraction.

56. Each exponential expression below is a power in which the base is a fraction. They are named in the usual way. That is:

$\left(\frac{3}{4}\right)^2$ is called "$\frac{3}{4}$ to the <u>second</u> power" or "$\frac{3}{4}$ squared".

$\left(\frac{1}{2}\right)^3$ is called "$\frac{1}{2}$ to the <u>third</u> power" or "$\frac{1}{2}$ cubed".

$\left(\frac{5}{3}\right)^6$ is called "$\frac{5}{3}$ to the ____________ power".

Answer: sixth

57. We found the value of the power below by completing the multiplication.

$$\left(\frac{1}{2}\right)^3 = \left(\frac{1}{2}\right)\left(\frac{1}{2}\right)\left(\frac{1}{2}\right) = \frac{1}{8}$$

Complete these:

a) $\left(\frac{5}{4}\right)^2 = \left(\frac{5}{4}\right)\left(\frac{5}{4}\right) =$ ____

b) $\left(\frac{2}{3}\right)^4 = \left(\frac{2}{3}\right)\left(\frac{2}{3}\right)\left(\frac{2}{3}\right)\left(\frac{2}{3}\right) =$ ____

Answers: a) $\frac{25}{16}$ b) $\frac{16}{81}$

58. We found the value of the power below by completing the multiplication.

$$\left(-\frac{2}{5}\right)^3 = \left(-\frac{2}{5}\right)\left(-\frac{2}{5}\right)\left(-\frac{2}{5}\right) = -\frac{8}{125}$$

Complete these:

a) $\left(-\frac{7}{6}\right)^2 = \left(-\frac{7}{6}\right)\left(-\frac{7}{6}\right) =$ ______

b) $\left(-\frac{1}{2}\right)^5 = \left(-\frac{1}{2}\right)\left(-\frac{1}{2}\right)\left(-\frac{1}{2}\right)\left(-\frac{1}{2}\right)\left(-\frac{1}{2}\right) =$ ______

59. Don't confuse $\left(-\frac{3}{4}\right)^2$ with $-\left(\frac{3}{4}\right)^2$.

$$\left(-\frac{3}{4}\right)^2 \quad \text{means} \quad \left(-\frac{3}{4}\right)\left(-\frac{3}{4}\right) = \frac{9}{16}$$

$$-\left(\frac{3}{4}\right)^2 \quad \text{means} \quad -\left(\frac{3}{4}\right)\left(\frac{3}{4}\right) = -\frac{9}{16}$$

Following the pattern above, complete these:

a) $\left(-\frac{1}{2}\right)^4 =$ ________ b) $-\left(\frac{1}{2}\right)^4 =$ ________

Answers: a) $\frac{49}{36}$ b) $-\frac{1}{32}$

---

60. Any power whose exponent is "1" is equal to its base. That is:

$\left(\frac{4}{5}\right)^1 = \frac{4}{5}$ $\left(-\frac{2}{3}\right)^1 = -\frac{2}{3}$ $\left(-\frac{7}{2}\right)^1 =$ ______

Answers: a) $\frac{1}{16}$ b) $-\frac{1}{16}$

---

61. Complete these:

a) $\left(\frac{4}{3}\right)^3 =$ ______ c) $-\left(\frac{9}{10}\right)^2 =$ ______

b) $\left(-\frac{6}{7}\right)^1 =$ ______ d) $\left(-\frac{1}{3}\right)^4 =$ ______

Answer: $-\frac{7}{2}$

Answers: a) $\frac{64}{27}$ b) $-\frac{6}{7}$ c) $-\frac{81}{100}$ d) $\frac{1}{81}$

2-9 ALGEBRAIC EXPRESSIONS AND FRACTIONS

In this section, we will discuss evaluations of algebraic expressions in which a fraction is substituted for the variable. The proper order of operations is shown.

---

62. We evaluated the first expression when $x = \frac{5}{8}$. Notice that we did the multiplication first. Evaluate the second expression when $x = \frac{1}{4}$.

$$3x - 1 = 3\left(\frac{5}{8}\right) - 1 = \frac{15}{8} - 1 = \frac{15}{8} - \frac{8}{8} = \frac{7}{8}$$

$$2 - 3x = 2 - 3\left(\frac{1}{4}\right) = \underline{\qquad\qquad}$$

---

63. We evaluated the first expression when $y = \frac{4}{5}$. Notice that we performed the operation within the grouping first. Evaluate the second expression when $y = -\frac{1}{3}$.

$$3(y + 1) = 3\left(\frac{4}{5} + 1\right) = 3\left(\frac{9}{5}\right) = \frac{27}{5}$$

$$2(y - 1) = 2\left(-\frac{1}{3} - 1\right) = \underline{\qquad\qquad}$$

Answer: $\frac{5}{4}$

64. We evaluated the first expression when $t = \frac{2}{5}$. Notice that we simplified the numerator first. Evaluate the second expression when $t = \frac{1}{3}$.

$$\frac{3t}{7} = \frac{3\left(\frac{2}{5}\right)}{7} + \frac{\frac{6}{5}}{7} = \frac{6}{5}\left(\frac{1}{7}\right) = \frac{6}{35}$$

$$\frac{1}{5t} = \_\_\_\_\_\_\_\_\_\_$$

Answer: $-\frac{8}{3}$

---

65. We evaluated the first expression when $x = \frac{4}{3}$. Notice that we simplified the denominator first. Evaluate the second expression when $x = -\frac{3}{5}$.

$$\frac{3}{2x - 1} = \frac{3}{2\left(\frac{4}{3}\right) - 1} = \frac{3}{\frac{8}{3} - 1} = \frac{3}{\frac{8}{3} - \frac{3}{3}} = \frac{3}{\frac{5}{3}} = 3\left(\frac{3}{5}\right) = \frac{9}{5}$$

$$\frac{4x + 1}{5} = \_\_\_\_\_\_\_\_\_\_$$

Answer: $\frac{3}{5}$

---

66. We evaluated the first expression when $y = \frac{3}{8}$. Evaluate the second expression when $y = \frac{5}{4}$.

$$\frac{2(y + 1)}{5} = \frac{2\left(\frac{3}{8} + 1\right)}{5} = \frac{2\left(\frac{11}{8}\right)}{5} = \frac{\frac{11}{4}}{5} = \frac{11}{4}\left(\frac{1}{5}\right) = \frac{11}{20}$$

$$\frac{2}{3(y - 1)} = \_\_\_\_\_\_\_\_\_\_$$

Answer: $-\frac{7}{25}$

---

67. We evaluated the first expression when $t = \frac{5}{4}$. Evaluate the second expression when $t = \frac{1}{4}$.

$$3t - (1 - t) = 3\left(\frac{5}{4}\right) - \left(1 - \frac{5}{4}\right)$$

$$= \frac{15}{4} - \left(-\frac{1}{4}\right) = \frac{15}{4} + \frac{1}{4} = \frac{16}{4} = 4$$

$$5(t + 1) - 3t =$$

Answer: $\frac{8}{3}$

---

68. We evaluated the expression below when $x = \frac{1}{2}$.

$$3x^2 + 2 = 3\left(\frac{1}{2}\right)^2 + 2 = 3\left(\frac{1}{4}\right) + 2 = \frac{3}{4} + \frac{8}{4} = \frac{11}{4}$$

Continued on following page.

Answer: $\frac{11}{2}$

68. Continued

Evaluate each of these when $x = \frac{1}{2}$.

a) $2x^2 + 1$ = ____________________

b) $x^3 - 1$ = ____________________

69. Evaluate each expression below when $y = -\frac{1}{3}$.

a) $y^2 + 2y$ = ____________________

b) $2y^2 - y + 1$ = ____________________

Answers (68): a) $\frac{3}{2}$ b) $-\frac{7}{8}$

Answers (69): a) $-\frac{5}{9}$ b) $\frac{14}{9}$

2-10 IMPROPER FRACTIONS AND MIXED NUMBERS

In this section, we will show the relationship between improper fractions and mixed numbers.

70. A fraction is a proper fraction if its numerator is smaller than its denominator.

$$\frac{2}{3}, \frac{5}{11}, \text{ and } \frac{1}{4} \text{ are proper fractions.}$$

A fraction is an improper fraction if its numerator is equal to or larger than its denominator.

$$\frac{7}{4}, \frac{15}{5}, \text{ and } \frac{8}{8} \text{ are improper fractions.}$$

Which fractions below are improper fractions? ____________

a) $\frac{12}{6}$ b) $\frac{4}{5}$ c) $\frac{4}{4}$ d) $\frac{1}{3}$ e) $\frac{4}{3}$

71. Any addition of a whole number and a proper fraction can be written as a mixed number. For example:

$$1 + \frac{3}{4} \text{ can be written } 1\frac{3}{4}$$

Any improper fraction that does not equal a whole number can be converted to a mixed number. For example:

$$\frac{9}{2} = \frac{8}{2} + \frac{1}{2} = 4 + \frac{1}{2} = 4\frac{1}{2}$$

$$\frac{10}{6} = \frac{6}{6} + \frac{4}{6} = 1 + \frac{4}{6} = 1\frac{2}{3}$$

Convert each fraction to a mixed number.

a) $\frac{4}{3}$ = ______ b) $\frac{13}{5}$ = ______ c) $\frac{15}{6}$ = ______

Answer (70): a) (a), (c), and (e)

72. Any negative improper fraction is equal to a negative mixed number. That is:

$-\frac{5}{3} = -1\frac{2}{3}$ $\quad -\frac{18}{4} = -4\frac{1}{2}$ $\quad -\frac{10}{8} =$ ______

a) $1\frac{1}{3}$ b) $2\frac{3}{5}$ c) $2\frac{1}{2}$

$-1\frac{1}{4}$

SELF-TEST 5 (pages 62-71)

Add. Report each sum in lowest terms.

1. $\frac{5}{4} + \left(-\frac{3}{4}\right) =$

2. $-\frac{3}{2} + \left(-\frac{1}{6}\right) =$

3. $-\frac{2}{5} + \frac{7}{3} =$

Subtract. Report each difference in lowest terms.

4. $\frac{2}{9} - \frac{8}{9} =$

5. $-\frac{3}{8} - \left(-\frac{5}{12}\right) =$

6. $\frac{8}{5} - 4 =$

Find the value of each power.

7. $\left(-\frac{3}{4}\right)^3 =$

8. $-\left(\frac{1}{3}\right)^2 =$

Convert to a mixed number.

9. $-\frac{20}{12} =$

Evaluate each expression when $x = \frac{2}{3}$.

10. $3(1 - x) =$

11. $\frac{5x - 2}{6} =$

Evaluate each expression when $y = -\frac{1}{2}$.

12. $4y - (y - 1) =$

13. $y^3 - y^2 - y =$

ANSWERS:

1. $\frac{1}{2}$
2. $-\frac{5}{3}$
3. $\frac{29}{15}$
4. $-\frac{2}{3}$
5. $\frac{1}{24}$
6. $-\frac{12}{5}$
7. $-\frac{27}{64}$
8. $-\frac{1}{9}$
9. $-1\frac{2}{3}$
10. 1
11. $\frac{2}{9}$
12. $-\frac{1}{2}$
13. $\frac{1}{8}$

2-11 ADDITION AND SUBTRACTION OF DECIMALS

In this section, we will discuss the procedure for additions and subtractions involving signed decimal numbers.

---

73. A decimal is a number written with a decimal point. To add decimals, we write them in a column with the decimal points lined up. An example is shown. Do the other addition.

1.84 + 5.03 + 6.11 | 25.8 + 49.2 + 16.7

```
  1.84
  5.03
+ 6.11
 12.98
```

---

74. In a whole number, the decimal point is understood to be at the right of the number. That is:

4 means 4.

When adding decimals and whole numbers, it is sometimes helpful to attach 0's so that all numbers end in the same place. An example is shown. Do the other addition.

3.1 + 4 + .25 | 3.41 + 8 + .275

```
   3.1      3.10
   4        4.00
+   .25   +  .25
            7.35
```

Answer to 73:

```
  25.8
  49.2
+ 16.7
  91.7
```

---

75. To subtract decimals, we also write them in a column with the decimal points lined up. An example is shown. Do the other subtraction.

.947 - .135 | 16.3 - 12.9

```
  .947
- .135
  .812
```

Answer to 74:

```
   3.410
   8.000
+   .275
  11.685
```

---

76. When subtracting decimals, it is also helpful at times to attach 0's so that all numbers end in the same place. Two examples are shown.

6 - .148 | .578 - .4

```
  6.000        .578
-  .148      - .400
  5.852        .178
```

Following the examples, do these:

a) 5.1 - 3.75   b) 1 - .66

Answer to 75:

```
  16.3
- 12.9
   3.4
```

77. Decimal numbers also have an absolute value. For example:

$|.47| = .47$ $|-8.6| = 8.6$

To add two negative decimals, we use the same rule we used for integers and fractions. That is, <u>we add their absolute values</u>. <u>The sum is negative</u>. For example:

-10.2 + (-20.5) = -30.7 -6.6 + (-9.9) = ______

Answers: a) 5.10 − 3.75 = 1.35 b) 1.00 − .66 = .34

---

78. To add one positive and one negative decimal, we also use the same rule we used with integers and fractions. That is:

1. <u>Subtract their absolute values to get the absolute value of the sum</u>.
2. <u>Give the sum the same sign as the decimal with the larger absolute value</u>.

The rule is applied in the two examples below.

6.4 + (-4.1) = 2.3 — Since 6.4 - 4.1 = 2.3, the absolute value of the sum is 2.3. Since 6.4 has the larger absolute value, the sum is positive.

-.7 + .55 = -.15 — Since .70 - .55 = .15, the absolute value of the sum is .15. Since -.7 (or -.70) has the larger absolute value, the sum is negative.

Use the rule for these:

a) 5.6 + (-8.1) = ______ b) -.02 + .069 = ______

Answer: -16.5

---

79. Complete these:

a) 1 + (-.3) = ______ b) -.75 + (-2) = ______

Answers: a) -2.5 b) .049

---

80. Two decimals are a pair of opposites if their sum is 0. Therefore:

2.7 and -2.7 are opposites, since 2.7 + (-2.7) = 0.

.61 and -.61 are opposites, since .61 + (-.61) = 0.

Answers: a) .7 b) -2.75

Continued on following page.

80. Continued

Any subtraction involving signed decimals can be converted to an addition by adding the opposite of the second term to the first term. That is:

1.4 - 8.5 = 1.4 + (-8.5) = -7.1

.05 - (-.02) = .05 + .02 = .07

a) -.7 - .1 = ____ + ______ = ______

b) -9.2 - (-4.7) = ____ + ______ = ______

---

81. Complete each subtraction.

a) 3.7 - 9.9 = ________________________

b) -.8 - (-.7) = ________________________

c) 1 - (-.25) = ________________________

d) -.3 - 1 = ________________________

Answers to 80: a) -.7 + (-.1) = -.8 b) -9.2 + 4.7 = -4.5

Answers to 81: a) -6.2 b) -.1 c) 1.25 d) -1.3

2-12 MULTIPLICATION OF DECIMALS

In this section, we will discuss the procedure for multiplications involving signed decimal numbers.

---

82. The number of decimal places in a decimal is the number of digits written to the right of the decimal point. A whole number has no decimal places. For example:

6.4 has one decimal place
17.28 has two decimal places
.0407 has four decimal places
27 has zero decimal places

When multiplying decimals, the number of decimal places in the product is the sum of the number of the decimal places in the factors. Two examples are shown:

(.71)(.9) = .639 (.06)(.03) = .0018

.71 (2 decimal places)	.06 (2 decimal places)
x .9 (1 decimal place)	x .03 (2 decimal places)
.639 (3 decimal places)	.0018 (4 decimal places)

Following the examples, do these:

a) 57(.03) = ________ b) (.02)(.4) = ________

83. When the product ends with one or more final 0's, the 0 or 0's must be counted when placing the decimal point. An example is shown. Do the other multiplication.

240(.005) = 1.200 = 1.2 6(.005) = ______

```
  240
x .005
 1.200
```

a) 1.71

b) .008

---

84. Following the example, do the other multiplication.

```
   2 1 4    (0 decimal places)          . 3 2
 x 9.0 1    (2 decimal places)        x . 1 4
   2 1 4
  0 0 0
1 9 2 6
1 9 2 8.1 4 (2 decimal places)
```

.030

---

85. <u>When one factor is positive and the other negative, the product is negative</u>. An example is shown. Notice how we multiplied vertically to get the absolute value of the product. Do the other multiplication.

7(-.064) = -.448 (-.09)(.8)

```
.064
 x 7
.448
```

.0448

---

86. When both factors are negative, the product is positive. An example is shown. Do the other multiplication.

(-.9)(-.75) = .675 -8(-.6) = ______

```
 .75
 x .9
.675
```

-.072

---

87. Following the example, do the other multiplication.

2.4(-1.8) = -4.32 -3.7(55) = ______

```
  1.8
x 2.4
  7 2
3 6
4.3 2
```

4.8

-203.5

88. Three properties of multiplication are listed below.

1) When a decimal is multiplied by "1", the product is identical to the decimal.

$1(.25) = .25$ $-6.4(1) = -6.4$

2) When a decimal is multiplied by -1, the product is the opposite of the decimal.

$-1(4.6) = -4.6$ $(-.08)(-1) = .08$

3) When a decimal is multiplied by 0, the product is 0.

$0(.007) = 0$ $(-1.66)(0) = 0$

Using the properties, complete these:

a) $-1(2.75) =$ ________ c) $0(-16.9) =$ ________

b) $(-.86)(1) =$ ________ d) $(-.15)(-1) =$ ________

---

89. We squared a decimal and cubed a decimal below.

$$(.3)^2 = (.3)(.3) = .09$$

$$(-.5)^3 = (-.5)(-.5)(-.5) = -.125$$

Following the examples, do these:

a) $(-1.2)^2 =$ ________ b) $(.4)^3 =$ ________

Answers to 88: a) -2.75 b) -.86 c) 0 d) .15

Answers to 89: a) 1.44 b) .064

2-13 DIVISION OF DECIMALS

In this section, we will discuss the procedure for divisions involving signed decimal numbers.

---

90. In each division below, we divided a decimal by a whole number. Notice, in each case, that we placed the decimal point in the quotient directly above the decimal point in the dividend. (See the arrows.) Do the other division.

```
   4.3           .0 1 4
6)2 5.↑8      5)↑.0 7 0      8).0 5 6
 -2 4           -5
   1 8           2 0
  -1 8          -2 0
```

---

91. In the division on the next page, we divided 25.83 by the decimal 12.3. Notice the steps:

Step 1: We changed the divisor 12.3 to the whole number 123 by shifting the decimal point one place to the right. We also had to shift the decimal point one place to the right in 25.83.

Step 2: We divided 258.3 by 123 in the usual way.

Continued on following page.

Answer to 90: .007

91. Continued

Following the example, complete the division of .138 by .04 .

Step 1: 12.3)2 5.8 3 .04).1 3 8

Step 2:

```
       2.1
123)2 5 8.3        4)1 3.8
   -2 4 6
      1 2 3
     -1 2 3
```

---

92. In algebra, divisions involving decimal numbers are ordinarily written as fractions. For example:

$$.028 \div 7 \text{ is written } \frac{.028}{7}$$

When one number is positive and the other negative, the quotient is negative. An example is shown. Notice how we divided in the usual way to get the absolute value of the quotient. Complete the other division.

$\frac{-9.12}{4} = -2.28$ $\frac{3.42}{-.06} =$ ________

```
  2.2 8
4)9.1 2
 -8
  1 1
 -8
    3 2
   -3 2
```

Answer: 3.45

---

93. When both numbers are negative, the quotient is positive. An example is shown. Complete the other division.

$\frac{-4.8}{-1.2} = 4$ $\frac{-.056}{-.28} =$ ________

```
      4.
1.2)4.8
   -4 8
```

Answer: -57

---

94. The names of some places in decimal numbers are given below:

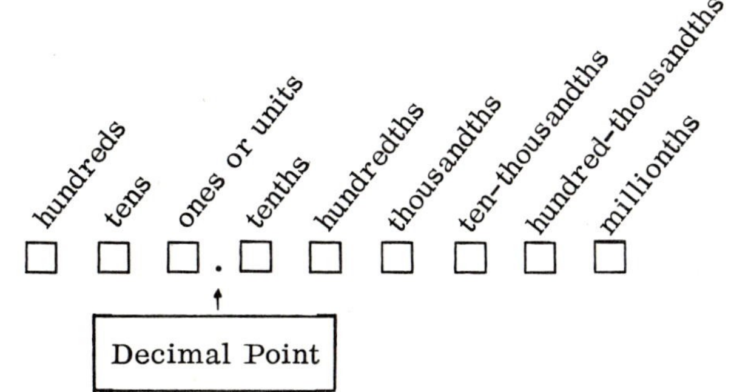

In .251687 : a) the "2" is in the ______________ place

b) the "1" is in the ______________ place

c) the "7" is in the ______________ place

Answer: .2

95. The following rules are used to round to a specific place.

> **Rules For Rounding**
>
> 1) If the digit in the place immediately to the right of the place rounded to is a 0, 1, 2, 3, or 4:
>
>   a) leave the digit in the place rounded to unchanged, and
>   b) drop all digits to the right of that place.
>
> 2) If the digit in the place immediately to the right of the place rounded to is a 5, 6, 7, 8, or 9:
>
>   a) add "1" to the digit in the place rounded to, and
>   b) drop all digits to the right of that place.
>
> Note: When deciding whether to add "1" or not to the digit in the place rounded to, we look only at the digit in the place immediately to its right. All other digits are ignored.

Some examples are shown below. We used the arrows as aids.

Rounding to tenths, 4.827 rounds to 4.8 .
(Because of the 2 in the hundredths place, we left the 8 unchanged.)

Rounding to thousandths, .096517 rounds to .097 .
(Because of the 5 in the ten-thousandths place, we added "1" to the 6 to get 7.)

Rounding to a whole number, 27.94 rounds to 28.
(Because of the 9 in the tenths place, we added "1" to the 7 to get 8.)

Following the examples, do these:

a) Round 1.65524 to hundredths. ____________

b) Round 256.48 to a whole number. ____________

c) Round .018704 to thousandths. ____________

a) tenths
b) thousandths
c) millionths

---

96. When rounding, we can get a 0 in the place rounded to. For example:

Rounding to hundredths, .7025 rounds to .70 .

Rounding to a whole number, 49.81 rounds to 50.

a) Round 1.03 to tenths. ____________

b) Round .05972 to thousandths. ____________

c) Round 9.96 to tenths. ____________

a) 1.66
b) 256
c) .019

97. The division at the right is non-terminating because it goes on and on. To show that the division process could continue, we write three periods "..." at the right side of the quotient.

```
     2.5 7 1 4 ...
 7)1 8.0 0 0 0
  -1 4
     4 0
    -3 5
       5 0
      -4 9
         1 0
         -7
           3 0
          -2 8
             2
```

To stop a non-terminating division, we ordinarily round the quotient to a specific place. For example:

Rounding to thousandths, 2.5714... rounds to 2.571.

Rounding to hundredths, 2.5714... rounds to 2.57.

Rounding to tenths, 2.5714... rounds to ________.

a) 1.0

b) .060

c) 10.0

---

98. When rounding a non-terminating quotient to a specific place, we continue dividing until we get a digit in the place immediately to the right of the specific place and then round back one place. An example is shown and discussed. Do the other division.

To round to the nearest whole number below, we divided until we got a digit in the tenths place and then rounded back.

```
    4 8 4.8   or  485
 9)4,3 6 4.0
  -3 6
     7 6
    -7 2
       4 4
      -3 6
         8 0
        -7 2
           8
```

Round to the nearest hundredths. That is, divide until you get a digit in the thousandths place and then round back.

```
 3).5 3 8
```

2.6

---

99. We rounded the quotient below to hundredths. Round the other quotient to tenths.

$$\frac{-7.56}{4.1} = -1.84 \qquad \frac{-10.9}{-.7}$$

```
       1.8 4 3   or  1.84
 4.1)7.5 6 0 0
    -4 1
     3 4 6
    -3 2 8
       1 8 0
      -1 6 4
         1 6 0
        -1 2 3
           3 7
```

.18 (from .179)

15.6 (from 15.57)

100. When a non-terminating division is performed on a calculator, the quotient shown contains many decimal places. Two examples are given below.

$47\overline{)130} = 2.765957$ $\qquad$ $.13\overline{).119} = .915385$

Ordinarily, such quotients are rounded to fewer places.

a) Round 2.765957 <u>to the nearest tenth</u>. ________

b) Round .915385 <u>to the nearest thousandth</u>. ________

Answers to 100: a) 2.8 $\quad$ b) .915

101. Three properties of division are extended to divisions involving decimals below.

1) When a decimal is divided by itself, the quotient is "1".

$\frac{.74}{.74} = 1$ $\qquad$ $\frac{-5.8}{-5.8} = 1$

2) When a decimal is divided by "1", the quotient is <u>identical</u> to the decimal.

$\frac{.025}{1} = .025$ $\qquad$ $\frac{-14.9}{1} = -14.9$

3) When a decimal is divided by -1, the quotient is <u>the opposite</u> of the decimal.

$\frac{5.7}{-1} = -5.7$ $\qquad$ $\frac{-.416}{-1} = .416$

Using the properties, complete these:

a) $\frac{-.08}{-.08} =$ ____ $\qquad$ b) $\frac{-6.27}{1} =$ ____ $\qquad$ c) $\frac{.509}{-1} =$ ____

Answers to 101: a) 1 $\quad$ b) -6.27 $\quad$ c) -.509

102. Two more properties of division are extended to divisions involving decimals below.

1) When 0 is divided by a decimal, the quotient is 0.

$\frac{0}{.86} = 0$ $\qquad$ $\frac{0}{-2.5} = 0$

2) Dividing a decimal by 0 is <u>IMPOSSIBLE</u> because division by 0 is impossible.

The division $\frac{18.7}{0}$ is not possible.

Using the properties, complete these:

a) $\frac{0}{-.4} =$ ____ $\qquad$ b) $\frac{-6.9}{0} =$ ____ $\qquad$ c) $\frac{0}{.668} =$ ____

Answers to 102: a) 0 $\quad$ b) impossible $\quad$ c) 0

2-14 FRACTION-DECIMAL CONVERSIONS

In this section, we will discuss the procedures for converting fractions to decimals and decimals to fractions.

---

103. To convert a fraction to a decimal, we divide. Following the example, complete the other conversion.

$$\frac{5}{8} = \begin{array}{r} .625 \\ 8\overline{)5.000} \\ \underline{-48} \\ 20 \\ \underline{-16} \\ 40 \\ \underline{-40} \end{array} \qquad\qquad \frac{3}{4} =$$

.75

104. Sometimes we get repeating decimals. Two examples are shown. The "..." means that the decimal goes on and on.

$$\frac{2}{3} = \begin{array}{r} .666... \\ 3\overline{)2.000} \\ \underline{-18} \\ 20 \\ \underline{-18} \\ 20 \\ \underline{-18} \\ 2 \end{array} \qquad\qquad \frac{4}{11} = \begin{array}{r} .3636... \\ 11\overline{)4.0000} \\ \underline{-33} \\ 70 \\ \underline{-66} \\ 40 \\ \underline{-33} \\ 70 \\ \underline{-66} \\ 4 \end{array}$$

For $\frac{2}{3}$, the digit "6" keeps repeating. For $\frac{4}{11}$, the pair of digits "36" keeps repeating. In such cases, the dots are often replaced by a bar over the repeating part. For example:

Instead of .666..., we write $.66\overline{6}$ or $.\overline{6}$.

Instead of .3636..., we write $.36\overline{36}$ or $.\overline{36}$.

In applied problems, we usually round a repeating pattern to a specific place. For example:

Rounded to hundredths, $\frac{2}{3} = .67$

Rounded to thousandths, $\frac{4}{11} =$ ________

.364

105. When an improper fraction is converted to a decimal, <u>the decimal is larger than "1"</u>. An example is shown. Complete the other conversion. <u>Round to hundredths</u>.

$$\frac{5}{2} = \begin{array}{r} 2.5 \\ 2\overline{)5.0} \\ \underline{-4} \\ 10 \\ \underline{-10} \end{array} \qquad\qquad \frac{7}{6} =$$

1.17

106. A decimal fraction is a fraction whose denominator is a number like 10, 100, or 1,000. A decimal fraction can be converted directly to a decimal. The decimal number contains as many decimal places as there are 0's in the denominator. For example:

$\frac{7}{10} = .7$ (10 has one "0"; .7 has one decimal place.)

$\frac{49}{100} = .49$ (100 has two 0's; .49 has two decimal places.)

$\frac{256}{1000} = .256$ (1000 has three 0's; .256 has three decimal places.)

Convert each fraction to a decimal:

a) $\frac{3}{10} =$ ______ b) $\frac{97}{100} =$ ______ c) $\frac{84}{1000} =$ ______

---

a) .3

b) .97

c) .084

107. Any decimal smaller than "1" can be converted directly to a decimal fraction. The denominator contains as many 0's as there are decimal places in the decimal. For example:

$.4 = \frac{4}{10}$ (.4 has one decimal place; 10 has one "0".)

$.61 = \frac{61}{100}$ (.61 has two decimal places; 100 has two 0's.)

$.854 = \frac{854}{1000}$ (.854 has three decimal places; 1000 has three 0's.)

Convert each decimal to a decimal fraction.

a) .9 = ______ b) .07 = ______ c) .249 = ______

---

a) $\frac{9}{10}$

b) $\frac{7}{100}$

c) $\frac{249}{1000}$

108. Any decimal larger than "1" can be converted to an improper decimal fraction by converting to a mixed number first. For example:

$$13.7 = 13\frac{7}{10} = 13 + \frac{7}{10} = \frac{130}{10} + \frac{7}{10} = \frac{137}{10}$$

$$2.56 = 2\frac{56}{100} = 2 + \frac{56}{100} = \frac{200}{100} + \frac{56}{100} = \frac{256}{100}$$

However, we can make the conversion directly. The number of decimal places tells us how many 0's to use in the denominator. That is:

$13.7 = \frac{137}{10}$ (One decimal place in 13.7, one "0" in 10.)

$2.56 = \frac{256}{100}$ (Two decimal places in 2.56, two 0's in 100.)

Convert each decimal to an improper decimal fraction.

a) 6.3 = ______ b) 4.19 = ______ c) 2.755 = ______

---

a) $\frac{63}{10}$ b) $\frac{419}{100}$ c) $\frac{2755}{1000}$

2-15 RATIONAL NUMBERS

In this section, we will define a rational number and show that all fractions, decimals, and integers fit the definitions.

---

109. A rational number is a number that can be expressed in the form $\frac{a}{b}$, where a and b are integers and b is not 0.

All of the following are rational numbers because they are a division of integers with a non-zero denominator.

$$\frac{1}{2} \qquad \frac{7}{5} \qquad \frac{-4}{5} \qquad \frac{9}{-2} \qquad \frac{256}{87}$$

Any negative fraction is a rational number because it can be written as a division of integers with a non-zero denominator. For example:

$-\frac{5}{6}$ can be written $\frac{-5}{6}$ or $\frac{5}{-6}$

$-\frac{13}{9}$ can be written ____ or ____

Answer: $\frac{-13}{9}$ or $\frac{13}{-9}$

110. Any positive decimal is a rational number because it can be written as a division of integers. For example:

$6.5 = \frac{65}{10}$ $\qquad$ $.037 = \frac{37}{1000}$

Any negative decimal is a rational number because it can be written as a division of integers. For example:

$-1.3 = -\frac{13}{10} = \frac{-13}{10}$ or $\frac{13}{-10}$

$-.49 = -\frac{49}{100} =$ ____ or ____

Answer: $\frac{-49}{100}$ or $\frac{49}{-100}$

111. Any integer is a rational number because it can be written as a division of itself and "1". For example:

$8 = \frac{8}{1}$ $\qquad$ $-3 = \frac{-3}{1}$ $\qquad$ a) $140 =$ ____ $\qquad$ b) $-79 =$ ____

Answer: a) $\frac{140}{1}$ b) $\frac{-79}{1}$

112. Rational numbers can be shown on a number line. For example, all of the numbers shown below are rational numbers.

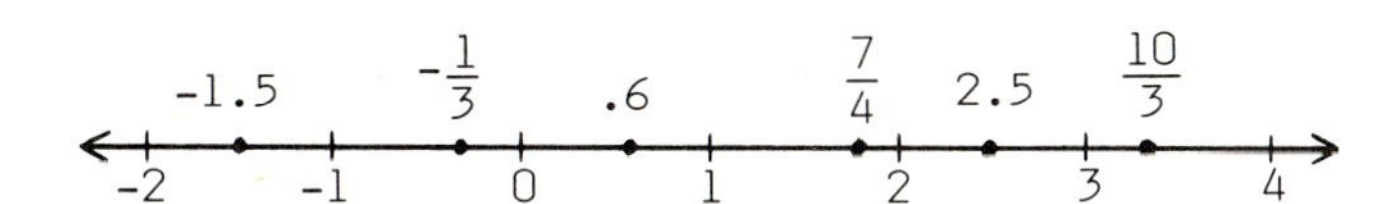

The order of rational numbers is shown on the number line. That is:

Since $-\frac{1}{3} < 0$, $-\frac{1}{3}$ is to the left of 0 on the number line.

Since $2.5 > 2$, 2.5 is to the ____ of 2 on the number line.

Answer: right

2-16 TRANSLATING PHRASES TO ALGEBRAIC EXPRESSIONS

In this section, we will translate English phrases to algebraic expressions that contain fractions and decimals.

---

113. An algebraic expression can contain a fraction or decimal. For example:

"some number plus $\frac{1}{2}$" translates to $n + \frac{1}{2}$.

"the difference between 4.5 and a number" translates to $4.5 - n$.

Using the letter n, translate each phrase to an algebraic expression.

a) 1.8 more than a number ______

b) some number reduced by $\frac{3}{4}$ ______

c) the square of a number, plus $\frac{1}{3}$ ______

d) the difference of 5.5 and a number ______

---

114. In the phrase "$\frac{1}{2}$ of a number", the word "of" means multiply. Therefore:

"$\frac{1}{2}$ of a number" translates to $\frac{1}{2}n$

Using the letter n, translate each phrase to an algebraic expression.

a) $\frac{2}{3}$ of a number ______

b) $\frac{3}{4}$ of the square of a number ______

c) 10.5 times the cube of a number ______

d) the product of a number and 6.8 ______

Answers to 113: a) $n + 1.8$ b) $n - \frac{3}{4}$ c) $n^2 + \frac{1}{3}$ d) $5.5 - n$

---

115. Following the examples, translate the other phrases to algebraic expressions. Use the letter x.

1.75 decreased by twice a number	$1.75 - 2x$
10 more than $\frac{1}{2}$ of a number	$\frac{1}{2}x + 10$

a) 3 times a number, minus 8.6 ______

b) $\frac{2}{3}$ of the square of a number, plus 5 ______

c) $\frac{1}{4}$ reduced by $\frac{1}{2}$ of a number ______

d) 7.5 reduced by 1.5 times a number ______

Answers to 114: a) $\frac{2}{3}n$ b) $\frac{3}{4}n^2$ c) $10.5n^3$ d) $6.8n$

116. Use the letter y to translate these phrases to algebraic expressions.

a) 2.75 divided by a number ____________

b) the quotient of the square of a number and 10.5 ____________

c) the cube of a number minus $\frac{1}{2}$ of the number ____________

Answers: a) $3x - 8.6$ b) $\frac{2}{3}x^2 + 5$ c) $\frac{1}{4} - \frac{1}{2}x$ d) $7.5 - 1.5x$

---

117. Using the letters x and y, translate these phrases to algebraic expressions.

a) the sum of two numbers minus $\frac{1}{3}$ ____________

b) 6.75 more than the difference of two numbers ____________

c) the product of two numbers, plus $\frac{4}{5}$ ____________

d) double the product of two numbers, minus 4.68 ____________

Answers: a) $\frac{2.75}{y}$ b) $\frac{y^2}{10.5}$ c) $y^3 - \frac{1}{2}y$

---

118. Parentheses are needed for these translations. Use x for one number and x and y for two numbers.

a) $\frac{1}{2}$ minus the difference of two numbers ____________

b) the product of 2.5, and the sum of two numbers ____________

c) the product of $\frac{3}{4}$, and the difference of the square of a number and $\frac{1}{2}$ ____________

d) 12.6 minus the sum of two numbers ____________

Answers: a) $x + y - \frac{1}{3}$ b) $x - y + 6.75$ c) $xy + \frac{4}{5}$ d) $2xy - 4.68$

Answers: a) $\frac{1}{2} - (x - y)$ b) $2.5(x + y)$ c) $\frac{3}{4}\left(x^2 - \frac{1}{2}\right)$ d) $12.6 - (x + y)$

SELF-TEST 6 (pages 72-86)

Perform each operation.

1. $-9.7 + 5 =$	2. $2.1 - 7.3 =$	3. $-.42 - (-.39) =$
4. $(-1.1)(-3.7) =$	5. $(.06)(-.94)$	6. $\frac{-76.8}{-3.2} =$

Convert to a decimal. If necessary, round to hundredths.

7. $\frac{19}{100} =$

8. $\frac{5}{6} =$

Divide. Round to tenths.

9. $\frac{-5.17}{.7} =$

Convert each decimal to a fraction.

10. $.03 =$	11. $6.9 =$	12. $.059 =$

Translate to an algebraic expression. Use x for one variable and x and y for two variables.

13. $\frac{1}{2}$ reduced by the product of $\frac{1}{4}$ and a number ______________

14. 1.5 times a number, plus the square of that number ______________

15. the product of two numbers minus $\frac{3}{5}$ ______________

16. 7.8 divided by the sum of two numbers ______________

ANSWERS:

1. -4.7	4. 4.07	7. $.19$	10. $\frac{3}{100}$	13. $\frac{1}{2} - \frac{1}{4}x$	16. $\frac{7.8}{x + y}$
2. -5.2	5. $-.0564$	8. $.83$	11. $\frac{69}{10}$	14. $1.5x + x^2$	
3. $-.03$	6. 24	9. -7.4	12. $\frac{59}{1000}$	15. $xy - \frac{3}{5}$	

2-17 LAWS OF EXPONENTS

In this section, we will discuss the laws of exponents for multiplication, division, and raising a power to a power. The discussion is limited to positive exponents.

---

119. Earlier we defined powers with positive exponents. For example:

$$2^4 = (2)(2)(2)(2) = 16$$
$$5^3 = (5)(5)(5) = 125$$
$$8^2 = (8)(8) = 64$$
$$7^1 = 7$$

Find the value of these:

a) $3^4 =$ ______ b) $2^3 =$ ______ c) $1^8 =$ ______

Answers: a) 81 b) 8 c) 1

120. We multiplied 4^2 and 4^3 below by substituting (4)(4) for 4^2 and (4)(4)(4) for 4^3. Since there are a total of five 4's, the exponent of the product is 5.

$$4^2 \cdot 4^3 = (4)(4) \cdot (4)(4)(4) = 4^5$$

We can get the exponent of the product by adding the exponents of the factors. That is:

$$4^2 \cdot 4^3 = 4^{2+3} = 4^5$$

Therefore, the general law of exponents for multiplying powers with the same base is:

$$\boxed{a^m \cdot a^n = a^{m+n}}$$

Two more examples of the above law are:

$$6^1 \cdot 6^3 = 6^{1+3} = 6^4$$
$$2^5 \cdot 2^2 = 2^{5+2} = 2^7$$

Using the law, complete these:

a) $5^4 \cdot 5^3 =$ ______ b) $9^1 \cdot 9^2 =$ ______ c) $3^5 \cdot 3^5 =$ ______

Answers: a) 5^7 b) 9^3 c) 3^{10}

121. To use the law below, we substituted 3^1 for 3.

$$3 \cdot 3^4 = 3^1 \cdot 3^4 = 3^{1+4} = 3^5$$

Use the same method for these:

a) $2 \cdot 2^5 =$ ______ b) $5^2 \cdot 5 =$ ______ c) $6 \cdot 6 =$ ______

Answers: a) 2^6 b) 5^3 c) 6^2

122. The law for multiplication applies only when the powers have the same base. It does not apply to either multiplication below.

$$2^3 \cdot 5^2 \qquad\qquad 6^4 \cdot 3^5$$

Use the law if it applies.

a) $6^4 \cdot 6^4 =$ ______ b) $7^2 \cdot 8^4 =$ ______ c) $(-3)^5 \cdot (-3)^2 =$ ______

---

123. The law also applies to multiplications with more than two factors. For example:

$$4^3 \cdot 4^2 \cdot 4^5 = 4^{3+2+5} = 4^{10}$$

Following the example, complete these:

a) $7^4 \cdot 7^6 \cdot 7^3 =$ ______ b) $5^3 \cdot 5 \cdot 5^3 \cdot 5^2 =$ ______

Answers to 122: a) 6^8 b) does not apply c) $(-3)^7$

---

124. Since each base below is also a power, each expression is a power of a power. The expressions are also a short way of writing a multiplication of identical factors.

$(5^3)^2 = 5^3 \cdot 5^3$ (There are two 5^3's)

$(6^2)^4 = 6^2 \cdot 6^2 \cdot 6^2 \cdot 6^2$ (There are four 6^2's)

To raise a power to a power, we can convert to a multiplication and use that law. For example:

$$(5^3)^2 = 5^3 \cdot 5^3 = 5^{3+3} = 5^6$$

$$(6^2)^4 = 6^2 \cdot 6^2 \cdot 6^2 \cdot 6^2 = 6^{2+2+2+2} = 6^8$$

However, it is simpler to raise a power to a power by multiplying exponents. That is:

$$(5^3)^2 = 5^{(3)(2)} = 5^6 \qquad\qquad (6^2)^4 = 6^{(2)(4)} = 6^8$$

Therefore, the general law of exponents for raising a power to a power is:

$$\boxed{(a^m)^n = a^{mn}}$$

Use the law of exponents to complete these:

a) $(8^5)^2 =$ ______ b) $(4^3)^4 =$ ______ c) $(7^5)^6 =$ ______

Answers to 123: a) 7^{13} b) 5^9

---

125. To do the multiplication below, we wrote each power as a multiplication of 4's and then factored out $\left(\frac{4 \cdot 4}{4 \cdot 4}\right)$ which equals "1".

$$\frac{4^5}{4^2} = \frac{4 \cdot 4 \cdot 4 \cdot 4 \cdot 4}{4 \cdot 4} = \left(\frac{4 \cdot 4}{4 \cdot 4}\right)(4 \cdot 4 \cdot 4) = 1(4 \cdot 4 \cdot 4) = 4 \cdot 4 \cdot 4 = 4^3$$

We can get the exponent of the quotient by subtracting exponents. That is:

$$\frac{4^5}{4^2} = 4^{5-2} = 4^3$$

Continued on following page.

Answers to 124: a) 8^{10} b) 4^{12} c) 7^{30}

125. Continued

Therefore, the general law of exponents for dividing powers with the same base is:

$$\frac{a^m}{a^n} = a^{m-n}$$

Two more examples of the above law are given below:

$$\frac{5^7}{5^4} = 5^{7-4} = 5^3 \qquad \frac{9^5}{9} = \frac{9^5}{9^1} = 9^{5-1} = 9^4$$

Using the law, complete these:

a) $\frac{2^6}{2^2}$ = ______ b) $\frac{8^2}{8}$ = ______ c) $\frac{6^{10}}{6^5}$ = ______

---

126. The law of exponents <u>does not apply</u> to $\frac{6^4}{3^2}$ because the powers have different bases. If it applies, use the law for these:

a) $\frac{4^8}{3^3}$ = ______ b) $\frac{7^6}{7^5}$ = ______ c) $\frac{2^4}{10^2}$ = ______

Answers (125): a) 2^4 b) $8^1 = 8$ c) 6^5

---

127. The three basic laws of exponents are summarized below:

$a^m \cdot a^n = a^{m+n}$	Multiplication
$\frac{a^m}{a^n} = a^{m-n}$	Division
$(a^m)^n = a^{mn}$	Raising a Power to a Power

Using the laws, complete these:

a) $7^4 \cdot 7^2$ = ______ d) $2^6 \cdot 2^3$ = ______

b) $\frac{7^4}{7^2}$ = ______ e) $\frac{2^6}{2^3}$ = ______

c) $(7^4)^2$ = ______ f) $(2^6)^3$ = ______

Answers (126): a) does not apply b) $7^1 = 7$ c) does not apply

Answers (127): a) 7^6 b) 7^2 c) 7^8 d) 2^9 e) 2^3 f) 2^{18}

2-18 ZERO AND NEGATIVE EXPONENTS

In this section, we will define zero and negative exponents and extend the laws of exponents to zero and negative exponents.

---

128. If we divide a power by itself, the quotient is "1". For example:

$$\frac{5^3}{5^3} = \frac{(5)(5)(5)}{(5)(5)(5)} = 1$$

Continued on following page.

128. Continued

If we use the law of exponents for division, we get:

$$\frac{5^3}{5^3} = 5^{3-3} = 5^0$$

Comparing the two quotients, we can see that $5^0 = 1$. Based on divisions like those above, we can define powers with "0" as the exponent.

$$\boxed{a^0 = 1 \qquad a \neq 0}$$

Note: The $a \neq 0$ means that $0^0 \neq 1$. 0^0 is meaningless.

Using the definition above, complete these:

a) $2^0 =$ ______ b) $10^0 =$ ______ c) $(-7)^0 =$ ______

---

129. The laws of exponents apply to powers with "0" as an exponent. Use the laws for these:

a) $2^4 \cdot 2^0 =$ ______ c) $8^0 \cdot 8 =$ ______

b) $\frac{6^7}{6^7} =$ ______ d) $\frac{10^5}{10^0} =$ ______

Answers (128): a) 1 b) 1 c) 1

---

130. If we use the law of exponents for this division, we get a negative exponent.

$$\frac{4^3}{4^5} = 4^{3-5} = 4^{-2}$$

We can do the same division by writing each power as a multiplication of 4's and then factoring out $\left(\frac{4 \cdot 4 \cdot 4}{4 \cdot 4 \cdot 4}\right)$ which equals "1".

$$\frac{4^3}{4^5} = \frac{4 \cdot 4 \cdot 4}{4 \cdot 4 \cdot 4 \cdot 4 \cdot 4} = \left(\frac{4 \cdot 4 \cdot 4}{4 \cdot 4 \cdot 4}\right)\left(\frac{1}{4 \cdot 4}\right) = 1\left(\frac{1}{4 \cdot 4}\right) = \frac{1}{4^2}$$

Comparing the two quotients, we can see this fact:

$$4^{-2} = \frac{1}{4^2}$$

Based on divisions like those above, we can define powers with negative exponents.

$$\boxed{a^{-n} = \frac{1}{a^n}}$$

Therefore: $9^{-2} = \frac{1}{9^2}$ $3^{-5} = \frac{1}{3^5}$ $5^{-8} =$ ______

Answers (129): a) 2^4 b) $6^0 = 1$ c) $8^1 = 8$ d) 10^5

---

131. By reversing the definition, we can convert each fraction below to a power with a negative exponent.

$\frac{1}{2^4} = 2^{-4}$ a) $\frac{1}{5^3} =$ ______ b) $\frac{1}{8^{10}} =$ ______

Answer (130): $\frac{1}{5^8}$

Answers (131): a) 5^{-3} b) 8^{-10}

132. Using the definition, we converted each power below to a fraction.

$7^{-2} = \frac{1}{7^2} = \frac{1}{49}$ $5^{-3} = \frac{1}{5^3} = \frac{1}{125}$

Convert each of these to a fraction:

a) $6^{-2} =$ ______   b) $4^{-3} =$ ______   c) $2^{-5} =$ ______

Answers: a) $\frac{1}{36}$ b) $\frac{1}{64}$ c) $\frac{1}{32}$

133. We converted a power to a fraction and a fraction to a power.

$5^{-1} = \frac{1}{5^1} = \frac{1}{5}$ $\frac{1}{7} = \frac{1}{7^1} = 7^{-1}$

Convert each power to a fraction and each fraction to a power.

a) $2^{-1} =$ ______   b) $\frac{1}{3} =$ ______   c) $\frac{1}{10} =$ ______   d) $12^{-1} =$ ______

Answers: a) $\frac{1}{2}$ b) 3^{-1} c) 10^{-1} d) $\frac{1}{12}$

134. The word names for some powers with negative exponents are given below.

8^{-2} is called "8 to the negative-two power" or "8 to the negative-two".

3^{-5} is called "3 to the negative-five power" or "3 to the negative-five".

Write the power corresponding to each word name.

a) 7 to the negative-nine power ____   b) 10 to the negative-one ____

Answers: a) 7^{-9} b) 10^{-1}

135. The laws of exponents also apply to powers with negative exponents. For example, two multiplications are shown below.

$6^{-3} \cdot 6^5 = 6^{-3+5} = 6^2$ $4^{-1} \cdot 4^{-5} = 4^{-1+(-5)} = 4^{-6}$

Use the law to complete these:

a) $2^{-4} \cdot 2^{-6} =$ ______   c) $8^{-9} \cdot 8 =$ ______

b) $7^5 \cdot 7^{-4} =$ ______   d) $10^3 \cdot 10^{-3} =$ ____

Answers: a) 2^{-10} b) $7^1 = 7$ c) 8^{-8} d) $10^0 = 1$

136. Two examples of raising a power to a power are shown below.

$(4^{-2})^3 = 4^{(-2)(3)} = 4^{-6}$ $(10^{-1})^7 = 10^{(-1)(7)} = 10^{-7}$

The same law can be used to raise a power to a negative power. For example:

$(5^4)^{-2} = 5^{4(-2)} = 5^{-8}$ $(7^{-3})^{-5} = 7^{(-3)(-5)} = 7^{15}$

Use the law to complete these:

a) $(3^{-4})^2 =$ ______   c) $(6^{-1})^{-5} =$ ______

b) $(8^2)^{-1} =$ ______   d) $(2^{-4})^{-3} =$ ______

Answers: a) 3^{-8} b) 8^{-2} c) 6^5 d) 2^{12}

137. Two divisions of powers are shown below. Notice how we did the subtraction of exponents.

$$\frac{4^{-3}}{4^5} = 4^{-3-5} = 4^{-3+(-5)} = 4^{-8}$$

$$\frac{2^{-4}}{2^{-7}} = 2^{-4-(-7)} = 2^{-4+7} = 2^3$$

Use the law to complete these:

a) $\frac{6^2}{6^4} =$ ______ c) $\frac{3^4}{3^{-5}} =$ ______

b) $\frac{9^{-1}}{9^3} =$ ______ d) $\frac{10^{-10}}{10^{-6}} =$ ______

---

a) 6^{-2} c) 3^9

b) 9^{-4} d) 10^{-4}

138. To do the divisions below, we substituted 2^1 for 2 and 5^1 for 5.

$$\frac{2}{2^5} = \frac{2^1}{2^5} = 2^{1-5} = 2^{1+(-5)} = 2^{-4}$$

$$\frac{5^{-4}}{5} = \frac{5^{-4}}{5^1} = 5^{-4-1} = 5^{-4+(-1)} = 5^{-5}$$

Complete these:

a) $\frac{7^{-1}}{7} =$ ______ c) $\frac{10}{10^{-8}} =$ ______

b) $\frac{3}{3^6} =$ ______ d) $\frac{6^{-3}}{6^{-4}} =$ ______

---

a) 7^{-2} b) 3^{-5} c) 10^9 d) 6^1 or 6

2-19 POWERS OF TEN

In this section, we will discuss powers of ten and the decimal-point-shift method for multiplying by powers of ten.

---

139. The table at the right shows the values of various powers of 10. You can see this fact:

The exponent equals the number of 0's in the ordinary number.

Using the above fact, convert each of these to a power of ten.

a) 10,000,000 = ______

b) 1,000,000,000 = ______

Power	Value
10^6 =	1,000,000
10^5 =	100,000
10^4 =	10,000
10^3 =	1,000
10^2 =	100
10^1 =	10
10^0 =	1

140. Convert each whole number to a power of ten and each power of ten to a whole number.

a) 10^3 = ________ c) 100 = ________

b) 1,000,000 = ________ d) 10^5 = ________

Answers (frame 139): a) 10^7 b) 10^9

---

141. Since $10^1 = 10$, $10^2 = 100$, and $10^3 = 1{,}000$, we can use the decimal-point-shift method to multiply by those powers. The exponent tells us how many places to shift the decimal point to the right. For example:

$10^1 \times .56 = .5{,}6 = 5.6$ (Shifted one place. Exponent is 1.)

$10^2 \times .39 = .39{,} = 39$ (Shifted two places. Exponent is 2.)

$10^3 \times 9.8 = 9.800{,} = 9{,}800$ (Shifted three places. Exponent is 3.)

Use the decimal-point-shift method for these:

a) $10^1 \times 470$ = ____ b) $10^2 \times 6.08$ = ____ c) $10^3 \times 6.5$ = ____

Answers (frame 140): a) 1,000 b) 10^6 c) 10^2 d) 100,000

---

142. The decimal-point-shift method can also be used to multiply by powers of ten with larger exponents. The number of places shifted depends on the exponent. For example:

$10^4 \times 2.56 = 2.5600{,} = 25{,}600$ (Shifted four places.)

$10^7 \times .0029 = .0029000{,} = 29{,}000$ (Shifted seven places.)

Use the decimal-point-shift method for these:

a) $10^6 \times .00098$ = ________ b) $10^8 \times .000125$ = ________

Answers (frame 141): a) 4,700 b) 608 c) 6,500

---

143. The decimal-point-shift method can also be used when the power of ten is the second factor. For example:

$775 \times 10^3 = 775.000{,} = 775{,}000$

$.094 \times 10^5 = .09400{,} = 9{,}400$

Use the decimal-point-shift method for these:

a) 1.5×10^1 = ____ b) $.27 \times 10^4$ = ____ c) $.00007 \times 10^6$ = ____

Answers (frame 142): a) 980 b) 12,500

---

144. Since $10^0 = 1$, when multiplying by 10^0, the product is identical to the other factor. That is:

$10^0 \times 55 = 55$ $1.39 \times 10^0 = 1.39$

Therefore, when multiplying by 10^0, we do not shift the decimal point. Complete these:

a) $10^0 \times .009$ = ________ b) 27.5×10^0 = ________

Answers (frame 143): a) 15 b) 2,700 c) 70

Answers (frame 144): a) .009 b) 27.5

145. Some powers of ten with negative exponents are shown in the table at the right. You can see this fact:

The absolute value of the exponent equals the number of decimal places in the ordinary number.

10^{-1}	= .1
10^{-2}	= .01
10^{-3}	= .001
10^{-4}	= .0001
10^{-5}	= .00001
10^{-6}	= .000001

Using the above fact, convert each of these to a power of ten.

a) .00000001 = ________

b) .0000000001 = __________

---

146. Convert each decimal to a power of ten and each power of ten to a decimal.

a) .001 = __________ c) 10^{-5} = __________

b) 10^{-2} = __________ d) .000001 = __________

Answers to 145: a) 10^{-8} b) 10^{-10}

---

147. Since $10^{-1} = .1$, $10^{-2} = .01$, and $10^{-3} = .001$, we can use the decimal-point-shift method to multiply by those powers. The absolute value of the exponent tells us how many places to shift the decimal point to the left. For example:

$10^{-1} \times 3.8$ = 3.8 = .38 (Shifted one place. Exponent is -1.)

$10^{-2} \times 7,500$ = 7,5 00. = 75 (Shifted two places. Exponent is -2.)

$10^{-3} \times .91$ = 000.91 = .00091 (Shifted three places. Exponent is -3.)

Use the decimal-point-shift method for these:

a) $10^{-1} \times 250$ = _____ b) $10^{-2} \times 487$ = _____ c) $10^{-3} \times 6.4$ = _____

Answers to 146: a) 10^{-3} b) .01 c) .00001 d) 10^{-6}

---

148. The decimal-point-shift method can also be used to multiply by powers of ten with larger negative exponents. The number of places shifted depends on the absolute value of the exponent. For example:

$10^{-4} \times 36$ = 0036. = .0036 (Shifted four places.)

$10^{-7} \times 90,000,000$ = 9 0,000,000. = 9 (Shifted seven places.)

Use the decimal-point-shift method for these:

a) $10^{-6} \times 391$ = __________ b) $10^{-8} \times 45,000,000$ = __________

Answers to 147: a) 25 b) 4.87 c) .0064

---

149. The same method can be used when the power of ten is the second factor. For example:

510×10^{-3} = 510. = .51

$1,200 \times 10^{-5}$ = 01,200. = .012

Use the decimal-point-shift method for these:

a) 58×10^{-1} = _____ b) 700×10^{-4} = _____ c) 2.4×10^{-6} = _____

Answers to 148: a) .000391 b) .45

150. Don't confuse the direction of the decimal-point-shift when multiplying by powers of ten.

If the exponent is positive, we shift to the right since we are multiplying by 10, 100, 1000, and so on. That is:

$$5.3 \times 10^3 = 5.300 = 5,300$$

If the exponent is negative, we shift to the left since we are multiplying by .1, .01, .001, and so on. That is:

$$5.3 \times 10^{-3} = .005.3 = .0053$$

Use the decimal-point-shift method for these:

a) 72×10^2 = ________ c) $.48 \times 10^4$ = ________

b) 72×10^{-2} = ________ d) $.48 \times 10^{-4}$ = ________

(Answers to previous frame: a) 5.8 b) .07 c) .0000024)

Answers: a) 7,200 b) .72 c) 4,800 d) .000048

2-20 SCIENTIFIC NOTATION

The very large and very small numbers that occur in science are frequently written in scientific notation. We will discuss scientific notation in this section.

151. A number is written in scientific notation when:

1) The first factor is a number between 1 and 10.
2) The second factor is a power of ten.

Some numbers are written in scientific notation below.

		Scientific Notation
6,200,000	=	6.2×10^6
17,000	=	1.7×10^4
.0084	=	8.4×10^{-3}
.000039	=	3.9×10^{-5}

(A number between 1 and 10) (A power of 10)

Which of the following are written in scientific notation? That is, which are multiplications of a number between 1 and 10 and a power of ten? ________

a) 25×10^8 b) 3.4×10^5 c) 9.1×10^{-3} d) $.67 \times 10^{-7}$

Only (b) and (c)

152. To convert a number written in scientific notation to its ordinary form, we simply perform the multiplication by the decimal-point-shift method. For example:

$2.7 \times 10^3 = 2.700. = 2,700$

$9.6 \times 10^{-5} = .00009.6 = .000096$

Convert to a whole number or decimal.

a) 3×10^4 = ________ c) 2.54×10^2 = ________

b) 3×10^{-4} = ________ d) 2.54×10^{-2} = ________

---

153. A number written in scientific notation can stand for a very large or very small number. For example:

$5.1 \times 10^7 = 5.1000000. = 51,000,000$

$3.9 \times 10^{-9} = .000000003.9 = .0000000039$

Convert to a whole number or decimal.

a) 7.13×10^9 = ________ b) 5×10^{-10} = ________

Answers to 152: a) 30,000 c) 254 b) .0003 d) .0254

---

154. To convert a number larger than "1" to scientific notation, we must find both the first factor and the exponent for the power of ten. Since the first factor must be a number between 1 and 10, we can find it by writing a caret (∧) after the first digit on the left. That is:

For 7∧3,000 , the first factor is 7.3 .

For 9∧00,000 , the first factor is 9 .

For 1∧8.4 , the first factor is 1.84 .

Having found the first factor, we can find the exponent of the power of ten by counting the number of places from the caret to the decimal point. That is:

7∧3,000.	$= 7.3 \times 10^4$	(Four places to the decimal point)
8∧00,000.	$= 8 \times 10^5$	(Five places to the decimal point)
9∧1.7	$= 9.17 \times 10^1$	(One place to the decimal point)

Write each number in scientific notation.

a) 47,000 = ______ x ______ c) 88.55 = ______ x ______

b) 200 = ______ x ______ d) 9,160,000 = ______ x ______

Answers to 153: a) 7,130,000,000 b) .0000000005

---

155. To convert a number smaller than "1" to scientific notation, we must again find both the first factor and the exponent for the power of ten. Since the first factor must be a number between 1 and 10, we can find it by writing a caret after the first non-zero digit in the number. That is:

For .06∧8 , the first factor is 6.8 .

For .1∧95 , the first factor is 1.95 .

For .00007∧ , the first factor is 7 .

Continued on following page.

Answers to 154: a) 4.7×10^4 b) 2×10^2 c) 8.855×10^1 d) 9.16×10^6

155. Continued

Having found the first factor, we can find the exponent of the power of ten by counting the number of places from the caret to the decimal point. Since we count to the left, the exponent is negative. That is:

.06‸8 = 6.8×10^{-2} (Two places to the decimal point)

.1‸95 = 1.95×10^{-1} (One place to the decimal point)

.00007‸ = 7×10^{-5} (Five places to the decimal point)

Write each number in scientific notation.

a) .00025 = ________ x ________ c) .06718 = ________ x ________

b) .4 = ________ x ________ d) .0000019 = ________ x ________

Answers: a) 2.5×10^{-4} b) 4×10^{-1} c) 6.718×10^{-2} d) 1.9×10^{-6}

---

156. In scientific notation, the exponent can be either positive or negative.

If the number is larger than "1", we count to the right from the caret to the decimal point. Therefore, the exponent is positive. For example:

9‸1,300 = 9.13×10^4

If the number is smaller than "1", we count to the left from the caret to the decimal point. Therefore, the exponent is negative. For example:

.009‸13 = 9.13×10^{-3}

Write each number in scientific notation.

a) 5,600 = ________ x ________ c) 300,000,000 = ________ x ________

b) .056 = ________ x ________ d) .000000003 = ________ x ________

Answers: a) 5.6×10^3 b) 5.6×10^{-2} c) 3×10^8 d) 3×10^{-9}

---

157. Scientific notation is used to express the very large and very small numbers that occur in science. For example:

The speed of light is 2.998×10^8 meters per second.

The diameter of a large molecule is 1.7×10^{-7} centimeter.

Using the decimal-point-shift method, convert each measurement above to an ordinary number.

a) 2.998×10^8 meters per second = ____________ meters per second.

b) 1.7×10^{-7} centimeter = ____________ centimeter

Answers: a) 299,800,000 b) .00000017

158. Two more measurements are given below. A light-year (the distance light travels in one year) is 5,870,000,000,000 miles. One cycle of a television broadcast signal takes .00000000481 second. Ordinarily measurements of that size would be expressed in scientific notation. Convert them to scientific notation below. a) 5,870,000,000,000 miles = ________________ miles b) .00000000481 second = ________________ second	
159. To multiply the two numbers in scientific notation below, we multiplied the two whole number factors and the two powers of ten. Complete the other multiplication. $(3 \times 10^4) \times (2 \times 10^5) = (3 \times 2) \times (10^4 \times 10^5) = 6 \times 10^9$ $(2 \times 10^{-3}) \times (4 \times 10^{-7}) =$ ________________	a) 5.87×10^{12} miles b) 4.81×10^{-9} second
160. In the multiplication below, the product is not in scientific notation because 142 is not a number between 1 and 10. We converted the product to scientific notation by writing 142 in scientific notation and then multiplying the two powers of ten. $20 \times (7.1 \times 10^6) = 142 \times 10^6 = (1.42 \times 10^2) \times 10^6 = 1.42 \times 10^8$ Use the same method to convert the product below to scientific notation. $.04 \times (5.3 \times 10^{-5}) = .212 \times 10^{-5} =$ ________________	8×10^{-10}
161. To divide the two numbers in scientific notation below, we divided the two whole numbers and the two powers of ten. Complete the other division. $\frac{8 \times 10^9}{2 \times 10^4} = \frac{8}{2} \times \frac{10^9}{10^4} = 4 \times 10^5$ $\frac{9 \times 10^{-8}}{3 \times 10^{-6}} = \frac{9}{3} \times \frac{10^{-8}}{10^{-6}} =$ ________________	2.12×10^{-6}
162. Following the example, convert the other quotient below to scientific notation. $\frac{3.01 \times 10^7}{7} = .43 \times 10^7 = (4.3 \times 10^{-1}) \times 10^7 = 4.3 \times 10^6$ $\frac{6.66 \times 10^{10}}{8.88 \times 10^{20}} = .75 \times 10^{-10} =$ ________________	3×10^{-2}
	7.5×10^{-11}

SELF-TEST 7 (pages 87-99)

Find the value of each power.

1. 5^{-1} =	2. 8^{0} =	3. 3^{-4} =

Use the laws of exponents for these.

4. $2^{-5} \cdot 2^{3}$ =	5. $6^{7} \cdot 6$ =	6. $(4^{3})^{-1}$ =
7. $(8^{-5})^{7}$ =	8. $\frac{5^6}{5^9}$ =	9. $\frac{3^{-9}}{3^{-4}}$ =

Convert to a whole number or decimal.

10. 10^{1} =	11. 10^{5} =	12. 10^{-2} =	13. 10^{-7} =

Convert to a power of ten.

14. 100 =	15. .001 =	16. 10,000,000 =	17. .000001 =

Convert to a whole number or decimal.

18. 4.5×10^{6} =	19. 2.1×10^{-3} =	20. 7.12×10^{-7} =

Convert to scientific notation.

21. 51,000 =	22. .0608 =	23. .00000085 =

ANSWERS:

1. $\frac{1}{5}$
2. 1
3. $\frac{1}{81}$
4. 2^{-2}
5. 6^{8}
6. 4^{-3}
7. 8^{-35}
8. 5^{-3}
9. 3^{-5}
10. 10
11. 100,000
12. .01
13. .0000001
14. 10^{2}
15. 10^{-3}
16. 10^{7}
17. 10^{-6}
18. 4,500,000
19. .0021
20. .000000712
21. 5.1×10^{4}
22. 6.08×10^{-2}
23. 8.5×10^{-7}

SUPPLEMENTARY PROBLEMS - CHAPTER 2

Assignment 4

Report each product in lowest terms.

1. $\left(-\frac{1}{3}\right)\left(\frac{1}{5}\right)$
2. $\left(\frac{3}{7}\right)\left(-\frac{7}{4}\right)$
3. $\left(-\frac{4}{3}\right)\left(-\frac{5}{2}\right)$
4. $\left(\frac{9}{16}\right)\left(\frac{10}{3}\right)$
5. $5\left(-\frac{7}{20}\right)$
6. $(-2)\left(\frac{3}{8}\right)$
7. $\left(-\frac{9}{5}\right)(-1)$
8. $\left(-\frac{1}{6}\right)(-12)$
9. $\left(-\frac{3}{2}\right)(0)$
10. $\left(-\frac{5}{6}\right)\left(-\frac{3}{10}\right)$
11. $\left(-\frac{21}{16}\right)\left(\frac{8}{9}\right)$
12. $\left(\frac{36}{5}\right)\left(-\frac{15}{4}\right)$

Complete the following:

13. $-\frac{4}{9} = -\frac{\square}{27}$
14. $-\frac{7}{4} = -\frac{\square}{40}$
15. $5 = \frac{\square}{20}$
16. $-3 = -\frac{\square}{6}$

Reduce each fraction to lowest terms.

17. $\frac{18}{60}$
18. $-\frac{32}{24}$
19. $-\frac{14}{42}$
20. $-\frac{55}{11}$

Report each quotient in lowest terms.

21. $\dfrac{\frac{1}{8}}{-\frac{3}{4}}$
22. $\dfrac{-\frac{5}{3}}{-\frac{3}{10}}$
23. $\dfrac{-\frac{7}{16}}{\frac{1}{4}}$
24. $\dfrac{-\frac{45}{4}}{-\frac{15}{8}}$
25. $\dfrac{-\frac{8}{7}}{2}$
26. $\dfrac{-15}{-\frac{3}{5}}$
27. $\dfrac{-\frac{2}{3}}{-1}$
28. $\dfrac{-\frac{9}{5}}{\frac{9}{5}}$

Assignment 5

Report each sum in lowest terms.

1. $\frac{11}{8} + \left(-\frac{3}{8}\right)$
2. $-\frac{5}{7} + \frac{5}{7}$
3. $-\frac{3}{4} + \left(-\frac{11}{4}\right)$
4. $-\frac{7}{3} + \left(-\frac{2}{3}\right)$
5. $-\frac{1}{5} + \left(-\frac{9}{20}\right)$
6. $\frac{7}{2} + \left(-\frac{5}{6}\right)$
7. $-\frac{9}{8} + \frac{11}{12}$
8. $-\frac{1}{6} + \left(-\frac{3}{10}\right)$
9. $-3 + \frac{5}{4}$
10. $-\frac{7}{3} + 5$
11. $-\frac{9}{5} + (-1)$
12. $-4 + \left(-\frac{5}{8}\right)$

Assignment 7

Find the value of each power.

1. 4^0 2. 8^{-1} 3. 7^{-2} 4. 6^{-3} 5. 3^{-4} 6. 2^{-6}

Use the laws of exponents for these:

7. $4 \cdot 4^3$ 8. $2^{-3} \cdot 2^2$ 9. $6^9 \cdot 6^{-1}$ 10. $7^{-4} \cdot 7^{-5}$ 11. $10^{-7} \cdot 10$

12. $\frac{5^4}{5}$ 13. $\frac{9^2}{9^3}$ 14. $\frac{2^8}{2^8}$ 15. $\frac{8^{-1}}{8^3}$ 16. $\frac{3^{-2}}{3^{-9}}$

17. $(3^4)^2$ 18. $(10^5)^4$ 19. $(4^{-1})^8$ 20. $(7^2)^{-6}$ 21. $(2^{-7})^{-3}$

Convert to a whole number or decimal.

22. 10^2 23. 10^{-3} 24. 10^1 25. 10^{-1} 26. 10^0

27. 10^{-5} 28. 10^6 29. 10^{-8} 30. 10^9 31. 10^{-10}

Convert to a power of ten.

32. 10 33. 10,000 34. .1 35. .001 36. 1

37. 1,000,000 38. .00001 39. 1,000 40. .00000001 41. 10,000,000,000

Convert to a whole number or decimal.

42. 2.8×10^1 43. 6.56×10^4 44. 7×10^7 45. 9.4×10^9

46. 4.1×10^{-1} 47. 1.14×10^{-3} 48. 2×10^{-6} 49. 8.8×10^{-8}

Convert to scientific notation.

50. 27.7 51. 519 52. 49,600 53. 6,000,000 54. 425,000,000

55. .314 56. .075 57. .000004 58. .0000639 59. .000000011

Write each answer in scientific notation.

60. $(4 \times 10^2) \times (2 \times 10^7)$ 61. $51 \times (7 \times 10^5)$ 62. $\frac{8 \times 10^{-7}}{2 \times 10^{-2}}$ 63. $\frac{2.46 \times 10^{-5}}{6}$

64. In chemistry, the quantity 602,000,000,000,000,000,000,000 molecules is called Avogadro's number. Write that number in scientific notation.

65. The wavelength of an x-ray is 7.82×10^{-7} centimeter. Write that length as a decimal number.

66. One second equals .000011574 day. Write that time in scientific notation.

67. The "half life" of a radioactive uranium 238 is 4.5×10^{10} years. Write that time as a whole number.

3 Solving Equations

In this chapter, we will discuss the algebraic principles and processes that are used to solve non-fractional and some fractional equations. Formula evaluation is discussed, including some evaluations that require equation-solving. Formula rearrangement is also discussed.

3-1 THE ADDITION AXIOM

In this section, we will discuss equations and the meaning of "solving an equation". Then we will show how the addition axiom for equations is used to solve equations.

---

1. An equation is a statement that two algebraic expressions are equal. Three examples are shown:

$$4 + 8 = 12 \qquad x + 2 = 5 \qquad 9 = 4 + y$$

Continued on following page.

1. Continued

An equation contains a left side (or left member) and a right side (or right member) connected by an equal sign. That is:

Left Side	=	Right Side

In $x + 2 = 5$: the left side is $x + 2$

the right side is 5

In $9 = 4 + y$: a) the left side is ______.

b) the right side is ______.

---

2. When an equation contains only numbers, it is either true or false.

The equation $4 + 8 = 12$ is true.

The equation $9 - 4 = 6$ is false.

When an equation contains a variable, it is neither true nor false because we do not know what the variable stands for.

The equation $x + 2 = 5$ is neither true nor false.

However, as soon as we substitute a number for the variable, the equation becomes either true or false.

a) If we substitute -1 for $\underline{x}$, is $x + 2 = 5$ true? ______

b) If we substitute 3 for $\underline{x}$, is $x + 2 = 5$ true? ______

Answers: a) 9 b) $4 + y$

---

3. When substituting a number for a variable, the number that makes the equation true is said to satisfy the equation. The number that satisfies an equation is called its solution or root.

a) Is 9 the solution or root of $y + 10 = 19$? ______

b) Is 7 the solution or root of $3 + d = 9$? ______

Answers: a) No b) Yes

---

4. The solution of $x + 4 = 10$ is 6. Instead of saying "the solution is 6", we simply say $x = 6$.

The solution of $4 + y = 9$ is 5. We say: $y = 5$.

The solution of $9 - t = 6$ is 3. We say: ______

Answers: a) Yes b) No

---

5. To solve an equation means to find its solution or root. There are various principles used to solve equations. One principle is THE ADDITION AXIOM FOR EQUATIONS. It says:

> IF WE ADD THE SAME QUANTITY TO BOTH SIDES OF AN EQUATION THAT IS TRUE, THE NEW EQUATION IS ALSO TRUE. That is:
>
> If: $A = B$
>
> Then: $A + C = B + C$

Continued on following page.

Answer: $t = 3$

5. Continued

When an equation contains a variable and a number on the same side, we use the addition axiom to isolate the variable and solve the equation. To isolate the variable, we "get rid of" the number by adding its opposite to both sides. Two examples are shown:

To isolate x below, we "get rid of" the 2 by adding its opposite (-2) to both sides.

$$x + 2 = 7$$
$$x + \underbrace{2 + (-2)}_{0} = 7 + (-2)$$
$$x + 0 = 5$$
$$x = 5$$

To isolate y below, we "get rid of" the -3 by adding its opposite (3) to both sides.

$$y - 3 = 6$$
$$y \underbrace{- 3 + 3}_{0} = 6 + 3$$
$$y + 0 = 9$$
$$y = 9$$

To check a solution, we substitute it in the original equation to see that it satisfies the equation. We checked 5 as the solution of $x + 2 = 7$. Check the other solution.

Check

$$x + 2 = 7$$
$$5 + 2 = 7$$
$$7 = 7$$

Check

Answer: $y - 3 = 6$, $9 - 3 = 6$, $6 = 6$

6. To solve the equation below, we added -10 to both sides. Solve the other equation by adding 3 to both sides.

$$7 = 10 + d$$
$$-10 + 7 = \underbrace{-10 + 10}_{0} + d$$
$$-3 = 0 + d$$
$$-3 = d$$
$$d = -3$$

$$-9 = -3 + t$$

Answer: $-9 = -3 + t$, $3 - 9 = \underbrace{3 - 3}_{0} + t$, $-6 = 0 + t$, $-6 = t$, $t = -6$

7. What number would we add to both sides to solve each equation below?

a) $12 + x = 4$ ________

b) $20 = p + 30$ ________

c) $V - 4 = -5$ ________

d) $2 = -9 + y$ ________

Answers: a) -12 b) -30 c) 4 d) 9

8. Use the addition axiom to solve each equation.

a) $y - 2 = -3$

b) $-20 = Q + 13$

Answers: a) $y = -1$ b) $Q = -33$

9. To solve the equation below, we added 2 to both sides.

$$x - 2 = 0$$
$$x \underbrace{- 2 + 2} = 0 + 2$$
$$x + 0 = 2$$
$$x = 2$$

Solve each of these equations:

a) $m + 5 = 0$

b) $0 = F - 7$

---

10. To solve the equation below, we added -1.2 to both sides. Solve and check the other equation.

$$x + 1.2 = 3.7$$
$$x + \underbrace{1.2 + (-1.2)} = 3.7 + (-1.2)$$
$$x + 0 = 2.5$$
$$x = 2.5$$

$$2.4 = m - 8.1$$

Check

$$x + 1.2 = 3.7$$
$$2.5 + 1.2 = 3.7$$
$$3.7 = 3.7$$

Check

Answers to 9: a) $m = -5$ b) $F = 7$

---

11. To solve the equation below, we added $-\frac{2}{5}$ to both sides. Solve and check the other equation.

$$x + \frac{2}{5} = \frac{3}{5}$$
$$x + \underbrace{\frac{2}{5} + \left(-\frac{2}{5}\right)} = \frac{3}{5} + \left(-\frac{2}{5}\right)$$
$$x + 0 = \frac{1}{5}$$
$$x = \frac{1}{5}$$

$$y + \frac{1}{4} = \frac{3}{4}$$

Check

$$x + \frac{2}{5} = \frac{3}{5}$$
$$\frac{1}{5} + \frac{2}{5} = \frac{3}{5}$$
$$\frac{3}{5} = \frac{3}{5}$$

Check

Answer to 10: $m = 10.5$

Check

$$2.4 = m - 8.1$$
$$2.4 = 10.5 - 8.1$$
$$2.4 = 2.4$$

12. Following the example, solve the other equation.

$$m - \frac{2}{3} = \frac{1}{6}$$

$$m \underbrace{- \frac{2}{3} + \frac{2}{3}}_{\downarrow} = \frac{1}{6} + \frac{2}{3}$$

$$m + 0 = \frac{1}{6} + \frac{4}{6}$$

$$m = \frac{5}{6}$$

$$d - \frac{1}{4} = \frac{5}{8}$$

Answer: $y = \frac{1}{2}$

Check

$$y + \frac{1}{4} = \frac{3}{4}$$

$$\frac{1}{2} + \frac{1}{4} = \frac{3}{4}$$

$$\frac{3}{4} = \frac{3}{4}$$

---

13. Following the example, solve the other equation.

$$x + 1 = \frac{2}{5}$$

$$x \underbrace{+ 1 + (-1)}_{\downarrow} = \frac{2}{5} + (-1)$$

$$x + 0 = \frac{2}{5} + \left(-\frac{5}{5}\right)$$

$$x = -\frac{3}{5}$$

$$y - 1 = \frac{7}{8}$$

Answer: $d = \frac{7}{8}$

---

14. Following the example, solve the other equation.

$$y - \frac{1}{2} = 0$$

$$y \underbrace{- \frac{1}{2} + \frac{1}{2}}_{\downarrow} = 0 + \frac{1}{2}$$

$$y + 0 = \frac{1}{2}$$

$$y = \frac{1}{2}$$

$$x + 3.5 = 0$$

Answer: $y = \frac{15}{8}$

Answer: $x = -3.5$

3-2 THE MULTIPLICATION AXIOM

In this section, we will show how the multiplication axiom for equations is used to solve equations.

---

15. We saw earlier that expressions like $3x$ or $-5y$ indicate a multiplication of a number and a variable. That is:

$3x$ means "3 times x".

$-5y$ means "-5 times y".

In expressions like those above, the numerical factor is called the coefficient of the variable. That is:

In $3x$, 3 is the coefficient of x.

In $-5y$, _____ is the coefficient of y.

16. A second principle used to solve equations is THE MULTIPLICATION AXIOM FOR EQUATIONS. It says:

> IF WE MULTIPLY BOTH SIDES OF AN EQUATION THAT IS TRUE BY THE SAME QUANTITY, THE NEW EQUATION IS ALSO TRUE. That is:
>
> If: $B = C$
>
> Then: $AB = AC$

The multiplication axiom was used to solve the equations below. Notice that we multiplied both sides by the reciprocal of the coefficient of the variable.

We multiplied both sides by $\frac{1}{5}$.

$$5x = 20$$
$$\frac{1}{5}(5x) = \frac{1}{5}(20)$$
$$1x = 4$$
$$x = 4$$

We multiplied both sides by $-\frac{1}{3}$.

$$-3y = 18$$
$$-\frac{1}{3}(-3y) = -\frac{1}{3}(18)$$
$$1y = -6$$
$$y = -6$$

We checked one solution below. Check the other solution.

Check

$$5x = 20$$
$$5(4) = 20$$
$$20 = 20$$

Check

Answer: -5

---

17. We used the multiplication axiom to solve two equations below.

$$6x = 24$$
$$\frac{1}{6}(6x) = \frac{1}{6}(24)$$
$$1x = 4$$
$$x = 4$$

$$-5y = 35$$
$$-\frac{1}{5}(-5y) = -\frac{1}{5}(35)$$
$$1y = -7$$
$$y = -7$$

You can see that using this axiom is the same as dividing the number by the coefficient of the variable. Therefore, we can use that shorter method to solve the same two equations.

$$6x = 24$$
$$x = \frac{24}{6}$$
$$x = 4$$

$$-5y = 35$$
$$y = \frac{35}{-5}$$
$$y = -7$$

Use the shorter method to solve these:

a) $7m = 56$ b) $-2y = 20$ c) $-10F = -40$

Answer:

$$-3y = 18$$
$$-3(-6) = 18$$
$$18 = 18$$

---

a) $m = 8$

b) $y = -10$

c) $F = 4$

18. In each solution below, we also divided by the coefficient of the variable.

$$-16 = 2t \qquad \frac{-16}{2} = t \qquad t = -8$$

$$-30 = -5V \qquad \frac{-30}{-5} = V \qquad V = 6$$

Use the shorter method to solve these:

a) $4 = 4x$ b) $-30 = 3t$ c) $-48 = -6m$

Answers: a) $x = 1$ b) $t = -10$ c) $m = 8$

19. When the number is 0, the solution is 0. Two examples are shown.

$$5x = 0 \qquad x = \frac{0}{5} \qquad x = 0$$

$$0 = -9y \qquad \frac{0}{-9} = y \qquad y = 0$$

Solve these:

a) $-4t = 0$ b) $0 = 7d$ c) $-8 = 8R$

Answers: a) $t = 0$ b) $d = 0$ c) $R = -1$

20. When the solution is a fraction, the fraction is always reduced to lowest terms. For example:

$$5x = 4 \qquad x = \frac{4}{5}$$

$$15 = 9m \qquad \frac{15}{9} = m \qquad \frac{5}{3} = m$$

Solve. Reduce to lowest terms.

a) $11R = 13$ b) $2 = 6y$ c) $30p = 45$

Answers: a) $R = \frac{13}{11}$ b) $y = \frac{1}{3}$ c) $p = \frac{3}{2}$

21. When a fractional solution is negative, we write the negative sign in front of the fraction. For example:

$$7y = -3 \qquad y = \frac{-3}{7} \qquad y = -\frac{3}{7}$$

$$8 = -6d \qquad \frac{8}{-6} = d \qquad -\frac{4}{3} = d$$

Solve. Reduce to lowest terms.

a) $-6x = 1$ b) $-2 = 4R$ c) $-8h = 10$

22. When both terms in the division are negative, the solution is positive. For example:

$$-3y = -1 \qquad\qquad -4 = -6m$$
$$y = \frac{-1}{-3} \qquad\qquad \frac{-4}{-6} = m$$
$$y = \frac{1}{3} \qquad\qquad \frac{2}{3} = m$$

Solve. Reduce to lowest terms.

a) $-9x = -2$ b) $-1 = -5p$ c) $-12d = -15$

Answers: a) $x = -\frac{1}{6}$ b) $R = -\frac{1}{2}$ c) $h = -\frac{5}{4}$

---

23. Any variable with a - in front of it has a coefficient of -1. That is:

$$-x = -1x \qquad\qquad -y = -1y$$

To solve each equation below, we began by writing the -1 coefficient explicitly.

$$-x = 10 \qquad\qquad -4 = -y$$
$$-1x = 10 \qquad\qquad -4 = -1y$$
$$x = \frac{10}{-1} \qquad\qquad \frac{-4}{-1} = y$$
$$x = -10 \qquad\qquad y = 4$$

Using the same method, solve these:

a) $-t = 7$ b) $-p = -9$ c) $10 = -V$

Answers: a) $x = \frac{2}{9}$ b) $p = \frac{1}{5}$ c) $d = \frac{5}{4}$

---

24. We used both the multiplication axiom and the shorter method to solve $1.2x = 4.8$ below.

$$1.2x = 4.8 \qquad\qquad 1.2x = 4.8$$
$$\frac{1}{1.2}(1.2x) = \frac{1}{1.2}(4.8) \qquad\qquad x = \frac{4.8}{1.2}$$
$$1\;x = \frac{4.8}{1.2} \qquad\qquad x = 4$$
$$x = 4$$

When decimals are involved, we usually use the shorter method. Use that method to solve these:

a) $-1.5y = 7.5$ b) $.06 = .3d$

Answers: a) $t = -7$ b) $p = 9$ c) $V = -10$

Answers: a) $y = -5$ b) $d = .2$

25. We used both the multiplication axiom and the shorter method to solve $\frac{3}{5}x = \frac{1}{2}$ below.

$$\frac{3}{5}x = \frac{1}{2} \qquad\qquad \frac{3}{5}x = \frac{1}{2}$$

$$\frac{5}{3}\left(\frac{3}{5}x\right) = \frac{5}{3}\left(\frac{1}{2}\right) \qquad\qquad x = \frac{\frac{1}{2}}{\frac{3}{5}}$$

$$1\ x = \frac{5}{6}$$

$$x = \frac{5}{6} \qquad\qquad x = \frac{1}{2}\left(\frac{5}{3}\right) = \frac{5}{6}$$

Ordinarily we use the axiom for such solutions to avoid the division of fractions. We used the axiom below. Notice that we multiplied both sides by $\frac{3}{2}$, the reciprocal of $\frac{2}{3}$. Solve the other equation by multiplying both sides by $\frac{2}{7}$.

$$\frac{2}{3}x = 5 \qquad\qquad -3 = \frac{7}{2}y$$

$$\frac{3}{2}\left(\frac{2}{3}x\right) = \frac{3}{2}(5)$$

$$1\ x = \frac{15}{2}$$

$$x = \frac{15}{2}$$

26. To solve the equation below, we multiplied both sides by $\frac{7}{5}$. Solve the other equation.

$y = -\frac{6}{7}$

$$\frac{5}{7}p = \frac{3}{4} \qquad\qquad \frac{1}{5} = \frac{2}{3}t$$

$$\frac{7}{5}\left(\frac{5}{7}p\right) = \frac{7}{5}\left(\frac{3}{4}\right)$$

$$1\ p = \frac{21}{20}$$

$$p = \frac{21}{20}$$

27. Following the example, solve the other equation.

$t = \frac{3}{10}$

$$4t = \frac{1}{2} \qquad\qquad \frac{7}{8} = 3m$$

$$\frac{1}{4}(4t) = \frac{1}{4}\left(\frac{1}{2}\right)$$

$$1\ t = \frac{1}{8}$$

$$t = \frac{1}{8}$$

$m = \frac{7}{24}$

28. Following the example, solve the other equation.

$$\frac{1}{3}x = 4$$
$$3\left(\frac{1}{3}x\right) = 3(4)$$
$$1\,x = 12$$
$$x = 12$$

$$\frac{1}{2}y = 1$$

Answer: $y = 2$

29. Solve each equation.

a) $2x = \frac{8}{9}$

b) $\frac{5}{4}F = 0$

Answers: a) $x = \frac{4}{9}$ b) $F = 0$

3-3 USING BOTH AXIOMS

Both axioms must be used to solve some equations. We will discuss solutions of that type in this section.

30. Both axioms are used to solve an equation like $3x + 5 = 26$. That is:

The addition axiom is used to isolate $3x$ by "getting rid of" the 5 on the left side.

The multiplication axiom is then used to solve the resulting equation.

The equation is solved and checked below.

$$3x + 5 = 26$$
$$3x + 5 + (-5) = 26 + (-5)$$
$$3x + 0 = 21$$
$$3x = 21$$
$$x = \frac{21}{3}$$
$$x = 7$$

Check

$$3x + 5 = 26$$
$$3(7) + 5 = 26$$
$$21 + 5 = 26$$
$$26 = 26$$

Using the same steps, solve each equation below.

a) $7y + 24 = 10$

b) $87 = 51 + 9F$

31. Following the example, solve the other equation.

$$4d - 13 = 15$$
$$4d \underbrace{- 13 + 13}_{\downarrow} = 15 + 13$$
$$4d + 0 = 28$$
$$4d = 28$$
$$d = \frac{28}{4}$$
$$d = 7$$

$83 = 10t - 7$

a) $y = -2$

b) $F = 4$

---

32. Following the example, solve the other equation.

$$20 - 5y = 65$$
$$\underbrace{(-20) + 20}_{\downarrow} - 5y = 65 + (-20)$$
$$0 - 5y = 45$$
$$-5y = 45$$
$$y = \frac{45}{-5}$$
$$y = -9$$

$79 = 23 - 8p$

$t = 9$

---

33. Following the example, solve the other equation.

$$3x + 18 = 0$$
$$3x + \underbrace{18 + (-18)}_{\downarrow} = 0 + (-18)$$
$$3x + 0 = -18$$
$$3x = -18$$
$$x = \frac{-18}{3}$$
$$x = -6$$

$28 + 7y = 0$

$p = -7$

---

34. Following the example, solve the other equation.

$$6p - 30 = 0$$
$$6p \underbrace{- 30 + 30}_{\downarrow} = 0 + 30$$
$$6p + 0 = 30$$
$$6p = 30$$
$$p = \frac{30}{6}$$
$$p = 5$$

$12 - 4d = 0$

$y = -4$

---

35. Solve. Reduce the solution to lowest terms.

a) $7t + 9 = 8$

b) $9 - 6n = 5$

$d = 3$

---

a) $t = -\frac{1}{7}$ b) $n = \frac{2}{3}$

36. To solve the equation below, we wrote the coefficient of x explicitly before using the addition axiom. Solve the other equation.

$$\begin{aligned} 7 - x &= 3 \\ 7 - 1x &= 3 \\ -7 + 7 - 1x &= 3 + (-7) \\ -1x &= -4 \\ x &= \frac{-4}{-1} \\ x &= 4 \end{aligned}$$

$2 - y = 8$

$y = -6$

37. We used both axioms to solve the equation below.

$$\begin{aligned} .4x + 1.1 &= 3.9 \\ .4x + 1.1 + (-1.1) &= 3.9 + (-1.1) \\ .4x &= 2.8 \\ x &= \frac{2.8}{.4} \\ x &= 7 \end{aligned}$$

Following the example, solve these:

a) $.6 - .2x = .8$

b) $1.2t - 5.3 = 2.5$

---

a) $x = -1$ b) $t = 6.5$

38. To solve the equation below, we began by writing the coefficient of m explicitly. Solve the other equation.

$$\begin{aligned} .7 - m &= .1 \\ .7 - 1m &= .1 \\ -.7 + .7 - 1m &= .1 + (-.7) \\ -1m &= -.6 \\ m &= \frac{-.6}{-1} \\ m &= .6 \end{aligned}$$

$5.49 = 2.18 - d$

$d = -3.31$

39. We used both axioms to solve the equation below. Solve the other equation.

$$\begin{aligned} \frac{2}{5}x + \frac{1}{5} &= \frac{4}{5} \\ \frac{2}{5}x + \frac{1}{5} + \left(-\frac{1}{5}\right) &= \frac{4}{5} + \left(-\frac{1}{5}\right) \\ \frac{2}{5}x &= \frac{3}{5} \\ \frac{5}{2}\left(\frac{2}{5}x\right) &= \frac{5}{2}\left(\frac{3}{5}\right) \\ x &= \frac{3}{2} \end{aligned}$$

$\frac{1}{2}y + \frac{1}{3} = 1$

$y = \frac{4}{3}$

40. We used both axioms to solve the equation below. Solve the other equation.

$$2x - \frac{1}{3} = \frac{1}{6}$$

$$2x - \frac{1}{3} + \frac{1}{3} = \frac{1}{6} + \frac{1}{3}$$

$$2x = \frac{1}{6} + \frac{2}{6}$$

$$2x = \frac{1}{2}$$

$$\frac{1}{2}(2x) = \frac{1}{2}\left(\frac{1}{2}\right)$$

$$x = \frac{1}{4}$$

$$1 - \frac{1}{6}x = 0$$

$x = 6$

SELF-TEST 8 (pages 104-116)

Solve each equation. Report fractional solutions in lowest terms.

1. $w - 5 = -8$	2. $p + 2 = 0$	3. $y + \frac{1}{8} = \frac{3}{4}$
4. $1.3x = 3.9$	5. $-7 = -28t$	6. $-6y = 15$
7. $10s + 19 = 15$	8. $6 - x = 7$	9. $\frac{1}{4}y + \frac{1}{3} = 1$

ANSWERS:

1. $w = -3$	4. $x = 3$	7. $s = -\frac{2}{5}$
2. $p = -2$	5. $t = \frac{1}{4}$	8. $x = -1$
3. $y = \frac{5}{8}$	6. $y = -\frac{5}{2}$	9. $y = \frac{8}{3}$

3-4 THE DISTRIBUTIVE PRINCIPLE

In this section, we will discuss the distributive principle of multiplication over addition and subtraction.

---

41. In the multiplication $3(4 + 2)$, the second factor is an addition. To evaluate, we can either add before multiplying or multiply before adding.

$$3(4 + 2) \quad\quad 3(4+2) = 3(4) + 3(2)$$
$$3\ (6) = 18 \quad\quad = 12 + 6 = 18$$

The "multiply before adding" method uses the DISTRIBUTIVE PRINCIPLE OF MULTIPLICATION OVER ADDITION. That principle is diagrammed below. Notice that we multiply both 4 and 2 by 3.

$$3(4 + 2) = 3(4) + 3(2) = 12 + 6$$

The process above is called multiplying by the distributive principle. Two more examples are shown.

$$4(x + 6) = 4(x) + 4(6) = 4x + 24$$
$$9(2 + y) = 9(2) + 9(y) = 18 + 9y$$

Following the examples, complete these:

a) $2(x + 7) = \_\_\_\_ + \_\_\_\_ = \_\_\_\_ + \_\_\_\_$

b) $5(8 + t) = \_\_\_\_ + \_\_\_\_ = \_\_\_\_ + \_\_\_\_$

Answers to 41:

a) $2(x) + 2(7) = 2x + 14$

b) $5(8) + 5(t) = 40 + 5t$

42. The DISTRIBUTIVE PRINCIPLE OF MULTIPLICATION OVER ADDITION is:

$$a(b + c) = ab + ac$$

Two more examples of multiplying by the distributive principle are shown below.

$$6(3x + 7) = 6(3x) + 6(7) = 18x + 42$$
$$2(5 + 8y) = 2(5) + 2(8y) = 10 + 16y$$

Following the examples, complete these:

a) $3(5x + 8) = \_\_\_\_ + \_\_\_\_ = \_\_\_\_ + \_\_\_\_$

b) $7(2 + 9y) = \_\_\_\_ + \_\_\_\_ = \_\_\_\_ + \_\_\_\_$

Answers to 42:

a) $3(5x) + 3(8) = 15x + 24$

b) $7(2) + 7(9y) = 14 + 63y$

43. We usually try to write the final product in one step. For example:

$$8(x + 4) = 8x + 32$$
$$5(6 + 8y) = 30 + 40y$$

Following the examples, complete these:

a) $5(x + 1) = \_\_\_\_ + \_\_\_\_$

b) $6(10 + y) = \_\_\_\_ + \_\_\_\_$

c) $4(7d + 3) = \_\_\_\_ + \_\_\_\_$

d) $9(4 + 6p) = \_\_\_\_ + \_\_\_\_$

44. A common error is forgetting to multiply the second term of the addition by the first factor. Some examples are shown.

Error (see arrow)	Correct
$3(x + 8) = 3x + 8$	$3(x + 8) = 3x + 24$
$7(1 + 6m) = 7 + 6m$	$7(1 + 6m) = 7 + 42m$

Avoiding the common error, complete these:

a) $2(x + 7) =$ ____________ c) $5(6 + y) =$ ____________

b) $7(5b + 1) =$ ____________ d) $8(5 + 7m) =$ ____________

Answers: a) $5x + 5$ b) $60 + 6y$ c) $28d + 12$ d) $36 + 54p$

---

45. There is also a DISTRIBUTIVE PRINCIPLE OF MULTIPLICATION OVER SUBTRACTION. It is diagrammed below.

$$a(b - c) = ab - ac$$

Two examples of multiplying by the distributive principle when the second factor is a subtraction are shown below.

$$9(x - 3) = 9(x) - 9(3) = 9x - 27$$

$$4(6 - 2y) = 4(6) - 4(2y) = 24 - 8y$$

Usually we write the final product in one step. For example:

$$9(x - 3) = 9x - 27$$

$$4(6 - 2y) = 24 - 8y$$

Complete these. Remember that the terms are subtracted, not added.

a) $6(m - 1) =$ ____________ c) $3(5x - 2) =$ ____________

b) $5(4 - h) =$ ____________ d) $10(1 - 4y) =$ ____________

Answers: a) $2x + 14$ b) $35b + 7$ c) $30 + 5y$ d) $40 + 56m$

---

46. Be careful with the signs for these.

a) $7(2 - x) =$ ____________ c) $4(2a + 1) =$ ____________

b) $6(y + 9) =$ ____________ d) $9(7 - 8b) =$ ____________

Answers: a) $6m - 6$ b) $20 - 5h$ c) $15x - 6$ d) $10 - 40y$

---

47. In each multiplication below, the first factor is negative. Notice how we wrote the final products to minimize the number of signs between terms. Both $-2x - 6$ and $-4t + 20$ have only one sign between terms.

$$-2(x + 3) = (-2)(x) + (-2)(3) = -2x + (-6) = -2x - 6$$

$$-4(t - 5) = (-4)(t) - (-4)(5) = -4t - (-20) = -4t + 20$$

Following the examples, do these:

a) $-3(y + 7) =$ ____________

b) $-5(2d + 1) =$ ____________

c) $-1(m - 9) =$ ____________

d) $-10(4 - 2d) =$ ____________

Answers: a) $14 - 7x$ b) $6y + 54$ c) $8a + 4$ d) $63 - 72b$

48. We can also multiply by the distributive principle when the addition or subtraction is the first factor. That is:	a) $-3y - 21$
	b) $-10d - 5$
	c) $-1m + 9$
	d) $-40 + 20d$

$$(a + b)c = ac + bc \quad \text{and} \quad (a - b)c = ac - bc$$

Two examples of the above principle are shown.

$$(2 + 3)x = 2x + 3x$$

$$(6 - 4)y = 6y - 4y$$

Following the examples, complete these:

a) $(7 + 5)x$ = ______________ b) $(8 - 3)y$ = ______________

a) $7x + 5x$ b) $8y - 3y$

3-5 COMBINING LIKE TERMS

In this section, we will show how like terms can be combined by factoring by the distributive principle. We will then use that process to solve some equations.

49. Two multiplications by the distributive principle over addition are shown below.

$$(5 + 4)x = 5x + 4x$$

$$(8 + 2)y = 8y + 2y$$

By reversing the process, we can break up each product into the original factor. Doing so is called factoring by the distributive principle. That is:

$5x + 4x$ can be factored to get $(5 + 4)x$

$8y + 2y$ can be factored to get __________

$(8 + 2)y$

50. In an algebraic expression, the parts separated by a plus sign or a minus sign are called terms. For example:

In $x + 2y - 3$, the terms are x, $2y$, and 3.

In $3x - 4y + 5$, the terms are $3x$, $4y$, and 5.

Though the definition will be stated more precisely later, for now we will define like terms as terms with the same variable. For example:

$3x$ and $9x$ are like terms.

$7y$ and $-5y$ are like terms.

Like terms can be combined by factoring by the distributive principle. That is:

$$3x + 9x = (3 + 9)x = 12x$$

$$7y + (-5y) = [7 + (-5)]y = 2y$$

Continued on following page.

50. Continued

Using the same process, complete these:

a) $8t + 7t = (8 + 7)t =$ ________

b) $-3R + 9R = (-3 + 9)R =$ ________

Answers: a) $15t$ b) $6R$

---

51. As you can see, we can combine like terms by simply adding their coefficients. That is:

$$7x + 5x = 12x, \quad \text{since} \quad 7 + 5 = 12$$

$$-8y + 3y = -5y, \quad \text{since} \quad -8 + 3 = -5$$

By simply adding coefficients, complete these:

a) $11m + 3m =$ ________ c) $6t + (-9t) =$ ________

b) $-2V + 7V =$ ________ d) $-5x + (-5x) =$ ________

Answers: a) $14m$ b) $5V$ c) $-3t$ d) $-10x$

---

52. Any variable without a coefficient has a coefficient of "1". That is:

$$x = 1x \qquad y = 1y$$

To do the additions below, we wrote the coefficient "1" explicitly.

$$7x + x = 7x + 1x = 8x$$

$$y + (-4y) = 1y + (-4y) = -3y$$

Complete these:

a) $x + 3x =$ ______ b) $-9t + t =$ ______ c) $m + m =$ ______

Answers: a) $4x$ b) $-8t$ c) $2m$

---

53. Since $0x$ means "0 times x", $0x = 0$. Therefore, each sum below is 0.

$$5x + (-5x) = 0x = 0$$

$$x + (-x) = 1x + (-1x) = 0x = 0$$

Two quantities are called <u>opposites</u> if their sum is 0. It should be obvious that two like terms are opposites <u>if their coefficients are opposites</u>. That is:

$5x$ and $(-5x)$ are opposites, since $5 + (-5) = 0$

x and $-x$ are opposites, since $1 + (-1) = 0$

Write the opposite of each term.

a) $8y$ ______ b) d ______ c) $-12t$ ______ d) $-V$ ______

Answers: a) $-8y$ b) $-d$ c) $12t$ d) V

---

54. Two multiplications by the distributive principle over subtraction are shown below.

$$(8 - 4)x = 8x - 4x$$

$$(2 - 7)y = 2y - 7y$$

We can also factor by the distributive principle to combine the like terms above. That is:

$$8x - 4x = (8 - 4)x = 4x$$

$$2y - 7y = (2 - 7)y = -5y$$

Continued on following page.

54. Continued

By simply subtracting coefficients, combine the like terms below:

a) $10R - 2R =$ ________ c) $-2x - 5x =$ ________

b) $5t - 15t =$ ________ d) $9y - 9y =$ ________

---

55. To do the subtractions below, we wrote the coefficient "1" explicitly.

$$x - 7x = 1x - 7x = -6x$$

$$3y - y = 3y - 1y = 2y$$

Complete these:

a) $10d - d =$ ______ b) $V - 4V =$ ______ c) $p - p =$ ______

Answers (54): a) $8R$ b) $-10t$ c) $-7x$ d) 0

---

56. As we saw earlier, $-x = -1x$ and $-y = -1y$. It is helpful to write the -1 coefficient explicitly in additions and subtractions like those below.

$$-x + 3x = -1x + 3x = 2x$$

$$4y + (-y) = 4y + (-1y) = 3y$$

$$-t - 6t = -1t - 6t = -7t$$

Following the examples, complete these:

a) $-m + 10m =$ ________ c) $6d + (-d) =$ ________

b) $-R - 2R =$ ________ d) $-p - p =$ ________

Answers (55): a) $9d$ b) $-3V$ c) 0

---

57. To solve the equations below, we combined like terms and then used the multiplication axiom.

$$5x + 2x = 21 \qquad\qquad 20 = y - 5y$$

$$7x = 21 \qquad\qquad 20 = -4y$$

$$x = \frac{21}{7} \qquad\qquad \frac{20}{-4} = y$$

$$x = 3 \qquad\qquad y = -5$$

Using the same steps, solve these:

a) $2t + t = 27$ b) $20 = 8d - 3d$

Answers (56): a) $9m$ b) $-3R$ c) $5d$ d) $-2p$

---

58. Solve: a) $45 = -6t + (-3t)$ b) $9x - x = 16$

Answers (57): a) $t = 9$ b) $d = 4$

Answers (58): a) $t = -5$ b) $x = 2$

59. Solve: a) $y + 4y = 3$ b) $-d - 2d = 7$

---

60. Solve. Reduce to lowest terms.

a) $-12 = 10V - 20V$ b) $-x + 7x = 3$

Answers (59): a) $y = \frac{3}{5}$ b) $d = -\frac{7}{3}$

---

61. The solution of the equation below is 0. Solve the other equation.

$$3x + 4x = 0$$
$$7x = 0$$
$$x = \frac{0}{7}$$
$$x = 0$$

$$y - 2y = 0$$

Answers (60): a) $V = \frac{6}{5}$ b) $x = \frac{1}{2}$

---

62. Following the example, solve the other equation.

$$1.5x + 3.1x = 13.8$$
$$4.6x = 13.8$$
$$x = \frac{13.8}{4.6}$$
$$x = 3$$

$$6.8y - 2.4y = -88$$

Answer (61): $y = 0$

---

63. Following the example, solve the other equation.

$$\frac{5}{6}x - \frac{2}{3}x = 5$$
$$\frac{5}{6}x - \frac{4}{6}x = 5$$
$$\frac{1}{6}x = 5$$
$$6\left(\frac{1}{6}x\right) = 6(5)$$
$$x = 30$$

$$\frac{1}{2}y + \frac{1}{4}y = 2$$

Answer (62): $y = -20$

---

64. Following the example, solve the other equation.

$$x - \frac{1}{3}x = 5$$
$$\frac{2}{3}x = 5$$
$$\frac{3}{2}\left(\frac{2}{3}x\right) = \frac{3}{2}(5)$$
$$x = \frac{15}{2}$$

$$y + \frac{1}{4}y = 10$$

Answer (63): $y = \frac{8}{3}$

65. Some expressions containing more than two terms can be simplified by combining numbers or combining like terms. Two examples are shown. $7 + 3x - 5 = 2 + 3x$ $-5d - 4 - d = -6d - 4$ Following the examples, simplify these: a) $10 + 7x + 5 =$ ________ c) $-6y + 4 + 2y =$ ________ b) $y + 9 + 3y =$ ________ d) $5 - t - 4 =$ ________	$y = 8$
66. We simplified two more expressions below. $4 - 6x - 7 = -3 - 6x$ $-y - 3y + 2 = -4y + 2$ Following the examples, simplify these: a) $10x - 7 - x =$ ________ c) $x - 7x + 4 =$ ________ b) $3 + 5y - 9 =$ ________ d) $-t + 6t - 7 =$ ________	a) $15 + 7x$ b) $4y + 9$ c) $-4y + 4$ d) $1 - t$
67. To solve the equation below, we began by simplifying the left side. Solve the other equation. $7 + 3x - 9 = 6$ $-2 + 3x = 6$ $2 - 2 + 3x = 6 + 2$ $3x = 8$ $x = \frac{8}{3}$ $4 - 5y - 3 = 2$	a) $9x - 7$ b) $-6 + 5y$ c) $-6x + 4$ d) $5t - 7$
68. To solve the equation below, we began by simplifying the right side. Solve the other equation. $10 = x - 5x + 4$ $10 = -4x + 4$ $-4 + 10 = -4x + 4 + (-4)$ $6 = -4x$ $\frac{6}{-4} = x$ $x = -\frac{3}{2}$ $9 = -y + 10 - y$	$y = -\frac{1}{5}$
	$y = \frac{1}{2}$

3-6 EQUATIONS WITH LIKE TERMS ON OPPOSITE SIDES

In this section, we will use both axioms to solve equations with like terms on opposite sides.

---

69. The equation $7x = 2x + 20$ has like terms on opposite sides. To solve it, we use both axioms.

The addition axiom is used to get both like terms on one side so that they can be combined.

The multiplication axiom is then used to solve the resulting equation.

The steps used to solve the equation are described below.

1) To get both like terms on the same side so that they can be combined, we add -2x to both sides.

$$7x = 2x + 20$$
$$-2x + 7x = \underbrace{-2x + 2x}_{0} + 20$$
$$5x = 0 + 20$$
$$5x = 20$$

2) Then we use the multiplication axiom to solve $5x = 20$.

$$x = \frac{20}{5} = 4$$

Complete the checking of 4 as the solution of the original equation at the right.

$$7x = 2x + 20$$
$$7(4) = 2(4) + 20$$
$$\_\_\_ = \_\_\_ + 20$$
$$\_\_\_ = \_\_\_$$

Answer: $28 = 8 + 20$

$28 = 28$

70. For equations like those above, we use the addition axiom to move the like term on the same side as the number to the other side. To do so, we add its opposite to both sides.

For $5y + 36 = 9y$, we move $5y$ by adding $-5y$ to both sides.

For $7d = 8d - 60$, we move $8d$ by adding $-8d$ to both sides.

Using the facts above, solve these equations.

a) $5y + 36 = 9y$

b) $7d = 8d - 60$

71. Following the example, solve the other equation.

$$7x = 40 - 3x$$
$$3x + 7x = 40 \underbrace{- 3x + 3x}$$
$$10x = 40 + 0$$
$$10x = 40$$
$$x = \frac{40}{10}$$
$$x = 4$$

$$11 - 4y = 4y$$

Answers: a) $y = 9$ b) $d = 60$

---

72. Following the example, solve the other equation.

$$y = 14 + 3y$$
$$-3y + y = 14 + 3y + (-3y)$$
$$-2y = 14$$
$$y = \frac{14}{-2}$$
$$y = -7$$

$$4 - x = 4x$$

Answer: $y = \frac{11}{8}$

---

73. Following the example, solve the other equation.

$$\frac{4}{9}x - \frac{2}{9} = \frac{5}{9}x$$
$$-\frac{4}{9}x + \frac{4}{9}x - \frac{2}{9} = \frac{5}{9}x + \left(-\frac{4}{9}x\right)$$
$$-\frac{2}{9} = \frac{1}{9}x$$
$$9\left(-\frac{2}{9}\right) = 9\left(\frac{1}{9}x\right)$$
$$-2 = x$$

$$\frac{1}{2}y + \frac{1}{4} = \frac{3}{4}y$$

Answer: $x = \frac{4}{5}$

---

74. Following the example, solve the other equation.

$$5.2x - 4.5 = 3.7x$$
$$-5.2x + 5.2x - 4.5 = 3.7x + (-5.2x)$$
$$-4.5 = -1.5x$$
$$x = \frac{-4.5}{-1.5}$$
$$x = 3$$

$$3.6t = 1.6t + 5$$

Answer: $y = 1$

---

75. The equation $7x + 9 = 4x + 21$ has a variable term and a number on each side. To solve it, we must get both like terms on one side and both numbers on the other side so that they can be combined. To do so, we must move a variable term from one side and a number from the other side. We have a choice of two pairs:

1) moving $4x$ and 9 or 2) moving $7x$ and 21

Continued on following page.

Answer: $t = 2.5$

75. Continued

With either choice, we must use the addition axiom twice. We solved below by moving 4x and 9.

To move 4x, we add -4x to both sides.

$$7x + 9 = 4x + 21$$

$$-4x + 7x + 9 = -4x + 4x + 21$$

$$3x + 9 = 21$$

To move 9, we add -9 to both sides.

$$3x + 9 + (-9) = 21 + (-9)$$

$$3x = 12$$

$$x = \frac{12}{3} = 4$$

Check the solution at the right.

$$7x + 9 = 4x + 21$$

---

76. By moving either possible pair, solve these:

a) $3y + 8 = 9y + 2$

b) $8t + 1 = 4t - 9$

Answer to 75:

$$7(4) + 9 = 4(4) + 21$$

$$28 + 9 = 16 + 21$$

$$37 = 37$$

---

77. Following the example, solve the other equation.

$$30 - 5d = 20 - 9d$$

$$30 - 5d + 5d = 20 - 9d + 5d$$

$$30 = 20 - 4d$$

$$-20 + 30 = -20 + 20 - 4d$$

$$10 = -4d$$

$$d = -\frac{10}{4}$$

$$d = -\frac{5}{2}$$

$$15 - 2h = 8 - h$$

Answers to 76:

a) $y = 1$

b) $t = -\frac{5}{2}$

Answer to 77:

$h = 7$

78. Following the example, solve the other equation.

$$\frac{7}{8}t - \frac{1}{4} = \frac{1}{2}t + \frac{5}{8}$$

$$\frac{7}{8}t - \frac{1}{4} + \frac{1}{4} = \frac{1}{2}t + \frac{5}{8} + \frac{1}{4}$$

$$\frac{7}{8}t = \frac{1}{2}t + \frac{7}{8}$$

$$-\frac{1}{2}t + \frac{7}{8}t = -\frac{1}{2}t + \frac{1}{2}t + \frac{7}{8}$$

$$\frac{3}{8}t = \frac{7}{8}$$

$$\frac{8}{3}\left(\frac{3}{8}t\right) = \frac{8}{3}\left(\frac{7}{8}\right)$$

$$t = \frac{7}{3}$$

$$x + \frac{1}{3} = \frac{2}{3}x + 1$$

$x = 2$

SELF-TEST 9 (pages 117-127)

Multiply:

1. $4(1 - 2d) =$
2. $-3(x + 5) =$

Combine like terms:

3. $4t - t =$
4. $-m + 7 - 6m =$

Solve each equation. Report fractional solutions in lowest terms.

5. $(-2d) + 9d = 0$
6. $2 = r - 5r$
7. $7t + 8 = 3t$
8. $w = 10 - 5w$
9. $3.3p - 4.8 = 2.1p$
10. $\frac{1}{3}x + \frac{1}{2} = x$
11. $11 - 2P = P + 11$
12. $6h - 13 = 22 - 9h$

ANSWERS:

1. $4 - 8d$
2. $-3x - 15$
3. $3t$
4. $-7m + 7$
5. $d = 0$
6. $r = -\frac{1}{2}$
7. $t = -2$
8. $w = \frac{5}{3}$
9. $p = 4$
10. $x = \frac{3}{4}$
11. $P = 0$
12. $h = \frac{7}{3}$

3-7 REMOVING GROUPING SYMBOLS

In this section, we will discuss the procedures for removing grouping symbols to simplify expressions.

---

79. If a grouping is preceded by a + sign or no sign, we can simply drop the grouping symbols. For example:

$$+(x - 4) \text{ can be written } x - 4$$

$$(3y + 1) \text{ can be written } 3y + 1$$

Using the fact above, we simplified the expressions below. Simplify the other expressions.

$$7 + (x - 4) = 7 + x - 4 = 3 + x$$

$$(3y + 1) - y = 3y + 1 - y = 2y + 1$$

a) $4t + (3t - 8) =$ ______________________

b) $10 + (7 - 2d) =$ ______________________

Answers: a) $7t - 8$ b) $17 - 2d$

80. If a grouping is preceded by a - sign, we <u>cannot</u> simply drop the grouping symbols. To get rid of the grouping symbols, we can substitute -1 for the - sign and multiply by the distributive principle. That is:

$$-(3y + 2) = -1(3y + 2) = (-1)(3y) + (-1)(2) = -3y + (-2) = -3y - 2$$

$$-(6x - 4) = -1(6x - 4) = (-1)(6x) - (-1)(4) = -6x - (-4) = -6x + 4$$

As you can see, we can remove the grouping symbols above <u>by simply changing the sign of each term</u>. That is:

$$-(3y + 2) = -3y - 2$$

$$-(6x - 4) = -6x + 4$$

Remove the grouping symbols from these <u>by simply changing the sign of each term</u>.

a) $-(7x + 1) =$ __________ c) $-(t + 10) =$ __________

b) $-(4y - 5) =$ __________ d) $-(8 - d) =$ __________

Answers: a) $-7x - 1$ b) $-4y + 5$ c) $-t - 10$ d) $-8 + d$

81. To remove the grouping symbols below, we changed the sign of each term in $(3y + 2)$. We then simplified.

$$5 - (3y + 2) = 5 - 3y - 2 = 3 - 3y$$

Following the example, remove the grouping symbols and simplify.

a) $2d - (4d + 7) =$ ______________ = ______________

b) $7 - (p + 1) =$ ______________ = ______________

Answers: a) $2d - 4d - 7 = -2d - 7$ b) $7 - p - 1 = 6 - p$

82. To remove the grouping symbols below, we changed the sign of each term in $(6x - 4)$. We then simplified.

$$7 - (6x - 4) = 7 - 6x + 4 = 11 - 6x$$

Following the examples, remove the grouping symbols and simplify.

a) $5 - (x - 3)$ = ______________ = ______________

b) $6y - (5 - 2y)$ = ______________ = ______________

Answers to 82: a) $5 - x + 3 = 8 - x$ b) $6y - 5 + 2y = 8y - 5$

83. $3(x + 5)$ and $4(2y - 6)$ are instances of the distributive principle. To remove the grouping symbols, we multiply.

$$3(x + 5) = 3x + 15$$

$$4(2y - 6) = 8y - 24$$

Remove the grouping symbols in these.

a) $5(4d + 1)$ = __________ b) $7(5 - p)$ = __________

Answers to 83: a) $20d + 5$ b) $35 - 7p$

84. When an instance of the distributive principle is preceded by a + sign or no sign, we can remove the grouping symbols by multiplying. For example:

$$3(x + 4) + 7 = 3x + 12 + 7 = 3x + 19$$

$$10 + 6(2y - 3) = 10 + 12y - 18 = -8 + 12y$$

Following the examples, remove the grouping symbols and simplify.

a) $2(4d - 6) - 3d$ = ______________________________

b) $12 + 4(1 - 3x)$ = ______________________________

Answers to 84: a) $5d - 12$ b) $16 - 12x$

85. When an instance of the distributive principle is preceded by a − sign, we multiply first and then remove the grouping symbols in the usual way. <u>It is very helpful to draw brackets around the instance of the distributive principle before multiplying</u>. For example:

$10 - 2(x + 3)$	$= 10 - [2(x + 3)]$	Drawing brackets
	$= 10 - [2x + 6]$	Multiplying
	$= 10 - 2x - 6$	Removing the grouping symbols
	$= 4 - 2x$	Simplifying

Following the example, simplify these. <u>Begin by drawing brackets around the instance of the distributive principle</u>.

a) $5 - 3(d + 1)$ b) $t - 2(4t + 5)$

86. To simplify the expression below, we began by drawing brackets around $5(x - 1)$.

$$6 - 5(x - 1) = 6 - [5(x - 1)]$$
$$= 6 - [5x - 5]$$
$$= 6 - 5x + 5$$
$$= 11 - 5x$$

Following the example, simplify these. Draw brackets.

a) $20 - 4(d - 2)$ b) $10y - 3(2y - 9)$

Answers: a) $2 - 3d$ b) $-7t - 10$

---

87. $-2(x + 6)$ and $-5(3y - 4)$ are instances of the distributive principle. To remove the grouping symbols, we multiply.

$$-2(x + 6) = -2x - 12$$
$$-5(3y - 4) = -15y + 20$$

Remove the grouping symbols by multiplying.

a) $-3(7p + 1)$ = ________ b) $-4(10 - t)$ = ________

Answers: a) $28 - 4d$ b) $4y + 27$

---

88. Following the example, simplify the other expressions.

$$-3(t + 4) + 5 = -3t - 12 + 5 = -3t - 7$$

a) $-2(3y + 1) - 7$ = ________________

b) $-5(4 - d) - 2d$ = ________________

Answers: a) $-21p - 3$ b) $-40 + 4t$

Answers: a) $-6y - 9$ b) $-20 + 3d$

3-8 EQUATIONS WITH GROUPING SYMBOLS

In this section, we will solve equations that contain grouping symbols.

---

89. To solve the equation below, we removed the grouping symbols, simplified the left side, and then used the axioms. Solve the other equation.

$$3 + (2y + 1) = 9$$
$$3 + 2y + 1 = 9$$
$$4 + 2y = 9$$
$$-4 + 4 + 2y = 9 + (-4)$$
$$2y = 5$$
$$y = \frac{5}{2}$$

$$(t - 3) - 8 = 0$$

90. To solve the equation below, we removed the grouping symbols, simplified the left side, and then used the axioms. Solve the other equation. $5x - (2x + 1) = 4$ $5x - 2x - 1 = 4$ $3x - 1 = 4$ $3x - 1 + 1 = 4 + 1$ $3x = 5$ $x = \frac{5}{3}$ $F = 11 - (6F + 7)$	$t = 11$
91. To solve the equation below, we removed the grouping symbols, simplified the left side, and then used the axioms. Solve the other equation. $6y - (2y - 3) = 1$ $6y - 2y + 3 = 1$ $4y + 3 = 1$ $4y + 3 + (-3) = 1 + (-3)$ $4y = -2$ $y = -\frac{2}{4}$ $y = -\frac{1}{2}$ $x = 10 - (5x - 1)$	$F = \frac{4}{7}$
92. Following the example, solve the other equation. $4t - (t + 2) = 0$ $4t - t - 2 = 0$ $3t - 2 = 0$ $3t - 2 + 2 = 0 + 2$ $3t = 2$ $t = \frac{2}{3}$ $8p - (2p - 1) = 0$	$x = \frac{11}{6}$
93. To solve the equation below, we began by multiplying by the distributive principle. Solve the other equation. $3(x + 2) = 21$ $3x + 6 = 21$ $3x + 6 + (-6) = 21 + (-6)$ $3x = 15$ $x = 5$ $9(2b - 4) = 18$	$p = -\frac{1}{6}$
	$b = 3$

94. Following the example, solve the other equation.

$$5(3 + 2y) = 20y$$
$$15 + 10y = 20y$$
$$15 + 10y + (-10y) = 20y + (-10y)$$
$$15 = 10y$$
$$\frac{15}{10} = y$$
$$y = \frac{3}{2}$$

$$3(7 - m) = m$$

---

95. Following the example, solve the other equation.

$m = \frac{21}{4}$

$$5(x - 1) = 2x - 7$$
$$5x - 5 = 2x - 7$$
$$-2x + 5x - 5 = -2x + 2x - 7$$
$$3x - 5 = -7$$
$$3x - 5 + 5 = -7 + 5$$
$$3x = -2$$
$$x = -\frac{2}{3}$$

$$2y + 9 = 3(y + 2)$$

---

96. To solve the equation below, we multiplied by the distributive principle on both sides. Solve the other equation.

$y = 3$

$$7(d + 2) = 2(3 + d)$$
$$7d + 14 = 6 + 2d$$
$$-2d + 7d + 14 = 6 + 2d + (-2d)$$
$$5d + 14 = 6$$
$$5d + 14 + (-14) = 6 + (-14)$$
$$5d = -8$$
$$d = -\frac{8}{5}$$

$$5(t + 3) = 3(t - 5)$$

---

97. To solve the equation below, we simplified the expression on the left side and then used the axioms. Solve the other equation.

$t = -15$

$$2(m + 4) - 3 = 10$$
$$2m + 8 - 3 = 10$$
$$2m + 5 = 10$$
$$2m + 5 + (-5) = 10 + (-5)$$
$$2m = 5$$
$$m = \frac{5}{2}$$

$$4t = 5 + 3(2t - 3)$$

$t = 2$

98. To solve the equation below, we began by simplifying the expression on the left side. Solve the other equation.

$$\begin{aligned} 10 - 3(x + 4) &= 16 \\ 10 - [3(x + 4)] &= 16 \\ 10 - [3x + 12] &= 16 \\ 10 - 3x - 12 &= 16 \\ -2 - 3x &= 16 \\ 2 - 2 - 3x &= 16 + 2 \\ -3x &= 18 \\ x &= -6 \end{aligned}$$

$$9y - 7(y + 2) = 4$$

$y = 9$

99. To solve the equation below, we began by simplifying the expression on the right side. Solve the other equation.

$$\begin{aligned} 35 &= 47 - 6(4 - x) \\ 35 &= 47 - [6(4 - x)] \\ 35 &= 47 - [24 - 6x] \\ 35 &= 47 - 24 + 6x \\ 35 &= 23 + 6x \\ -23 + 35 &= -23 + 23 + 6x \\ 12 &= 6x \\ x &= 2 \end{aligned}$$

$$20 = t - 3(5 - 2t)$$

$t = 5$

3-9 FRACTIONAL EQUATIONS

To solve a fractional equation containing only one fraction, we begin by clearing the fraction. We will discuss the method in this section.

---

100. An algebraic fraction is a fraction that contains a variable in either its numerator or denominator or both. Some examples are:

$$\frac{x}{3} \qquad \frac{5}{2y} \qquad \frac{m+1}{4} \qquad \frac{t}{t-1}$$

Just like numerical fractions, algebraic fractions stand for a division. That is:

$\frac{x}{3}$ means: x divided by 3 .

$\frac{t}{t-1}$ means: _____ divided by _____

---

t divided by $t - 1$

101. When the denominator of an algebraic fraction is "1", the fraction equals its numerator. That is:

$$\frac{x}{1} = x \qquad \frac{4y}{1} = 4y \qquad \frac{t+7}{1} = \underline{\hspace{2cm}}$$

Answer: t + 7

102. When the numerator and denominator of an algebraic fraction are identical, the fraction equals "1". That is:

$$\frac{x}{x} = 1 \qquad \frac{5y}{5y} = 1 \qquad \frac{m-3}{m-3} = \underline{\hspace{1cm}}$$

Answer: 1

103. In each example below, we multiplied an algebraic fraction by its denominator. Notice that each product is <u>the numerator of the original fraction</u>.

$$3\left(\frac{x}{3}\right) = \frac{(3)(x)}{3} = \left(\frac{3}{3}\right)(x) = (1)(x) = x$$

$$y\left(\frac{5}{y}\right) = \frac{(y)(5)}{y} = \left(\frac{y}{y}\right)(5) = (1)(5) = 5$$

The "cancelling" shortcut below can be used for the same multiplications.

$$\overset{1}{\cancel{3}}\left(\frac{x}{\underset{1}{\cancel{3}}}\right) = \frac{x}{1} = x \qquad \overset{1}{\cancel{y}}\left(\frac{5}{\underset{1}{\cancel{y}}}\right) = \frac{5}{1} = 5$$

Use the shortcut for these:

a) $7\left(\frac{d}{7}\right) = \underline{\hspace{2cm}}$ b) $m\left(\frac{8}{m}\right) = \underline{\hspace{2cm}}$

Answers: a) d b) 8

104. In each example below, we used the "cancelling" shortcut to multiply an algebraic fraction by its denominator. Notice that each product is <u>the numerator of the original fraction</u>.

$$\overset{1}{\cancel{5}}\left(\frac{4y}{\underset{1}{\cancel{5}}}\right) = \frac{4y}{1} = 4y \qquad \overset{1}{\cancel{2x}}\left(\frac{9}{\underset{1}{\cancel{2x}}}\right) = \frac{9}{1} = 9$$

Use the shortcut for these:

a) $10\left(\frac{3t}{10}\right) = \underline{\hspace{2cm}}$ b) $7m\left(\frac{6}{7m}\right) = \underline{\hspace{2cm}}$

Answers: a) 3t b) 6

105. In each example below, we multiplied an algebraic fraction by its denominator. Notice again that each product is <u>the numerator of the original fraction</u>.

$$\overset{1}{\cancel{8}}\left(\frac{x+5}{\underset{1}{\cancel{8}}}\right) = \frac{x+5}{1} = x + 5 \qquad \overset{1}{(\cancel{t-5})}\left(\frac{t}{\underset{1}{\cancel{t-5}}}\right) = \frac{t}{1} = t$$

Use the same method for these:

a) $y\left(\frac{y-3}{y}\right) = \underline{\hspace{2cm}}$ b) $(x+8)\left(\frac{4}{x+8}\right) = \underline{\hspace{2cm}}$

Answers: a) y - 3 b) 4

106. Whenever an algebraic fraction is multiplied by its denominator, the product is identical to its numerator. That is:

$$7x\left(\frac{1}{7x}\right) = 1 \qquad V\left(\frac{V-7}{V}\right) = V - 7 \qquad (x+4)\left(\frac{5}{x+4}\right) = 5$$

Using the above fact, complete these:

a) $6\left(\frac{5R}{6}\right) =$ ____ b) $5\left(\frac{1+h}{5}\right) =$ ____ c) $(10 - x)\left(\frac{x}{10-x}\right) =$ ____

Answers: a) 5R b) 1 + h c) x

107. To solve the fractional equation below, we began by <u>clearing the fraction</u>. To do so, we used the multiplication axiom, multiplying both sides by 5, <u>the denominator of the fraction</u>. Solve the other equation by multiplying both sides by 8.

$$\frac{x}{5} = 7$$
$$5\left(\frac{x}{5}\right) = 5(7)$$
$$x = 35$$

Check

$$\frac{x}{5} = 7$$
$$\frac{35}{5} = 7$$
$$7 = 7$$

$$4 = \frac{t}{8}$$

Answer:

$$8(4) = 8\left(\frac{t}{8}\right)$$
$$32 = t$$
$$t = 32$$

108. To clear the fraction below, we multiplied both sides by <u>y</u>, <u>the denominator of the fraction</u>. To solve the other equation, begin by multiplying both sides by R.

$$\frac{12}{y} = 3$$
$$y\left(\frac{12}{y}\right) = y(3)$$
$$12 = 3y$$
$$\frac{12}{3} = y$$
$$y = 4$$

$$6 = \frac{42}{R}$$

Answer:

$$R(6) = R\left(\frac{42}{R}\right)$$
$$6R = 42$$
$$R = 7$$

109. To solve the equation below, we began by multiplying both sides by 3 to clear the fraction. To solve the other equation, begin by multiplying both sides by 2.

$$\frac{2m}{3} = 6$$
$$3\left(\frac{2m}{3}\right) = 3(6)$$
$$2m = 18$$
$$m = \frac{18}{2}$$
$$m = 9$$

Check

$$\frac{2m}{3} = 6$$
$$\frac{2(9)}{3} = 6$$
$$\frac{18}{3} = 6$$
$$6 = 6$$

$$12 = \frac{3x}{2}$$

Answer: $x = 8$

110. To solve the equation below, we began by multiplying both sides by 3t. To solve the other equation, begin by multiplying both sides by 9q.

$$\frac{7}{3t} = 2 \qquad\qquad 1 = \frac{5}{9q}$$

$$3t\left(\frac{7}{3t}\right) = 3t(2)$$

$$7 = 6t$$

$$t = \frac{7}{6}$$

Answer: $q = \frac{5}{9}$

---

111. To solve the equation below, we began by multiplying both sides by 2. Solve the other equation.

$$\frac{x + 3}{2} = 5 \qquad\qquad 1 = \frac{y - 3}{4}$$

$$2\left(\frac{x+3}{2}\right) = 2(5)$$

$$x + 3 = 10$$

$$x + 3 + (-3) = 10 + (-3)$$

$$x = 7$$

Answer: $y = 7$

---

112. To solve the equation below, we began by multiplying both sides by 4. Solve the other equation.

$$\frac{2k + 7}{4} = 1 \qquad\qquad \frac{x - 1}{x} = 3$$

$$4\left(\frac{2k+7}{4}\right) = 4(1)$$

$$2k + 7 = 4$$

$$2k + 7 + (-7) = 4 + (-7)$$

$$2k = -3$$

$$k = -\frac{3}{2}$$

Answer: $x = -\frac{1}{2}$

---

113. To solve the equation below, we began by multiplying both sides by $(x + 3)$. Notice that we then had to multiply by the distributive principle on the right side. Solve the other equation.

$$\frac{7}{x + 3} = 5 \qquad\qquad 4 = \frac{y}{y - 1}$$

$$(x + 3)\left(\frac{7}{x + 3}\right) = 5(x + 3)$$

$$7 = 5x + 15$$

$$-15 + 7 = 5x + 15 + (-15)$$

$$-8 = 5x$$

$$x = -\frac{8}{5}$$

114. Following the example, solve the other equation.

$$\frac{3 - x}{7} = 2 \qquad\qquad 1 = \frac{5}{10 - d}$$

$$7\left(\frac{3 - x}{7}\right) = 7(2)$$

$$3 - x = 14$$

$$-3 + 3 - x = 14 + (-3)$$

$$-1x = 11$$

$$x = \frac{11}{-1}$$

$$x = -11$$

Answer: $y = \frac{4}{3}$

---

115. Following the example, solve the other equation.

$$\frac{2m}{7} = 0 \qquad\qquad \frac{y + 5}{9} = 0$$

$$7\left(\frac{2m}{7}\right) = 7(0)$$

$$2m = 0$$

$$m = \frac{0}{2}$$

$$m = 0$$

Answer: $d = 5$

Answer: $y = -5$

SELF-TEST 10 (pages 128-138)

Solve each equation. Report each solution in lowest terms.

1. $2h + (7h + 3) = 0$
2. $5 = 2E - (4E + 1)$
3. $1 - (2 - 3t) = t$
4. $2w + 7 = 3(5 - 4w)$
5. $2t = 9 - 5(2t + 3)$
6. $5x - 2(3x - 1) = 1$
7. $\frac{7}{5y} = 1$
8. $2 = \frac{10 - b}{7}$
9. $\frac{t}{t + 1} = 3$

ANSWERS:

1. $h = -\frac{1}{3}$
2. $E = -3$
3. $t = \frac{1}{2}$
4. $w = \frac{4}{7}$
5. $t = -\frac{1}{2}$
6. $x = 1$
7. $y = \frac{7}{5}$
8. $b = -4$
9. $t = -\frac{3}{2}$

3-10 FORMULA EVALUATION

In this section we will discuss formula evaluation, including the type that requires solving an equation.

Note: Since we want to emphasize the numerical part of formula evaluation rather than a method for dealing with the "units" in formulas, "units" are frequently avoided in this section.

---

116. The relationship stated below contains three variables.

The distance traveled by a moving object can be found by multiplying its average velocity by the time traveled.

The relationship is usually written as a formula with letters as abbreviations for the variables. That is:

$$\boxed{s = vt}$$

where: s = distance traveled

v = average velocity

t = time traveled

By substituting numbers for the variables, we can use the formula to solve applied problems. For example:

If the average velocity is 50 miles per hour and the time traveled is 5 hours, the distance traveled is 250 miles, since:

$$s = vt = 50(5) = 250 \text{ miles}$$

If the average velocity is 300 kilometers per hour and the time traveled is 10 hours, the distance traveled is 3,000 kilometers, since:

$$s = vt = 300(10) = \underline{\qquad\qquad} \text{ kilometers}$$

Answer: 3,000 kilometers

117. When it makes sense to use the same letter for more than one variable in a formula, subscripts are used. Either letters or numbers can be used as the subscripts. For example, the formula below shows the relationship between total force and the three contributing forces in a situation. The letter "T" and the numbers 1, 2, and 3 are used as subscripts.

$$\boxed{F_T = F_1 + F_2 + F_3}$$

where: F_T = total force

F_1 = first force

F_2 = second force

F_3 = third force

Let's use the formula to find F_T when $F_1 = 10$, $F_2 = 20$, and $F_3 = 40$.

$$F_T = F_1 + F_2 + F_3 = 10 + 20 + 40 = \underline{\qquad}$$

Answer: 70

118. a) The formula relating degrees-Celsius (C) and degrees-Kelvin (K) is $\boxed{C = K - 273°}$. Find the number of degrees-Celsius corresponding to 400° Kelvin.

$$C = K - 273° = \_\_\_\_\_\_$$

b) The formula for the area of a triangle is $\boxed{A = \frac{1}{2}bh}$, where A = area, b = length of the base, and h = height. Find the area of a triangle when b = 10 inches and h = 6 inches.

$$A = \frac{1}{2}bh = \_\_\_\_\_\_$$

---

119. Complete each evaluation.

a) In $\boxed{P = 2L + 2W}$, find P when L = 25 and W = 10.

$$P = 2L + 2W = \_\_\_\_\_\_$$

b) In $\boxed{C = \frac{5}{9}(F - 32)}$, find C when F = 50.

$$C = \frac{5}{9}(F - 32) = \_\_\_\_\_\_$$

Answers to 118: a) C = 127° b) A = 30 square inches

---

120. Complete each evaluation.

a) In $\boxed{V = \frac{4st}{a}}$, find V when s = 5, t = 3, and a = 10.

$$V = \frac{4st}{a} = \_\_\_\_\_\_$$

b) In $\boxed{F = \frac{9C}{5} + 32}$, find F when C = 10.

$$F = \frac{9C}{5} + 32 = \_\_\_\_\_\_$$

Answers to 119: a) P = 70 b) C = 10

---

121. Complete each evaluation.

a) In $\boxed{A = \frac{B}{B + 1}}$, find A when B = 99.

$$A = \frac{B}{B + 1} = \_\_\_\_\_\_$$

b) In $\boxed{m = \frac{y_2 - y_1}{x_2 - x_1}}$, find m when $y_2 = 18$, $y_1 = 10$, $x_2 = 10$, and $x_1 = 6$.

$$m = \frac{y_2 - y_1}{x_2 - x_1} = \_\_\_\_\_\_$$

Answers to 120: a) V = 6 b) F = 50

122. To complete the evaluation below, we had to solve an equation after substituting. Complete the other evaluation.

In the formula below, find V when M = 150.

$$\boxed{M = V + 100}$$

$$150 = V + 100$$

$$150 + (-100) = V + 100 + (-100)$$

$$V = 50$$

In the formula below, find R when E = 150 and I = 15.

$$\boxed{E = IR}$$

a) A = .99, from $\frac{99}{100}$

b) m = 2

---

123. In this frame and the following frames, we will give an example that requires equation-solving and ask you to do a similar evaluation.

In the formula below, find L when V = 40, W = 2, and H = 4.

$$\boxed{V = LWH}$$

$$40 = L(2)(4)$$

$$40 = 8L$$

$$\frac{40}{8} = L$$

$$L = 5$$

In the formula below, find b when A = 28 and h = 8.

$$\boxed{A = \frac{1}{2}bh}$$

R = 10

---

124. In the formula below, find r when e = 20, E = 50, and I = 2.

$$\boxed{e = E - Ir}$$

$$20 = 50 - 2r$$

$$-30 = -2r$$

$$\frac{-30}{-2} = r$$

$$r = 15$$

In the formula below, find Q when D = 100, P = 5, and R = 10.

$$\boxed{D = P(Q + R)}$$

b = 7

---

125. In the formula below, find s when v = 50 and t = 3.

$$\boxed{v = \frac{s}{t}}$$

$$50 = \frac{s}{3}$$

$$3(50) = \not{3}\left(\frac{s}{\not{3}}\right)$$

$$s = 150$$

In the formula below, find V_2 when $P_1 = 25$, $P_2 = 5$, and $V_1 = 4$.

$$\boxed{P_1 = \frac{P_2V_2}{V_1}}$$

Q = 10

126. In the formula below, find v_o when $v_{av} = 50$ and $v_f = 70$.

$$\boxed{v_{av} = \frac{v_o + v_f}{2}}$$

$$50 = \frac{v_o + 70}{2}$$

$$2(50) = \cancel{2}\left(\frac{v_o + 70}{\cancel{2}}\right)$$

$$100 = v_o + 70$$

$$v_o = 30$$

In the formula below, find v_1 when $a = 20$, $v_2 = 100$, and $t = 3$.

$$\boxed{a = \frac{v_2 - v_1}{t}}$$

Answer: $V_2 = 20$

127. In the formula below, find y_2 when $m = 2$, $y_1 = 4$, $x_2 = 12$, and $x_1 = 9$.

$$\boxed{m = \frac{y_2 - y_1}{x_2 - x_1}}$$

$$2 = \frac{y_2 - 4}{12 - 9}$$

$$2 = \frac{y_2 - 4}{3}$$

$$3(2) = \cancel{3}\left(\frac{y_2 - 4}{\cancel{3}}\right)$$

$$6 = y_2 - 4$$

$$y_2 = 10$$

In the formula below, find t_2 when $v = 50$, $s_2 = 200$, $s_1 = 100$, and $t_1 = 2$.

$$\boxed{v = \frac{s_2 - s_1}{t_2 - t_1}}$$

Answer: $v_1 = 40$

Answer: $t_2 = 4$

3-11 THE MULTIPLICATION AXIOM AND FORMULAS

In this section, we will use the multiplication axiom to rearrange formulas.

128. To complete the evaluation below, we had to solve the equation $50 = 10W$.

Find W when $A = 50$ and $L = 10$.

$$A = LW$$

$$50 = 10W$$

$$W = \frac{50}{10}$$

$$W = 5$$

Continued on following page.

128. Continued

However, we can use algebraic principles to rearrange the above formula to solve for W. That is:

Rearranging $A = LW$, we get $W = \frac{A}{L}$

Using the rearranged formula, we can do the same evaluation without solving an equation.

Find W when $A = 50$ and $L = 10$.

$$W = \frac{A}{L} = \frac{50}{10} = 5$$

The main purpose of formula rearrangement is to avoid the need for equation-solving when doing a formula evaluation. A rearrangement is especially useful when the same evaluation has to be repeated a number of times.

---

129. Before using the multiplication axiom to rearrange formulas, we must extend some algebraic principles to literal expressions. For example:

Just as $\frac{1}{5}(3) = \frac{3}{5}$, $\frac{1}{R}(C) = \frac{C}{R}$

Following the example, complete these:

a) $\frac{1}{2a}(v^2) =$ ______ b) $\frac{1}{I^2}(P) =$ ______ c) $\frac{1}{Q+R}(D) =$ ______

---

130. When any expression is divided by itself, the quotient is "1". For example:

$\frac{I^2}{I^2} = 1$ a) $\frac{2a}{2a} =$ ______ b) $\frac{p+q}{p+q} =$ ______

Answers to 129: a) $\frac{v^2}{2a}$ b) $\frac{P}{I^2}$ c) $\frac{D}{Q+R}$

---

131. Two quantities are a pair of reciprocals if their product is "1".

Since $\frac{1}{LW}(LW) = \frac{LW}{LW} = 1$, the reciprocal of LW is $\frac{1}{LW}$.

Since $\frac{1}{c+d}(c+d) = \frac{c+d}{c+d} = 1$, the reciprocal of $c + d$ is $\frac{1}{c+d}$.

You can see that the reciprocal of a literal expression is "1" divided by the expression. That is:

The reciprocal of Ir^2 is $\frac{1}{Ir^2}$

The reciprocal of $v_1 - v_2$ is $\frac{1}{v_1 - v_2}$

Write the reciprocal of each expression.

a) I^2 ______ b) $2a$ ______ c) $AK(t_2 - t_1)$ ______

Answers to 130: a) 1 b) 1

132. In any multiplication, the coefficient of a variable is <u>the other factor or factors</u>. For example:

In 1.5t, the coefficient of t is 1.5.

In LW, the coefficient of L is W.

In $.24I^2Rt$, the coefficient of R is $.24I^2t$.

In $P(Q + R)$, the coefficient of P is $(Q + R)$.

Following the examples, complete these:

a) In IR, the coefficient of R is ______.

b) In LWH, the coefficient of W is ______.

c) In $ms(t_2 - t_1)$, the coefficient of m is ______.

Answers: a) $\frac{1}{I^2}$ b) $\frac{1}{2a}$ c) $\frac{1}{AK(t_2 - t_1)}$

---

133. We used the multiplication axiom to solve for h below. That is, we multiplied both sides by $\frac{1}{.433}$, the reciprocal of the coefficient of "h". Solve for t in the other formula by multiplying both sides by $\frac{1}{10}$.

$$P = .433h$$
$$\frac{1}{.433}(P) = \frac{1}{.433}(.433h)$$
$$\frac{P}{.433} = 1h$$
$$h = \frac{P}{.433}$$

$$v = 10t$$

Answers: a) I b) LH c) $s(t_2 - t_1)$

---

134. To solve for R below, we multiplied both sides by $\frac{1}{I^2}$, the reciprocal of the coefficient of R. Solve for V_1 in the other formula.

$$P = I^2R$$
$$\frac{1}{I^2}(P) = \frac{1}{I^2}(I^2R)$$
$$\frac{P}{I^2} = 1R$$
$$R = \frac{P}{I^2}$$

$$P_1V_1 = P_2V_2$$

Answer: $t = \frac{v}{10}$, from:

$$\frac{1}{10}(v) = \frac{1}{10}(10t)$$
$$\frac{v}{10} = 1t$$

---

135. To solve for s below, we multiplied both sides by $\frac{1}{2a}$, the reciprocal of the coefficient of "s". Solve for L in the other formula.

$$v^2 = 2as$$
$$\frac{1}{2a}(v^2) = \frac{1}{2a}(2as)$$
$$\frac{v^2}{2a} = 1s$$
$$s = \frac{v^2}{2a}$$

$$V = LWH$$

Answer: $V_1 = \frac{P_2V_2}{P_1}$, from:

$$\frac{1}{P_1}(P_1V_1) = \frac{1}{P_1}(P_2V_2)$$
$$1V_1 = \frac{P_2V_2}{P_1}$$

136. To solve for P below, we multiplied both sides by $\frac{1}{Q + R}$, the reciprocal of the coefficient of P. Solve for b in the other formula.

$$D = P(Q + R) \qquad\qquad a = b(c - d)$$

$$\left(\frac{1}{Q + R}\right)(D) = P(Q + R)\left(\frac{1}{Q + R}\right)$$

$$\frac{D}{Q + R} = P \cdot 1$$

$$P = \frac{D}{Q + R}$$

Answer: $L = \frac{V}{WH}$, from:

$$\frac{1}{WH}(V) = LWH\left(\frac{1}{WH}\right)$$

$$\frac{V}{WH} = L \cdot 1$$

---

137. Two solutions from the preceding frames are shown below.

If $P = .433h$,

$$h = \frac{P}{.433}$$

If $V = LWH$,

$$L = \frac{V}{WH}$$

As you can see from the above examples, there is a shortcut for the multiplication axiom. That is, to solve for each letter, we can simply divide the other side of the formula by the coefficient of that letter. For example:

In $E = IR$: The coefficient of I is R.

Therefore, $I = \frac{E}{R}$

In $2Pt^2 = rw$: The coefficient of P is $2t^2$.

Therefore, P =

Answer: $b = \frac{a}{c - d}$, from:

$$\frac{1}{c-d}(a) = b(c-d)\left(\frac{1}{c-d}\right)$$

$$\frac{a}{c - d} = b \cdot 1$$

---

138. Two more solutions from the preceding frames are shown below.

If $D = P(Q + R)$,

$$P = \frac{D}{Q + R}$$

If $a = b(c - d)$,

$$b = \frac{a}{c - d}$$

Again you can see that we can solve for each letter by simply dividing the other side of the formula by the coefficient of the letter.

In $2RF = W(R - r)$: The coefficient of W is $(R - r)$.

Therefore, $W = \frac{2RF}{R - r}$

In $H = ms(t_2 - t_1)$: The coefficient of s is $m(t_2 - t_1)$.

Therefore, s =

Answer: $P = \frac{rw}{2t^2}$

Answer: $s = \frac{H}{m(t_2 - t_1)}$

139. When rearranging a formula, a capital letter should not be changed to a small letter and a small letter should not be changed to a capital letter. That is:

Don't change "A" to "a".

Don't change "e" to "E".

Use the shortcut of the multiplication axiom for these:

a) Solve for F.

$$d^2Fr = m_1m_2$$

$$F =$$

b) Solve for m.

$$H = ms(t_2 - t_1)$$

$$m =$$

---

a) $F = \frac{m_1m_2}{d^2r}$ b) $m = \frac{H}{s(t_2 - t_1)}$

3-12 THE ADDITION AXIOM AND FORMULAS

In this section, we will use the addition axiom to rearrange formulas.

---

140. Two terms are opposites if their sum is 0. Therefore:

Since $2ab + (-2ab) = 0$, $2ab$ and $-2ab$ are opposites.

Since $CD + (-CD) = 0$, CD and $-CD$ are opposites.

Write the opposite of each term.

a) 7TV ______ b) -5pq ______ c) AB ______ d) -cd ______

---

141. To solve for A below, we added L to both sides. Solve for P in the other formula by adding D to both sides.

$$C = A - L$$

$$L + C = A \underbrace{- L + L}$$

$$L + C = A + 0$$

$$A = L + C$$

$$N = P - D$$

Answers to 140: a) -7TV b) 5pq c) -AB d) cd

---

142. Using the same steps, solve for the indicated letter in each formula.

a) Solve for K.

$$C = K - 273$$

b) Solve for E.

$$e = E - Ir$$

Answer to 141: P = D + N

143. To solve for C below, we added -M to both sides. To solve for V below, we added -100 to both sides.

$$R = C + M \qquad F = V + 100$$

$$R + (-M) = C + \underbrace{M + (-M)}_{0} \qquad F + (-100) = V + \underbrace{100 + (-100)}_{0}$$

$$R + (-M) = C + 0 \qquad F + (-100) = V + 0$$

$$C = R + (-M) \qquad V = F + (-100)$$

In each solution above, the right side is an addition in which <u>the second term is negative</u>. Ordinarily, additions of that type are converted to a subtraction. That is:

Instead of $C = R + (-M)$, we write $C = R - M$.

Instead of $V = F + (-100)$, we write ________________.

Answers (frame 142): a) $K = C + 273$ b) $E = e + Ir$

---

144. Solve for the indicated letter. Write each solution as a subtraction.

a) Solve for E.

$$H = E + PV$$

b) Solve for R.

$$D = R + S + T$$

Answer (frame 143): $V = F - 100$

---

145. To solve for P below, we used the addition axiom to isolate PV. Then we used the multiplication axiom. Solve for <u>m</u> in the other formula.

$$H = E + PV$$

$$H + (-E) = (-E) + E + PV$$

$$H - E = PV$$

$$P = \frac{H - E}{V}$$

$$I_a = I_c + md^2$$

Answers (frame 144): a) $E = H - PV$ b) $R = D - S - T$

---

146. The opposite of a subtraction can be obtained by interchanging the two terms. That is:

The opposite of $(5 - 3)$ is $(3 - 5)$, since:

$5 - 3 = 2$ and $3 - 5 = -2$

The same procedure was used to get each opposite below.

The opposite of $(2x - 1)$ is $(1 - 2x)$.

The opposite of $(a - b)$ is $(b - a)$.

The opposite of $(at - v_2)$ is $(v_2 - at)$.

Write the opposite of each subtraction.

a) $4y - 1$ ____________

b) $p - q$ ____________

c) $a^2 - c^2$ ____________

d) $v_f^2 - 2gs$ ____________

Answer (frame 145): $m = \frac{I_a - I_c}{d^2}$

147. If we replace each side of an equation with its opposite, the new equation is equivalent to the original. For example:

$-5x = 15$ and $5x = -15$ are equivalent because the solution of each is -3.

Replacing each side of an equation by its opposite is called the oppositing principle for equations. That principle was used to solve for y below.

$$-y = 10 - 2x$$

$$y = 2x - 10$$

The same principle can be used to solve for D and v_1 below. That is:

$-D = N - L$ $\qquad$ $-v_1 = at - v_2$

$D = L - N$ $\qquad$ $v_1 = v_2 - at$

Solve for t in each formula below by replacing each side with its opposite.

a) $-t = a - b$ $\qquad$ b) $-t = cd - pq$

$t =$ ______________ $\qquad$ $t =$ ______________

Answers: a) $1 - 4y$ b) $q - p$ c) $c^2 - a^2$ d) $2gs - v_f^2$

---

148. Use the oppositing principle for these.

a) Solve for I. $\qquad$ b) Solve for a.

$-I = E_1 - E_2$ $\qquad$ $-a = bF_1 - cF_2$

Answers: a) $t = b - a$ b) $t = pq - cd$

---

149. In solving for L below, we had to use the oppositing principle for the final step. Solve for D in the other formula.

$C = A - L$ $\qquad$ $N = P - D$

$C + (-A) = (-A) + A - L$

$C - A = -L$

$L = A - C$

Answers: a) $I = E_2 - E_1$ b) $a = cF_2 - bF_1$

---

150. In solving for T_C below, we used the oppositing principle for the final step. Solve for R_1 in the other formula.

$T_K - T_C = 273$ $\qquad$ $R_t - R_1 = R_2$

$(-T_K) + T_K - T_C = 273 + (-T_K)$

$-T_C = 273 - T_K$

$T_C = T_K - 273$

Answer: $D = P - N$

Answer: $R_1 = R_t - R_2$

3-13 FRACTIONAL FORMULAS

In this section, we will discuss the method for rearranging formulas that contain a fraction on one side.

---

151. In each example below, we used the cancelling shortcut to multiply a literal fraction by its denominator. Notice that each product is the numerator of the original fraction.

$$\not{R}\left(\frac{E}{\not{R}}\right) = E \qquad\qquad \not{dt^2}\left(\frac{ac}{\not{dt^2}}\right) = ac$$

Using the fact above, complete these:

a) $2\left(\frac{gt^2}{2}\right) =$ ________ b) $P_1\left(\frac{P_2V_2}{P_1}\right) =$ ________

Answers: a) gt^2 b) P_2V_2

152. In the expression below, we multiplied a fraction by its denominator. Notice again that the product is the numerator of the original fraction.

$$\not{t}\left(\frac{v_2 - v_1}{\not{t}}\right) = v_2 - v_1$$

a) $2\left(\frac{v_0 + v_f}{2}\right) =$ ________ b) $AB\left(\frac{A - B}{AB}\right) =$ ________

Answers: a) $v_0 + v_f$ b) $A - B$

153. To clear the fraction in the formula below, we multiplied both sides by its denominator. Use the same method to clear the fractions in the other formula.

$$v = \frac{s}{t} \qquad a)\ H = \frac{D^2N}{2.5} \qquad b)\ F = \frac{m_1m_2}{rd^2}$$

$$t(v) = t\left(\frac{s}{t}\right)$$

$$tv = s$$

Answers: a) $2.5H = D^2N$ b) $Frd^2 = m_1m_2$

154. Following the example, clear the fraction in the other formula.

$$v_{av} = \frac{v_o + v_f}{2} \qquad\qquad a = \frac{v_2 - v_1}{t}$$

$$2(v_{av}) = \not{2}\left(\frac{v_o + v_f}{\not{2}}\right)$$

$$2v_{av} = v_o + v_f$$

Answer: $ta = v_2 - v_1$

155. To solve for B below, we simply cleared the fraction. Solve for a in the other formula.

$$D = \frac{B}{N} \qquad\qquad h = \frac{a}{s^2}$$

$$N(D) = N\left(\frac{B}{N}\right)$$

$$DN = B$$

$$B = DN$$

$a = hs^2$

156. To solve for s below, we cleared the fraction and then used the multiplication axiom. Solve for N in the other formula.

$$P = \frac{Fs}{t} \qquad\qquad H = \frac{D^2N}{2.5}$$

$$t(P) = t\left(\frac{Fs}{t}\right)$$

$$Pt = Fs$$

$$s = \frac{Pt}{F}$$

$N = \frac{2.5H}{D^2}$

157. We solved for R below. Solve for r in the other formula.

$$I = \frac{E}{R} \qquad\qquad F = \frac{m_1m_2}{rd^2}$$

$$R(I) = R\left(\frac{E}{R}\right)$$

$$IR = E$$

$$R = \frac{E}{I}$$

$r = \frac{m_1m_2}{Fd^2}$

158. We solved for t below. Solve for r in the other formula.

$$a = \frac{v_2 - v_1}{t} \qquad\qquad m = \frac{c - d}{2r}$$

$$t(a) = t\left(\frac{v_2 - v_1}{t}\right)$$

$$at = v_2 - v_1$$

$$t = \frac{v_2 - v_1}{a}$$

$r = \frac{c - d}{2m}$

159. We solved for F_o below. Solve for v_2 in the other formula.

$$w = \frac{F_o + F_i}{t} \qquad\qquad a = \frac{v_2 - v_1}{t}$$

$$t(w) = t\left(\frac{F_o + F_i}{t}\right)$$

$$tw = F_o + F_i$$

$$tw + (-F_i) = F_o + F_i + (-F_i)$$

$$tw - F_i = F_o$$

$$F_o = tw - F_i$$

$v_2 = at + v_1$

160. Notice how we used the oppositing principle to solve for v_1 below. Solve for d in the other formula.

$$a = \frac{v_2 - v_1}{t} \qquad\qquad m = \frac{b - d}{r}$$

$$t(a) = t\left(\frac{v_2 - v_1}{t}\right)$$

$$at = v_2 - v_1$$

$$at + (-v_2) = -v_2 + v_2 - v_1$$

$$at - v_2 = -v_1$$

$$v_1 = v_2 - at$$

$d = b - mr$

161. To solve for h below, we began by multiplying both sides by 2 to get rid of the $\frac{1}{2}$. Then we proceeded in the usual way. Solve for a in the other formula.

$$A = \frac{1}{2}bh \qquad\qquad s = \frac{1}{2}at^2$$

$$2(A) = 2\left(\frac{1}{2}bh\right)$$

$$2A = bh$$

$$h = \frac{2A}{b}$$

$a = \frac{2s}{t^2}$

162. To solve for h below, we again began by multiplying both sides by 2 to get rid of the $\frac{1}{2}$. Solve for v^2 in the other formula.

$$A = \frac{1}{2}h(b_1 + b_2) \qquad\qquad E = \frac{1}{2}(m_1 + m_2)v^2$$

$$2(A) = 2\left[\frac{1}{2}h(b_1 + b_2)\right]$$

$$2A = h(b_1 + b_2)$$

$$h = \frac{2A}{b_1 + b_2}$$

$$v^2 = \frac{2E}{m_1 + m_2}$$

SELF-TEST 11 (pages 139-153)

1. In $A = \frac{1}{2}h(b + c)$, find A when h = 8, b = 11, and c = 14.

2. In $m = \frac{y_2 - y_1}{x_2 - x_1}$, find m when $y_2 = 20$, $y_1 = 4$, $x_2 = 14$, and $x_1 = 10$.

3. In $E = I(R - r)$, find r when E = 40, I = 10, and R = 15.

4. In $v = \frac{s}{t_2 - t_1}$, find t_2 when v = 120, s = 480, and $t_1 = 2$.

5. Solve for F_1.

$$d_1F_1 = d_2F_2$$

6. Solve for h.

$$r = h(w - v)$$

7. Solve for F_2.

$$F_t = F_1 + F_2$$

8. Solve for Q.

$$P - Q = R$$

9. Solve for H.

$$L = \frac{V}{HW}$$

10. Solve for d_1.

$$v = \frac{d_1 - d_2}{t}$$

ANSWERS:

1. $A = 100$
2. $m = 4$
3. $r = 11$
4. $t_2 = 6$
5. $F_1 = \frac{d_2F_2}{d_1}$
6. $h = \frac{r}{w - v}$
7. $F_2 = F_t - F_1$
8. $Q = P - R$
9. $H = \frac{V}{LW}$
10. $d_1 = vt + d_2$

SUPPLEMENTARY PROBLEMS - CHAPTER 3

Assignment 8

Solve each equation. Report each solution in lowest terms.

1. $y + 9 = 5$
2. $t - 3 = 8$
3. $29 = x - 12$
4. $54 = N + 1$
5. $G - 1 = 0$
6. $5 + d = 0$
7. $x - 2.5 = 4.1$
8. $7.5 = p + 5.7$
9. $d + \frac{1}{6} = \frac{5}{6}$
10. $x - \frac{3}{4} = \frac{1}{8}$
11. $V + 1 = \frac{4}{7}$
12. $m + \frac{2}{3} = 0$
13. $9p = 4$
14. $-8R = -15$
15. $7F = -7$
16. $2 = 14y$
17. $-35 = 21w$
18. $-3h = 0$
19. $-x = 8$
20. $36.6 = 12.2y$
21. $\frac{3}{4}p = 1$
22. $2x = -\frac{4}{5}$
23. $\frac{2}{3}y = \frac{5}{6}$
24. $\frac{7}{8}H = 0$
25. $3x + 5 = 17$
26. $2r + 15 = 9$
27. $11 - 15m = 21$
28. $7 - 12s = 4$
29. $3 = 9 + 4H$
30. $-1 = 1 + 2A$
31. $4y + 7 = 0$
32. $4x - 1 = 3$
33. $3c - 5 = -6$
34. $2 - 11p = 2$
35. $5 - t = 9$
36. $4.5 - m = 0$
37. $.7A - 1.4 = 5.6$
38. $\frac{1}{3}y + \frac{1}{4} = 1$
39. $2x - \frac{1}{4} = \frac{1}{8}$

Assignment 9

Do each multiplication.

1. $3(x + 7)$
2. $9(2 + 3y)$
3. $10(5R + 1)$
4. $7(x - 2)$
5. $5(3 - 4y)$
6. $8(1 - 5d)$
7. $-4(m + 3)$
8. $-1(7 - 6P)$

Combine like terms.

9. $x + 5x$
10. $7y + (-y)$
11. $-5m + (-2m)$
12. $9d - 4d$
13. $r - 7r$
14. $2 + 5p - 3$
15. $6V - 7 - V$
16. $-a - 9 - 2a$

Solve each equation. Report each solution in lowest terms.

17. $4R + (-6R) = 8$
18. $13 = -3y + (-2y)$
19. $9P - 3P = 2$
20. $25 = 2x - 12x$
21. $5w - 2w = 0$
22. $7h - h = 20$
23. $6 = F - 9F$
24. $.4x + .8x = 6$
25. $\frac{3}{4}y - \frac{3}{8}y = \frac{1}{2}$
26. $6r - 5 = 2r$
27. $b = 3b + 4$
28. $5x - 4 = 11x$
29. $a = 1 - a$
30. $10 - V = 9V$
31. $1.9t - 2.4 = 1.1t$
32. $\frac{1}{8}x + \frac{1}{4} = \frac{7}{8}x$
33. $5N + 1 = 2N + 2$
34. $2c + 7 = 7 + 9c$
35. $E + 1 = 3 - E$
36. $2h - 7 = 3h + 5$
37. $2 + 3y = 10 - y$
38. $15 - 6w = 25 - 2w$
39. $5 - k = 1 - 5k$
40. $\frac{2}{3}d - \frac{1}{2} = \frac{1}{6}d - \frac{1}{3}$

Assignment 10

Solve each equation. Report each answer in lowest terms.

1. $a + (3a + 2) = 1$
2. $4R = (5 - 2R) + 3$
3. $18 - (2s + 7) = 12$
4. $x = 1 - (5x + 2)$
5. $6 - (2 + 5w) = 4$
6. $3b - (b + 5) = 0$
7. $G - (2G - 3) = 5$
8. $9 - (2 - 5h) = 3h$
9. $2r - (7r - 3) = 1$
10. $4t - (1 - t) = 0$
11. $4(x + 3) = 5$
12. $7y = 5(4 + y)$
13. $2 = 4(P - 1)$
14. $5(3F - 4) = 7F$
15. $2(d + 4) = 3(d + 2)$
16. $E + 11 = 4(5 + 4E)$
17. $2(x - 4) = 5x - 2$
18. $3(V + 1) = 2(V - 1)$
19. $3w + 5(w + 2) = 50$
20. $7 + 4(3d - 1) = 0$
21. $25 - 2(4x + 5) = 45$
22. $8y - 3(2y + 7) = 9$
23. $3 = 9 - 2(1 - 4E)$
24. $4 - 5(2r - 7) = 3r$
25. $\frac{w}{6} = 9$
26. $\frac{24}{x} = 8$
27. $15 = \frac{3y}{5}$
28. $4 = \frac{5G}{3}$
29. $\frac{1}{2v} = 5$
30. $1 = \frac{7}{2d}$
31. $\frac{3p}{8} = 1$
32. $\frac{5t}{8} = 0$
33. $\frac{h + 9}{2} = 3$
34. $x = \frac{3x - 1}{4}$
35. $\frac{4 - 2y}{3} = 2y$
36. $\frac{4w + 3}{6} = 0$
37. $\frac{8}{F + 3} = 2$
38. $\frac{a}{a - 5} = 6$
39. $2 = \frac{1 - t}{3}$

Assignment 11

1. In $\boxed{v = at}$, find v when $t = 12$ and $a = 8$.
2. In $\boxed{E = e + ir}$, find E when $e = 24$, $i = 4$, and $r = 12$.
3. In $\boxed{B = 180 - (A + C)}$, find B when $A = 25$ and $C = 85$.
4. In $\boxed{V = \frac{1}{3}Bh}$, find V when $B = 40$ and $h = 30$.
5. In $\boxed{F = \frac{9}{5}C + 32}$, find F when $C = 20$.
6. In $\boxed{a = \frac{F}{m}}$, find a when $F = 40$ and $m = 8$.
7. In $\boxed{W = \frac{P - 2L}{2}}$, find W when $P = 84$ and $L = 30$.
8. In $\boxed{I = \frac{E}{R - r}}$, find I when $E = 120$, $R = 50$, and $r = 20$.
9. In $\boxed{P = EI}$, find I when $E = 120$ and $P = 600$.

10. In $\boxed{A = \frac{1}{2}bh}$, find h when A = 60 and b = 15.

11. In $\boxed{Z = A - N}$, find N when Z = 82 and A = 93.

12. In $\boxed{V = LWH}$, find W when V = 400, L = 10, and H = 5.

13. In $\boxed{v = V - at}$, find t when v = 30, V = 90, and a = 15.

14. In $\boxed{F = P(A - a)}$, find A when F = 200, P = 50, and a = 16.

15. In $\boxed{L = \frac{A}{W}}$, find W when L = 8 and A = 48.

16. In $\boxed{P = \frac{pv}{V}}$, find v when p = 9, V = 3, and P = 12.

17. In $\boxed{E = \frac{I - i}{R}}$, find i when E = 1, I = 5, and R = 2.

18. In $\boxed{v = \frac{s}{t_2 - t_1}}$, find t_2 when v = 20, s = 60, and t_1 = 5.

Complete each rearrangement.

19. Solve for s.

$P = 4s$

20. Solve for L.

$A = LW$

21. Solve for r.

$C = 2\pi r$

22. Solve for R.

$E^2 = PR$

23. Solve for h.

$V = \pi r^2 h$

24. Solve for v.

$d = v(t_2 - t_1)$

25. Solve for M.

$F(p - a) = 2Mm$

26. Solve for V_2.

$T_1V_2 = T_2V_1$

27. Solve for F_t.

$F_1 = F_t - F_2$

28. Solve for P_2.

$P_1 = P_2 - EI$

29. Solve for Q.

$P = Q + R$

30. Solve for W.

$P = 2L + 2W$

31. Solve for R.

$-R = Q - AB$

32. Solve for F_1.

$F_t - F_1 = F_2$

33. Solve for D.

$B = C - D$

34. Solve for k.

$w = ah - ks$

35. Solve for W.

$F = \frac{W}{d}$

36. Solve for f.

$t = \frac{1}{f}$

37. Solve for c.

$T = \frac{Q}{cm}$

38. Solve for G.

$W = \frac{GKN}{R^2}$

39. Solve for h.

$V = \frac{1}{3}Bh$

40. Solve for r.

$i = \frac{E - e}{r}$

41. Solve for G_1.

$W = \frac{G_1 + G_2}{P}$

42. Solve for d_2.

$v = \frac{d_1 - d_2}{t}$

Word Problems

4

In this chapter, we will introduce a method for solving word problems and then use it to solve word problems of various types. Percents and word problems involving percents are discussed. Some formulas for the perimeter, area, and volume of basic geometric figures are reviewed and then used to solve word problems.

4-1 TRANSLATING ENGLISH SENTENCES TO EQUATIONS

In this section, we will discuss the process of translating English sentences to equations.

1. In the sentence "The sum of a number and 3 is 7.", the word "is" means "equal". Using $\underline{x}$ for the number, we translated the sentence to an equation below.

The sum of a number and 3	is	7
↓	↓	↓
$x + 3$	$=$	7

Continued on following page.

1. Continued

Following the example, translate each of these to an equation. Use $\underline{x}$ for the number.

a) What number increased by 10 is 25?

____

b) 8 more than 3 times a number is 29.

____

---

2. We translated "The difference of a number and 5 is 9." to an equation below.

The difference of a number and 5	is	9
↓	↓	↓
$x - 5$	$=$	9

Using $\underline{x}$ for the number, translate these to equations.

a) What number minus 7 is 12?

____

b) 50 reduced by four times a number is 14.

____

Answers to 1: a) $x + 10 = 25$ b) $3x + 8 = 29$

---

3. We translated "The product of a number and 6 is 24." to an equation below.

The product of a number and 6	is	24
↓	↓	↓
$6x$	$=$	24

Using $\underline{x}$ for the number, translate these to an equation.

a) The product of 3 and a number equal the number plus 7.

____

b) 100 minus the product of a number and 5 is 15 times the number.

____

Answers to 2: a) $x - 7 = 12$ b) $50 - 4x = 14$

---

4. Following the example, translate to equations.

Three-fourths of what number is 36? $\frac{3}{4}x = 36$

a) One-third of what number is $\frac{7}{8}$? ________________

b) Two-fifths of a number equals the number plus 3. ________________

Answers to 3: a) $3x = x + 7$ b) $100 - 5x = 15x$

---

5. Following the example, translate to equations.

Three times a number, plus 8 is 14. $3x + 8 = 14$

a) Double a number, minus 10 equals the number plus 100. ________________

b) One-half a number, plus 24 equals two-fifths of the number. ________________

Answers to 4: a) $\frac{1}{3}x = \frac{7}{8}$ b) $\frac{2}{5}x = x + 3$

6. Following the example, translate to equations.

The product of 5, and a number decreased by 3 is 27. $5(x - 3) = 27$

a) If the sum of a number and 2 is multiplied by 7, we get 35. ____________

b) 10 minus the difference of a number and 3 equals 14. ____________

Answers: a) $2x - 10 = x + 100$ b) $\frac{1}{2}x + 24 = \frac{2}{5}x$

---

7. The phrase "the result is" also means "equals". Following the example, translate to equations.

If twice a number is decreased by 7, the result is 15. $2x - 7 = 15$

a) If the difference of a number and 2 is multiplied by 3, the result is 24. ____________

b) If a number is added to one-half the number, the result is 15. ____________

Answers: a) $7(x + 2) = 35$ b) $10 - (x - 3) = 14$

---

8. Following the example, translate to equations.

The quotient of a number and 3 is 4. $\frac{x}{3} = 4$

a) The quotient of double a number and 7 is 6. ____________

b) 18 divided by 3 times a number equals 2. ____________

Answers: a) $3(x - 2) = 24$ b) $x + \frac{1}{2}x = 15$

---

9. Following the example, translate to equations. Use n as the variable.

If the sum of a number and 2 is divided by 3, the result is 4. $\frac{n + 2}{3} = 4$

a) If the difference of twice a number and 5 is divided by 7, the result is 2. ____________

b) If 20 is divided by the sum of triple a number and 7, the result is 3. ____________

Answers: a) $\frac{2x}{7} = 6$ b) $\frac{18}{3x} = 2$

---

10. Following the examples, translate to equations. Use y as the variable.

If the difference of a number and 2 is divided by the number, the result is 10. $\frac{y - 2}{y} = 10$

a) If the sum of a number and 4 is divided by the number, the result is 5. ____________

b) If double a number is divided by the difference of that number and 3, the result is 6. ____________

Answers: a) $\frac{2n - 5}{7} = 2$ b) $\frac{20}{3n + 7} = 3$

Answers: a) $\frac{y + 4}{y} = 5$ b) $\frac{2y}{y - 3} = 6$

4-2 NUMBER PROBLEMS

In this section, we will introduce a method for solving word problems. All of the problems in this section involve an "unknown number".

11.

Four Steps For Solving Word Problems
1. Represent the unknown or unknowns with a letter or algebraic expression.
2. Translate the problem to an equation.
3. Solve the equation.
4. Check the solution in the original words of the problem.

The four steps are used to solve the problem below.

Problem: If we subtract 5 from double a number, we get 11. Find the number.

Step 1: Represent the unknown with a letter.

Let x equal the unknown number.

Step 2: Translate the problem to an equation.

$$2x - 5 = 11$$

Step 3: Solve the equation.

$$2x - 5 = 11$$

$$2x = 16$$

$$x = 8$$

Step 4: Check the solution in the original words of the problem.

Double 8 is 16. If we subtract 5 from 16, we get 11. The solution checks.

Let's use the four steps to solve this one.

Problem: What number plus 10 equals 13?

Represent the unknown with a letter.

Let x equal the unknown number.

a) Translate the problem to an equation.

b) Solve the equation.

c) Check the solution in the original words of the problem.

12. Let's use the four steps to solve this one.

Problem: If 3 times a number is subtracted from 50, the result is 20. Find the number.

Represent the unknown with a letter.

Let x equal the unknown number.

a) Translate the problem to an equation.

b) Solve the equation.

c) Check the solution in the original words of the problem.

Answers:
a) $x + 10 = 13$
b) $x = 3$
c) 3 plus 10 does equal 13.

---

13. Use the four steps to solve each of these.

a) What number minus $\frac{1}{3}$ equals $\frac{1}{2}$?

b) $\frac{3}{4}$ of a number is 36. Find the number.

Answers:
a) $50 - 3x = 20$
b) $x = 10$
c) 3 times 10 = 30. If 30 is subtracted from 50, we do get 20.

---

14. Solve each problem.

a) Five times a number, plus 10 equals 25. Find the number.

b) A number minus 7 equals twice the number. Find the number.

Answers:
a) $\frac{5}{6}$, from: $x - \frac{1}{3} = \frac{1}{2}$
b) 48, from: $\frac{3}{4}x = 36$

---

15. Solve each problem.

a) If we add $\frac{1}{2}$ of a number to the number itself, we get 18. Find the number.

b) If the difference of a number and 2 is multiplied by 3, we get 21. Find the number.

Answers:
a) 3, from: $5x + 10 = 25$
b) -7, from: $x - 7 = 2x$

16. Solve each problem

a) If 5 is subtracted from three times a number, the result is the number plus 11. Find the number.

b) If 3 times a number is divided by 2, the result is 9. Find the number.

a) 12, from:

$$x + \frac{1}{2}x = 18$$

b) 9, from:

$$3(x - 2) = 21$$

---

17. Solve each problem.

a) If the sum of a number and 10 is divided by 3, the result is 8. Find the number.

b) If double a number is divided by the difference of the number and 2, the result is 3. Find the number.

a) 8 b) 6

---

a) 14 b) 6

4-3 CONSECUTIVE INTEGER PROBLEMS

In this section, we will discuss word problems about consecutive integers.

---

18. Consecutive integers are integers that are next to each other, like 5 and 6. The larger is one more than the smaller. Therefore, if the smaller is $\underline{x}$, the larger is $\underline{x + 1}$. The problem below involves consecutive integers.

<u>Problem</u>: The sum of two consecutive integers is 27. What are the integers?

Let's use the four problem-solving steps to solve this problem.

1. <u>Represent the unknowns with a letter or algebraic expression</u>.

   Let $\underline{x}$ be the smaller integer and $\underline{x + 1}$ the larger integer.

2. <u>Translate to an equation</u>.

Smaller integer	+	larger integer	=	27
↓	↓	↓	↓	↓
x	+	(x + 1)	=	27

Continued on following page.

18. Continued

3. Solve the equation.

$$x + (x + 1) = 27$$
$$2x + 1 = 27$$
$$2x = 26$$
$$x = 13 \quad (\text{and } x + 1 = 14)$$

4. Check the solution in the original wording of the problem.

Our answers are: $x = 13$ and $x + 1 = 14$. 13 and 14 are consecutive integers, and their sum is 27. Therefore, our answer checks.

Using the same steps, solve these:

a) The sum of two consecutive integers is 53. What are the integers?

b) Find two consecutive integers whose sum is 111.

---

a) 26 and 27

b) 55 and 56

19. Consecutive odd integers are odd integers that are next to each other, like 7 and 9. The larger is 2 more than the smaller. Therefore, if the smaller is x, the larger is $x + 2$. We used this fact to solve the problem below. Solve the other problem.

The sum of two consecutive odd integers is 48. What are the two integers?

$$x + (x + 2) = 48$$
$$2x + 2 = 48$$
$$2x = 46$$
$$x = 23$$
$$(\text{and } x + 2 = 25)$$

The two consecutive odd integers are 23 and 25.

Find two consecutive odd integers whose sum is 164.

---

81 and 83

20. Consecutive even integers are even integers that are next to each other, like 12 and 14. The larger is 2 more than the smaller. Therefore, if the smaller is $\underline{x}$, the larger is $\underline{x+2}$. We used this fact to solve one problem below. Solve the other problem.

The sum of two consecutive even integers is 86. Find the two integers.

$$x + (x + 2) = 86$$
$$2x + 2 = 86$$
$$2x = 84$$
$$x = 42$$

The two consecutive even integers are 42 and 44.

Find two consecutive even integers whose sum is 150.

---

74 and 76

21. Three consecutive integers are integers like 5, 6, and 7. If the smallest is $\underline{x}$, the middle one is $\underline{x+1}$ and the largest is $\underline{x+2}$. We used this fact to solve one problem below. Solve the other problem.

The sum of three consecutive integers is 48. What are the integers?

$$x + (x + 1) + (x + 2) = 48$$
$$3x + 3 = 48$$
$$3x = 45$$
$$x = 15$$

The three consecutive integers are 15, 16, and 17.

Find three consecutive integers whose sum is 126.

41, 42, and 43

22. Three consecutive odd integers are integers like 7, 9, and 11. If the smallest is $\underline{x}$, the middle one is $\underline{x+2}$ and the largest is $\underline{x+4}$. We used this fact to solve one problem below. Solve the other problem.

Find three consecutive odd integers whose sum is 69.

$$x + (x + 2) + (x + 4) = 69$$
$$3x + 6 = 69$$
$$3x = 63$$
$$x = 21$$

The three consecutive odd integers are 21, 23, and 25.

The sum of three consecutive odd integers is 111. What are the integers?

35, 37, and 39

23. Three consecutive even integers are integers like 6, 8, and 10. If the smallest is x, the middle one is $x + 2$ and the largest is $x + 4$. We used this fact to solve one problem below. Solve the other problem.

The sum of three consecutive even integers is 72. Find the integers.

$$x + (x + 2) + (x + 4) = 72$$
$$3x + 6 = 72$$
$$3x = 66$$
$$x = 22$$

The three consecutive even integers are 22, 24, and 26.

Find three consecutive even integers whose sum is 156.

---

24. To solve the problem below, we used x and $x + 1$ for the two consecutive integers. Solve the other problem.

50, 52, and 54

When the smaller of two consecutive integers is added to four times the larger, the result is 104. Find the integers.

$$x + 4(x + 1) = 104$$
$$x + 4x + 4 = 104$$
$$5x + 4 = 104$$
$$5x = 100$$
$$x = 20$$

The integers are 20 and 21.

When the larger of two consecutive integers is added to double the smaller, the result is 76. Find the integers.

---

25. To solve the problem below, we used x, $x + 1$, and $x + 2$ for the three consecutive integers. Solve the other problems.

25 and 26

When the largest of three consecutive integers is added to twice the smallest, the result is 13 less than four times the middle one. Find the integers.

$$(x + 2) + 2x = 4(x + 1) - 13$$
$$3x + 2 = 4x + 4 - 13$$
$$3x + 2 = 4x - 9$$
$$x = 11$$

The integers are 11, 12, and 13.

If the middle of three consecutive integers is added to 100, the result is 31 more than the sum of the smallest and twice the largest. Find the integers.

33, 34, and 35

26. To solve the problem below, we used $\underline{x}$, $\underline{x + 2}$, and $\underline{x + 4}$ for the three consecutive odd integers. Solve the other problem.

If 11 is added to the largest of three consecutive odd integers, the answer equals the sum of the first two integers. Find the integers.

$$(x + 4) + 11 = x + (x + 2)$$

$$x + 15 = 2x + 2$$

$$x = 13$$

The integers are 13, 15, and 17.

If the middle and largest of three consecutive even integers are added, the result is 46 more than the smallest. Find the integers.

40, 42, and 44

4-4 PART-WHOLE PROBLEMS

In this section, we will discuss word problems in which a whole is divided into two or three parts.

27. The problem below involves dividing a whole into two parts.

Problem: A wire 50 centimeters long is cut into two parts. The larger part is 10 centimeters longer than the smaller part. How long is each part?

We drew a diagram for the problem below. In the diagram, we let $\underline{x}$ equal the length of the smaller part. Therefore, the length of the larger part is $\underline{x + 10}$.

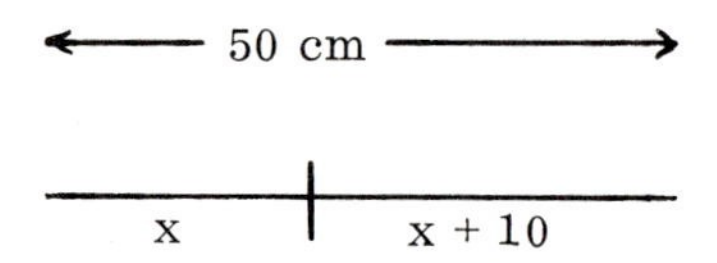

We translated to an equation, solved the equation, and checked the solution below.

$$\text{smaller} + \text{larger} = 50$$

$$x + (x + 10) = 50$$

$$2x + 10 = 50$$

$$2x = 40$$

$$x = 20 \quad (\text{and } x + 10 = 30)$$

Check: The smaller part is 20 cm and the larger part is 30 cm. 30 is 10 longer than 20 and 30 + 20 equals 50. Therefore, the solution checks.

Continued on following page.

27. Continued

Following the example, solve these:

a) A rope 90 feet long is cut into two parts. One part is 14 feet longer than the other. Find the length of the two parts.

b) A man has a board 48 inches long. He wants to cut it into two pieces so that one piece will be 8 inches longer than the other. How long should the shorter piece be?

---

a) 38 ft. and 52 ft.

b) 20 in.

28. The problem below also involves a whole (graduation class) which is divided into two parts (males and females). To solve it, we used $\underline{x}$ for the number of males and $\underline{x + 27}$ for the number of females.

Problem: There were 27 more females than males in a graduation class. If the total number of graduates was 207, find the number of males and females.

$$x + (x + 27) = 207$$

$$2x + 27 = 207$$

$$2x = 180$$

$$x = 90$$

Therefore, the number of males is 90 and the number of females is $90 + 27 = 117$.

Following the example, solve these:

a) There were two opposing candidates in an election. 483 votes were cast. Candidate A received 63 more votes than candidate B. How many votes did candidate B get?

b) A total of 52 points were scored in a football game. One team scored 18 points more than the other. What was the score of the game?

29. To solve the problem below, we used $\underline{x}$ for the number of adults and $\underline{2x}$ for the number of children. Solve the other problem.

There were twice as many children as adults at a movie. If the total number attending was 369, how many children attended?

$$x + 2x = 369$$

$$3x = 369$$

$$x = 123$$

The number of children attending was $2(123) = 246$.

Mary did three times as many sit-ups as Joan. If the total number of sit-ups for both was 68, how many sit-ups did Joan do?

a) 210 votes

b) 35 to 17

---

30. To solve the problem below, we used $\underline{x}$ for the first piece, $\underline{2x}$ for the second piece, and $3(2x) = \underline{6x}$ for the third piece.

A 270 cm wire is cut into three pieces. The second piece is twice as long as the first. The third piece is three times as long as the second. How long is each piece?

$$x + 2x + 3(2x) = 270$$

$$x + 2x + 6x = 270$$

$$9x = 270$$

$$x = 30$$

The three pieces are 30 cm, 60 cm, and 180 cm.

A 350m rope is cut into three pieces. The second piece is twice as long as the first. The third piece is four times as long as the first. How long is each piece?

17 sit-ups

---

31. To solve the problem below, we used $\underline{x}$, $\underline{2x}$, and $\underline{x - 5}$ for the three branches. Solve the other problem.

An electric current of 83 amperes is branched off into three circuits. The second branch carries twice as much current as the first. The third branch carries 5 amperes less current than the first. Find the amount of current in each branch.

$$x + 2x + (x - 5) = 83$$

$$4x - 5 = 83$$

$$4x = 88$$

$$x = 22$$

Therefore, the amount of current in the three branches is 22 amperes, 44 amperes, and 17 amperes.

The flow capacity of a pipeline is 800 liters per minute. The pipeline separates into three branches. The second branch has 3 times the capacity of the first branch. The third branch has a capacity of 50 liters more than the first. Find the capacity of each branch.

50m, 100m, and 200m

---

150 liters, 450 liters, and 200 liters

SELF-TEST 12 (pages 157-169)

Translate each sentence to an equation. Use x for the variable.

1. 100 minus three times a number is 31. ____________
2. Four times a number, minus 10 equals 74. ____________
3. What number minus $\frac{1}{4}$ equals $\frac{5}{8}$? ____________
4. If the difference between a number and 5 is multiplied by 3, the result is 60. ____________
5. 48 divided by twice a number equals 3. ____________
6. If a number is divided by the difference of that number and 2, the result is 5. ____________

Solve each problem.

7. If a number is added to one-third the number, the result is 36. Find the number.

8. If the sum of a number and 5 is divided by 4, we get 21. Find the number.

9. The sum of two consecutive odd integers is 76. Find the two integers.

10. The sum of three consecutive integers is 162. What are the three integers?

11. On an algebra test, the highest grade was 52 points more than the lowest grade. The sum of the two grades was 140. Find the lowest grade.

12. A 150 ft rope is cut into 3 pieces. The second piece is twice as long as the first. The third piece is three times as long as the first. How long is each piece?

ANSWERS:

1. $100 - 3x = 31$
2. $4x - 10 = 74$
3. $x - \frac{1}{4} = \frac{5}{8}$
4. $3(x - 5) = 60$
5. $\frac{48}{2x} = 3$
6. $\frac{x}{x - 2} = 5$
7. 27
8. 79
9. 37 and 39
10. 53, 54, and 55
11. 44 points
12. 25 ft, 50 ft, and 75 ft

4-5 CONVERTING PERCENTS TO FRACTIONS AND DECIMALS

In this section, we will define percents and show the methods for converting percents to fractions and decimals.

---

32. The word percent means per 100 or ___/100. That is, 42% means 42 per 100 or 42/100. Therefore, any percent can be written as a hundredths decimal fraction with the percent-value as its numerator. For example:

$63\% = \frac{63}{100}$ $\quad 7.5\% = \frac{7.5}{100}$ $\quad 240\% = \frac{240}{100}$

Convert each percent to a hundredths decimal-fraction.

a) 1% = ____ b) 150% = ____ c) 8.75% = ____

---

33. Some percents can be converted to simple fractions by reducing the decimal fraction to lowest terms. For example:

$50\% = \frac{50}{100} = \frac{1}{2}$ $\quad 80\% = \frac{80}{100} = \frac{4}{5}$

Convert each percent to a simple fraction.

a) 25% = ____ b) 10% = ____ c) 60% = ____

Answers to 32: a) $\frac{1}{100}$ b) $\frac{150}{100}$ c) $\frac{8.75}{100}$

---

34. Any multiple of 100% equals a whole number. That is:

$100\% = \frac{100}{100} = 1$ $\quad 200\% = \frac{200}{100} = 2$ $\quad 500\% = \frac{500}{100} =$ ____

Answers to 33: a) $\frac{1}{4}$ b) $\frac{1}{10}$ c) $\frac{3}{5}$

---

35. To convert each percent below to a decimal, we converted it to a decimal fraction and then divided the numerator by 100. (Note: To divide by 100, we can simply shift the decimal point two places to the left.)

$83\% = \frac{83}{100} = .83. = .83$

$172\% = \frac{172}{100} = 1.72. = 1.72$

$6.5\% = \frac{6.5}{100} = .06.5 = .065$

The same conversions can be made directly by shifting the decimal point two places to the left and dropping the percent sign. That is:

$83\% = .83. = .83$

$172\% = 1.72. = 1.72$

$6.5\% = .06.5 = .065$

Convert each percent to a decimal.

a) 7% = ____ b) 215% = ____ c) 12.75% = ____

Answer to 34: 5

36. After converting some percents to decimal numbers, we can drop one or two final 0's after the decimal point. For example:

$$30\% = .30. = .30 = .3$$

$$400\% = 4.00. = 4.00 = 4$$

Convert each percent to a decimal number.

a) 90% = ______ b) 200% = ______ c) 130% = ______

Answers (frame 35): a) .07 b) 2.15 c) .1275

---

37. To convert a mixed-number percent to a decimal number, we convert it to a decimal-number percent first. For example:

$12\frac{1}{2}\% = 12.5\% = .125$ $9\frac{3}{4}\% = 9.75\% = .0975$

Convert each percent to a decimal number.

a) $11\frac{3}{4}\%$ = ______ b) $5\frac{1}{2}\%$ = ______

Answers (frame 36): a) .9 b) 2 c) 1.3

---

38. Some common percent-fraction-decimal equivalents are given in the table below. Because of their usefulness, <u>they should be memorized</u>.

$25\% = \frac{1}{4} = .25$	$33\frac{1}{3}\% = \frac{1}{3} = .33$	$20\% = \frac{1}{5} = .2$	$10\% = \frac{1}{10} = .1$
$50\% = \frac{1}{2} = .5$	$66\frac{2}{3}\% = \frac{2}{3} = .67$	$40\% = \frac{2}{5} = .4$	$30\% = \frac{3}{10} = .3$
$75\% = \frac{3}{4} = .75$		$60\% = \frac{3}{5} = .6$	$70\% = \frac{7}{10} = .7$
		$80\% = \frac{4}{5} = .8$	$90\% = \frac{9}{10} = .9$

<u>Note</u>: The decimal equivalents for $33\frac{1}{3}\%$ and $66\frac{2}{3}\%$ are rounded <u>to the nearest hundredth</u>.

From memory, convert each percent to a fraction.

a) 25% = ______ b) 80% = ______ c) $66\frac{2}{3}\%$ = ______

d) 90% = ______ e) 50% = ______ f) $33\frac{1}{3}\%$ = ______

Answers (frame 37): a) .1175 b) .055

Answers (frame 38): a) $\frac{1}{4}$ b) $\frac{4}{5}$ c) $\frac{2}{3}$ d) $\frac{9}{10}$ e) $\frac{1}{2}$ f) $\frac{1}{3}$

4-6 CONVERTING DECIMALS AND FRACTIONS TO PERCENTS

In this section, we will discuss the methods for converting decimals and fractions to percents.

39. Because percent means hundredths, any hundredths decimal fraction can be converted directly to a percent. For example:

$$\frac{9}{100} = 9\% \qquad \frac{60}{100} = 60\% \qquad \frac{225}{100} = \_\_\_\_$$

Answer: 225%

40. We converted the decimals below to percents by converting them to a hundredths decimal fraction first.

$$.05 = \frac{5}{100} = 5\% \qquad 1.37 = \frac{137}{100} = 137\%$$

$$.4 = \frac{4}{10} = \frac{40}{100} = 40\% \qquad 1.8 = \frac{18}{10} = \frac{180}{100} = 180\%$$

The same conversions can be made directly by shifting the decimal point two places to the right and attaching the percent sign. That is:

.05 = .05‸ = 5% 1.37 = 1.37‸ = 137%

.4 = .40‸ = 40% 1.8 = 1.80‸ = 180%

Convert to a percent.

a) .83 = ____ b) .01 = ____ c) 2.96 = ____

d) .1 = ____ e) .9 = ____ f) 2.3 = ____

Answers: a) 83% b) 1% c) 296% d) 10% e) 90% f) 230%

41. Any whole number can be converted to a percent by converting it to a hundredths decimal fraction first. For example:

$$1 = \frac{100}{100} = 100\% \qquad 3 = \frac{300}{100} = 300\%$$

The decimal-point-shift method can be used for the same conversions. That is:

1 = 1.00‸ = 100% 3 = 3.00‸ = 300%

Convert to a percent.

a) 2 = ____ b) 4 = ____ c) 7 = ____

Answers: a) 200% b) 400% c) 700%

42. We used the decimal-point-shift method for the conversions below.

.497 = .49‸7 = 49.7% .0875 = .08‸75 = 8.75%

Convert to a percent.

a) .065 = ____ b) .1325 = ____ c) .0915 = ____

43. Some fractions can be converted to percents by converting them to equivalent <u>hundredths</u> decimal fractions. For example:

$\frac{1}{2} = \frac{50}{100} = 50\%$ $\frac{3}{4} = \frac{75}{100} = 75\%$

Using the same method, complete these conversions.

a) $\frac{3}{5} = \frac{\square}{100} =$ ______ b) $\frac{1}{10} = \frac{\square}{100} =$ ______

Answers: a) 6.5% b) 13.25% c) 9.15%

---

44. We can convert fractions to percents by converting them to decimals first. To do so, we divide. An example is shown. Complete the other conversion.

$\frac{1}{4} = 4\overline{)1.00}$ = .25 = 25%
-8
20
-20

$\frac{4}{5} = 5\overline{)4.}$ =

Answers: a) $\frac{60}{100} = 60\%$ b) $\frac{10}{100} = 10\%$

---

45. The division method is generally used to convert fractions to percents. A calculator can be used for the divisions. Two examples are shown. Use a calculator for the other conversions.

$\frac{51}{68} = .75 = 75\%$ $\frac{3}{16} = .1875 = 18.75\%$

a) $\frac{34}{40} =$ ______ = ______ b) $\frac{10}{16} =$ ______ = ______

Answer: $5\overline{)4.0}$ = .8 = 80%

---

46. When the quotient is larger than "1", the percent is larger than 100%. An example is shown. Use a calculator for the other conversion.

$\frac{31}{25} = 1.24 = 124\%$ $\frac{47}{20} =$ ______ = ______

Answers: a) .85 = 85% b) .625 = 62.5%

---

47. When one or both terms of the fraction is a decimal number, we convert it to a percent in the usual way. An example is shown. Use a calculator for the other conversion.

$\frac{8.4}{17.5} = .48 = 48\%$ $\frac{5.65}{4.52} =$ ______ = ______

Answer: 2.35 = 235%

---

48. When the division is non-terminating, we round to a specific place. For example:

To round <u>to the nearest whole number percent</u>, we round <u>to two decimal places</u> before converting.

$\frac{211}{375} = .56266667 = .56 = 56\%$

To round <u>to the nearest tenth of a percent</u>, we round <u>to three decimal places</u> before converting.

$\frac{39}{550} = .07090909 = .071 = 7.1\%$

Continued on following page.

Answer: 1.25 = 125%

48. Continued

a) Convert to the nearest whole number percent.

$\frac{18}{170} =$ ______

b) Convert to the nearest tenth of a percent.

$\frac{15.9}{17.5} =$ ______

a) 11%   b) 90.9%

4-7 THE THREE TYPES OF PERCENT PROBLEMS

In this section, we will discuss the three types of percent problems. We will show how they can be solved by converting to equations and then solving the equations.

Note: You will find it helpful to use a calculator for some problems in this section and the next section.

49. Two examples of the first type of percent problems are shown below. We converted each to an equation.

What is 40% of 200 ?
↓ ↓ ↓ ↓ ↓
x = 40% · 200

75% of 96 is what?
↓ ↓ ↓ ↓ ↓
75% · 96 = x

Translate each of these to an equation.

a) What is 50% of 138 ?

______

b) 63% of 837 is what?

______

50. Another example of the first type of percent problems is translated to an equation. Translate the other problem to an equation.

Find 8.5% of $500.
↓ ↓ ↓
8.5% · $500 = x

Find 145% of 925.

______

a) x = 50% · 138

b) 63% · 837 = x

51. To solve an equation involving a percent, we must convert the percent to a fraction or decimal. Below, for example, we solved for x by converting $33\frac{1}{3}$% to a fraction to multiply. Use the same method for the other problem.

145% · 925 = x

Find $33\frac{1}{3}$% of 600.

$x = 33\frac{1}{3}\% \cdot 600$

$x = \frac{1}{3}(600) = 200$

25% of 48 is what?

x = 12, from $\frac{1}{4}(48)$

52. Ordinarily we have to convert to a decimal when multiplying by a percent. An example is shown. Do the other problem.

Find 2.4% of 120.

$x = 2.4\% \cdot 120$

$x = .024(120)$

$x = 2.88$

What is 18% of \$7,500 ?

Answer: $x = \$1{,}350$, from $.18(\$7{,}500)$

53. Two examples of the second type of percent problems are shown below. We translated each to an equation.

17 is what percent of 80 ?
↓ ↓ ↓ ↓ ↓
$17 = x \cdot 80$

What percent of 200 is 159 ?
↓ ↓ ↓ ↓ ↓
$x \cdot 200 = 159$

Translate each problem to an equation.

a) 20 is what percent of 50 ? ________

b) What percent of 900 is 650 ? ________

Answers: a) $20 = x \cdot 50$ b) $x \cdot 900 = 650$

54. To solve equations like those in the last frame, we divide the other number by the coefficient of x and then convert the fraction to a percent. An example is shown. Solve the other equation.

$x \cdot 90 = 60$

$$x = \frac{60}{90} = \frac{2}{3} = 66\frac{2}{3}\%$$

$20 = x \cdot 50$

Answer: 40%, from $\frac{20}{50}$

55. Ordinarily we have to use the division method to solve the second type of percent problem. An example is shown. Do the other problem. Round to the nearest whole number percent.

What percent of 70 is 98 ?

$x \cdot 70 = 98$

$$x = \frac{98}{70} = 140\%$$

14.9 is what percent of 84.7 ?

Answer: $x = 18\%$, from $\frac{14.9}{84.7}$

56. Two examples of the third type of percent problems are shown below.

25 is 12% of what?
↓ ↓ ↓ ↓ ↓
$25 = 12\% \cdot x$

80% of what is 62 ?
↓ ↓ ↓ ↓ ↓
$80\% \cdot x = 62$

Translate each problem below to an equation.

a) 50% of what is 20 ? ________

b) 75 is 140% of what? ________

Answers: a) $50\% \cdot x = 20$ b) $75 = 140\% \cdot x$

57. To solve equations like those in the last frame, we can sometimes convert the percent to a fraction. An example is shown. Solve the other equation.

$$30 = 75\% \cdot x$$

$$30 = \frac{3}{4}x$$

$$\frac{4}{3}(30) = \frac{4}{3}\left(\frac{3}{4}x\right)$$

$$40 = x$$

$$50\% \cdot x = 35$$

---

x = 70

58. Usually we have to convert the percent to a decimal to solve the third type of percent problem. An example is shown. Do the other problem. Round to the nearest whole number.

22 is 2.5% of what?

$$22 = 2.5\% \cdot x$$

$$22 = .025x$$

$$x = \frac{22}{.025} = 880$$

12% of what is 200 ?

---

x = 1,667, from: $.12x = 200$

59. Do these.

a) Find $7\frac{1}{2}\%$ of \$600.

b) 39.6 is what percent of 28.3 ? (Round to the nearest whole number percent.)

---

a) \$45 b) 140%

60. Do these.

a) What percent of 9,500 is 427 ? (Round to the nearest tenth of a percent.)

b) 75 is 45% of what? (Round to the nearest whole number.)

---

a) 4.5% b) 167

4-8 WORD PROBLEMS INVOLVING PERCENTS

In this section, we will discuss a method for solving word problems involving percents.

---

61. To solve the problem below, we reworded it, translated the rewording to an equation, and then solved the equation. Use the same method for the other problem.

A team won 60% of its games in one season. If they played 80 games, how many games did they win?

Rewording: What is 60% of 80 ?

Equation: $x = 60\% \cdot 80$

Solution: $x = .6(80)$

$x = 48$

Therefore, they won 48 games in that season.

On a 40-item test, a student had 90% of the items correct. How many did she have correct?

Rewording: ______________________

Equation: ______________________

Solution: ______________________

What is 90% of 40?

$x = 90\% \cdot 40$

$x = .9(40) = 36$ items correct

62. We used the rewording-equation method to solve the problem below. Use the same method for the other problem. Round to the nearest whole number percent.

Of 268 motors tested, 37 were defective. What percent were defective?

Rewording: 37 is what percent of 268 ?

Equation: $37 = x \cdot 268$

Solution: $x = \frac{37}{268} = 14\%$

Therefore, 14% of the motors were defective.

In a class of 44 students, 8 students received a grade of A. What percent received a grade of A?

Rewording: ______________________

Equation: ______________________

Solution: ______________________

63. We used the rewording-equation method to solve the problem below. Use the same method for the other problem.

How much money must be invested at 5% annual interest to earn $500 interest annually?

Rewording: $500 is 5% of what?

Equation: $\$500 = 5\% \cdot x$

Solution: $x = \frac{\$500}{.05} = \$10,000$

Therefore, $10,000 must be invested to earn $500 annually.

A stainless steel alloy contains 8% nickel. How many pounds of the alloy can be made from 560 pounds of nickel?

Rewording: ____________________

Equation: ____________________

Solution: ____________________

8 is what percent of 44?

$8 = x \cdot 44$

$x = \frac{8}{44} = 18\%$

---

64. Use the rewording hints to solve the problems in this frame and the next frame.

a) If the sales tax rate is 4%, how much tax is paid for a $425.95 TV set. (Round to the nearest cent.)

(Hint: 4% of $425.95 is what?)

b) If a student got 26 items correct on a 32-item test, what percent grade did he receive? (Round to the nearest whole number percent.)

(Hint: 26 is what percent of 32?)

560 is 8% of what?

$560 = 8\% \cdot x$

$x = \frac{560}{.08} = 7,000$ pounds

---

65. a) An alloy weighing 32.7 kilograms contains 25.3 kilograms of aluminum. What percent of the alloy is aluminum? (Round to the nearest whole number percent.

(Hint: 25.3 is what percent of 32.7?)

b) A basketball team won 44% of its games. If it won 11 games, how many games did it play?

(Hint: 11 is 44% of what?)

a) $17.04

b) 81%

66. Make up your own <u>rewording</u> hints for the problems in this frame and the next frame.

a) If the annual interest on a $500 loan is $60, what is the interest rate?

b) Mary's weekly gross pay is $240. If 22% of her check is withheld, how much is withheld?

Answers: a) 77% b) 25 games

---

67. a) A jewelry store pays $20 for a necklace and sells it for $45. Therefore, the markup is $25. What is the percent markup?

b) How much money must be invested at 8% annual interest to earn $1,600 interest annually?

Answers: a) 12% b) $52.80

Answers: a) 125% b) $20,000

SELF-TEST 13 (pages 170-180)

Convert each percent to a fraction in lowest terms.

1. 20% = ______ 2. $66\frac{2}{3}\%$ = ______

Convert each percent to a decimal number.

3. 15.3% = ______ 4. 6.75% = ______

Convert each decimal to a percent.

5. .085 = ______ 6. 1.8 = ______

Convert each fraction to a percent.

7. $\frac{3}{4}$ = ______ 8. $\frac{36}{32}$ = ______

9. Convert $\frac{2.63}{74.7}$ to the nearest tenth of a percent. ______

10. Convert $\frac{8,190}{3,620}$ to the nearest whole number percent. ______

Continued on following page.

SELF-TEST 13 (pages 170-180) - continued

11. What is 6.25% of $2,500 ?

12. 240 is what percent of 400 ?

13. 20% of what is 70 ?

14. A student got 61 items correct on a 72-item test. Find her percent grade. (Round to the nearest whole number percent.)

15. How many games would a team have to win to win 80% of the games of a 30-game schedule?

16. An alloy contains 18% chromium. How much of the alloy can be made from 150 pounds of chromium? (Round to the nearest whole number.)

ANSWERS:

1. $\frac{1}{5}$	5. 8.5%	9. 3.5%	13. 350
2. $\frac{2}{3}$	6. 180%	10. 226%	14. 85%
3. .153	7. 75%	11. $156.25	15. 24 games
4. .0675	8. 112.5%	12. 60%	16. 833 pounds

4-9 RECTANGLES AND SQUARES

In this section, we will review the formulas for the perimeter and area of rectangles and squares. Some word problems are included.

68. Lengths can be measured with units from either the English system or the metric system. Some common units from each system and their abbreviations are shown in the table below. Notice that periods are not used in the abbreviations.

	Unit	Abbreviation
English System	inches	in
	feet	ft
	yards	yd
Metric System	centimeters	cm
	meters	m

Continued on following page.

68. Continued

A rectangle is a four-sided figure with opposite sides equal and four right angles. (A right angle contains 90°.)

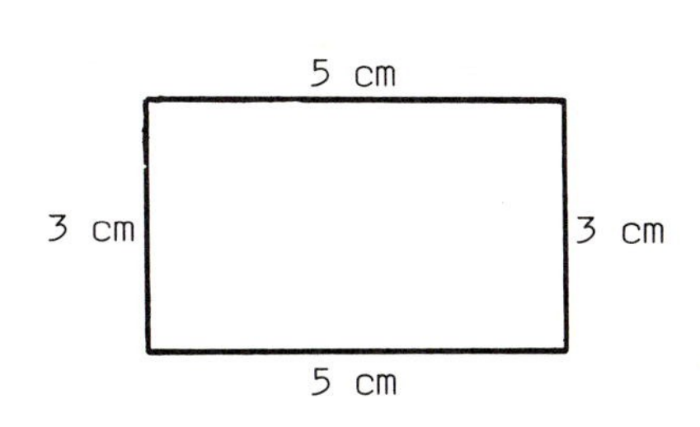

The longer sides are called the length (or L).

The shorter sides are called the width (or W).

The perimeter (P) of a rectangle is the distance around it. We can find the perimeter by adding the length of its four sides. That is, $P = L + W + L + W = 2L + 2W$. Therefore, we use the following formula:

$$\boxed{P = 2L + 2W}$$

When using the formula, we treat the units as if they are variables. That is, for the rectangle above:

$$P = 2(5 \text{ cm}) + 2(3 \text{ cm})$$
$$= 10 \text{ cm} + 6 \text{ cm}$$
$$= (10 + 6) \text{ cm}$$
$$= \_\_\_\_\_ \text{ cm}$$

---

69. The area of a geometric figure is the number of unit squares needed to fill it. Some unit squares are shown below. Following the patterns, the abbreviations for "a square foot" and "a square yard" are ft^2 and yd^2.

Answer: 16 cm

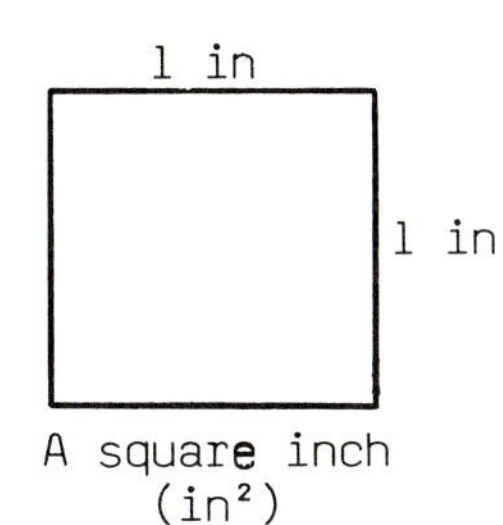

A square inch (in^2)

1 cm
1 cm

A square centimeter (cm^2)

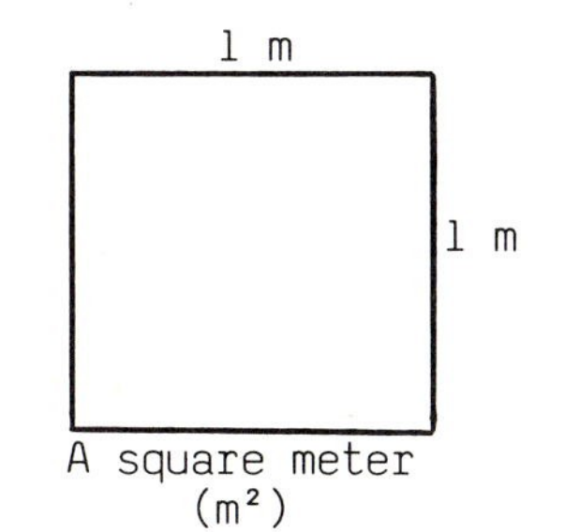

A square meter (m^2)

It takes 6 square centimeters to fill up the rectangle at the right. Therefore, its area is 6 cm^2. Notice that we can find its area by multiplying its length and width. That is:

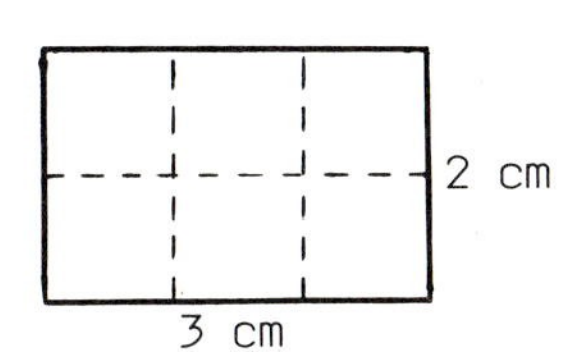

$$\boxed{A = LW}$$

When using the formula, we also treat the units as if they were variables. That is, for the rectangle above:

$$A = (3 \text{ cm})(2 \text{ cm})$$
$$= (3)(2)(\text{cm})(\text{cm})$$
$$= \_\_\_\_\_ \text{ cm}^2$$

70. Use the formulas to find the perimeter and area of the rectangle at the right.

a) $P = 2L + 2W$ b) $A = LW$

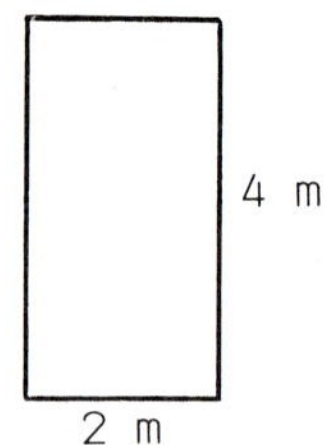

Answer: 6 cm^2

---

71. To find the width of a rectangle if its area is 40 ft^2 and its length is 10 ft, we can use either the equation-solving method or the rearrangement method. For example:

Equation-Solving Method	Rearrangement Method
$A = LW$	$A = LW$
$40 = 10W$	$W = \frac{A}{L}$
$W = \frac{40}{10} = 4 \text{ ft}$	$W = \frac{40}{10} = 4 \text{ ft}$

Using either method, solve the problems below.

a) Find the width of a rectangle if its area is 54 in^2 and its length is 9 in.

b) Find the length of a rectangle if its area is 20 cm^2 and its width is 2 cm.

Answers (70): a) $P = 12 \text{ m}$ b) $A = 8 \text{ m}^2$

---

72. To find the length of a rectangle if its perimeter is 16 cm and its width is 3 cm, we can also use either the equation-solving method or the rearrangement method. For example:

Equation-Solving Method	Rearrangement Method
$P = 2L + 2W$	$P = 2L + 2W$
$16 = 2L + 2(3)$	$P - 2W = 2L$
$16 = 2L + 6$	$L = \frac{P - 2W}{2}$
$10 = 2L$	$L = \frac{16 - 2(3)}{2}$
$L = \frac{10}{2} = 5 \text{ cm}$	$L = \frac{16 - 6}{2} = \frac{10}{2} = 5 \text{ cm}$

Continued on following page.

Answers (71): a) $W = 6 \text{ in}$ b) $L = 10 \text{ cm}$

72. Continued

Using either method, solve the problems below.

a) Find the length of a rectangle if its perimeter is 30 yd and its width is 5 yd.

b) Find the width of a rectangle if its perimeter is 18 m and its length is 7 m.

73. To solve the problem below, we used the formula $P = 2L + 2W$. We let $W = x$ and $L = x + 20$. Use the same method for the other problem. Draw an appropriate figure and label the sides.

a) $L = 10$ yd

b) $W = 2$ m

The perimeter of a rectangle is 160 cm. The length is 20 cm greater than the width. Find the dimensions.

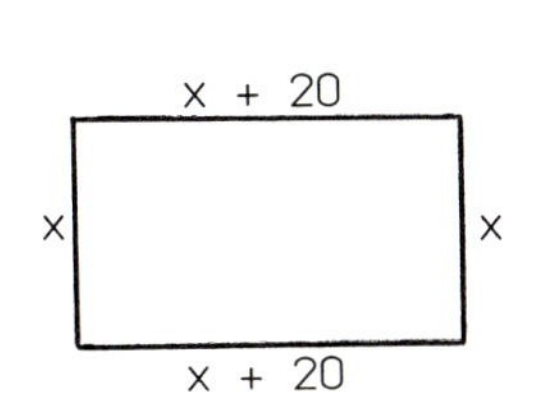

$$P = 2L + 2W$$
$$160 = 2(x + 20) + 2(x)$$
$$160 = 2x + 40 + 2x$$
$$160 = 4x + 40$$
$$120 = 4x$$
$$x = 30$$

Therefore, the width is 30 cm and the length is 30 + 20 = 50 cm.

The perimeter of a rectangle is 250 ft. The length is 25 ft greater than the width. Find the dimensions.

$W = 50$ ft; $L = 75$ ft

74. To solve the problem below, we also used the formula $P = 2L + 2W$. We let $L = x$ and $W = x - 15$. Use the same method for the other problem. Draw an appropriate figure and label the sides.

The perimeter of a rectangle is 170 m. The width is 15 m less than the length. Find the dimensions.

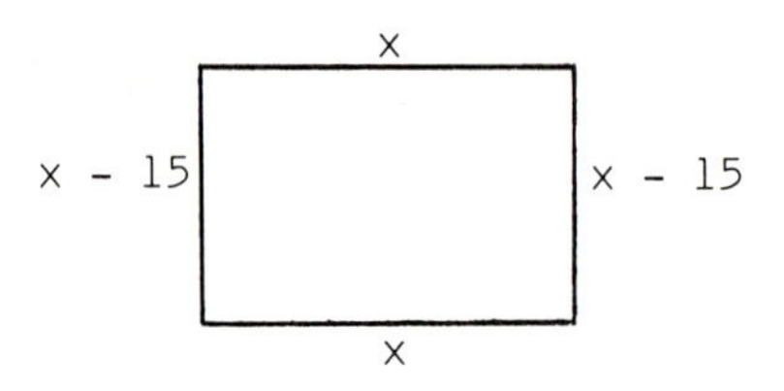

$$P = 2L + 2W$$
$$170 = 2(x) + 2(x - 15)$$
$$170 = 2x + 2x - 30$$
$$170 = 4x - 30$$
$$200 = 4x$$
$$x = 50$$

Therefore the length is 50 m and the width is $50 - 15 = 35$ m.

The perimeter of a rectangle is 236 cm. The width is 22 cm less than the length. Find the dimensions.

---

75. A square is a special type of rectangle in which all four sides (s) are equal. Therefore:

s
s s
s

The <u>perimeter</u> of a square is $s + s + s + s$. That is:

$$\boxed{P = 4s}$$

The <u>area</u> of a square is (s)(s). That is:

$$\boxed{A = s^2}$$

Using the formulas, find the perimeter and area of a square of its side is 10 cm.

a) $P = 4s$

b) $A = s^2$

$L = 70$ cm; $W = 48$ cm

76. To find the side of a square if its perimeter is 12 in, we can use either method below.

Equation-Solving Method

$$P = 4s$$
$$12 = 4s$$
$$s = \frac{12}{4} = 3 \text{ in}$$

Rearrangement Method

$$P = 4s$$
$$s = \frac{P}{4}$$
$$s = \frac{12}{4} = 3 \text{ in}$$

Using either method, solve each problem

a) Find the side of a square if its perimeter is 100 m.

b) Find the side of a square if its perimeter is 48 ft.

Answers (frame 75): a) $P = 40$ cm b) $A = 100 \text{ cm}^2$

Answers (frame 76): a) $s = 25$ m b) $s = 12$ ft

4-10 PARALLELOGRAMS AND TRIANGLES

In this section, we will review the formulas for the area of parallelograms and triangles, and the angle-sum formula for triangles. Some word problems are included.

77. A parallelogram is a four-sided figure with two pairs of parallel sides. To show that the area of a parallelogram is equal to its base (b) times its height (h), we can cut off part of the parallelogram and use it to form a rectangle as we have done below.

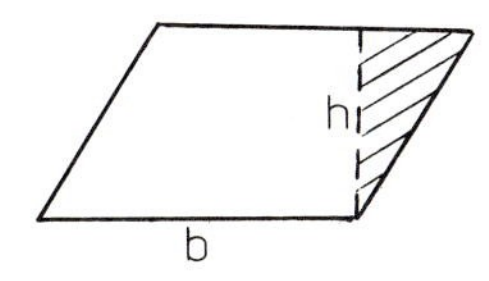

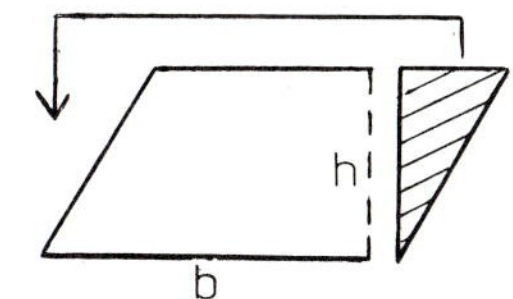

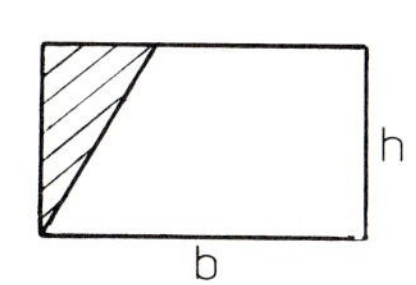

Since the areas of the parallelogram and rectangle are equal, we can use the following formula for the area of a parallelogram.

$$\boxed{A = bh}$$

Using the formula, find the area of each parallelogram.

a)

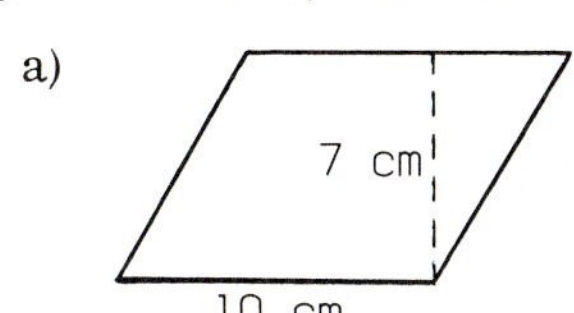

b)

3 in

10.5 in

78. To find the base (or height) of a parallelogram when its area and height (or base) are given, we can use either the equation-solving method or the rearrangement method. Use either method to solve the problems below.

a) Find the base of a parallelogram if its area is 30 m^2 and its height is 6 m.

b) Find the height of a parallelogram if its area is 100 ft^2 and its base is 25 ft.

a) $A = 70\ cm^2$

b) $A = 31.5\ in^2$

---

79. As you can see from the figure at the right, a triangle is simply half a parallelogram. Since the area of the parallelogram is bh, the area of the triangle is half of bh. That is:

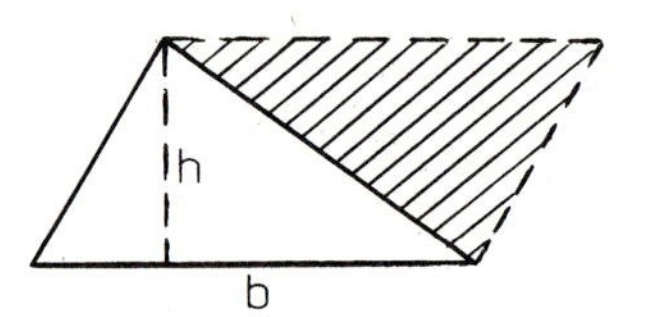

$$A = \frac{1}{2}bh$$

Using the formula above, find the area of each triangle.

a)

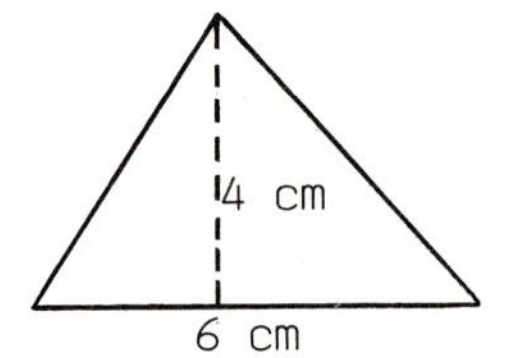

b)

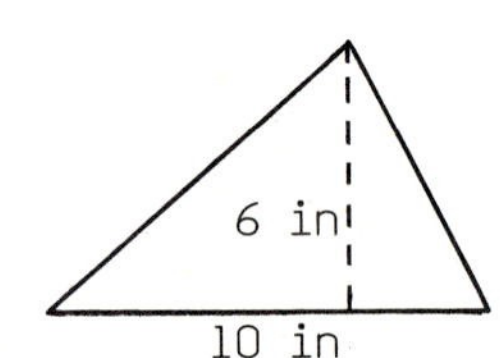

a) $b = 5$ m

b) $h = 4$ ft

---

80. To find the height of a triangle if its area is 36 m^2 and its base is 8 m, we can use either method below.

Equation-Solving Method

$$A = \frac{1}{2}bh$$

$$36 = \frac{1}{2}(8)h$$

$$36 = 4h$$

$$h = \frac{36}{4} = 9\ m$$

Rearrangement Method

$$A = \frac{1}{2}bh$$

$$2A = bh$$

$$h = \frac{2A}{b}$$

$$h = \frac{2(36)}{8} = \frac{72}{8} = 9\ m$$

Continued on following page.

a) $A = 12\ cm^2$

b) $A = 30\ in^2$

80. Continued

Using either method, solve the problems below.

a) Find the height of a triangle if its area is 80 cm^2 and its base is 10 cm.

b) Find the base of a triangle if its area is 100 in^2 and its height is 20 in.

---

a) h = 16 cm

b) b = 10 in

---

81. A right triangle is a triangle that contains a right angle (90°). The right angle is marked with a small box. In a right triangle:

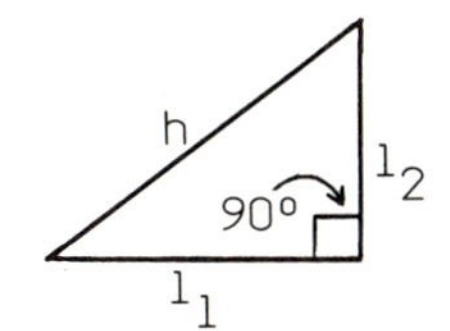

1) The longest side (the one opposite the right angle) is called the hypotenuse (h).

2) The two shorter sides are called the legs (l_1 and l_2).

In the right triangle shown:

1) leg 1 (or l_1) is the base.

2) leg 2 (or l_2) is the height.

Since the area of a triangle is half the base times the height, we use the formula below to find the area of a right triangle.

$$A = \frac{1}{2}l_1l_2$$

Using the formula, find the area of each right triangle.

a)

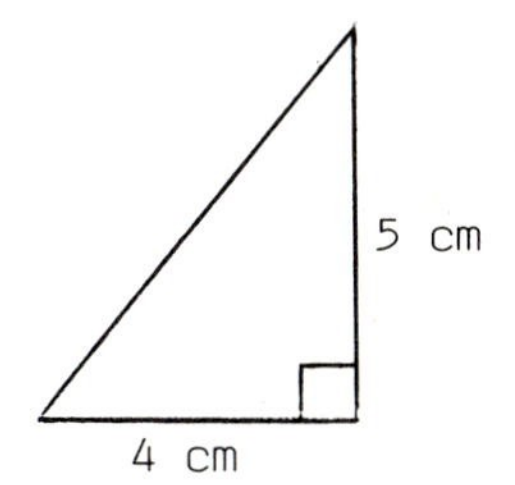

b)

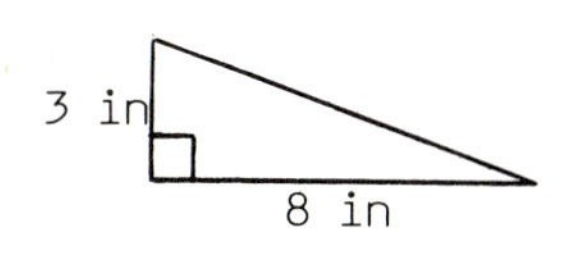

82. Using either the equation-solving method or the rearrangement method, solve each problem.

a) In a right triangle, find 1_2 if A = 50 m^2 and 1_1 = 20 m.

b) In a right triangle, find 1_1 if A = 48 ft^2 and 1_2 = 8 ft.

a) A = 10 cm^2

b) A = 12 in^2

---

83. In any triangle, the sum of the three angles is 180°. If the three angles are A, B, and C, we can state the angle-sum fact with the formula below.

$$\angle A + \angle B + \angle C = 180°$$

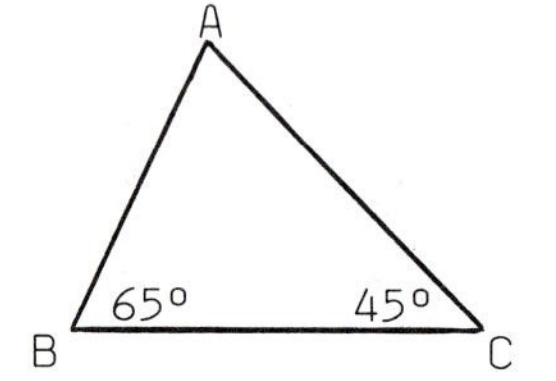

To find the size of $\angle A$ in the triangle shown, we can use either method below.

Equation-Solving Method	Rearrangement Method
$\angle A + \angle B + \angle C = 180°$	$\angle A + \angle B + \angle C = 180°$
$\angle A + 65° + 45° = 180°$	$\angle A = 180° - \angle B - \angle C$
$\angle A + 110° = 180°$	$\angle A = 180° - 65° - 45°$
$\angle A = 70°$	$\angle A = 115° - 45° = 70°$

Using either method, find the size of the missing angle in each triangle.

a) 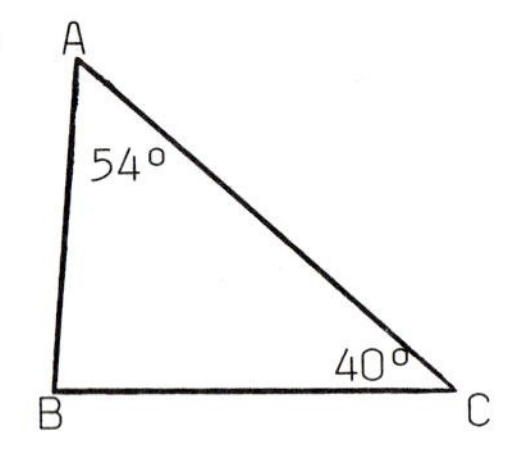

b) 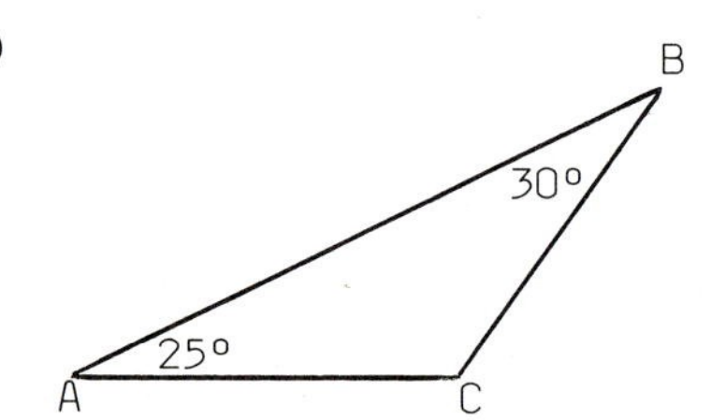

a) 1_2 = 5m

b) 1_1 = 12 ft

---

84. Find the missing angle in each right triangle. (Note: Each right angle is 90°.)

a) 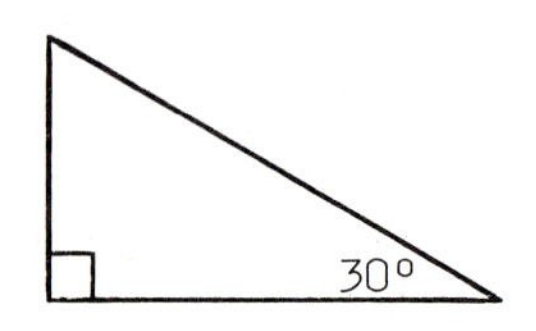

b) 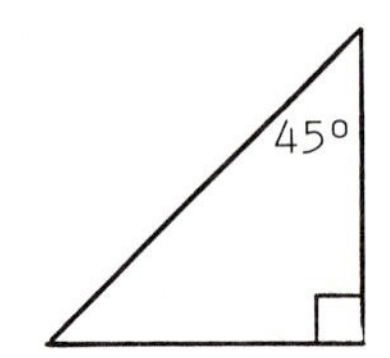

a) $\angle B$ = 86°

b) $\angle C$ = 125°

85. We used the angle-sum formula to solve the problem below. To do so, we let the first angle equal $\underline{x}$, the second angle equal 3x, and the third angle equal x + 30. Use the same method for the other problem.

The second angle of a triangle is three times as large as the first. The third angle is 30° larger than the first. How large are the angles?

1st angle = x

2nd angle = 3x

3rd angle = x + 30

$$x + 3x + (x + 30) = 180$$
$$5x + 30 = 180$$
$$5x = 150$$
$$x = 30$$

Therefore, we get these sizes for the angles:

1st angle = 30°

2nd angle = 90°

3rd angle = 60°

The second angle of a triangle is twice as large as the first. The third angle is 20° less than the first. How large are the angles?

a) 60° b) 45°

---

86. Following the example, solve the other problem.

The second angle of a triangle is twice as large as the first. The third angle is 60° less than the sum of the other two angles. Find the size of the three angles.

1st angle = x

2nd angle = 2x

3rd angle = x + 2x - 60

= 3x - 60

$$x + 2x + (3x - 60) = 180$$
$$6x - 60 = 180$$
$$6x = 240$$
$$x = 40$$

Therefore, the sizes of the angles are:

1st angle = 40°

2nd angle = 80°

3rd angle = 60°

The second angle of a triangle is four times as large as the first. The third angle is 40° less than the sum of the other two angles. How large are the angles?

1st angle = 50°

2nd angle = 100°

3rd angle = 30°

1st angle = 22°
2nd angle = 88°
3rd angle = 70°

4-11 CIRCLES

In this section, we will review the formulas for the circumference and area of a circle and use them to solve some word problems.

87. Point O is the center of the circle at the right.

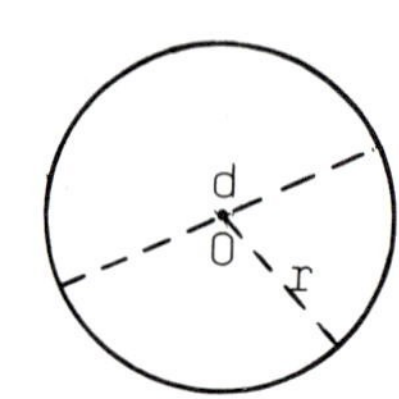

The line d is a diameter of the circle.

The line r is a radius of the circle.

As you can see, a diameter is twice as long as a radius. That is:

$$\boxed{d = 2r}$$

Using the formula above, solve each problem below.

a) Find the diameter of a circle if its radius is 10 cm.

b) Find the radius of a circle if its diameter is 12 in.

88. The distance around a circle is called its circumference. If we divide the circumference of any circle by its diameter, we always get the same number. The greek letter π (pronounced pie) is used for that number. Though π stands for the unending number 3.1415925636..., we will use 3.14 for π. Therefore:

$$\frac{C}{d} = \pi \quad \text{or} \quad \frac{C}{d} = 3.14$$

Solving for C, we get the following circumference formula.

$$\boxed{C = \pi d} \quad \text{or} \quad \boxed{C = 3.14d}$$

Note: If you use the calculator value for π instead of 3.14, you will get slightly different answers.

Using the above formula, solve each problem below.

a) Find the circumference of a circle if its diameter is 10 cm.

b) Find the diameter of a circle if its circumference is 9.42 in.

a) d = 20 cm

b) r = 6 in

89. Since $d = 2r$, we can also state a circumference formula in terms of the radius. That is:

$$C = \pi d$$

$$C = \pi(2r) \quad \text{or} \quad \boxed{C = 2\pi r}$$

Using the formula above, solve each problem below. Use 3.14 for π.

a) Find the circumference of a circle if its radius is 4 in.

b) Find the radius of a circle if its circumference is 12.56 cm.

a) $C = 31.4$ cm

b) $d = 3$ in

---

90. To find the area of a circle, we can use the formula below.

$$\boxed{A = \pi r^2} \quad \text{or} \quad \boxed{A = 3.14r^2}$$

Using the formula above, solve each problem.

a) Find the area of a circle if its radius is 2 in.

b) Find the area of a circle if its radius is 10 cm.

a) $C = 25.12$ in

b) $r = 2$ cm

---

a) $A = 12.56\ in^2$ b) $A = 314\ cm^2$

4-12 VOLUMES

In this section, we will review the formulas for the volume of a rectangular solid, a cube, and a cylinder. The formulas are used to solve some word problems.

---

91. The volume of a geometric figure is the number of unit cubes needed to fill it up. Some unit cubes are shown below. Following the pattern, the abbreviations for "a cubic meter", "a cubic foot", and "a cubic yard" are m^3, ft^3, and yd^3.

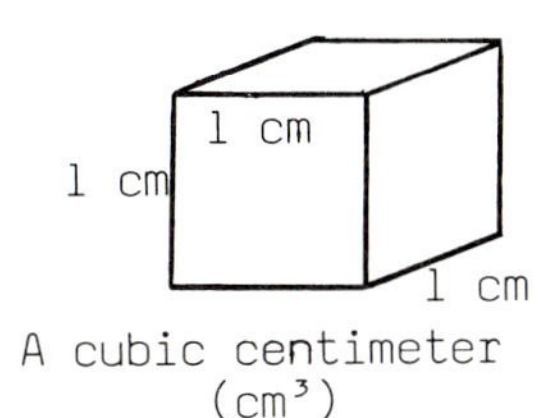

A cubic centimeter (cm^3)

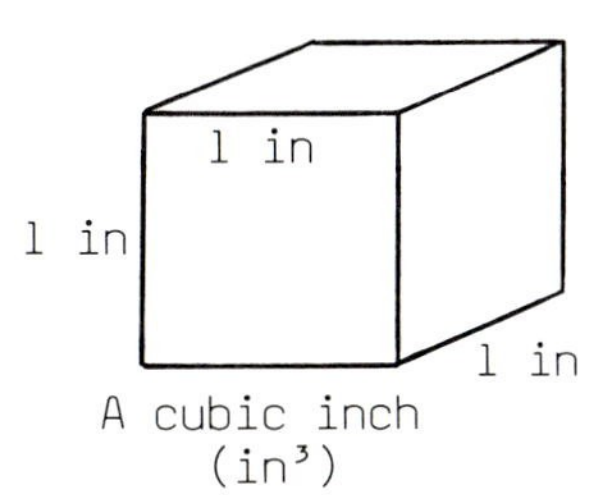

A cubic inch (in^3)

Continued on following page.

91. Continued

It takes 24 cubic centimeters to fill up the rectangular solid at the right. Therefore, its volume is 24 cm^3. Notice that we can find its volume by multiplying its length, width, and height. That is:

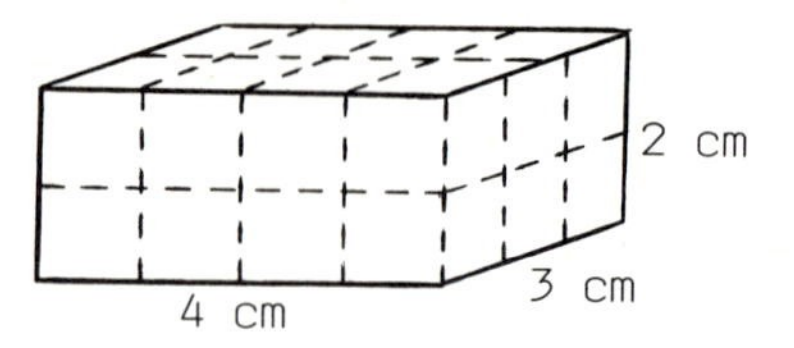

$$V = LWH$$

Using the formula for the rectangular solid above, we get:

$$V = (4\ cm)(3\ cm)(2\ cm)$$
$$= (4)(3)(2)(cm)(cm)(cm)$$
$$= \_\_\_\_\_\_\_\_ cm^3$$

Answer: 24 cm^3

---

92. Use the formula to find the volume of each solid.

a)

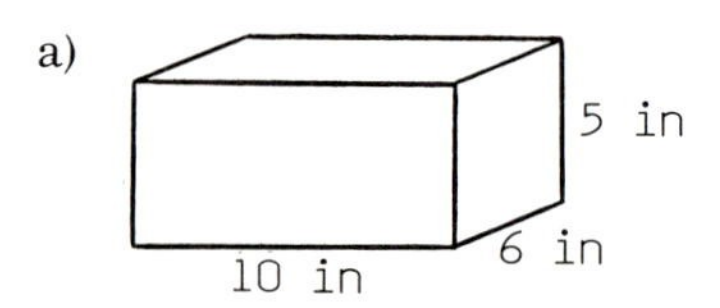

b)

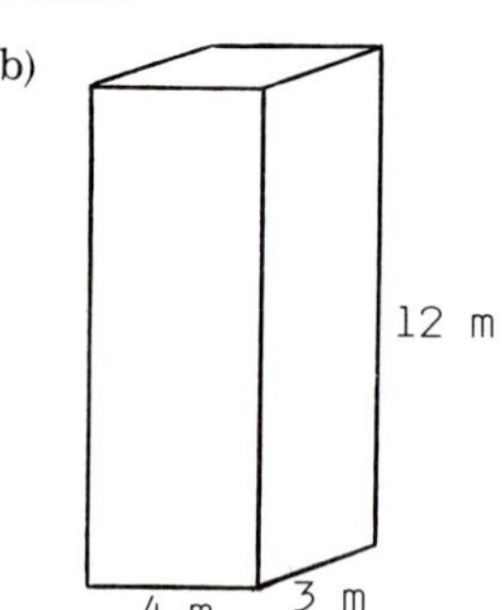

Answers: a) $V = 300\ in^3$ b) $V = 144\ m^3$

---

93. To find the width of a rectangular solid when its volume is 60 in^3, its length is 5 in, and its height is 6 in, we can use either method below.

<u>Equation-Solving Method</u>

$$V = LWH$$
$$60 = (5)(W)(6)$$
$$60 = 30W$$
$$W = \frac{60}{30} = 2\ in$$

<u>Rearrangement Method</u>

$$V = LWH$$
$$W = \frac{V}{LH}$$
$$W = \frac{60}{(5)(6)} = \frac{60}{30} = 2\ in$$

Using either method, solve each problem.

a) Find the <u>height</u> of a rectangular solid if $V = 120\ m^3$, $L = 4\ m$, and $W = 3\ m$.

b) Find the <u>length</u> of a rectangular solid if $V = 24\ yd^3$, $W = 2\ yd$, and $H = 2\ yd$.

94. A cube is a special type of rectangular solid in which all the sides are equal. Therefore, the volume of a cube equals (s)(s)(s). That is:

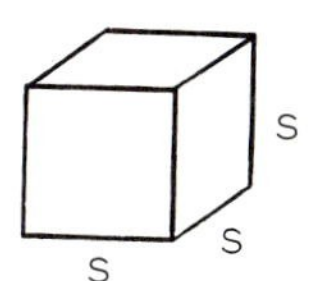

$$V = s^3$$

Using the formula above, we found the volume of one cube below. Find the volume of the other cube.

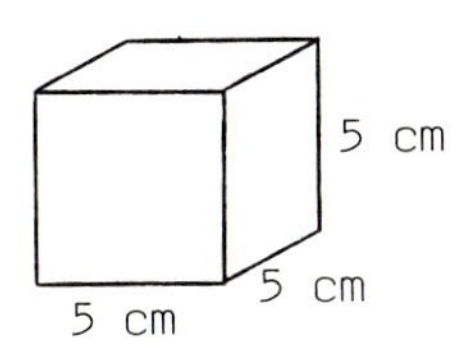

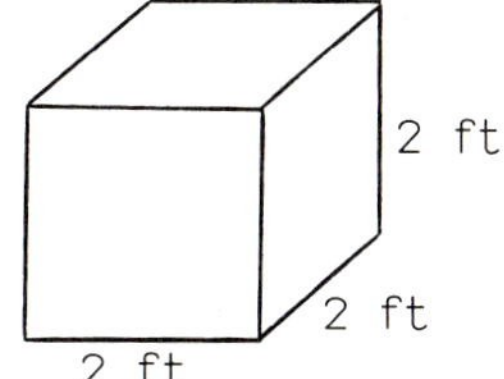

$V = s^3$

$V = (5 \text{ cm})^3$

$V = (5^3)(\text{cm})^3$

$V = 125 \text{ cm}^3$

a) H = 10 m

b) L = 6 yd

---

95. The formula for the volume of a rectangular solid is the same as multiplying the area of the base by the height. That is:

Area of the rectangular base = LW

Height = H

Volume = (LW)(H) or LWH

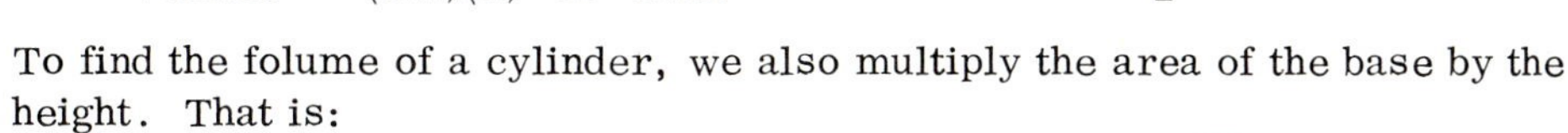

To find the folume of a cylinder, we also multiply the area of the base by the height. That is:

Area of the circular base = πr^2

Height = h

Volume = $(\pi r^2)(h) = \pi r^2 h$

or

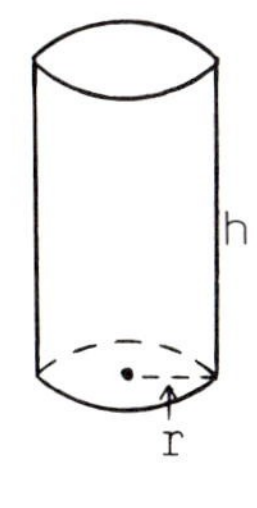

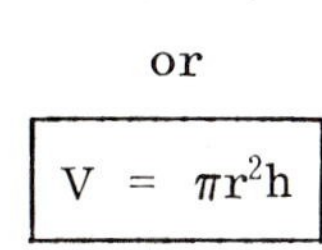

$$V = \pi r^2 h$$

We used the formula above to find the volume of one cylinder below. Find the volume of the other cylinder. Use 3.14 for π.

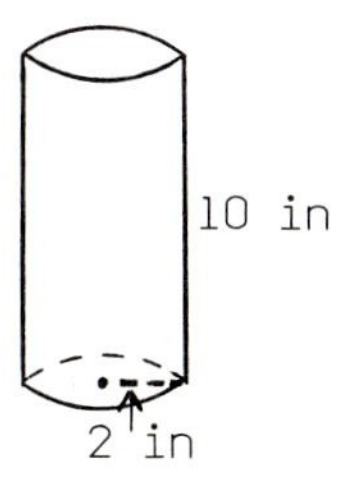

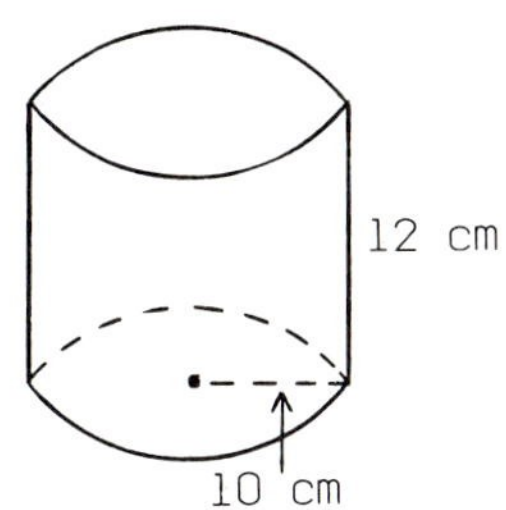

$V = \pi r^2 h$

$V = (3.14)(2^2)(10)$

$V = (3.14)(4)(10)$

$V = (12.56)(10)$

$V = 125.6 \text{ in}^3$

$V = 8 \text{ ft}^3$

$V = 3,768 \text{ cm}^3$

SUMMARY OF FORMULAS

Rectangle	$P = 2L + 2W$
	$A = LW$
Square:	$P = 4s$
	$A = s^2$
Parallelogram:	$A = bh$
Triangle:	$\angle A + \angle B + \angle C = 180°$
	$A = \frac{1}{2}bh$
Right Triangle	$A = \frac{1}{2} l_1 l_2$
Circle:	$d = 2r$
	$C = \pi d$ or $C = 2\pi r$
	$A = \pi r^2$
Rectangular Solid:	$V = LWH$
Cube:	$V = s^3$
Cylinder:	$V = \pi r^2 h$

SELF-TEST 14 (pages 180-195)

1. Find the perimeter.

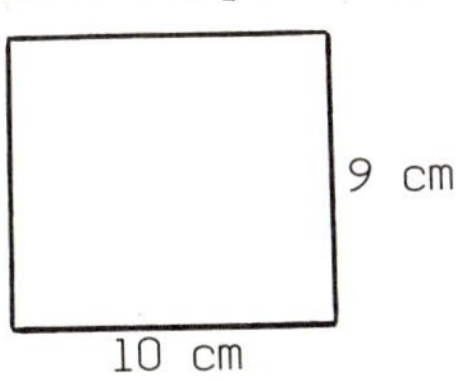

2. Find the area.

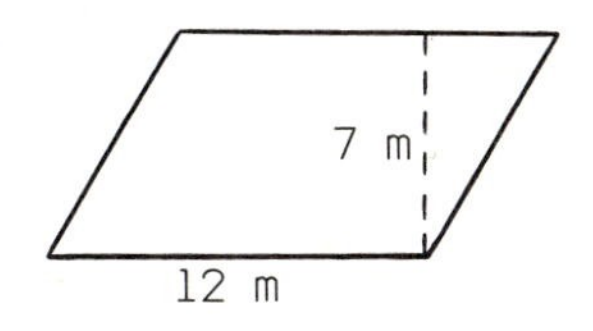

3. Find the area.

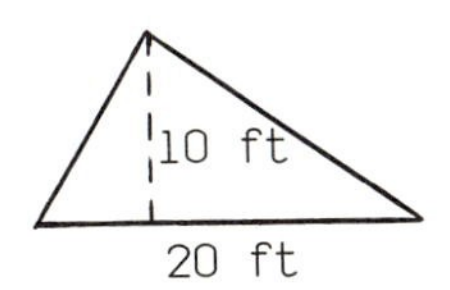

4. Find the circumference.

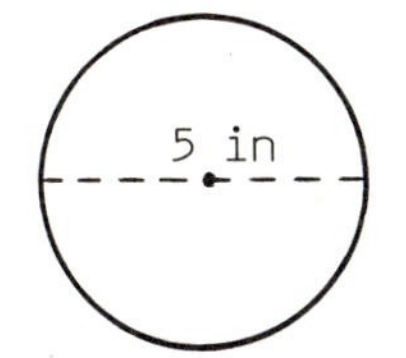

5. Find the volume.

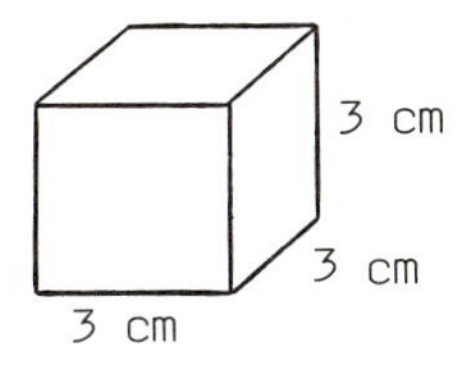

6. Find the volume.

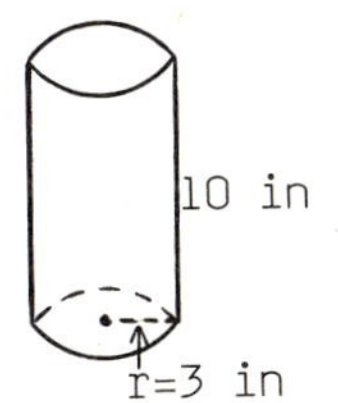

7. Find the length of a rectangle if its area is 48 cm^2 and its width is 6 cm.

8. Find 1_1 in a right triangle if A is 40 m^2 and 1_2 is 10 m.

9. Find the radius of a circle if its circumference is 37.68 m.

10. Find the height of a rectangular solid if its volume is 30 cm^3, its length is 3 cm, and its width is 2 cm.

11. The perimeter of a rectangle is 104 m. The width is 12 m less than the length. Find the dimensions.

12. The second angle of a triangle is four times as large as the first. The third angle is 30° larger than the first. How large are the three angles?

ANSWERS:

1. P = 38 cm
2. A = 84 m^2
3. A = 100 ft^2
4. C = 15.7 in
5. V = 27 cm^3
6. V = 282.6 in^3
7. L = 8 cm
8. 1_1 = 8 m
9. r = 6 m
10. H = 5 cm
11. L = 32 m, W = 20 m
12. 25°, 100°, and 55°

SUPPLEMENTARY PROBLEMS - CHAPTER 4

Assignment 12

Translate each sentence to an equation.

1. The difference between a number and 15 is 32.
2. Four times a number equals the number plus 45.
3. One-fifth of a number equals the number minus 16.
4. Double a number, plus 17 equals 91.
5. If the sum of a number and 7 is multiplied by 10, the result is 120.
6. If one-half of a number is subtracted from the number, we get 24.
7. If the product of a number and 3 is divided by 5, the result is 30.
8. If the difference of a number and 12 is divided by the number, we get 3.

Solve each problem.

9. If 42 is added to double a number, the result is 94. Find the number.
10. If the difference between a number and 12 is multiplied by 5, we get 20. Find the number.
11. If a number is divided by the difference of the number and 9, the result is 4. Find the number.
12. The sum of two consecutive even integers is 234. Find the two integers.
13. The sum of three consecutive integers is 54. Find the three integers.
14. If 100 is added to the middle of three consecutive integers, the result is 21 more than the sum of the largest and three times the smallest.
15. A wire 125 centimeters long is divided into two parts. The larger part is 35 centimeters longer than the smaller part. How long is each part?
16. There were three times as many children as adults at a circus performance. If the total number attending was 3,196, how many adults attended?
17. An electric current of 118 amperes is branched off into three circuits. The second branch carries 14 amperes more current than the first. The third branch carries 10 amperes less current than the first. Find the amount of current in each branch.

Assignment 13

Convert each percent to a fraction in lowest terms.

1. 75% 2. 40% 3. $33\frac{1}{3}\%$ 4. 90% 5. 25%

Convert each percent to a decimal number or whole number.

6. 19% 7. 3.9% 8. 8.75% 9. 147% 10. 300%

Convert each decimal number or whole number to a percent.

11. .81 12. .105 13. .0425 14. 1.3 15. 5

Convert each fraction to a percent.

16. $\frac{1}{2}$ 17. $\frac{4}{5}$ 18. $\frac{3}{10}$ 19. $\frac{2}{3}$ 20. $\frac{59}{100}$

Convert to the nearest whole number percent.

21. $\frac{14}{40}$ 22. $\frac{243}{185}$

Convert to the nearest tenth of a percent.

23. $\frac{76}{1,830}$ 24. $\frac{1.81}{4.26}$

Solve these problems.

25. 20 is what percent of 50 ?
26. What is 75% of 400 ?
27. 30% of what is 15 ?
28. Find 8.5% of $250.
29. 56 if 8% of what?
30. What percent of 300 is 12 ?
31. How much sales tax is paid for a $9,470 car if the sales tax rate is 5%?
32. If a student got 19 problems correct on a 25-problem test, what was his percent grade?
33. How much money must be invested at 12% interest to earn $1,500 interest annually?
34. If 58.45 grams of sodium chloride contains 35.46 grams of chlorine, what percent is chlorine? (Round to the nearest tenth of a percent.)
35. If silver solder contains 75% silver, how much solder can be made from 500 grams of silver? (Round to the nearest gram.)
36. An ore contains 1.37% copper. How much copper is there in 2,000 kilograms of the ore?

Assignment 14

Find the perimeter and area of each figure.

1.

3 cm
5 cm

2.

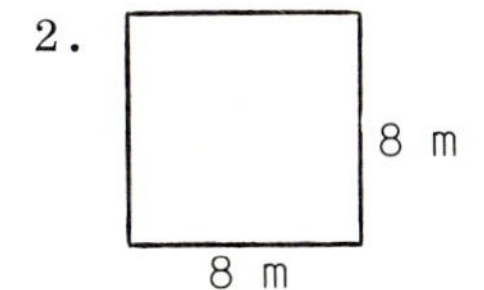

Find the circumference and area of each circle.

3.

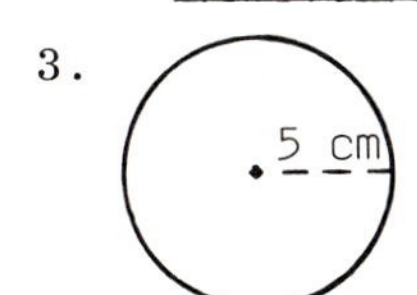

4.

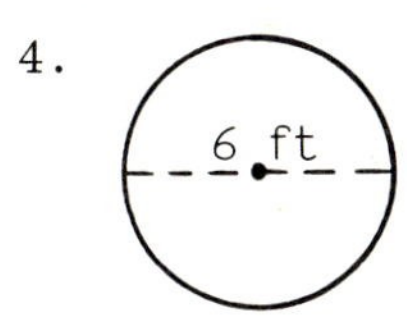

Find the area of each figure.

5.

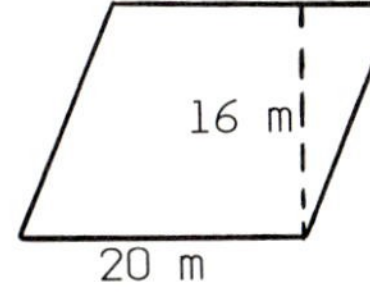

6.

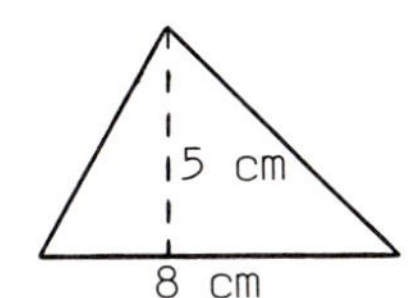

7.

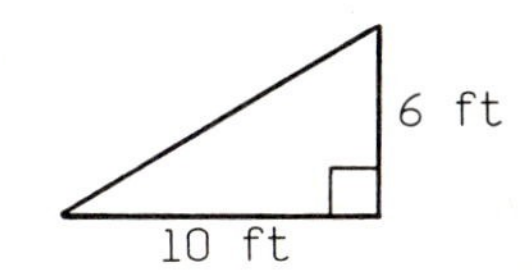

Find the unknown angle in each triangle.

8.

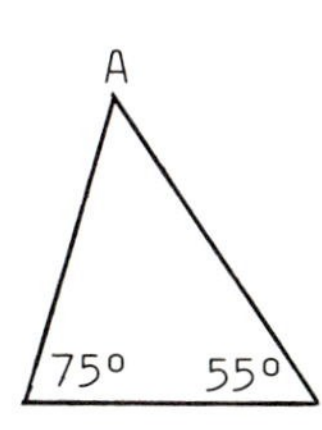

9.

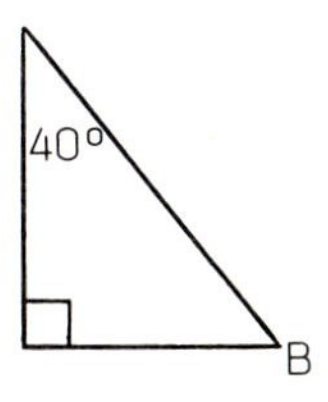

10.

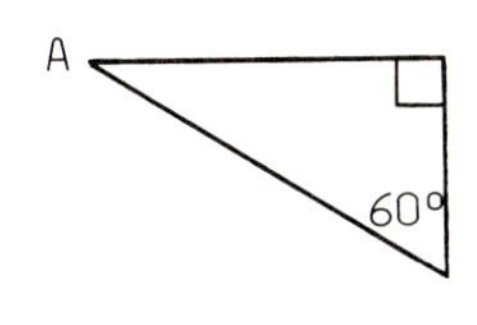

Find the volume of each figure.

11.

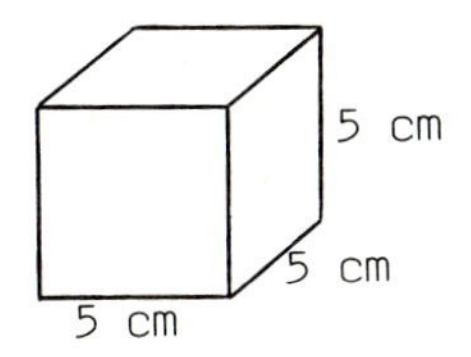

12.

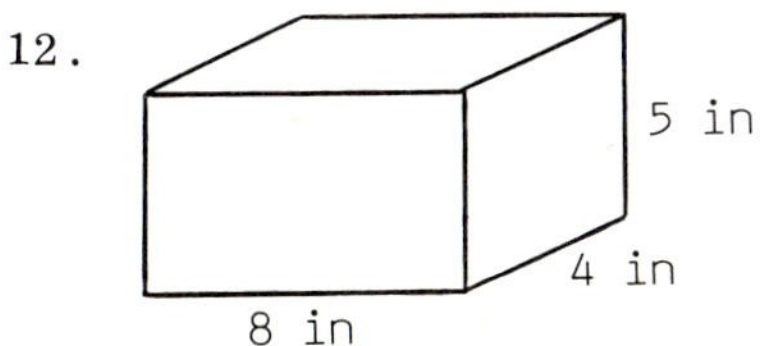

13.

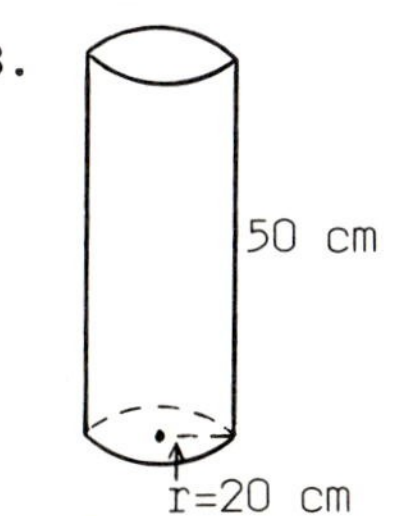

14. Find the width of a rectangle if its perimeter is 30 cm and its length is 8 cm.

15. Find the length of a rectangle if its area is 80 ft^2 and its width is 4 ft.

16. Find the side of a square whose perimeter is 48 m.

17. Find the height of a parallelogram if its area is 100 cm^2 and its base is 20 cm.

18. Find the base of a triangle if its area is 40 yd^2 and its height is 8 yd.

19. Find the area of a right triangle if $1_1 = 16$ in and $1_2 = 8$ in.

20. Find 1_2 in a right triangle if its area is 120 m^2 and 1_1 is 12 m.

21. Find the radius of a circle if its diameter is 14 cm.

22. Find the diameter of a circle if its circumference is 25.12 in.

23. Find the width of a rectangular solid if $V = 100$ ft^3, $L = 5$ ft, and $H = 5$ ft.

24. The perimeter of a rectangle is 210 m. The length is 25 m greater than the width. Find the dimensions.

25. The perimeter of a rectangle is 240 ft. The width is 20 ft less than the length. Find the dimensions.

26. The second angle of a triangle is three times as large as the first. The third angle is 60° larger than the first. How large are the angles?

27. The second angle of a triangle is twice as large as the first. The third angle is 12° less the sum of the other two. How large are the angles?

Polynomials

5

In this chapter, we will extend the definition of powers to include a variable as the base and then use the laws of exponents with powers of that type. We will define polynomials and show the procedure for adding, subtracting, multiplying, and dividing polynomials. Special products of binomials are discussed.

5-1 EXPONENTS

In earlier chapters, we defined powers with a numerical base. In this section, we will extend the definitions to powers with a variable as the base.

1, In 5^3, 5 is the base and ^{3} is the exponent. The expression is a short way of writing a multiplication of identical factors. That is:

$5^3 = (5)(5)(5)$ (The exponent tells us to multiply three 5's.)

The base can also be a variable. When it is, the expression also stands for a multiplication of identical factors. That is:

$x^2 = (x)(x)$ (The exponent tells us to multiply two x's.)

$y^4 = (y)(y)(y)(y)$ (The exponent tells us to multiply four y's.)

Write each expression as a multiplication of identical factors.

a) t^3 = __________ b) d^6 = __________

---

a) $(t)(t)(t)$

b) $(d)(d)(d)(d)(d)(d)$

2. Any multiplication of identical factors can be written in exponential form. For example:

$(5)(5)(5) = 5^3$ (Since there are three 5's, the exponent is 3.)

$(x)(x)(x)(x) = x^4$ (Since there are four x's, the exponent is 4.)

Write each multiplication in exponential form.

a) $(y)(y) =$ ______ b) $(m)(m)(m)(m)(m) =$ ______

---

3. Any exponential expression is called a power of the base. For example:

Answers to 2: a) y^2 b) m^5

5^2, 5^3, and 5^6 are called powers of 5.

y^3, y^4, and y^7 are called powers of y.

The following language is used to distinguish powers of the same base.

x^2 is called x to the second power or x squared

x^3 is called x to the third power or x cubed

x^5 is called x to the fifth power

Similarly: a) t^4 is called t to the ________ power.

b) m^{10} is called m to the ________ power.

---

4. Any power whose exponent is "1" is equal to its base. That is:

Answers to 3: a) fourth b) tenth

$5^1 = 5$ $x^1 = x$

Any power whose exponent is 0 is equal to "1". That is:

$5^0 = 1$ $x^0 = 1$

Using the definitions above, complete these:

a) $3^1 =$ _____ b) $y^0 =$ _____ c) $9^0 =$ _____ d) $m^1 =$ _____.

---

5. Any number or variable can be written as a power whose exponent is "1" For example:

Answers to 4: a) 3 b) 1 c) 1 d) m

$2 = 2^1$ $t = t^1$ a) $8 =$ ______ b) $y =$ ______

---

6. The definition of any power with a negative exponent is given below.

Answers to 5: a) 8^1 b) y^1

$$a^{-n} = \frac{1}{a^n}$$

Therefore: $9^{-2} = \frac{1}{9^2}$ $x^{-3} = \frac{1}{x^3}$ $y^{-7} =$ ______

---

7. By reversing the definition, we can convert each fraction below to a power with a negative exponent.

Answer to 6: $\frac{1}{y^7}$

$\frac{1}{2^4} = 2^{-4}$ $\frac{1}{y^9} = y^{-9}$ a) $\frac{1}{7^5} =$ ______ b) $\frac{1}{x^{10}} =$ ______

Answers to 7: a) 7^{-5} b) x^{-10}

8. We converted each power below to a fraction.

$$5^{-1} = \frac{1}{5^1} = \frac{1}{5} \qquad x^{-1} = \frac{1}{x^1} = \frac{1}{x}$$

We converted each fraction below to a power.

$$\frac{1}{9} = \frac{1}{9^1} = 9^{-1} \qquad \frac{1}{y} = \frac{1}{y^1} = y^{-1}$$

Convert each power to a fraction and each fraction to a power.

a) 2^{-1} = ______ b) $\frac{1}{4}$ = ______ c) $\frac{1}{t}$ = ______ d) v^{-1} = ______

---

9. The word names for some powers with negative exponents are given below.

5^{-2} is called "5 to the negative two power" or "5 to the negative two".

y^{-5} is called "y to the negative five power" or "y to the negative five".

Write the power corresponding to each word name.

a) 3 to the negative four power = ______ b) t to the negative one = ______

Answers (8): a) $\frac{1}{2}$ b) 4^{-1} c) t^{-1} d) $\frac{1}{v}$

Answers (9): a) 3^{-4} b) t^{-1}

5-2 LAWS OF EXPONENTS

In an earlier chapter, we discussed the laws of exponents for powers with a numerical base. In this section, we will extend the use of those laws to powers whose base is a variable.

---

10. The following law of exponents is used to multiply powers with the same base. Two examples are shown:

$$\boxed{a^m \cdot a^n = a^{m+n}}$$

$$x^2 \cdot x^5 = x^{2+5} = x^7 \qquad y^{-5} \cdot y^3 = y^{-5+3} = y^{-2}$$

Using the above law, complete these:

a) $m^9 \cdot m^{-6}$ = ______ b) $t^{-1} \cdot t^{-2}$ = ______

---

11. To perform the multiplication below, we substituted x^1 for x.

$$x \cdot x^{-3} = x^1 \cdot x^{-3} = x^{-2}$$

Complete these:

a) $y^4 \cdot y$ = ______ b) $t^{-8} \cdot t$ = ______ c) $d \cdot d$ = ______

Answers (10): a) m^3 b) t^{-3}

Answers (11): a) y^5 b) t^{-7} c) d^2

12. If the product is a base to the first power, the product equals the base.

$$x^{-2} \cdot x^{3} = x^{1} = x$$

If the product is a base to the zero power, the product is "1".

$$t^{5} \cdot t^{-5} = t^{0} = 1$$

Complete these:

a) $m^{7} \cdot m^{-6} =$ ______ b) $R^{-9} \cdot R^{9} =$ ______

Answers: a) m b) 1

13. The same law is used for multiplications of more than two powers of the same base. That is:

$$y^{4} \cdot y^{-7} \cdot y^{9} \cdot y^{-1} = y^{4+(-7)+9+(-1)} = y^{5}$$

Complete these:

a) $x^{3} \cdot x^{7} \cdot x^{-4} =$ ______ b) $m^{-9} \cdot m \cdot m^{6} \cdot m^{-1} =$ ______

Answers: a) x^{6} b) m^{-3}

14. The law of exponents for multiplication applies only if the powers have the same base. It does not apply to the multiplications below.

$$x^{3} \cdot y^{4} \qquad b \cdot a^{-2} \cdot c^{5}$$

Four multiplications like those above, we usually write the powers side by side in alphabetical order. That is:

$$x^{3} \cdot y^{4} = x^{3}y^{4} \qquad b \cdot a^{-2} \cdot c^{5} = a^{-2}bc^{5}$$

The multiplications below contain expressions like those above. Notice how we used the law of exponents twice in each.

$$(x^{2}y^{3})(x^{4}y^{-1}) = x^{2+4} \cdot y^{3+(-1)} = x^{6}y^{2}$$

$$(ab^{-5})(a^{-3}b) = a^{1+(-3)} \cdot b^{-5+1} = a^{-2}b^{-4}$$

Following the examples, complete these:

a) $(m^{-4}p^{2})(m^{7}p^{-6}) =$ ______ b) $(c^{4}d)(cd^{-3}) =$ ______

Answers: a) $m^{3}p^{-4}$ b) $c^{5}d^{-2}$

15. We used the law of exponents three times for the multiplication below.

$$(xy^{-3}z^{5})(x^{4}y^{2}z) = x^{1+4} \cdot y^{-3+2} \cdot z^{5+1} = x^{5}y^{-1}z^{6}$$

Following the example, do these:

a) $(m^{4}pq^{-7})(mp^{-3}q^{-1}) =$ ____________

b) $(a^{2}b^{-1}c)(a^{2}bc^{-3})(a^{-1}b^{-6}c^{7}) =$ ____________

Answers: a) $m^{5}p^{-2}q^{-8}$ b) $a^{3}b^{-6}c^{5}$

16. The following law of exponents is used to divide powers with the same base. Two examples are shown.

$$\frac{a^{m}}{a^{n}} = a^{m-n}$$

$$\frac{x^{5}}{x^{2}} = x^{5-2} = x^{3} \qquad \frac{y^{3}}{y^{7}} = y^{3-7} = y^{-4}$$

Continued on following page.

16. Continued

Using the previous law, complete these:

a) $\frac{t^{10}}{t^4}$ = ______ b) $\frac{m^2}{m^9}$ = ______

---

17. The same law of exponents is used for the divisions below.

$$\frac{x^{-3}}{x^5} = x^{-3-5} = x^{-3+(-5)} = x^{-8}$$

$$\frac{y^{-4}}{y^{-7}} = y^{-4-(-7)} = y^{-4+7} = y^3$$

Following the examples, complete these:

a) $\frac{d^3}{d^{-1}}$ = ______ b) $\frac{t^{-8}}{t^{-3}}$ = ______

Answers to 16: a) t^6 b) m^{-7}

---

18. To perform the divisions below, we substituted x^1 for x and y^1 for y.

$$\frac{x}{x^3} = \frac{x^1}{x^3} = x^{1-3} = x^{-2}$$

$$\frac{y^{-5}}{y} = \frac{y^{-5}}{y^1} = y^{-5-1} = y^{-6}$$

Following the examples, complete these:

a) $\frac{d}{d^{-2}}$ = ______ b) $\frac{R^{-1}}{R}$ = ______

Answers to 17: a) d^4 b) t^{-5}

---

19. To simplify the expression below, we used the law of exponents for division twice.

$$\frac{x^3y^4}{x^5y} = \left(\frac{x^3}{x^5}\right)\left(\frac{y^4}{y}\right) = x^{-2}y^3$$

Following the example, simplify these:

a) $\frac{a^9b^8}{ab^6}$ = ______ b) $\frac{p^6q^2}{p^7q}$ = ______

Answers to 18: a) d^3 b) R^{-2}

---

20. To simplify the expression below, we used the law of exponents for division three times.

$$\frac{a^4b^6c^3}{a^3bc^8} = \left(\frac{a^4}{a^3}\right)\left(\frac{b^6}{b}\right)\left(\frac{c^3}{c^8}\right) = ab^5c^{-5}$$

Following the example, simplify these:

a) $\frac{x^9y^5z^8}{xy^4z^5}$ = ______ b) $\frac{m^2p^{10}q^3}{mp^5q^5}$ = ______

Answers to 19: a) a^8b^2 b) $p^{-1}q$

Answers to 20: a) x^8yz^3 b) mp^5q^{-2}

21. The following law of exponents is used to raise a power to a power. Two examples are shown.

$$(a^m)^n = a^{mn}$$

$(x^4)^3 = x^{(4)(3)} = x^{12}$ $\qquad$ $(y^{-2})^5 = y^{(-2)(5)} = y^{-10}$

Using the above law, complete these:

a) $(m^2)^6 =$ ______ $\quad$ b) $(t^{-1})^4 =$ ______ $\quad$ c) $(d^{-6})^7 =$ ______

Answers: a) m^{12} b) t^{-4} c) d^{-42}

22. The same law of exponents is used to raise a power to a negative power. For example:

$(a^4)^{-2} = a^{(4)(-2)} = a^{-8}$ $\qquad$ $(c^{-5})^{-6} = c^{(-5)(-6)} = c^{30}$

Following the examples, complete these:

a) $(x^4)^{-1} =$ ______ $\quad$ b) $(y^{-1})^{-3} =$ ______ $\quad$ c) $(p^{-7})^{-5} =$ ______

Answers: a) x^{-4} b) y^3 c) p^{35}

23. To raise the expression below to a power, we converted to a multiplication.

$$(x^4y^5)^3 = (x^4y^5)(x^4y^5)(x^4y^5) = x^{12}y^{15}$$

However, it is simpler to multiply each exponent by 3. That is:

$$(x^4y^5)^3 = x^{(4)(3)}y^{(5)(3)} = x^{12}y^{15}$$

Based on the above example, we get the following law of exponents.

$$(a^mb^n)^p = a^{mp}b^{np}$$

Using the law, complete these:

a) $(c^2d^4)^5 =$ ______ $\quad$ b) $(b^{-1}t^{-3})^2 =$ ______

Answers: a) $c^{10}d^{20}$ b) $b^{-2}t^{-6}$

24. Another example of the same law is shown below.

$$(x^4y^{-3})^{-2} = x^{(4)(-2)}y^{(-3)(-2)} = x^{-8}y^6$$

Following the example, do these:

a) $(P^{-2}R^9)^{-3} =$ ______ $\quad$ b) $(c^{-5}t^{-1})^{-1} =$ ______

Answers: a) P^6R^{-27} b) c^5t

25. Notice how we substituted x^1 for <u>x</u> below.

$$(xy^{-2})^4 = (x^1y^{-2})^4 = x^4y^{-8}$$

Following the example, do these:

a) $(p^{-1}q)^8 =$ ______ $\quad$ b) $(ab^4)^{-6} =$ ______

Answers: a) $p^{-8}q^8$ b) $a^{-6}b^{-24}$

26. The four basic laws of exponents are restated below.

$$a^m \cdot a^n = a^{m+n}$$

$$\frac{a^m}{a^n} = a^{m-n}$$

$$(a^m)^n = a^{mn}$$

$$(a^m b^n)^p = a^{mp} b^{np}$$

Using the laws, complete these:

a) $x^5 \cdot x^2 =$ ______

b) $\frac{x^5}{x^2} =$ ______

c) $(x^5)^2 =$ ______

d) $y^{-2} \cdot y^3 =$ ______

e) $\frac{y^{-2}}{y^3} =$ ______

f) $(y^{-2})^3 =$ ______

Answers: a) x^7 b) x^3 c) x^{10} d) y e) y^{-5} f) y^{-6}

27. Using the laws of exponents, complete these:

a) $(xy^2)(x^2y) =$ ______

b) $\frac{s^9t^5}{s^2t^4} =$ ______

c) $(cd^2)^4 =$ ______

d) $(a^{-1}b^5)(a^{-2}b^{-3}) =$ ______

e) $\frac{p^3q^6}{p^7q} =$ ______

f) $(m^{-1}t^3)^{-4} =$ ______

Answers: a) x^3y^3 b) s^7t c) c^4d^8 d) $a^{-3}b^2$ e) $p^{-4}q^5$ f) m^4t^{-12}

5-3 POLYNOMIALS IN ONE VARIABLE

In this section, we will discuss polynomials in one variable and related terminology.

28. Each term below has the form ax^n. That is, each contains one power with a numerical coefficient.

$3x^4$ $-2y^5$ $\frac{1}{2}t^3$ $-8.5d^2$

Each term below can be written in the form ax^n by writing the coefficient, the exponent, or both explicitly. That is:

$x^5 = 1x^5$ $4x = 4x^1$ $d = 1d^1$

$-y^3 = -1y^3$ $-7t = -7t^1$ $-m = -1m^1$

Write each of these in the form ax^n.

a) $p^2 =$ ______ b) $-6V =$ ______ c) $-x =$ ______

Answers: a) $1p^2$ b) $-6V^1$ c) $-1x^1$

29. Numbers like 3 and -5 can be written in the form ax^n. To do so, we use the zero power of some variable. For example: 3 can be written $3x^0$, since $x^0 = 1$ -5 can be written $-5y^0$, since $y^0 =$ ______	
30. Each term in the expressions below can be written explicitly in the form ax^n. That is: $2x^3 - x^2 + x - 1$ can be written $2x^3 - 1x^2 + 1x^1 - 1x^0$ $y^5 - 8y^3 - y + 5$ can be written $1y^5 - 8y^3 - 1y^1 + 5y^0$ However, when writing expressions like those above, we do not usually write "1" and -1 coefficients, "1" exponents, and zero powers. That is: $1t^3 - 1t^2 + 8t^1 - 3t^0$ is usually written $t^3 - t^2 + 8t - 3$ $-2b^5 + 1b^4 - 1b^1 + 9b^0$ is usually written ____________	1
31. A <u>polynomial in one variable</u> is an expression containing only terms of the form ax^n, where the numerical coefficient <u>a</u> can be any number and the exponent <u>n</u> is either a positive integer or 0. Some examples are shown. $3x$ $y + 1.7$ $2a^2 - 5a + 6$ $t^4 - 4t^3 + \frac{1}{2}t^2 - 1$ A polynomial can contain one or more terms. There are special names for polynomials containing one, two, or three terms. A <u>one-term</u> polynomial is called a <u>monomial</u>. Some examples are: -5 x $\frac{1}{3}y$ $-t^2$ $7d^4$ A <u>two-term</u> polynomial is called a <u>binomial</u>. Some examples are: $2x + 9$ $y^4 - 1.5y$ $2t^5 + 3t^3$ A <u>three-term</u> polynomial is called a <u>trinomial</u>. Some examples are: $x^2 + 8x - 7$ $4p^4 - p^3 + 8p^2$ State whether each expression is a monomial, binomial, or trinomial. a) 8 ________ d) $5m^6 - 6m^5$ ________ b) $x^4 - 2.5$ ________ e) $-\frac{1}{2}p$ ________ c) $2y^3 + y + 7$ ________ f) $a^4 - 3a^2 + 1$ ________	$-2b^5 + b^4 - b + 9$
32. There are no special names for polynomials with more than three terms. $2x^3 - x^2 + 4x - 6$ is called a polynomial with <u>four</u> terms. $x^4 - 5x^3 + 2x^2 - x + 9$ is called a polynomial with ________ terms.	a) monomial b) binomial c) trinomial d) binomial e) monomial f) trinomial
	five

33. Polynomials in one variable are usually written in descending order. That is, they are written so that the powers are in order, with the largest exponent at the left. For example:

$$3x^4 + 5x^3 + 6x^2 + 2x + 9$$

Let's put each polynomial below in descending order.

$2 + 5y^4 + 3y^2 = 5y^4 + 3y^2 + 2$

a) $8b + b^2 + 1 + 4b^3 =$ ______________

b) $5y^3 + y - 6y^4 + 2y^2 =$ ______________

---

34. The degree of a term in a polynomial is its exponent. As an example, let's identify the degree of each term of $5x^3 + 7x + 2$.

The degree of $5x^3$ is 3.

The degree of $7x$ is 1, since $7x = 7x^1$.

The degree of 2 is 0, since $2 = 2x^0$.

Identify the degree of each term of $2y^5 - 7y^2 - 4y + 6$.

a) $2y^5$ ______ b) $-7y^2$ ______ c) $-4y$ ______ d) 6 ______

Answers to 33: a) $4b^3 + b^2 + 8b + 1$ b) $-6y^4 + 5y^3 + 2y^2 + y$

---

35. The degree of a polynomial is the largest exponent in the polynomial. For example:

The degree of $5x^3 - 2x^2 + 4$ is 3.

a) The degree of $7a^5 - a^3 + 4a$ is ______.

b) The degree of $-2y^7 + 4y^6 - 5y^3 + 2y$ is ______.

Answers to 34: a) 5 b) 2 c) 1 d) 0

---

36. We identified the numerical coefficient of each term of the polynomial below.

$$5x^4 - 3x^2 + x - 4$$

The coefficient of the first term is 5.

The coefficient of the second term is -3.

The coefficient of the third term is 1. $(x = 1x^1)$

The coefficient of the fourth term is -4. $(-4 = -4x^0)$

Identify the coefficient of each term in the polynomial below.

$$6y^5 - y^3 - 7y + 1$$

a) first term ______ c) third term ______

b) second term ______ d) fourth term ______

Answers to 35: a) 5 b) 7

Answers to 36: a) 6 b) -1 c) -7 d) 1

37. In the polynomial below, there is no term with x^3. We say that we have a missing term.

$$3x^4 + x^2 - 5x - 1$$

When there is a missing term, we can write it with a 0 coefficient or leave a space. For example:

$$3x^4 + 0x^3 + x^2 - 5x - 1$$

$$3x^4 + \qquad x^2 - 5x - 1$$

But we usually do not write missing terms or leave a space. That is:

Instead of $y^2 + 0y + 5$, we write $y^2 + 5$.

Instead of $2t^3 - \quad 4t + 9$, we write ______________.

Answer: $2t^3 - 4t + 9$

5-4 POLYNOMIALS IN MORE THAN ONE VARIABLE

In this section, we will discuss polynomials in more than one variable and related terminology.

38. The terms below are of the form ax^ny^m or $ax^ny^mz^p$. That is, each contains a numerical coefficient with two or more powers.

$5x^2y^3$ $\qquad$ $-7a^6b^5c^4$

Each term below can be written in the form ax^ny^m or $ax^ny^mz^p$ by writing the coefficients or exponents explicitly.

$c^8d^2 = 1c^8d^2$ $\qquad$ $4ab^3 = 4a^1b^3$

$-x^5y^2z^4 = -1x^5y^2z^4$ $\qquad$ $-9cx^2y = 9c^1x^2y^1$

Write each of these in the form ax^ny^m or $ax^ny^mz^p$.

a) p^7q = __________ $\qquad$ b) $-a^4mn$ = __________

Answers: a) $1p^7q^1$ b) $-1a^4m^1n^1$

39. Numbers like 2 and -7 can be written in the form ax^ny^m or $ax^ny^mz^p$ by using some powers with 0 exponents. That is:

$2 = 2x^0y^0$, since x^0 and $y^0 = 1$

$-7 = -7a^0b^0c^0$, since a^0, b^0, and c^0 = ______

Answer: 1

40. A polynomial in more than one variable is an expression containing only terms of the form ax^ny^m or $ax^ny^mz^p$, where a can be any number and the exponents are positive integers or 0. Some examples are shown.

$x^4y^2 - 5x^2y^3 + 1$ $\qquad$ $5pq$ $\qquad$ $a^4b^2c + abc^3$

State whether each polynomial is a monomial (one term), a binomial (two terms) or a trinomial (three terms).

a) $p^4q^3 - p^2q$ ______________ $\qquad$ c) $x^2y^7z^5 - xyz^8$ ______________

b) $7a^4b^3c$ ______________ $\qquad$ d) $m^4t - m^2 - 6$ ______________

41. Polynomials in more than one variable are usually written in descending order for one of the variables.

The polynomial below is written in descending order for x.

$$x^4y - 2x^2y^3 + 1$$

The polynomial below is written in descending order for b.

$$ab^2c^5 + a^4bc - a^2$$

Write the polynomial below in descending order for m.

$m^2t^5 - 3 - 5mt^4 + m^3t$ = ____________________

a) binomial
b) monomial
c) binomial
d) trinomial

---

42. The degree of a term in a polynomial in more than one variable is the sum of the exponents of the variables. Some examples are shown.

Term	Degree	
$2a^3b^2$	5	(from 3 + 2)
$m^4p^7q^3$	14	(from 4 + 7 + 3)
xy	2	(from 1 + 1)
$-4ab^4c$	6	(from 1 + 4 + 1)

Write the degree of each term.

a) $4d^5p^2$ ______ b) $-x^2yz^2$ ______ c) pqr^8 ______

$m^3t + m^2t^5 - 5mt^4 - 3$

---

43. The degree of a polynomial is the degree of the term with the highest degree.

The degree of $4x^2y + 3xy^3 - 7xy$ is 4,

since the degree of $3xy^3$ is 4.

Similarly: a) The degree of $a^4b^3 - 3a^3b^2 + 4ab$ is ______.

b) The degree of $x^5yz + x^4y^3z^2 - xyz^3$ is ______.

a) 7
b) 5
c) 10

---

44. Identify the numerical coefficient of each term in the polynomial below.

$$-3x^3y^2 + 5xy^3z^4 - xy + y^2 - 7x + 5$$

a) first term ______ d) fourth term ______
b) second term ______ e) fifth term ______
c) third term ______ f) sixth term ______

a) 7
b) 9

---

a) -3 d) 1
b) 5 e) -7
c) -1 f) 5

5-5 COMBINING LIKE TERMS

In this section, we will define like terms and show how they can be combined by addition and subtraction.

---

45. Two monomials with one power are like terms if they contain the same variable to the same power. The pairs below are like terms.

$5y^3$ and $4y^3$ $\qquad$ $8t^5$ and $-6t^5$

Two monomials with one power are unlike terms if they contain the same variable to different powers or different variables. The pairs below are unlike terms.

$2x^2$ and $5x^3$ $\qquad$ $-3t^5$ and $7y^5$ $\qquad$ $4x$ and $6y^2$

Which of the following are pairs of like terms? ____________

a) $5m^4$ and $2m^4$ $\qquad$ c) $-4b^2$ and $-2b^2$

b) $-3p^5$ and $4p^2$ $\qquad$ d) $7y$ and $-4y^3$

---

46. Two monomials with more than one power are like terms if they contain exactly the same variables to the same powers. Otherwise, they are unlike terms. Some examples are shown.

Only (a) and (c)

Like Terms	Unlike Terms
$3x^2y^3$ and $5x^2y^3$	$2p^4q^3$ and $5p^2q^3$
$2ab^4c^5$ and $-7ab^4c^5$	$-4c^2dt^5$ and $9c^2t^5$

Which of the following pairs are like terms? ____________

a) $8xy^2$ and $2x^2y$ $\qquad$ c) $3abc^3$ and $-9abc^3$

b) $-3m^4t^6$ and $4m^4t^6$ $\qquad$ d) $5pq^2r^7$ and $7pr^7$

---

47. When the terms in an addition are like terms, we can factor by the distributive principle to combine them. Doing so is the same as adding their numerical coefficients. That is:

Only (b) and (c)

$$5y^3 + 2y^3 = (5 + 2)y^3 = 7y^3$$

$$-ab^4 + 7ab^4 = (-1 + 7)ab^4 = 6ab^4$$

Combine by adding the numerical coefficients.

a) $4t^2 + (-7t^2) =$ ________ $\qquad$ b) $x^3y + x^3y =$ ________

---

a) $-3t^2$ $\qquad$ b) $2x^3y$

48. When the terms in a subtraction are like terms, we can also factor by the distributive principle to combine them. Doing so is the same as subtracting their numerical coefficients. That is:

$$3a^4b - 5a^4b = (3 - 5)a^4b = -2a^4b$$

$$4x^7 - x^7 = (4 - 1)x^7 = 3x^7$$

Combine by subtracting numerical coefficients.

a) $7y^5 - 6y^5 =$ ________ b) $mp^3t^2 - 9mp^3t^2 =$ ________

Answers to 48: a) $1y^5 = y^5$ b) $-8mp^3t^2$

49. When the numerical coefficients in a subtraction are identical, we get 0 as the difference. That is:

$$3y^2 - 3y^2 = (3 - 3)y^2 = 0y^2 = 0$$

$$a^8b^7 - a^8b^7 = (1 - 1)a^8b^7 = 0a^8b^7 = 0$$

Do each subtraction.

a) $10mt^7 - 10mt^7 =$ ______ b) $x^8 - x^8 =$ ______

Answers to 49: a) 0 b) 0

50. When the terms in an addition or subtraction are unlike, they cannot be combined. For example, neither expression below can be simplified by combining terms.

$$2x^4 + 3y^7 \qquad 4b^3c^2 - bc$$

If possible, simplify each expression by combining like terms.

a) $3xy + y =$ ________ d) $2p^2 - x^2 =$ ________

b) $p^4 - 8p^4 =$ ________ e) $a^4b^3c - a^4b^3 =$ ________

c) $6d^2t + d^2t =$ ________ f) $mpq^3 - mpq^3 =$ ________

Answers to 50: a) Not possible b) $-7p^4$ c) $7d^2t$ d) Not possible e) Not possible f) 0

51. In the polynomial below, there are two pairs of like terms: ($4y^3$ and y^3) and ($2y$ and $6y$). We simplified the polynomial by combining like terms.

$$4y^3 + 2y + y^3 + 6y = 5y^3 + 8y$$

Following the example, simplify these:

a) $x^4 + 7x^2 + 2x^4 + 5x^2 =$

b) $4a^3b^2 + 5 + a^3b^2 - 1 =$

Answers to 51: a) $3x^4 + 12x^2$ b) $5a^3b^2 + 4$

52. We simplified the polynomial below by combining like terms.

$$3x^2 - 5x - x^2 - 2x = 2x^2 - 7x$$

Following the example, simplify these polynomials.

a) $5t^4 + 3t^2 - 2t^4 - t^2 =$

b) $6a^5b^4 - ab^2 + a^5b^4 - ab^2 =$

53. When combining like terms to simplify a polynomial, we do not ordinarily write "1" and -1 coefficients. Simplify these. a) $2t^2 + t - t^2 - 2t =$ b) $5x^3y^2 - 7x^2y^3 - 4x^3y^2 + 6x^2y^3 =$	a) $3t^4 + 2t^2$ b) $7a^5b^4 - 2ab^2$
54. When combining like terms below, we got $0x^2$ or 0. Notice that the 0 is not written in the final simplification. $3x^2 + x - 3x^2 - 5x = 0x^2 - 4x = 0 - 4x = -4x$ Simplify these polynomials. a) $5t^3 + 7 - 5t^3 - 2 =$ b) $p^2q + pq^3 + p^2q - pq^3 =$	a) $t^2 - t$ b) $x^3y^2 - x^2y^3$
55. There are three pairs of like terms in the polynomial below. Notice how we simplified it. $3x^2 - 5x + 1 + 2x^2 + 5x - 7 = 5x^2 + 0x - 6 = 5x^2 - 6$ Simplify each polynomial. a) $y^2 + 2y - 9 + y^2 - y - 3 =$ b) $at^4 - a^2t^2 + 3 - at^4 + 4a^2t^2 - 3 =$	a) 5 b) $2p^2q$
56. There are only two pairs of like terms in the polynomial below. Notice how we wrote the terms in the simplified polynomial in descending order. $5a^3 + 2a^2 - a + a^2 - 3a + 8 = 5a^3 + 3a^2 - 4a + 8$ Simplify this polynomial. Write the terms in descending order. $3t^5 + t^4 - 1 - t^5 - t^4 + 4t =$	a) $2y^2 + y - 12$ b) $3a^2t^2$
57. We simplified the polynomial below and wrote the terms in descending order for $\underline{x}$. $3xy^2 - 9 + 4x^2y - xy^2 - 1 = 4x^2y + 2xy^2 - 10$ Simplify this polynomial. Write the terms in descending order for $\underline{b}$. $bc - 5b^3c^2 + 7b^2 + 2bc - b^3c^2 =$	$2t^5 + 4t - 1$
	$-6b^3c^2 + 7b^2 + 3bc$

SELF-TEST 15 (pages 199-213)

Use the laws of exponents for these.

1. $(y^5)(y)(y^{-3}) =$ ______
2. $(abc)(ab^2c)(a^{-3}bc) =$ ______
3. $\frac{p^6q^3}{p^2q^8} =$ ______
4. $\frac{m^4s^2t^3}{m^3s^2t^{-1}} =$ ______
5. $(x^3y^{-2})^3 =$ ______
6. $(kt^4)^{-2} =$ ______

Write each polynomial in descending order for x.

7. $3x - x^3 + 7 - 2x^2$

______

8. $d^3x + 2d - 4d^2x^3 + dx^5$

______

State the degree of each term.

9. $4xy$ ______
10. $-5r^2st^3$ ______

State whether each expression is a monomial, binomial, or trinomial.

11. $2y - 1$ ______
12. $3ab^2 + a^2b + 5$ ______

Simplify by combining like terms.

13. $7x^3 + 6x + x^3 - 3x =$ ______
14. $a^2b + ab^4 + a^2b - ab^4 =$ ______

Simplify and write in descending order for x.

15. $3x^2 + x - x^3 - 4x + 2x^3 - 1 =$ ______
16. $xy - 4x^3y^2 + 7x^2 + 3xy - x^3y^2 =$ ______

ANSWERS:

1. y^3
2. $a^{-1}b^4c^3$
3. p^4q^{-5}
4. mt^4
5. x^9y^{-6}
6. $k^{-2}t^{-8}$
7. $-x^3 - 2x^2 + 3x + 7$
8. $dx^5 - 4d^2x^3 + d^3x + 2d$
9. Degree: 2
10. Degree: 6
11. Binomial
12. Trinomial
13. $8x^3 + 3x$
14. $2a^2b$
15. $x^3 + 3x^2 - 3x - 1$
16. $-5x^3y^2 + 7x^2 + 4xy$

5-6 ADDITION OF POLYNOMIALS

In this section, we will discuss the procedure for adding polynomials both horizontally and vertically.

58. When writing an addition of polynomials horizontally, the polynomials are written in parentheses. For example: The addition of $2x^2 + 5$ and $3x^2 - 1$ is written $(2x^2 + 5) + (3x^2 - 1)$. To perform the addition, we simply drop the parentheses and combine like terms. That is: $(2x^2 + 5) + (3x^2 - 1) = 2x^2 + 5 + 3x^2 - 1 = 5x^2 + 4$ Following the example, do the following addition. $(ay^2 - 2by) + (2ay^2 - 3by) =$	
59. You should be able to perform an addition without rewriting the addition with the parentheses dropped. That is: $(y^4 - 2y^2 + 1) + (y^4 + 3y^2 - 9) = 2y^4 + y^2 - 8$ Following the example, complete this addition. $(4ax^2 - bx - 1) + (ax^2 + bx - 3) =$	$3ay^2 - 5by$
60. We added three polynomials below. Complete the other addition. $(x^2 + 5) + (3x^2 - 7x + 2) + (2x - 10) = 4x^2 - 5x - 3$ $(by^3 - 2cy^2 - y) + (2by^3 - cy^2) + (2cy^2 + y) =$	$5ax^2 - 4$
61. Do each addition. Write each sum in descending order. a) $(y^3 - 2) + (3y^4 - y^2 - 1) + (y^4 + 5) =$ b) $(4t^2 - 3t + 1) + (t^3 - 1) + (t^3 - 4t^2) =$	$3by^3 - cy^2$
62. Do this addition. Write the sum in descending order for $\underline{x}$. $(ax^2 - 3x) + (bx^3 + ax^2) + (bx^3 + 3x) =$	a) $4y^4 + y^3 - y^2 + 2$ b) $2t^3 - 3t$
63. To add polynomials vertically, we line up like terms in columns and then find the sum of the columns. An example is shown. Complete the other addition. Add $2x^2 - 1$ and $5x^2 - 3$. Add $y^3 + 4y$ and $-2y^3 - y$. $\begin{array}{r} 2x^2 - 1 \\ \underline{5x^2 - 3} \\ 7x^2 - 4 \end{array}$	$2bx^3 + 2ax^2$
	$\begin{array}{r} y^3 + 4y \\ \underline{-2y^3 - y} \\ -y^3 + 3y \end{array}$

64. Notice how we left space for the missing terms below. Complete the other addition.

Add $2x^2 + 5x - 1$, $3x + 5$, and $x^3 - 4x^2 - 7$.

$$\begin{array}{r} 2x^2 + 5x - 1 \\ 3x + 5 \\ x^3 - 4x^2 \quad\quad - 7 \\ \hline x^3 - 2x^2 + 8x - 3 \end{array}$$

Add $y + 6$, $y^2 - 6y + 1$, and $y^3 + 5y - 5$.

$$\begin{array}{r} y + 6 \\ y^2 - 6y + 1 \\ y^3 \quad\quad + 5y - 5 \\ \hline y^3 + y^2 \quad\quad + 2 \end{array}$$

65. Following the example, complete the other addition.

Add $bc + 1$, $ac^2 - 2bc - 1$, and $2ac^2 + bc - 3$.

$$\begin{array}{r} bc + 1 \\ ac^2 - 2bc - 1 \\ 2ac^2 + bc - 3 \\ \hline 3ac^2 \quad\quad - 3 \end{array}$$

Add $ax^4 - 7$, $ax^4 + 3bx^2 - 1$, and $-2ax^4 - bx^2 + 6$.

$$\begin{array}{r} ax^4 \quad\quad - 7 \\ ax^4 + 3bx^2 - 1 \\ -2ax^4 - bx^2 + 6 \\ \hline 2bx^2 - 2 \end{array}$$

5-7 SUBTRACTION OF POLYNOMIALS

We will begin this section by defining the opposites of polynomials. Then we will discuss the procedure for subtracting polynomials both horizontally and vertically.

---

66. Two monomials are a pair of opposites if their sum is 0. For example:

Since $3x^2 + (-3x^2) = 0$: the opposite of $3x^2$ is $-3x^2$.
the opposite of $-3x^2$ is $3x^2$.

Since $ay^3 + (-ay^3) = 0$: the opposite of ay^3 is $-ay^3$.
the opposite of $-ay^3$ is ay^3.

As you can see, we can get the opposite of a monomial by changing the sign of its numerical coefficient. Write the opposite of each monomial.

a) $9b^7$ ________ b) $-4cx^4$ ________ c) p^2qr^3 ________

---

a) $-9b^7$ b) $4cx^4$ c) $-p^2qr^3$

67. Two polynomials are a pair of opposites if each pair of like terms is a pair of opposites. For example:

Since $(2x^2 + 5) + (-2x^2 - 5) = 0x^2 + 0 = 0$:

the opposite of $2x^2 + 5$ is $-2x^2 - 5$.
the opposite of $-2x^2 - 5$ is $2x^2 + 5$.

Since $(y^3 - 2y - 7) + (-y^3 + 2y + 7) = 0y^3 + 0y + 0 = 0$:

the opposite of $y^3 - 2y - 7$ is $-y^3 + 2y + 7$.
the opposite of $-y^3 + 2y + 7$ is $y^3 - 2y - 7$.

As you can see, we can get the opposite of a polynomial by changing the sign of each term. Write the opposite of each polynomial.

a) $3b^5 - 6b^4$ ________ c) $x^3 - 7x + 2$ ________

b) $5cy^2 - 1$ ________ d) $-3ad^4 + 8bd^2 - d$ ________

---

68. To subtract polynomials, we add the opposite of the second polynomial. That is:

Opposite

$(3x^2 - 5) - (x^2 + 4) = (3x^2 - 5) + (-x^2 - 4) = 2x^2 - 9$

$(5y + 7) - (6y - 3) = (5y + 7) + (-6y + 3) = -y + 10$

Following the examples, complete these subtractions.

a) $(y - 1) - (3y - 5) = (y - 1) + (\qquad) =$

b) $(2t^4 + t^2) - (t^4 + t^2) = (2t^4 + t^2) + (\qquad) =$

Answers to 67:
a) $-3b^5 + 6b^4$
b) $-5cy^2 + 1$
c) $-x^3 + 7x - 2$
d) $3ad^4 - 8bd^2 + d$

---

69. To subtract below, we added the opposite of the second polynomial.

Opposite

$(7ax^4 - bx^2) - (ax^4 - 3bx^2) = (7ax^4 - bx^2) + (-ax^4 + 3bx^2) = 6ax^4 + 2bx^2$

Following the example, complete this subtraction.

$(p^2q^3 + 1) - (4p^2q^3 - 3) = (p^2q^3 + 1) + (\qquad) =$

Answers to 68:
a) $+ (-3y + 5) = -2y + 4$
b) $+ (-t^4 - t^2) = t^4$

---

70. Another subtraction of polynomials is shown below. Notice again that we added the opposite of the second polynomial.

Opposite

$(5y^3 - 2y^2 + 1) - (y^3 + 2y^2 + 1) = (5y^3 - 2y^2 + 1) + (-y^3 - 2y^2 - 1)$
$= 4y^3 - 4y^2$

Following the example, complete this subtraction.

$(m^2 - m + 4) - (m^2 - 2m - 1) = (m^2 - m + 4) + (\qquad)$
$=$

Answer to 69: $+(-4p^2q^3+3) = -3p^2q^3+4$

Answer to 70: $+(-m^2+2m+1) = m+5$

71. Do each subtraction. Write each answer in descending order.

a) $(2x^2 - 5x) - (x^4 + 3x) =$

b) $(6y - 1) - (-3y^3 + 6y - 1) =$

---

72. Do this subtraction. Write the answer in descending order for x.

$(4x^3y - 2xy) - (x^2y^2 + xy) =$

Answers to 71: a) $-x^4 + 2x^2 - 8x$ b) $3y^3$

---

73. To subtract polynomials vertically, we line up like terms in columns and then add the opposite of the bottom polynomial. An example is shown. Complete the other subtraction.

Answer to 72: $4x^3y - x^2y^2 - 3xy$

Subtract $2y - 7$ from $5y + 1$.

$$\begin{array}{r} 5y + 1 \\ (-)\ \underline{2y - 7} \end{array} \quad \text{becomes} \quad \begin{array}{r} 5y + 1 \\ \underline{-2y + 7} \\ 3y + 8 \end{array}$$

Subtract $x^3 + x^2$ from $3x^3 - 2x^2$.

$$\begin{array}{r} 3x^3 - 2x^2 \\ (-)\ \underline{x^3 + x^2} \end{array} \quad \text{becomes}$$

---

74. Following the example, complete the other subtraction.

Answer to 73:

$$\begin{array}{r} 3x^3 - 2x^2 \\ \underline{-x^3 - x^2} \\ 2x^3 - 3x^2 \end{array}$$

Subtract $2x^2 - 3x + 5$ from $7x^2 - x - 3$.

$$\begin{array}{r} 7x^2 - x - 3 \\ (-)\ \underline{2x^2 - 3x + 5} \end{array} \quad \text{becomes} \quad \begin{array}{r} 7x^2 - x - 3 \\ \underline{-2x^2 + 3x - 5} \\ 5x^2 + 2x - 8 \end{array}$$

Subtract $t^4 - t^2 - 1$ from $t^4 + t^2 + 1$.

$$\begin{array}{r} t^4 + t^2 + 1 \\ (-)\ \underline{t^4 - t^2 - 1} \end{array} \quad \text{becomes}$$

---

75. Following the example, complete the other subtraction.

Answer to 74:

$$\begin{array}{r} t^4 + t^2 + 1 \\ \underline{-t^4 + t^2 + 1} \\ 2t^2 + 2 \end{array}$$

Subtract $2xy - 5$ from $x^2y + 4xy$.

$$\begin{array}{r} x^2y + 4xy \\ (-)\ \underline{2xy - 5} \end{array} \quad \text{becomes} \quad \begin{array}{r} x^2y + 4xy \\ \underline{-2xy + 5} \\ x^2y + 2xy + 5 \end{array}$$

Subtract $ay^2 + 3$ from $4ay^2 - by + 1$.

$$\begin{array}{r} 4ay^2 - by + 1 \\ (-)\ \underline{ay^2 + 3} \end{array} \quad \text{becomes}$$

Answer to 75:

$$\begin{array}{r} 4ay^2 - by + 1 \\ \underline{-ay^2 - 3} \\ 3ay^2 - by - 2 \end{array}$$

5-8 MULTIPLYING BY A MONOMIAL

In this section, we will discuss the procedure for multiplying and squaring monomials. We will also show how the distributive principle can be used to multiply other polynomials by a monomial.

---

76. To perform the multiplications below, we multiplied the numerical coefficients and used the law of exponents for multiplication.

$$(2x)(7x) = (2)(7)(x)(x) = 14x^2$$

$$(-y)(5y) = (-1)(5)(y)(y) = -5y^2$$

Using the same method, complete these:

a) $(-3d)(8d) =$ ________ b) $(10t)(t) =$ ________

Answers: a) $-24d^2$ b) $10t^2$

77. To perform the multiplications below, we also multiplied the numerical coefficients and used the law of exponents for multiplication.

$$(-5x^2)(3x^4) = (-5)(3)(x^2)(x^4) = -15x^6$$

$$(2ab^2)(8a^4b) = (2)(8)(a)(a^4)(b^2)(b) = 16a^5b^3$$

Using the same method, do these:

a) $(4t^3)(3t^4) =$ ________ b) $(-6x^2y^2)(-5xy^7) =$ ________

Answers: a) $12t^7$ b) $30x^3y^9$

78. We used the law of exponents when possible in the multiplication below.

$$(apq^3)(pq^2r^5) = (a)(p)(p)(q^3)(q^2)(r^5) = ap^2q^5r^5$$

Following the example, complete these:

a) $(9x^3y^4)(7bx) =$ ________________

b) $(-b^2c)(5ab^2)(c^4d) =$ ________________

Answers: a) $63bx^4y^4$ b) $-5ab^4c^5d$

79. To square a monomial, we multiply the monomial by itself. For example:

$$(3y)^2 = (3y)(3y) = 9y^2$$

$$(x^2y^3)^2 = (x^2y^3)(x^2y^3) = x^4y^6$$

Following the examples, do these:

a) $(-5t)^2 = (-5t)(-5t) =$ ____________

b) $(2a^5b^4)^2 = (2a^5b^4)(2a^5b^4) =$ ____________

Answers: a) $25t^2$ b) $4a^{10}b^8$

80. We used the same method to square the monomial below.

$$(4x^3y^4)^2 = (4x^3y^4)(4x^3y^4) = 16x^6y^8$$

As you can see, squaring a monomial is the same as <u>squaring the numerical coefficient and doubling each exponent</u>. Using the shorter method, complete these:

a) $(4d)^2$ = ________ c) $(3x^2y)^2$ = ________

b) $(-6v^3)^2$ = ________ d) $(c^3d^6f^9)^2$ = ________

Answers: a) $16d^2$ b) $36v^6$ c) $9x^4y^2$ d) $c^6d^{12}f^{18}$

---

81. To multiply a binomial by a monomial, we use the distributive principle. Each term in the binomial is multiplied by the monomial. For example:

$$3(x + 5) = 3(x) + 3(5) = 3x + 15$$

$$7(4y - 1) = 7(4y) - 7(1) = 28y - 7$$

The distributive principle is also used when the monomial is a variable. For example:

$$x(x + 2) = x(x) + x(2) = x^2 + 2x$$

$$y(y - 7) = y(y) - y(7) = y^2 - 7y$$

Following the examples, complete these multiplications.

a) $m(m + 5)$ = ________ b) $t(t - 6)$ = ________

Answers: a) $m^2 + 5m$ b) $t^2 - 6t$

---

82. Notice that we did not write the coefficient "1" in each product below.

$$x(x + 1) = x(x) + x(1) = x^2 + x$$

$$y(y - 1) = y(y) - y(1) = y^2 - y$$

Following the examples, complete these multiplications.

a) $p(p + 1)$ = ________ b) $t(t - 1)$ = ________

Answers: a) $p^2 + p$ b) $t^2 - t$

---

83. Following the examples, complete the other multiplications.

$$b(3b + 4) = b(3b) + b(4) = 3b^2 + 4b$$

$$6y(y - 1) = 6y(y) - 6y(1) = 6y^2 - 6y$$

a) $3m(m + 1)$ = ________ b) $x(2x - 8)$ = ________

Answers: a) $3m^2 + 3m$ b) $2x^2 - 8x$

---

84. Following the examples, complete the other multiplications.

$$4x(2x + 1) = 4x(2x) + 4x(1) = 8x^2 + 4x$$

$$3y(5y - 6) = 3y(5y) - 3y(6) = 15y^2 - 18y$$

a) $7x(3x + 8)$ = ________ b) $9y(4y - 1)$ = ________

Answers: a) $21x^2 + 56x$ b) $36y^2 - 9y$

85. To multiply $4y^3 + 5$ by y^2 below, we multiplied each term in the binomial by y^2.

$$y^2(4y^3 + 5) = y^2(4y^3) + y^2(5) = 4y^5 + 5y^2$$

Complete these by multiplying each term in the binomial by the monomial.

a) $a^3(2a + 7) =$ ______

b) $5t^2(t^3 - 2t) =$ ______

---

86. To multiply $2x^4 - x^2 + 7$ by $5x$ below, we multiplied each term in the trinomial by $5x$.

$$5x(2x^4 - x^2 + 7) = 5x(2x^4) - 5x(x^2) + 5x(7)$$

$$= 10x^5 - 5x^3 + 35x$$

Complete these by multiplying each term in the trinomial by the monomial.

a) $p^4(p^3 + 2p^2 - 3p) =$ ______

b) $2t(t^4 - 5t^2 - 8) =$ ______

Answers to 85: a) $2a^4 + 7a^3$ b) $5t^5 - 10t^3$

---

87. Notice how we multiplied each term in the binomial below by xy^2.

$$xy^2(3x^3y - 2x) = xy^2(3x^3y) - xy^2(2x)$$

$$= 3x^4y^3 - 2x^2y^2$$

Complete these by multiplying each term in the other polynomial by the monomial.

a) $5a^3b^2(3a^4b^2 + a^2b) =$ ______

b) $pq(p^2q - pq^5 - 4) =$ ______

Answers to 86: a) $p^7 + 2p^6 - 3p^5$ b) $2t^5 - 10t^3 - 16t$

Answers to 87: a) $15a^7b^4 + 5a^5b^3$ b) $p^3q^2 - p^2q^6 - 4pq$

5-9 MULTIPLYING BY A BINOMIAL OR TRINOMIAL

In this section, we will show how the distributive principle is used to multiply by a binomial or trinomial either horizontally or vertically.

---

88. We used the distributive principle below to multiply a monomial and a binomial when the binomial is on the left. The arrows show that 5 is multiplied by each term ($\underline{x}$ and 4) in the binomial.

$$(x + 4)(5) = x(5) + 4(5) = 5x + 20$$

Continued on following page.

88. Continued

The same form of the distributive principle can be used to multiply a binomial by a binomial. An example is shown. The arrows show that $(x + 3)$ is multiplied by each term (x and 2) in the first binomial.

$$(x + 2)(x + 3) = x(x + 3) + 2(x + 3)$$

On the right side above, both $x(x + 3)$ and $2(x + 3)$ are multiplications of a binomial by a monomial. We completed those multiplications and simplified below.

$$\begin{aligned}(x + 2)(x + 3) &= x(x + 3) + 2(x + 3)\\ &= x^2 + 3x + 2x + 6\\ &= x^2 + 5x + 6\end{aligned}$$

Following the example, complete this multiplication.

$$\begin{aligned}(a + 6)(a - 4) &= \underline{\qquad\qquad}\\ &= \underline{\qquad\qquad}\\ &= \underline{\qquad\qquad}\end{aligned}$$

---

89. We used the same method to multiply two binomials below. Notice that we multiplied $x + 5$ by both $2x$ and -1.

$$\begin{aligned}(2x - 1)(x + 5) &= 2x(x + 5) - 1(x + 5)\\ &= 2x^2 + 10x - x - 5\\ &= 2x^2 + 9x - 5\end{aligned}$$

Following the example, complete this multiplication.

$$\begin{aligned}(3y - 4)(2y - 1) &= \underline{\qquad\qquad}\\ &= \underline{\qquad\qquad}\\ &= \underline{\qquad\qquad}\end{aligned}$$

Answer to 88:

$$\begin{aligned}&= a(a - 4) + 6(a - 4)\\ &= a^2 - 4a + 6a - 24\\ &= a^2 + 2a - 24\end{aligned}$$

---

90. We used the same method to multiply a trinomial by a binomial below. Notice that we multiplied $x^2 + 4x + 5$ by both x and 3.

$$\begin{aligned}(x + 3)(x^2 + 4x + 5) &= x(x^2 + 4x + 5) + 3(x^2 + 4x + 5)\\ &= x^3 + 4x^2 + 5x + 3x^2 + 12x + 15\\ &= x^3 + 7x^2 + 17x + 15\end{aligned}$$

Following the example, complete this multiplication.

$$\begin{aligned}(y + 2)(3y^2 - 5y + 1) &= \underline{\qquad\qquad}\\ &= \underline{\qquad\qquad}\\ &= \underline{\qquad\qquad}\end{aligned}$$

Answer to 89:

$$\begin{aligned}&= 3y(2y - 1) - 4(2y - 1)\\ &= 6y^2 - 3y - 8y + 4\\ &= 6y^2 - 11y + 4\end{aligned}$$

Answer to 90:

$$\begin{aligned}&= y(3y^2 - 5y + 1) + 2(3y^2 - 5y + 1)\\ &= 3y^3 - 5y^2 + y + 6y^2 - 10y + 2\\ &= 3y^3 + y^2 - 9y + 2\end{aligned}$$

91. We can multiply polynomials vertically. As an example, we multiplied $3x + 4$ and $x + 2$ below. Notice that we multiplied the top row $(3x + 4)$ by both 2 and $\underline{x}$ and then added.

$$\begin{array}{rl} 3x + 4 & \\ \underline{x + 2} & \\ 6x + 8 & \text{Multiplying the top row by 2} \\ \underline{3x^2 + \ \ 4x\qquad} & \text{Multiplying the top row by } \underline{x} \\ 3x^2 + 10x + 8 & \text{Adding} \end{array}$$

Following the example, do these:

a) Multiply $x + 3$ and $x + 5$

$$\begin{array}{r} x + 3 \\ \underline{x + 5} \end{array}$$

b) Multiply $2y - 1$ and $3y + 4$

$$\begin{array}{r} 2y - 1 \\ \underline{3y + 4} \end{array}$$

---

92. To multiply $5x - 1$ and $x - 6$ below, we multiplied the top row by -6 and $\underline{x}$ and then added. Do the other multiplication.

$$\begin{array}{r} 5x - 1 \\ \underline{x - 6} \\ -30x + 6 \\ \underline{5x^2 - \quad x\qquad} \\ 5x^2 - 31x + 6 \end{array}$$

$$\begin{array}{r} 4y + 3 \\ \underline{2y - 1} \end{array}$$

Answers to 91:

a) $$\begin{array}{r} x + 3 \\ \underline{x + 5} \\ 5x + 15 \\ \underline{x^2 + 3x\qquad} \\ x^2 + 8x + 15 \end{array}$$

b) $$\begin{array}{r} 2y - 1 \\ \underline{3y + 4} \\ 8y - 4 \\ \underline{6y^2 - 3y\qquad} \\ 6y^2 + 5y - 4 \end{array}$$

---

93. To multiply $3y^2 - 4y + 1$ and $y - 2$, we multiplied the top row by -2 and $\underline{y}$ and then added. Do the other multiplication.

$$\begin{array}{r} 3y^2 - 4y + 1 \\ \underline{y - 2} \\ -6y^2 + 8y - 2 \\ \underline{3y^3 - \ \ 4y^2 + \ \ y\qquad} \\ 3y^3 - 10y^2 + 9y - 2 \end{array}$$

$$\begin{array}{r} m^2 - 3m - 2 \\ \underline{4m - 1} \end{array}$$

Answer to 92: $8y^2 + 2y - 3$

---

94. To multiply $x^2 + 4x - 3$ and $2x^2 - x + 3$ below, we multiplied the top row by 3, $-x$, and $2x^2$ and then added. Do the other multiplication.

$$\begin{array}{r} x^2 + \ \ 4x - 3 \\ \underline{2x^2 - \quad x + 3} \\ 3x^2 + 12x - 9 \\ -x^3 - 4x^2 + \ \ 3x\qquad \\ \underline{2x^4 + 8x^3 - 6x^2\qquad\qquad} \\ 2x^4 + 7x^3 - 7x^2 + 15x - 9 \end{array}$$

$$\begin{array}{r} 3y^2 - \ \ y + 4 \\ \underline{y^2 + 2y - 1} \end{array}$$

Answer to 93: $4m^3 - 13m^2 - 5m + 2$

---

95. Sometimes we have to leave space for a missing term so that like terms are lined up in columns. Below, for example, we left space for a y-term in $4y^2 - 12$ and space for a y^2-term in $2y^3 - 6y$. Do the other multiplication.

$$\begin{array}{r} y^2 - \ \ 3 \\ \underline{2y + \ \ 4} \\ 4y^2 \qquad - 12 \\ \underline{2y^3 \qquad\quad - 6y\qquad} \\ 2y^3 + 4y^2 - 6y - 12 \end{array}$$

$$\begin{array}{r} 3x^3 - 5x \ + 4 \\ \underline{x^2 + 1} \end{array}$$

Answer to 94: $3y^4 + 5y^3 - y^2 + 9y - 4$

Answer to 95: $3x^5 - 2x^3 + 4x^2 - 5x + 4$

SELF-TEST 16 (pages 214-223)

Add. Write each sum in descending order for x.

1. $(5x + 1) + (4x^2 - x - 3) + (x^2 + 4)$

2. Add $dx^2 - 1$, $kx^4 - 2dx^2 + 1$, and $dx^2 - rx$

Subtract. Write each difference in descending order for t.

3. $(2t^3 - 3t^2 - t) - (t^3 - 3t^2 + 3)$

4. Subtract $bt^2 - 2at$ from $ht^4 + bt^2 - 1$

Multiply or square.

5. $(-2xy^3)(6x^2y)$

6. $(-3ab)(5a^2b)(-2a^3b^2)$

7. $(-4tw^3)^2$

8. $4y(y^2 + 3y - 2)$

9. $2p^2w(3p^3w^3 - pw + 5)$

10. $(x + 4)(x^2 + 3x - 2)$

11.
$$\begin{array}{r} 2y^2 - y + 5 \\ \underline{3y - 1} \end{array}$$

ANSWERS:

1. $5x^2 + 4x + 2$
2. $kx^4 - rx$
3. $t^3 - t - 3$
4. $ht^4 + 2at - 1$
5. $-12x^3y^4$
6. $30a^6b^4$
7. $16t^2w^6$
8. $4y^3 + 12y^2 - 8y$
9. $6p^5w^4 - 2p^3w^2 + 10p^2w$
10. $x^3 + 7x^2 + 10x - 8$
11. $6y^3 - 5y^2 + 16y - 5$

5-10 MULTIPLYING BINOMIALS

In this section, we will discuss the FOIL method for multiplying two binomials.

---

96. The first two steps used to multiply $(x + 2)$ and $(x + 6)$ by the distributive-principle method are shown below.

$$(x + 2)(x + 6) = x(x + 6) + 2(x + 6)$$
$$= x^2 + 6x + 2x + 12$$

In the second step above, there are four terms in the product. We can skip the first step and write those four terms directly as we have done below.

$$\underset{I\ \ L}{\overset{F\ \ O}{(x + 2)(x + 6)}} = \underset{F}{x^2} + \underset{O}{6x} + \underset{I}{2x} + \underset{L}{12}$$

Note: 1) To get F, we multiplied the first terms: $(x)(x) = x^2$

2) To get O, we multiplied the outside terms: $(x)(6) = 6x$

3) To get I, we multiplied the inside terms: $(2)(x) = 2x$

4) To get L, we multiplied the last terms: $(2)(6) = 12$

Using the FOIL method, write the four terms of each product below.

a) $(y + 3)(y + 5)$ = ____ (F) + ____ (O) + ____ (I) + ____ (L)

b) $(2t + 3)(4t + 1)$ = ____ (F) + ____ (O) + ____ (I) + ____ (L)

---

97. In the FOIL method, we multiply both terms in the second binomial:

1) by the first term in the first binomial

2) by the second term in the first binomial

Using the FOIL method, write each four-term product.

a) $(x + 7)(x + 5)$ = ____ + ____ + ____ + ____

b) $(5t + 1)(2t + 8)$ = ____ + ____ + ____ + ____

Answers to 96:

a) $y^2 + 5y + 3y + 15$

b) $8t^2 + 2t + 12t + 3$

---

98. After writing the four-term product below, we combined the like terms ($4x$ and $8x$) to get a trinomial product.

$$(x + 8)(x + 4) = x^2 + 4x + 8x + 32$$
$$= x^2 + 12x + 32$$

Continued on following page.

Answers to 97:

a) $x^2 + 5x + 7x + 35$

b) $10t^2 + 40t + 2t + 8$

98. Continued

Following the example, write each four-term product and then combine the like terms to get a trinomial product.

a) $(y + 5)(y + 9) =$ ______ + ______ + ______ + ______

$=$ ______

b) $(2t + 1)(4t + 5) =$ ______ + ______ + ______ + ______

$=$ ______

---

a) $y^2 + 9y + 5y + 45$

$y^2 + 14y + 45$

b) $8t^2 + 10t + 4t + 5$

$8t^2 + 14t + 5$

99. We used the FOIL method below and then combined like terms.

$$(x + 5)(x - 2) = x^2 - 2x + 5x - 10$$
$$= x^2 + 3x - 10$$

Use the FOIL method and then combine like terms.

a) $(y + 2)(y - 8) =$ ______

$=$ ______

b) $(5m + 1)(m - 4) =$ ______

$=$ ______

---

a) $y^2 - 8y + 2y - 16$

$y^2 - 6y - 16$

b) $5m^2 - 20m + m - 4$

$5m^2 - 19m - 4$

100. We used the FOIL method below and then combined like terms.

$$(x - 4)(x + 9) = x^2 + 9x - 4x - 36$$
$$= x^2 + 5x - 36$$

Use the FOIL method and then combine like terms.

a) $(x - 1)(x + 2) =$ ______

$=$ ______

b) $(3d - 4)(4d + 3) =$ ______

$=$ ______

---

a) $x^2 + 2x - x - 2$

$x^2 + x - 2$

b) $12d^2 + 9d - 16d - 12$

$12d^2 - 7d - 12$

101. We used the FOIL method below and then combined like terms.

$$(m - 2)(m - 3) = m^2 - 3m - 2m + 6$$
$$= m^2 - 5m + 6$$

Use the FOIL method and then combine like terms.

a) $(x - 9)(x - 1) =$ ______

$=$ ______

b) $(3y - 1)(2y - 4) =$ ______

$=$ ______

102. We combined like terms to get a trinomial product below. Complete the other multiplications.

$(3 - 2y)(5 + y) = 15 + 3y - 10y - 2y^2 = 15 - 7y - 2y^2$

a) $(1 + 4x)(6 - x)$ = ____________

b) $(2 - 3t)(3 - 2t)$ = ____________

Answers: a) $x^2 - x - 9x + 9$, $x^2 - 10x + 9$ b) $6y^2 - 12y - 2y + 4$, $6y^2 - 14y + 4$

---

103. We combined like terms to get a trinomial product below. Complete the other multiplications.

$(t^2 + 3)(t^2 - 6) = t^4 - 6t^2 + 3t^2 - 18 = t^4 - 3t^2 - 18$

a) $(x^3 - 4)(x^3 + 5)$ = ____________

b) $(2y^4 - 1)(3y^4 - 1)$ = ____________

Answers: a) $6 + 23x - 4x^2$ b) $6 - 13t + 6t^2$

---

104. Following the example, complete the other multiplications.

$(p - 3q)(2p + q) = 2p^2 + pq - 6pq - 3q^2$
$= 2p^2 - 5pq - 3q^2$

a) $(4x + y)(2x + y)$ = ____________
= ____________

b) $(a - 2b)(a - 3b)$ = ____________
= ____________

Answers: a) $x^6 + x^3 - 20$ b) $6y^8 - 5y^4 + 1$

---

105. Combine like terms in each product below.

a) $(b^2 + c)(b^2 - 3c)$ = ____________
= ____________

b) $(2x^2 - y^2)(x^2 - 4y^2)$ = ____________
= ____________

Answers: a) $8x^2 + 4xy + 2xy + y^2$, $8x^2 + 6xy + y^2$ b) $a^2 - 3ab - 2ab + 6b^2$, $a^2 - 5ab + 6b^2$

---

106. In some multiplications, there are no like terms. Therefore, we are unable to simplify the product. For example:

$(a + b)(c + d) = ac + ad + bc + bd$

Following the example, do these:

a) $(c + d)(p - q)$ = ____________

b) $(2a - b)(5c + 3d)$ = ____________

Answers: a) $b^4 - 3b^2c + b^2c - 3c^2$, $b^4 - 2b^2c - 3c^2$ b) $2x^4 - 8x^2y^2 - x^2y^2 + 4y^4$, $2x^4 - 9x^2y^2 + 4y^4$

---

107. There are no like terms in the product below.

$(c^2 - d^2)(p^3 + q^3) = c^2p^3 + c^2q^3 - d^2p^3 - d^2q^3$

Following the example, do these:

a) $(2a + b^2)(x^3 - 3y)$ = ____________

b) $(2x - 4xy)(y^4 - 5x^2)$ = ____________

Answers: a) $cp - cq + dp - dq$ b) $10ac + 6ad - 5bc - 3bd$

Answers: a) $2ax^3 - 6ay + b^2x^3 - 3b^2y$ b) $2xy^4 - 10x^3 - 4xy^5 + 20x^3y$

5-11 MULTIPLYING THE SUM AND DIFFERENCE OF TWO TERMS

In this section, we will discuss multiplications of the sum and difference of the same two terms.

---

108. In the multiplication below, both binomials contain x and 3. One binomial is a sum; the other binomial is a difference.

$$(x + 3)(x - 3) = x^2 - 3x + 3x - 9 = x^2 - 9$$

Since $-3x + 3x = 0$, the product above simplifies to the binomial $x^2 - 9$. Following the example, complete this multiplication.

$(y + 6)(y - 6) =$ ____________________

Answer: $y^2 - 36$

109. In the multiplication below, both binomials contain 2t and 5. One binomial is a sum; the other binomial is a difference.

$$(2t + 5)(2t - 5) = 4t^2 - 10t + 10t - 25 = 4t^2 - 25$$

Since $-10t + 10t = 0$, the product above simplifies to the binomial $4t^2 - 25$. Following the example, complete this multiplication.

$(3m + 2)(3m - 2) =$ ____________________

Answer: $9m^2 - 4$

110. Here are the multiplications done in the last two frames.

$$(y + 6)(y - 6) = y^2 - 36$$

$$(3m + 2)(3m - 2) = 9m^2 - 4$$

You can see that each product is the square of the first term minus the square of the second term. That is:

$$\boxed{(A + B)(A - B) = A^2 - B^2}$$

We used the pattern above for the multiplication below.

$$(x + 4)(x - 4) = (x)^2 - (4)^2 = x^2 - 16$$

Using the pattern, write each binomial product.

a) $(t + 9)(t - 9) =$ __________ b) $(V + 1)(V - 1) =$ __________

Answer: a) $t^2 - 81$ b) $V^2 - 1$

111. We used the same pattern for the multiplication below.

$$(4x + 5)(4x - 5) = (4x)^2 - (5)^2 = 16x^2 - 25$$

Using the pattern, write each binomial product.

a) $(6y + 1)(6y - 1) =$ __________ b) $(2t + 7)(2t - 7) =$ __________

Answer: a) $36y^2 - 1$ b) $4t^2 - 49$

112. Using the pattern, write each binomial product.

a) $(m + 2)(m - 2) =$ __________ c) $(5x + 1)(5x - 1) =$ __________

b) $(10 + d)(10 - d) =$ __________ d) $(9 + 2t)(9 - 2t) =$ __________

113. Two more examples of the same pattern are shown below.

$$(x + y)(x - y) = x^2 - y^2$$

$$(2p + q)(2p - q) = 4p^2 - q^2$$

Following the examples, complete these:

a) $(c + d)(c - d) =$ ________ b) $(3x + 5y)(3x - 5y) =$ ________

Answers: a) $m^2 - 4$ b) $100 - d^2$ c) $25x^2 - 1$ d) $81 - 4t^2$

---

114. Another example of the same pattern is shown below.

$$(x^2 + 2y^3)(x^2 - 2y^3) = x^4 - 4y^6$$

Following the example, complete these:

a) $(p + q^2)(p - q^2) =$ ________ c) $(a^3 + b^2)(a^3 - b^2) =$ ________

b) $(3x + y^3)(3x - y^3) =$ ________ d) $(2m^4 + 6t^5)(2m^4 - 6t^5) =$ ________

Answers: a) $c^2 - d^2$ b) $9x^2 - 25y^2$

Answers: a) $p^2 - q^4$ b) $9x^2 - y^6$ c) $a^6 - b^4$ d) $4m^8 - 36t^{10}$

5-12 SQUARING BINOMIALS

In this section, we will discuss a method for squaring binomials.

---

115. To square a binomial, we multiply the binomial by itself. We can use the FOIL method to do so. For example:

$$(x + 4)^2 = (x + 4)(x + 4)$$
$$= x^2 + 4x + 4x + 16$$
$$= x^2 + 8x + 16$$

There is a shortcut that can be used to square $(x + 4)$. To see the shortcut, let's examine $x^2 + 8x + 16$.

1) The first term (x^2) is the square of x.

2) The second term $(8x)$ is double the product of the two terms of the binomial. That is: $8x = 2(x)(4)$

3) The third term (16) is the square of 4.

Let's use the shortcut to square $(y + 8)$.

a) Squaring y, we get ______.

b) Doubling the product of y and 8, we get ______.

c) Squaring 8, we get ______.

d) Therefore, $(y + 8)^2 =$ ____________________

116.	Let's use the shortcut to square $(2x+7)$. a) Squaring $2x$, we get _____. b) Doubling the product of $2x$ and 7, we get _____. c) Squaring 7, we get _____. d) Therefore, $(2x+7)^2$ = ____________	a) y^2 b) $16y$ c) 64 d) $y^2 + 16y + 64$
117.	Use the shortcut for these. Be sure to <u>double</u> the product of the two terms to get the middle term of the trinomial. a) $(m+3)^2$ = ____________ b) $(5d+2)^2$ = ____________	a) $4x^2$ b) $28x$ c) 49 d) $4x^2 + 28x + 49$
118.	We used the FOIL method to square $y - 9$ below. Notice that we can also use the shortcut to get $y^2 - 18y + 81$. However, <u>we subtract the second term.</u> $(y-9)^2 = (y-9)(y-9)$ $= y^2 - 9y - 9y + 81$ $= y^2 - 18y + 81$ Let's use the shortcut to square $(4x - 5)$. a) Squaring $4x$, we get _____. b) Doubling the product of $4x$ and 5, we get _____. c) Squaring 5, we get _____. d) Therefore, $(4x-5)^2$ = ____________	a) $m^2 + 6m + 9$ b) $25d^2 + 20d + 4$
119.	Use the shortcut for these. Be sure to <u>subtract</u> the middle term of the trinomial. a) $(m-2)^2$ = ____________ b) $(3y-4)^2$ = ____________	a) $16x^2$ b) $40x$ c) 25 d) $16x^2 - 40x + 25$
120.	In the example below, the number is the first term in the binomial. The same method is used. $(3+5d)^2 = 9 + 30d + 25d^2$ Following the example, complete these: a) $(12+t)^2$ = ____________ b) $(1-4m)^2$ = ____________	a) $m^2 - 4m + 4$ b) $9y^2 - 24y + 16$
121.	Let's use the shortcut to square $(2p + q)$. a) Squaring $2p$, we get _____. b) Doubling the product of $2p$ and q, we get _____. c) Squaring q, we get _____. d) Therefore, $(2p+q)^2$ = ____________	a) $144 + 24t + t^2$ b) $1 - 8m + 16m^2$

122. Let's use the shortcut to square $(t^2 - 4y)$.

a) Squaring t^2, we get ______.

b) Doubling the product of t^2 and $4y$, we get ______.

c) Squaring $4y$, we get ______.

d) Therefore, $(t^2 - 4y)^2 =$ ____________

Answers: a) $4p^2$ b) $4pq$ c) q^2 d) $4p^2 + 4pq + q^2$

---

123. Use the shortcut for these:

a) $(c - d)^2 =$ ____________

b) $(2x + 3y)^2 =$ ____________

c) $(4p^2 - q^3) =$ ____________

d) $(a^2b + 1)^2 =$ ____________

Answers: a) t^4 b) $8t^2y$ c) $16y^2$ d) $t^4 - 8t^2y + 16y^2$

---

a) $c^2 - 2cd + d^2$ b) $4x^2 + 12xy + 9y^2$ c) $16p^4 - 8p^2q^3 + q^6$ d) $a^4b^2 + 2a^2b + 1$

5-13 DIVIDING BY A MONOMIAL

In this section, we will discuss the procedure for dividing a polynomial by a monomial.

---

124. To divide a monomial by a number, we divide the numerical coefficient by the number. That is:

$\frac{12x^3}{6} = 2x^3$ a) $\frac{-6y}{2} =$ ______ b) $\frac{-20t^2}{-5} =$ ______

---

125. To divide monomials containing powers of the same variable or variables, we divide the numerical coefficients and use the law of exponents for division. For example:

$$\frac{10x^5}{5x^2} = \left(\frac{10}{5}\right)\left(\frac{x^5}{x^2}\right) = 2x^3$$

$$\frac{-8a^3b^6}{2ab^5} = \left(\frac{-8}{2}\right)\left(\frac{a^3}{a}\right)\left(\frac{b^6}{b^5}\right) = -4a^2b$$

Using the same method, complete each division.

a) $\frac{12b^2t^7}{3bt^3} =$ ______ b) $\frac{-18m^5}{-6m^4} =$ ______

Answers: a) $-3y$ b) $4t^2$

---

126. When the coefficient of the quotient is "1" or -1, the coefficient is not ordinarily written. That is:

$\frac{4x^7}{4x^2} = 1x^5$ or x^5 $\frac{-8m^5v^8}{8m^3v^7} =$ ____________

Answers: a) $4bt^4$ b) $3m$

$-1m^2v = -m^2v$

127. When the coefficient of the denominator is "1" or -1, it is helpful to write the "1" or -1 explicitly. For example:

$$\frac{5x^3}{x} = \frac{5x^3}{1x} = 5x^2 \qquad \frac{7a^4t^5}{-a^3t^2} = \frac{7a^4t^5}{-1a^3t^2} = \_\_\_\_\_\_$$

---

128. When the quotient contains a power whose exponent is zero, that power is not ordinarily written since it equals "1". For example:

$$\frac{12b^3x^4}{4b^3x} = 3b^0x^3 = 3(1)x^3 = 3x^3$$

Using the same method, complete these.

a) $\frac{10t^5}{-2t^5} = \_\_\_\_\_\_$ b) $\frac{a^2b}{a^2b} = \_\_\_\_\_\_$

Answer (127): $-7at^3$

---

129. When dividing, we sometimes get a fractional coefficient in the quotient. For example:

$$\frac{5x^2}{2x} = \frac{5}{2}x \qquad \frac{x^3y^2}{3xy} = \frac{1}{3}x^2y$$

Following the examples, complete these.

a) $\frac{4m^6}{3m^2} = \_\_\_\_\_\_$ b) $\frac{2a^2b^2}{4ab^2} = \_\_\_\_\_\_$

Answers (128): a) -5, from $-5t^0$ b) 1, from a^0b^0

---

130. To divide a binomial by a number, we divide each term by the number. That is:

$$\frac{8x^2 + 6}{2} = \frac{8x^2}{2} + \frac{6}{2} = 4x^2 + 3$$

$$\frac{9ay^3 - 12}{3} = \frac{9ay^3}{3} - \frac{12}{3} = \_\_\_\_\_\_$$

Answers (129): a) $\frac{4}{3}m^4$ b) $\frac{1}{2}a$

---

131. To divide a trinomial by a number, we divide each term by the number. For example:

$$\frac{5x^2 - 10x + 15}{10} = \frac{5x^2}{10} - \frac{10x}{10} + \frac{15}{10} = \frac{1}{2}x^2 - x + \frac{3}{2}$$

$$\frac{8p^4 + 4p^2 - 20}{8} = \frac{8p^4}{8} + \frac{4p^2}{8} - \frac{20}{8} = \_\_\_\_\_\_$$

Answer (130): $3ay^3 - 4$

---

132. In each division below, we also divide each term in the numerator by the denominator. That is:

a) $\frac{6y^2 + 4y}{2y} = \frac{6y^2}{2y} + \frac{4y}{2y} = \_\_\_\_\_\_$

b) $\frac{ax^6 - bx^5 - x^4}{x^2} = \frac{ax^6}{x^2} - \frac{bx^5}{x^2} - \frac{x^4}{x^2} = \_\_\_\_\_\_$

Answer (131): $p^4 + \frac{1}{2}p^2 - \frac{5}{2}$

133. Complete each division.

a) $\dfrac{3t^4 + t}{t}$ = ________ b) $\dfrac{3m^5 - 9m^3}{6m^3}$ = ________

a) $3y + 2$

b) $ax^4 - bx^3 - x^2$

---

134. Complete: a) $\dfrac{2y^3 + 6y^2 - 4y}{2y}$ = ________

b) $\dfrac{x^5y^2 - x^3y^3 + x^2y}{x^2y}$ = ________

a) $3t^3 + 1$

b) $\frac{1}{2}m^2 - \frac{3}{2}$

---

a) $y^2 + 3y - 2$ b) $x^3y - xy^2 + 1$

5-14 DIVIDING BY A BINOMIAL

To divide a trinomial or larger polynomial by a binomial, we use a procedure similar to long division. We will discuss the method in this section.

---

135. The two basic steps needed to divide $x^2 + 6x + 8$ by $x + 2$ are shown below.

1) Dividing the first term x^2 of the dividend by the first term of the divisor.

$$\begin{array}{r|l} x \leftarrow & \text{Dividing } x^2 \text{ by } x\text{: } \dfrac{x^2}{x} = x \\ x + 2\overline{)x^2 + 6x + 8} & \\ \underline{x^2 + 2x} \leftarrow & \text{Multiplying the divisor by } \underline{x} \\ 4x \leftarrow & \text{Subtracting } x^2 + 2x \text{ from } x^2 + 6x \end{array}$$

2) Bringing down the next term 8 of the dividend and dividing the first term $4x$ of $4x + 8$ by the first term $\underline{x}$ of the divisor.

$$\begin{array}{r|l} x + 4 \leftarrow & \text{Dividing } 4x \text{ by } x\text{: } \dfrac{4x}{x} = 4 \\ x + 2\overline{)x^2 + 6x + 8} & \\ \underline{x^2 + 2x} & \\ 4x + 8 & \\ \underline{4x + 8} \leftarrow & \text{Multiplying the divisor by } 4 \\ 0 \leftarrow & \text{Subtracting } 4x + 8 \text{ from } 4x + 8 \end{array}$$

Therefore, $x^2 + 6x + 8$ divided by $x + 2$ equals $x + 4$ with a 0 remainder. To check the division, multiply the divisor and quotient below.

$(x + 2)(x + 4)$ = ________

$x^2 + 6x + 8$

136. Using the steps from the last frame, divide $x^2 + 8x + 15$ by $x + 3$ at the right.

$$x + 3\overline{)x^2 + 8x + 15}$$

Therefore: $\dfrac{x^2 + 8x + 15}{x + 3} =$ ________

$$\begin{array}{r} x + 5 \\ x + 3\overline{)x^2 + 8x + 15} \\ \underline{x^2 + 3x} \\ 5x + 15 \\ \underline{5x + 15} \\ 0 \end{array}$$

$$\frac{x^2 + 8x + 15}{x + 3} = x + 5$$

137. We completed one division below. Notice how subtracting $-2x$ from $-5x$ is the same as adding $+2x$ and $-5x$. Complete the other division.

$$\begin{array}{r} x - 3 \\ x - 2\overline{)x^2 - 5x + 6} \\ \underline{x^2 - 2x} \\ -3x + 6 \\ \underline{-3x + 6} \\ 0 \end{array}$$

$$y - 6\overline{)y^2 - 11y + 30}$$

$$\begin{array}{r} y - 5 \\ y - 6\overline{)y^2 - 11y + 30} \\ \underline{y^2 - 6y} \\ -5y + 30 \\ \underline{-5y + 30} \\ 0 \end{array}$$

138. We completed one division below. Notice how subtracting $-5y$ from $-y$ is the same as adding $+5y$ and $-y$. Complete the other division.

$$\begin{array}{r} y + 4 \\ y - 5\overline{)y^2 - y - 20} \\ \underline{y^2 - 5y} \\ 4y - 20 \\ \underline{4y - 20} \\ 0 \end{array}$$

$$t - 3\overline{)t^2 + t - 12}$$

$$\begin{array}{r} t + 4 \\ t - 3\overline{)t^2 + t - 12} \\ \underline{t^2 - 3t} \\ 4t - 12 \\ \underline{4t - 12} \\ 0 \end{array}$$

139. Following the example, complete the other division.

$$\begin{array}{r} 2x + 1 \\ 3x - 4\overline{)6x^2 - 5x - 4} \\ \underline{6x^2 - 8x} \\ 3x - 4 \\ \underline{3x - 4} \\ 0 \end{array}$$

$$4x - 1\overline{)8x^2 + 2x - 1}$$

$$\begin{array}{r} 2x + 1 \\ 4x - 1\overline{)8x^2 + 2x - 1} \\ \underline{8x^2 - 2x} \\ 4x - 1 \\ \underline{4x - 1} \\ 0 \end{array}$$

140. There is a remainder of 44 in the division below. Notice how we wrote the remainder as the fraction $\dfrac{44}{m - 5}$ when writing the quotient. Complete the other division.

$$\begin{array}{r} m + 8 \\ m - 5\overline{)m^2 + 3m + 4} \\ \underline{m^2 - 5m} \\ 8m + 4 \\ \underline{8m - 40} \\ 44 \end{array}$$

$$t - 2\overline{)t^2 + 5t - 1}$$

$$\frac{m^2 + 3m + 4}{m - 5} = m + 8 + \frac{44}{m - 5}$$

$\dfrac{t^2 + 5t - 1}{t - 2} =$ ________

$$\begin{array}{r} t + 7 \\ t - 2\overline{)t^2 + 5t - 1} \\ \underline{t^2 - 2t} \\ 7t - 1 \\ \underline{7t - 14} \\ 13 \end{array}$$

$$\frac{t^2 + 5t - 1}{t - 2} = t + 7 + \frac{13}{t - 2}$$

SELF-TEST 17 (pages 224-234)

Find each product or square.

1. $(a + b)(c - d)$
2. $(3x - 2)(2x - 1)$
3. $(a^2 + 2b)(a^2 - 2b)$
4. $(2x - 3y)^2$

Do each division.

5. $\dfrac{12p^3q^4}{-3pq^4}$
6. $\dfrac{x^3y + xy^3}{xy}$
7. $\dfrac{12t^4 - 4t^3 + 8t^2}{4t^2}$
8. $m + 3\overline{)m^2 - 3m - 18}$
9. $3y - 4\overline{)6y^2 + y - 12}$

ANSWERS:

1. $ac - ad + bc - bd$
2. $6x^2 - 7x + 2$
3. $a^4 - 4b^2$
4. $4x^2 - 12xy + 9y^2$
5. $-4p^2$
6. $x^2 + y^2$
7. $3t^2 - t + 2$
8. $m - 6$
9. $2y + 3$

SUPPLEMENTARY PROBLEMS - CHAPTER 5

Assignment 15

Multiply.

1. $x^2 \cdot x^{-3}$
2. $y^4 \cdot y \cdot y^{-2}$
3. $(a^2b)(ab^3)$
4. $(pr^{-5})(pr)$
5. $(c^{-2}d^{-3})(cd^{-1})$
6. $(s^3v^2w)(svw^4)$
7. $(m^{-2}pq^3)(mpq^{-1})$
8. $(a^2d)(a^{-1}c)(cd^{-4})$

Divide.

9. $\frac{p^5}{p}$
10. $\frac{q^2}{q^6}$
11. $\frac{c^4d^2}{c^5d}$
12. $\frac{x^{-1}y^2}{x^3y^{-1}}$
13. $\frac{a^{-5}b}{a^3b^4}$
14. $\frac{m^2p^9q^3}{mp^2q^4}$
15. $\frac{x^2y^{-3}z}{x^3y^{-4}z^5}$
16. $\frac{ast^8}{a^{-2}s^2t}$

Raise to the indicated power.

17. $(xy^2)^3$
18. $(t^{-4}v^{-3})^2$
19. $(a^{-1}b)^5$
20. $(c^{-1}d^4)^{-2}$

Write each polynomial in descending order for $\underline{x}$.

21. $5x - 2x^2 - 1 + 3x^3$
22. $b^3x^2 + bx + b^2x^3 + 5b$
23. $cx + 2x^5 - 4y - dx^3$

State the degree of each term.

24. $-5y$
25. $3ab$
26. $10x^4$
27. $-3s^4t$
28. $6cd^3t^2$

State the degree of each polynomial.

29. $x^2 - 5x + 7$
30. $b^2y^3 + b^5y^2 - 3by$
31. $x^3yz^4 - x^2y^3z^4 - xy^5z$

State whether each polynomial is a monomial, binomial, or trinomial.

32. $y^2 - 9$
33. $2pq^2$
34. $x^3 - x + 2$
35. $6m^5 + m$

Simplify by combining like terms if possible.

36. $8t^3 + t^3$
37. $5ab + a$
38. $cd^2 - 5cd^2$
39. $4x^2y^3z - x^2y^3$

Simplify by combining like terms.

40. $y^3 + 8y - 3y^3 - y$
41. $3ax^4 - 2bx^2 - 2ax^4 + bx^2$
42. $3t^2 - 4t + 1 + t^2 + 4t - 5$

Simplify and write in descending order for $\underline{x}$.

43. $3x + 7x^2 - 4x - 1$
44. $xy - 3x^3y^2 + 5y^2 + 3xy$
45. $2xy^2 - 5 + 3x^2y - xy^2 - 1$

Assignment 16

Add. Write each sum in descending order for $\underline{x}$.

1. $(x^2 + 2) + (x^2 - 1)$
2. $(x^3 - 2x - 3) + (2x^3 + x + 3)$
3. $(3x^4 - x^2 + 1) + (x^4 + 2x^2 - 2)$
4. $(x^2 + x - 7) + (3x + 2) + (5x^2 - 4x)$
5. $(ax^2 + b) + (rx - 2b) + (kx^4 - rx)$

Subtract. Write each difference in descending order for $\underline{y}$.

6. $(3y^2 + 2) - (y^2 - 2)$
7. $(y^3 - 4y + 3) - (2y^3 - 7y + 1)$
8. $(2y + 3) - (3y - y^2)$
9. $(2xy^4 - 1) - (xy^4 + 2)$
10. $(3by^2 + dy + 2) - (by^2 + 2)$
11. $(ky^3 - ty + w) - (ky^3 - 2ty - w)$

Multiply these monomials.

12. $(4x)(x)$
13. $-V(7V)$
14. $(6y)(8y)$
15. $(9b)(-7b)$
16. $(3rt^2)(-4r^3t)$
17. $(bx^2y)(axy)$
18. $(-dk)(5d^2k^3)(-2dk^2)$
19. $(6p^3s)(-3a^2s)(ap)$

Square these monomials.

20. $(3V)^2$ 21. $(-7h)^2$ 22. $(12d)^2$ 23. $(-20T)^2$

24. $(4cd)^2$ 25. $(-2x^3)^2$ 26. $(s^2tw^4)^2$ 27. $(-5h^3p)^2$

Multiply by a monomial.

28. $x(x + 1)$ 29. $y(3y - 7)$ 30. $5d(d - 1)$

31. $10a(3a + 7)$ 32. $8h(5h - 6)$ 33. $4x(x^2 - 3)$

34. $a^2(2a + b)$ 35. $my(3y^3 - 2m)$ 36. $2t(t^2 - t + 5)$

37. $xy(x^2 + xy + y^2)$ 38. $3mv^2(m^2v - 2v^2 + m)$ 39. $5h^2s(3hs^2 - h^2s - 2hs)$

Multiply by a binomial or trinomial.

40. $(x + 4)(x - 5)$ 41. $(2y - 3)(4y + 1)$ 42. $(t + 2)(2t^2 - 3t - 4)$

43. $\begin{array}{r} d - 1 \\ \underline{2d + 5} \end{array}$ 44. $\begin{array}{r} 3x + 7 \\ \underline{x - 5} \end{array}$ 45. $\begin{array}{r} y^2 - 2y + 3 \\ \underline{y^2 - 1} \end{array}$ 46. $\begin{array}{r} 2p^2 - 3p + 5 \\ \underline{p^2 + 2p - 3} \end{array}$

Assignment 17

Multiply.

1. $(x + 5)(x + 3)$ 2. $(b + 1)(b - 4)$ 3. $(t - 2)(t + 3)$ 4. $(y - 7)(y - 8)$

5. $(2a + 1)(3a + 2)$ 6. $(5h + 2)(h - 2)$ 7. $(2m - 7)(2m + 5)$ 8. $(3p - 1)(4p - 5)$

9. $(c + d)(p + q)$ 10. $(m + 4)(t - 2)$ 11. $(b - 3d)(2s + w)$ 12. $(a^2 - b^2)(m - 2t)$

13. $(2m - 3p)(3m + p)$ 14. $(4d + 3k)(5d - k)$ 15. $(x^2 - y)(x^2 - 2y)$ 16. $(3r^2 + s^2)(4r^2 + 5s^2)$

Multiply.

17. $(a + 8)(a - 8)$ 18. $(R - 1)(R + 1)$ 19. $(4x + 3)(4x - 3)$ 20. $(7y - 10)(7y + 10)$

21. $(m + n)(m - n)$ 22. $(7a + 2b)(7a - 2b)$ 23. $(c^2 - d^2)(c^2 + d^2)$ 24. $(pq - 5)(pq + 5)$

Square.

25. $(x + 3)^2$ 26. $(y - 9)^2$ 27. $(3a + 4)^2$ 28. $(10 - 3d)^2$

29. $(2a + b)^2$ 30. $(3m - 4t)^2$ 31. $(x^2 - 6y^2)^2$ 32. $(5d^2 + 7h)^2$

Divide.

33. $\frac{16y^4}{2y}$ 34. $\frac{-5a^2b}{5ab}$ 35. $\frac{3c^4df^2}{-c^3df}$ 36. $\frac{12x^5 - 9x}{3x}$

37. $\frac{4m^2p^2 + 3mp^2}{mp}$ 38. $\frac{20t^2 - 12t + 16}{4}$ 39. $\frac{6x^4y + 2x^3z - 4x^2}{2x^2}$ 40. $\frac{m^7s^4 - m^6s^3 - m^4s^2}{m^3s^2}$

41. $x + 8\overline{)x^2 + 14x + 48}$ 42. $d + 3\overline{)d^2 - 2d - 15}$ 43. $a - 4\overline{)a^2 - 10a + 24}$

44. $2y + 5\overline{)6y^2 + 7y - 20}$ 45. $5r - 6\overline{)10r^2 + 3r - 18}$ 46. $4t - 3\overline{)20t^2 - 23t + 6}$

Factoring

6

In this chapter, we will discuss various methods of factoring polynomials. The principle of zero products is introduced. Some equations and word problems are solved by factoring and then using the principle of zero products.

6-1 FACTORING MONOMIALS

In this section, we will define factoring and discuss the factoring process for monomials.

---

1. Factoring is a process in which an expression is written as a multiplication. For example:

10	can be factored into	(2)(5)
18x	can be factored into	(3)(6x)
$24y^3$	can be factored into	$(6y)(4y^2)$

The expressions on the right above are called factors of the expressions on the left. That is:

2 and 5 are factors of 10

3 and 6x are factors of 18x

6y and $4y^2$ are factors of ________

$24y^3$

2. Some numbers can only be factored into themselves and "1". Numbers of that type are called prime numbers. For example:

$3 = (1)(3)$ $\qquad$ $7 = (1)(7)$

Most numbers, however, can be factored into more than one pair of factors. For example, 18 can be factored into three pairs of factors. They are:

$18 = (1)(18)$

$18 = (2)(9)$

$18 = (3)(6)$

24 can be factored into four pairs of factors. Do so below.

$24 = (\quad)(\quad)$ $\qquad$ $24 = (\quad)(\quad)$

$24 = (\quad)(\quad)$ $\qquad$ $24 = (\quad)(\quad)$

---

3. Terms containing a variable can also be factored. For example:

5x can be factored into (5)(x)

y can be factored into (1)(y)

Factor each of these.

a) $7p = (\quad)(\quad)$ $\qquad$ b) $t = (\quad)(\quad)$

Answers: (1)(24) (3)(8) (2)(12) (4)(6)

---

4. We factored 12x in various ways below.

$12x = (2)(6x)$

$12x = (3)(4x)$

$12x = (4)(3x)$

$12x = (6)(2x)$

Write the missing factor in each blank.

a) $6t = (2)(\quad)$ $\qquad$ b) $14m = (7m)(\quad)$ $\qquad$ c) $32p = (\quad)(4)$

Answers: a) (7)(p) b) (1)(t)

---

5. We factored $16x^2$ and ab^2 in various ways below.

$16x^2 = (16)(x^2)$ $\qquad$ $ab^2 = (a)(b^2)$

$16x^2 = (8)(2x^2)$ $\qquad$ $ab^2 = (ab)(b)$

$16x^2 = (4x)(4x)$ $\qquad$ $ab^2 = (b)(ab)$

Write the missing factor in each blank.

a) $20d^2 = (5)(\quad)$ $\qquad$ b) $y^2 = (\quad)(y)$ $\qquad$ c) $p^2q = (p)(\quad)$

Answers: a) (2)(3t) b) (7m)(2) c) (8p)(4)

---

6. We factored x^2y^2 and $12p^7t^3$ in various ways below.

$x^2y^2 = (x^2)(y^2)$ $\qquad$ $12p^7t^3 = (6)(2p^7t^3)$

$x^2y^2 = (x)(xy^2)$ $\qquad$ $12p^7t^3 = (4p^2t^2)(3p^5t)$

$x^2y^2 = (xy)(xy)$ $\qquad$ $12p^7t^3 = (p^4t)(12p^3t^2)$

Write the missing factor in each blank.

a) $m^4y^4 = (my^3)(\quad)$ $\qquad$ c) $10p^5q^3 = (5pq)(\quad)$

b) $p^2q^6 = (\quad)(p^2q^4)$ $\qquad$ d) $18a^4b = (\quad)(2a^3)$

Answers: a) (5)(4d²) b) (y)(y) c) (p)(pq)

7. Any term can be factored into itself and "1". For example:

$3x = (3x)(1)$ $\quad$ $p^3q = (p^3q)(1)$

$y^2 = (1)(y^2)$ $\quad$ $8t^6v^6 = (1)(8t^6v^6)$

Write the missing factor in each blank.

a) $d^8 = (\quad)(d^8)$ $\quad$ b) $24xy^9 = (24xy^9)(\quad)$

Answers to previous frame:
a) $(my^3)\underline{(m^3y)}$
b) $\underline{(q^2)}(p^2q^4)$
c) $(5pq)\underline{(2p^4q^2)}$
d) $\underline{(9ab)}(2a^3)$

---

a) $(1)(d^8)$ $\quad$ b) $(24xy^9)(1)$

6-2 COMMON MONOMIAL FACTORS

In this section, we will show how the distributive principle can be used to factor a common monomial factor out of polynomials containing two or more terms.

---

8. We used the distributive principle for each multiplication below.

$4(x + 3) = 4x + 12$

$5(y - 2) = 5y - 10$

By reversing the process, we can use the distributive principle to factor out the common numerical factor from each product. That is:

$4x + 12$ can be factored into $4(x + 3)$

$5y - 10$ can be factored into $5(y - 2)$

Following the examples, factor each of these:

a) $3t + 18$ = __________ $\quad$ b) $6d - 24$ = __________

---

9. When factoring, we can sometimes factor out more than one numerical factor. For example, we can factor out 2, 3, or 6 from $6x^2 + 12$. That is:

$6x^2 + 12 = 2(3x^2 + 6)$

$6x^2 + 12 = 3(2x^2 + 4)$

$6x^2 + 12 = 6(x^2 + 2)$

When factoring out a common numerical factor, <u>we always factor out the largest possible factor</u>. Therefore, the proper factoring for $6x^2 + 12$ is $6(x^2 + 2)$. Factor out the largest possible numerical factor from these:

a) $8a + 24$ = __________ $\quad$ c) $12t^2 + 18$ = __________

b) $9b^2 - 45$ = __________ $\quad$ d) $16y - 12$ = __________

Answers to frame 8:
a) $3(t + 6)$
b) $6(d - 4)$

---

a) $8(a + 3)$
b) $9(b^2 - 5)$
c) $6(2t^2 + 3)$
d) $4(4y - 3)$

10. Notice how we got "1" as the second term in each binomial factor below.

$$6x^2 + 2 = 2(3x^2 + 1)$$

$$5y - 5 = 5(y - 1)$$

Following the examples, factor these:

a) $3m + 3 =$ ______________ b) $20p^2 - 10 =$ ______________

Answers: a) $3(m + 1)$ b) $10(2p^2 - 1)$

11. To check factorings of this type, we can multiply by the distributive principle. For example:

$$8x + 12 = 4(2x + 3) \text{ is correct,}$$

$$\text{since } 4(2x + 3) = 8x + 12$$

Factor and check each factoring.

a) $16x + 24 =$ ______________ c) $9m^2 + 9 =$ ______________

b) $36y^2 - 12 =$ ______________ d) $15d - 6 =$ ______________

Answers: a) $8(2x + 3)$ b) $12(3y^2 - 1)$ c) $9(m^2 + 1)$ d) $3(5d - 2)$

12. We factored out a common numerical factor from the trinomial below.

$$12t^2 - 18t - 6 = 6(2t^2 - 3t - 1)$$

Factor each trinomial.

a) $16m^2 + 4m - 8 =$ ______________________

b) $9y^3 - 12y + 3 =$ ______________________

Answers: a) $4(4m^2 + m - 2)$ b) $3(3y^3 - 4y + 1)$

13. We were able to factor a common first power out of the polynomial below.

$$x^3 - 3x^2 + x = x(x^2 - 3x + 1)$$

Factor out a first power.

a) $4t^2 - 7t =$ ______________________

b) $y^5 + 9y^3 - y =$ ______________________

Answers: a) $t(4t - 7)$ b) $y(y^4 + 9y^2 - 1)$

14. When factoring a common power out of a polynomial, <u>we always factor out the largest possible power</u>. The largest possible power is the smallest power that occurs. For example, the largest possible power that can be factored out below is y^2.

$$2y^6 - 7y^4 + y^2 = y^2(2y^4 - 7y^2 + 1)$$

Factor out the largest possible power.

a) $3x^5 + 7x^3 =$ ______________________

b) $a^8 + 4a^6 - a^4 =$ ______________________

Answers: a) $x^3(3x^2 + 7)$ b) $a^4(a^4 + 4a^2 - 1)$

15. In the example below, we factored out the largest possible numerical factor and power.

$$10t^5 - 30t^2 = 10t^2(t^3 - 3)$$

Factor out the largest possible numerical factor and power.

a) $8m^9 + 4m^5 =$ ______________________

b) $6x^3 - 12x^2 - 9x =$ ______________________

16. In the example below, we factored out the largest possible power of each variable.

$$3x^2y^2 - xy^3 = xy^2(3x - y)$$

Factor out the largest possible power of each variable.

a) $c^4d + 7c^2d$ = ____________

b) $2p^5q^3 - 5p^4q - p^3q^2$ = ____________

a) $4m^5(2m^4 + 1)$

b) $3x(2x^2 - 4x - 3)$

---

17. In the example below, we factored out the largest possible numerical factor and the largest possible powers of the variables.

$$12x^2y - 20x^3y = 4x^2y(3 - 5x)$$

Factor out the largest numerical factor and powers.

a) $6p^4q^4 + 3p^2q^2$ = ____________

b) $8c^5d - 4c^4d^2 + 4c^3d^3$ = ____________

a) $c^2d(c^2 + 7)$

b) $p^3q(2p^2q^2 - 5p - q)$

---

a) $3p^2q^2(2p^2q^2 + 1)$ b) $4c^3d(2c^2 - cd + d^2)$

6-3 FACTORING BY GROUPING

In this section, we will discuss the process called factoring by grouping.

---

18. We multiplied by the distributive principle below. To do so, we multiplied $(x + 3)$ by both x and 2.

$$(x + 2)(x + 3) = x(x + 3) + 2(x + 3)$$

By reversing the process, we can use the distributive principle to factor out $(x + 3)$ which is a common binomial factor.

$$x(x + 3) + 2(x + 3) = (x + 2)(x + 3)$$

Following the example, factor out the common binomial factor from these.

a) $y(y - 1) + 5(y - 1)$ = ____________

b) $2t(4t + 7) + 1(4t + 7)$ = ____________

---

19. We cannot factor out a common monomial factor from $x^2 + 5x + 3x + 15$. However, we can factor $x^2 + 5x$ and $3x + 15$ and then factor out a common binomial factor. We get:

$$x^2 + 5x + 3x + 15 = x(x + 5) + 3(x + 5)$$
$$= (x + 3)(x + 5)$$

Following the example, factor the expression below.

$y^2 + 2y + 5y + 10$ = ____________

= ____________

a) $(y + 5)(y - 1)$

b) $(2t + 1)(4t + 7)$

$y(y + 2) + 5(y + 2)$

$(y + 5)(y + 2)$

20. The process used in the last frame is called factoring by grouping. We used that process to factor one expression below. Use it to factor the other expression.

$$3x^2 - 12x + 2x - 8 = 3x(x - 4) + 2(x - 4)$$
$$= (3x + 2)(x - 4)$$

$2y^2 - 16y + 5y - 40 =$ ________________

$=$ ________________

---

21. Following the example, factor the other expression.

Answer to frame 20: $2y(y - 8) + 5(y - 8)$; $(2y + 5)(y - 8)$

$$5m^2 - 15mt + 2mt - 6t^2 = 5m(m - 3t) + 2t(m - 3t)$$
$$= (5m + 2t)(m - 3t)$$

$8k^2 + 6kq + 12kq + 9q^2 =$ ________________

$=$ ________________

---

22. Notice how we factored -4 out of $(-4x - 12)$ below. Following the example, factor the other expression.

Answer to frame 21: $2k(4k + 3q) + 3q(4k + 3q)$; $(2k + 3q)(4k + 3q)$

$$x^2 + 3x - 4x - 12 = (x^2 + 3x) + (-4x - 12)$$
$$= x(x + 3) - 4(x + 3)$$
$$= (x - 4)(x + 3)$$

$y^2 + 5y - 2y - 10 = (y^2 + 5y) + (-2y - 10)$

$=$ ________________

$=$ ________________

---

23. Notice how we factored -3 out of $(-3t + 12)$ below. Following the example, factor the other expression.

Answer to frame 22: $y(y + 5) - 2(y + 5)$; $(y - 2)(y + 5)$

$$t^2 - 4t - 3t + 12 = (t^2 - 4t) + (-3t + 12)$$
$$= t(t - 4) - 3(t - 4)$$
$$= (t - 3)(t - 4)$$

$m^2 - 7m - 2m + 14 = (m^2 - 7m) + (-2m + 14)$

$=$ ________________

$=$ ________________

Answer to frame 23: $m(m - 7) - 2(m - 7)$; $(m - 2)(m - 7)$

6-4 FACTORING TRINOMIALS OF THE FORM: $x^2 + bx + c$

In this section, we will discuss the procedure for factoring trinomials in which the numerical coefficient of the first term is "1".

---

24. When multiplying binomials, some products can be simplified to a trinomial. Three examples are shown below.

#1 $(x + 2)(x + 3) = x^2 + 5x + 6$

#2 $(y - 3)(y - 4) = y^2 - 7y + 12$

#3 $(t + 2)(t - 5) = t^2 - 3t - 10$

Notice that each product is a trinomial of the form $x^2 + bx + c$. In each case, b is the sum and c is the product of the numbers in the factors. That is:

For #1, $b = 2 + 3 = 5$ and $c = (2)(3) = 6$

For #2, $b = (-3) + (-4) = -7$ and $c = (-3)(-4) = 12$

For #3, $b = 2 + (-5) = -3$ and $c = (2)(-5) = -10$

Using these facts, we can write the product for $(y + 5)(y + 4)$.

a) Since $5 + 4 = 9$, b = ________

b) Since $(5)(4) = 20$, c = ________

c) Therefore, $(y + 5)(y + 4)$ = ______________

---

25. Let's use the same facts to write the product for $(t - 8)(t - 3)$.

a) Since $-8 + (-3) = -11$, b = ________

b) Since $(-8)(-3) = 24$, c = ________

c) Therefore, $(t - 8)(t - 3)$ = ______________

Answers: a) 9 b) 20 c) $y^2 + 9y + 20$

---

26. Let's use the same facts to write the product for $(x + 5)(x - 7)$.

a) Since $5 + (-7) = -2$, b = ________

b) Since $(5)(-7) = -35$, c = ________

c) Therefore, $(x + 5)(x - 7)$ = ______________

Answers: a) -11 b) 24 c) $t^2 - 11t + 24$

---

27. By adding the numbers to get b and multiplying the numbers to get c, find each product below.

a) $(m + 1)(m + 7)$ = ______________

b) $(x - 6)(x - 5)$ = ______________

c) $(y - 3)(y + 8)$ = ______________

d) $(t + 1)(t - 10)$ = ______________

Answers: a) -2 b) -35 c) $x^2 - 2x - 35$

28. To factor a trinomial of the form $x^2 + bx + c$, we can use the same facts about b and c. As an example, we will factor the trinomial below. Notice that both b and c are positive.

$$x^2 + 6x + 8$$

Since the product of the numbers in the binomials must be 8 and their sum must be 6, both numbers must be positive. The possible pairs of positive factors for 8 are (1 and 8) and (2 and 4). The pair whose sum is 6 is (2 and 4). Therefore:

$$x^2 + 6x + 8 = (x + 2)(x + 4)$$

Using the same method, factor these.

a) $y^2 + 6y + 5$ = ______________________

b) $m^2 + 15m + 36$ = ______________________

Answers (previous frame):
a) $m^2 + 8m + 7$
b) $x^2 - 11x + 30$
c) $y^2 + 5y - 24$
d) $t^2 - 9t - 10$

---

29. In the trinomial below, b is negative and c is positive.

$$x^2 - 7x + 10$$

Since the product of the numbers in the binomials must be 10 and their sum must be -7, both numbers must be negative. The possible pairs of negative factors for 10 are (-1 and -10) and (-2 and -5). The pair whose sum is -7 is (-2 and -5). Therefore:

$$x^2 - 7x + 10 = (x - 2)(x - 5)$$

Using the same method, factor these.

a) $t^2 - 3t + 2$ = ______________________

b) $m^2 - 10m + 16$ = ______________________

Answers to 28:
a) $(y + 1)(y + 5)$
b) $(m + 3)(m + 12)$

---

30. In the trinomial below, c is negative. Let's factor the trinomial.

$$x^2 + 5x - 6$$

Since the product of the numbers in the binomials must be -6, one number must be positive and the other negative. The possible pairs of factors for -6 are (-1 and 6), (-6 and 1), (-2 and 3), and (-3 and 2). The pair whose sum is 5 is (-1 and 6). Therefore:

$$x^2 + 5x - 6 = (x - 1)(x + 6)$$

Using the same method, factor these.

a) $m^2 + 4m - 5$ = ______________________

b) $y^2 - 5y - 14$ = ______________________

Answers to 29:
a) $(t - 1)(t - 2)$
b) $(m - 2)(m - 8)$

Answers to 30:
a) $(m - 1)(m + 5)$
b) $(y + 2)(y - 7)$

31. To check the factoring of a trinomial, multiply the binomial factors to see whether you obtain the original trinomial. For example:

$$y^2 + 7y - 18 = (y - 2)(y + 9) \text{ is correct, since:}$$

$$(y - 2)(y + 9) = y^2 + 7y - 18$$

Check each factoring. State whether it is "correct" or "incorrect".

a) $x^2 - 8x + 12 = (x - 2)(x - 6)$ ______

b) $t^2 - 5t - 14 = (t - 2)(t + 7)$ ______

---

a) Correct

b) Incorrect, since:

$(t - 2)(t + 7) =$

$t^2 + 5t - 14$

32. Remember that c is the key to factoring a trinomial. That is:

1) If c is positive, the numbers in the binomials have the same sign. b tells us whether both are positive or negative.

If b is positive, both are positive.

If b is negative, both are negative.

2) If c is negative, the numbers in the binomials have different signs.

Factor each trinomial below and check your results.

a) $x^2 + 9x + 20 =$ ______

b) $y^2 - 3y - 18 =$ ______

c) $t^2 - 9t + 8 =$ ______

d) $m^2 + 4m - 12 =$ ______

---

a) $(x + 4)(x + 5)$

b) $(y + 3)(y - 6)$

c) $(t - 1)(t - 8)$

d) $(m - 2)(m + 6)$

33. Some trinomials of the form $x^2 + bx + c$ cannot be factored into binomials in which the numbers are integers. An example is discussed below.

$$x^2 + 4x + 6$$

The possible pairs of factors are (1 and 6) and (2 and 3). Neither pair has 4 as its sum.

Factor if possible.

a) $t^2 + 7t + 12 =$ ______

b) $x^2 - 11x + 16 =$ ______

c) $R^2 - R - 11 =$ ______

d) $m^2 + m - 2 =$ ______

---

a) $(t + 3)(t + 4)$

b) Not possible

c) Not possible

d) $(m - 1)(m + 2)$

34. We factored some trinomials containing two variables below. Notice that the same general method is used.

$$x^2 + 5xy + 6y^2 = (x + 2y)(x + 3y)$$

$$a^2 - 3ab - 28b^2 = (a + 4b)(a - 7b)$$

Factor these and check your results.

a) $p^2 - 7pq + 6q^2 =$ ______

b) $m^2 + 6mt - 16t^2 =$ ______

35. We factored two more trinomials containing two variables below.

$$x^2y^2 - 7xy + 12 = (xy - 3)(xy - 4)$$

$$a^2b^2 + 4ab - 32 = (ab - 4)(ab + 8)$$

Factor these and check your results.

a) $p^2q^2 + 11pq + 10$ = ________________

b) $m^2v^2 - 5mv - 14$ = ________________

a) $(p - q)(p - 6q)$

b) $(m - 2t)(m + 8t)$

---

36. We factored some trinomials containing higher powers below.

$$x^4 + 11x^2 + 30 = (x^2 + 5)(x^2 + 6)$$

$$p^6q^6 - 3p^3q^3 - 4 = (p^3q^3 + 1)(p^3q^3 - 4)$$

Factor these and check your results.

a) $y^6 - 10y^3 + 16$ = ________________

b) $a^4b^4 + a^2b^2 - 6$ = ________________

a) $(pq + 1)(pq + 10)$

b) $(mv + 2)(mv - 7)$

---

a) $(y^3 - 2)(y^3 - 8)$ b) $(a^2b^2 + 3)(a^2b^2 - 2)$

SELF-TEST 18 (pages 237-246)

Factor out the largest common factor.

1. $9x + 15$ = ________
2. $7t - 7$ = ________
3. $24y^2 + 18y$ = ________
4. $36p^2 - 9p$ = ________
5. $8x^3y^3 + 12xy$ = ________
6. $16a^5 - 8a^4 + 20a^3$ = ________

Factor by grouping.

7. $3m^2 - 9m + 2m - 6$ = ________
8. $d^2 + 5d - 3d - 15$ = ________

Factor each trinomial.

9. $x^2 + 7x + 12$ = ________
10. $w^2 - 9w + 20$ = ________
11. $t^2 + 5t - 14$ = ________
12. $m^2 - m - 30$ = ________
13. $a^2 - 5ab + 6b^2$ = ________
14. $x^2y^2 + 2xy - 15$ = ________

ANSWERS:
1. $3(3x + 5)$
2. $7(t - 1)$
3. $6y(4y + 3)$
4. $9p(4p - 1)$
5. $4xy(2x^2y^2 + 3)$
6. $4a^3(4a^2 - 2a + 5)$
7. $(3m + 2)(m - 3)$
8. $(d - 3)(d + 5)$
9. $(x + 3)(x + 4)$
10. $(w - 4)(w - 5)$
11. $(t + 7)(t - 2)$
12. $(m - 6)(m + 5)$
13. $(a - 2b)(a - 3b)$
14. $(xy + 5)(xy - 3)$

6-5 FACTORING TRINOMIALS OF THE FORM: $ax^2 + bx + c$

In this section, we will discuss the procedure for factoring trinomials in which the coefficient of the first term is a number other than "1".

---

37. The steps needed to multiply $(4x + 3)$ and $(x + 2)$ are shown below.

$$(4x + 3)(x + 2) = 4x^2 + 8x + 3x + 6 = 4x^2 + 11x + 6$$

Notice these points about the trinomial product:

$4x^2$ is the product of $4x$ and $\underline{x}$, the first terms.

6 is the product of 3 and 2, the numbers.

Using the facts above, if we multiplied $(2y - 4)$ and $(3y + 1)$:

a) The first term of the trinomial product would be ________.

b) The last term of the trinomial product would be ________.

---

38. The facts in the last frame are used to factor the trinomial below.

Answers to frame 37: a) $6y^2$, from $(2y)(3y)$ b) -4, from $(-4)(1)$

$$3x^2 + 14x + 8$$

1) Since the product of the first terms must be $3x^2$, the only possible pair of first terms is $(3x$ and $x)$.
2) Since the product of the numbers must be 8, the possible pairs of numbers are (1 and 8) and (2 and 4).
3) Therefore, the possible pairs of binomial factors are:

A: $(3x + 1)(x + 8)$

B: $(3x + 8)(x + 1)$

C: $(3x + 2)(x + 4)$

D: $(3x + 4)(x + 2)$

4) Only one pair of binomials is correct. It is the pair that produces $14x$ as the middle term of the trinomial product.

a) Which pair of binomial factors has $14x$ as the middle term of its product? Pair ______

b) Therefore: $3x^2 + 14x + 8 = (\qquad)(\qquad)$

Answers to frame 38: a) Pair C b) $(3x + 2)(x + 4)$

39. Let's factor the trinomial: $2y^2 - 13y + 15$

1) The only possible pair of first terms is (2y and y).
2) The possible pairs of numbers are (-1 and -15) and (-3 and -5).
3) The possible pairs of binomial factors are:

A: $(2y - 1)(y - 15)$

B: $(2y - 15)(y - 1)$

C: $(2y - 3)(y - 5)$

D: $(2y - 5)(y - 3)$

The correct pair of factors is the one with $-13y$ as the middle term of its product.

a) Which pair has $-13y$ as the middle term of its product? Pair ____

b) Therefore: $2y^2 - 13y + 15 = (\quad)(\quad)$

---

40. Let's factor the trinomial: $7t^2 - 2t - 5$

1) The only possible pair of first terms is (7t and t).
2) The possible pairs of numbers are (1 and -5) and (-1 and 5).
3) The possible pairs of binomial factors are:

A: $(7t + 1)(t - 5)$

B: $(7t - 5)(t + 1)$

C: $(7t - 1)(t + 5)$

D: $(7t + 5)(t - 1)$

a) Which pair has $-2t$ as the middle term of its product? Pair ____

b) Therefore: $7t^2 - 2t - 5 = (\quad)(\quad)$

Answers to 39:

a) Pair C

b) $(2y - 3)(y - 5)$

---

41. When the coefficient of the first term is a number other than "1", factoring is a process of trial and error. Each possible pair of factors has to be checked by multiplying them. That is:

$2y^2 - y - 1 = (2y + 1)(y - 1)$ is correct, since:

$(2y + 1)(y - 1) = 2y^2 - y - 1$

Factor these:

a) $5x^2 + 17x + 6 =$ ____________________

b) $7m^2 - 16m + 4 =$ ____________________

c) $2t^2 - 5t - 7 =$ ____________________

Answers to 40:

a) Pair D

b) $(7t + 5)(t - 1)$

42. Let's factor the trinomial: $6x^2 + 19x + 8$

1) The possible pairs of first terms are (x and 6x) and (2x and 3x).
2) The possible pairs of numbers are (1 and 8) and (2 and 4).
3) The possible pairs of binomial factors are:

A: $(x + 1)(6x + 8)$	E: $(2x + 1)(3x + 8)$
B: $(x + 8)(6x + 1)$	F: $(2x + 8)(3x + 1)$
C: $(x + 2)(6x + 4)$	G: $(2x + 2)(3x + 4)$
D: $(x + 4)(6x + 2)$	H: $(2x + 4)(3x + 2)$

a) Which pair of factors has a product whose middle term is 19x? Pair ______

b) Therefore: $6x^2 + 19x + 8 = (\quad)(\quad)$

Answers: a) $(5x + 2)(x + 3)$ b) $(7m - 2)(m - 2)$ c) $(2t - 7)(t + 1)$

---

43. Factor these:

a) $6x^2 - 17x + 5 =$ ______________________

b) $15t^2 - 7t - 4 =$ ______________________

Answers: a) Pair E b) $(2x + 1)(3x + 8)$

---

44. We factored some trinomials containing two variables below. Notice that the same general method is used.

$$2x^2 + 7xy + 3y^2 = (2x + y)(x + 3y)$$

$$8t^2 - 2tv - 15v^2 = (4t + 5v)(2t - 3v)$$

Factor these:

a) $3a^2 - 5ab + 2b^2 =$ ______________________

b) $4p^2 + 8pq - 5q^2 =$ ______________________

Answers: a) $(2x - 5)(3x - 1)$ b) $(5t - 4)(3t + 1)$

---

45. We factored two more trinomials containing two variables below.

$$5x^2y^2 - 12xy + 7 = (5xy - 7)(xy - 1)$$

$$6a^2b^2 - 7ab - 3 = (3ab + 1)(2ab - 3)$$

Factor these:

a) $7p^2q^2 + 9pq + 2 =$ ______________________

b) $10t^2v^2 + 9tv - 9 =$ ______________________

Answers: a) $(3a - 2b)(a - b)$ b) $(2p + 5q)(2p - q)$

---

46. We factored two trinomials containing higher powers below.

$$3x^4 + 8x^2 + 5 = (3x^2 + 5)(x^2 + 1)$$

$$8a^6b^6 - 10a^3b^3 - 3 = (2a^3b^3 - 3)(4a^3b^3 + 1)$$

Factor these:

a) $2y^6 - 11y^3 + 5 =$ ______________________

b) $6p^4q^4 + 7p^2q^2 - 10 =$ ______________________

Answers: a) $(7pq + 2)(pq + 1)$ b) $(5tv - 3)(2tv + 3)$

47. The trinomial below cannot be factored into binomials containing integers.

$$2m^2 + 6m + 3$$

The possible pairs of factors for the trinomial are:

$$(2m + 3)(m + 1)$$

$$(2m + 1)(m + 3)$$

Neither pair has $6m$ as the middle term of its product.

Factor if possible:

a) $4y^2 + 13y + 3 =$ ______

b) $5a^2b^2 + 7ab - 2 =$ ______

Answers (frame 46): a) $(2y^3 - 1)(y^3 - 5)$ b) $(6p^2q^2 - 5)(p^2q^2 + 2)$

Answers: a) $(4y + 1)(y + 3)$ b) Not possible

6-6 FACTORING THE DIFFERENCE OF TWO PERFECT SQUARES

In this section, we will discuss perfect squares and their square roots. Then we will discuss the pattern used to factor the difference of two perfect squares.

48. A square root of a number N is a number whose square is N. That is:

Since $5^2 = 25$, 5 is a square root of 25.

Since $8^2 = 64$, 8 is a square root of 64.

Instead of saying the square root of, we used the symbol $\sqrt{}$. That is:

Since $4^2 = 16$, $\sqrt{16} = 4$ Since $10^2 = 100$, $\sqrt{100} =$ ______

49. Any whole number whose square root is a whole number is a perfect square. For example:

36 is a perfect square, since $\sqrt{36} = 6$

Which of the following are perfect squares? ______

a) 1 b) 7 c) 25 d) 63 e) 81

Answer (frame 48): 10

50. When a power is squared, the exponent of its square is always an even number. For example:

$$(x)^2 = x^2 \qquad (y^3)^2 = y^6 \qquad (t^8)^2 = t^{16}$$

Therefore, any power whose exponent is even is a perfect square. To find the square root of a perfect square power, we divide the exponent by 2. That is:

$$\sqrt{m^2} = m^{\frac{2}{2}} = m^1 = m \qquad \sqrt{b^{10}} = b^{\frac{10}{2}} = b^5$$

Find the square root of each perfect square below.

a) $\sqrt{q^2} =$ ______ b) $\sqrt{d^6} =$ ______ c) $\sqrt{x^{20}} =$ ______

Answer (frame 49): (a), (c), and (e)

51. Both $4x^2$ and $25y^8$ are perfect squares because their coefficients and powers are perfect squares. To find the square root of each, we find the square root of the coefficient and the power. That is:

$$\sqrt{4x^2} = 2x \qquad \sqrt{25y^8} = 5y^4$$

Find the square root of each perfect square below.

a) $\sqrt{9p^2}$ = ____ b) $\sqrt{16t^4}$ = ____ c) $\sqrt{100m^{12}}$ = ____

Answers: a) $q^1 = q$ b) d^3 c) x^{10}

---

52. When monomials like those below are squared, the exponent of each power is always an even number. That is:

$$(xy^2)^2 = x^2y^4 \qquad (a^3b^6)^2 = a^6b^{12}$$

Therefore, any literal monomial in which the exponent of each power is even is a perfect square. To find its square root, we divide each exponent by 2. For example:

$$\sqrt{p^{10}q^4} = p^{\frac{10}{2}}q^{\frac{4}{2}} = p^5q^2$$

Find the square root of each perfect square below.

a) $\sqrt{c^2d^6}$ = ______ b) $\sqrt{x^8y^{14}}$ = ______

Answers: a) $3p$ b) $4t^2$ c) $10m^6$

---

53. The monomial $49m^4v^2$ is a perfect square because its numerical coefficient and both powers are perfect squares. To find its square root, we find the square root of the coefficient and both powers. That is:

$$\sqrt{49m^4v^2} = 7m^2v$$

Find the square root of each perfect square below.

a) $\sqrt{64x^2y^8}$ = ______ b) $\sqrt{81p^{10}q^8}$ = ______

Answers: a) cd^3 b) x^4y^7

---

54. In the multiplication below, the product is the difference of two perfect squares.

$$(x + 4)(x - 4) = x^2 - 16$$

By simply reversing the two sides above, we can factor $x^2 - 16$. That is:

$$x^2 - 16 = (x + 4)(x - 4)$$

Notice these points about the factoring:

1) x is the square root of x^2.
2) 4 is the square root of 16.
3) One factor is a sum; the other factor is a difference.

Following the pattern above, complete these factorings.

a) $y^2 - 36 = (y + 6)(\qquad)$ b) $m^2 - 1 = (\qquad)(m - 1)$

Answers: a) $8xy^4$ b) $9p^5q^4$

55. In the multiplication below, the product is the difference of two perfect squares.

$$(3y + 2)(3y - 2) = 9y^2 - 4$$

By simply reversing the two sides, we can factor $9y^2 - 4$. That is:

$$9y^2 - 4 = (3y + 2)(3y - 2)$$

Note: 1) $3y$ is the square root of $9y^2$.

2) 2 is the square root of 4.

3) One factor is a sum; the other factor is a difference.

Following the pattern above, complete these factorings.

a) $25t^2 - 64 = (5t + 8)(\quad)$ b) $49y^2 - 1 = (\quad)(7y - 1)$

Answers: a) $(y + 6)(y - 6)$ b) $(m + 1)(m - 1)$

---

56. Following the pattern in the last two frames, factor these:

a) $p^2 - 16 =$ ______ c) $36x^2 - 81 =$ ______

b) $t^2 - 100 =$ ______ d) $64h^2 - 9 =$ ______

Answers: a) $(5t + 8)(5t - 8)$ b) $(7y + 1)(7y - 1)$

---

57. In the multiplication below, the product is the difference of two perfect squares.

$$(x^2 + 5)(x^2 - 5) = x^4 - 25$$

By reversing the two sides above, we can factor $x^4 - 25$. That is:

$$x^4 - 25 = (x^2 + 5)(x^2 - 5)$$

Notice that x^2 is the square root of x^4. Following the pattern above, factor these:

a) $y^6 - 9 =$ ______ b) $4d^8 - 49 =$ ______

Answers: a) $(p + 4)(p - 4)$ b) $(t + 10)(t - 10)$ c) $(6x + 9)(6x - 9)$ d) $(8h + 3)(8h - 3)$

---

58. We used the same pattern to factor $9x^2 - 25y^4$ below. Notice that $3x$ is the square root of $9x^2$ and $5y^2$ is the square root of $25y^4$.

$$9x^2 - 25y^4 = (3x + 5y^2)(3x - 5y^2)$$

Following the example, factor these:

a) $81m^2 - 16t^2 =$ ______

b) $100p^2 - 49q^6 =$ ______

Answers: a) $(y^3 + 3)(y^3 - 3)$ b) $(2d^4 + 7)(2d^4 - 7)$

Answers: a) $(9m + 4t)(9m - 4t)$ b) $(10p + 7q^3)(10p - 7q^3)$

59. We used the same pattern to factor $9a^2b^4 - 4$ below.

$$9a^2b^4 - 4 = (3ab^2 + 2)(3ab^2 - 2)$$

Following the example, factor these:

a) $x^2y^2 - 36$ = ____________

b) $16p^2q^8 - 1$ = ____________

---

60. The factoring pattern we have been using applies only to the difference of two perfect squares. It does not apply to the sum of two perfect squares. Therefore, it does not apply to the binomials below.

$25y^2 + 81$ $\qquad$ $a^6 + 4$

Use the pattern to factor these if it applies.

a) $m^2 - 1$ = ____________ $\qquad$ c) $64a^2 + 9$ = ____________

b) $v^{12} + 25$ = ____________ $\qquad$ d) $a^8 - 64$ = ____________

Answers to 59:

a) $(xy + 6)(xy - 6)$

b) $(4pq^4 + 1)(4pq^4 - 1)$

---

61. The factoring pattern applies only when both terms are perfect squares. Therefore, it does not apply to the binomials below.

$t^2 - 39$ $\qquad$ $x^3 - 36$ $\qquad$ $7a^2b^2 - 9$

Use the pattern to factor these if possible.

a) $y^2 - 84$ = ____________ $\qquad$ c) $16p^4 - 1$ = ____________

b) $4x^2 - 9$ = ____________ $\qquad$ d) $x^2y^3 - 100$ = ____________

Answers to 60:

a) $(m + 1)(m - 1)$

b) Does not apply

c) Does not apply

d) $(a^4 + 8)(a^4 - 8)$

Answers to 61:

a) Does not apply $\qquad$ c) $(4p^2 + 1)(4p^2 - 1)$

b) $(2x + 3)(2x - 3)$ $\qquad$ d) Does not apply

6-7 FACTORING PERFECT SQUARE TRINOMIALS

The square of a binomial is called a perfect square trinomial. In this section, we will discuss the method used to identify perfect square trinomials and the procedure used to factor perfect square trinomials.

---

62. We squared two binomials below. Each square is called a perfect square trinomial.

$$(x + 5)^2 = x^2 + 10x + 25$$

$$(3a - 4b)^2 = 9a^2 - 24ab + 16b^2$$

Continued on following page.

62. Continued

Notice these three characteristics of the perfect square trinomial.

1. The first and last terms are perfect squares.
2. Though the middle term may be positive or negative, the first and last terms are always positive.
3. The middle term is double the product of the square roots of the first and last terms. That is:

$$10x = 2 \cdot \sqrt{x^2} \cdot \sqrt{25} = (2)(x)(5)$$

$$24ab = 2 \cdot \sqrt{9a^2} \cdot \sqrt{16b^2} = (2)(3a)(4b)$$

Since perfect square trinomials have three characteristics, three tests are needed to identify one. The first test is this:

<u>The first and last terms must be perfect squares.</u>

In which trinomials below are the first and last terms perfect squares? __________

a) $y^2 - 8y + 16$

b) $x^2 + 4x + 8$

c) $4c^2 - 12cd + 7d^2$

d) $9a^2 + 12ab + 4b^2$

---

63. The second test needed to identify a perfect square trinomial is this:

<u>Though the middle term may be positive or negative, the first and last terms are always positive.</u>

Which of the following <u>cannot</u> be perfect square trinomials for the reason above? __________

a) $y^2 + 4y - 4$

b) $x^2 - 10x + 25$

c) $-c^2 + 6cd + 9d^2$

d) $p^2 - 4pq + 4q^2$

Only (a) and (d)

---

64. The third test needed to identify a perfect square trinomial is this:

<u>The middle term must be double the product of the square roots of the first and last terms.</u>

The third test was applied to each example below.

$x^2 + 14x + 49$ <u>is</u> a perfect square trinomial, since:

$$2 \cdot \sqrt{x^2} \cdot \sqrt{49} = 2(x)(7) = 14x$$

$a^2 - 3ab + b^2$ <u>is not</u> a perfect square trinomial, since:

$$2 \cdot \sqrt{a^2} \cdot \sqrt{b^2} = 2(a)(b) = 2ab$$

Use the above test to decide whether each trinomial below is a perfect square or not.

a) $x^2 - 2x + 1$ ________

b) $y^2 + 5y + 9$ ________

c) $c^2 - 7cd + d^2$ ________

d) $p^2 + 2pq + q^2$ ________

Both (a) and (c)

65. The third test was applied to each trinomial below.

$25y^2 + 10y + 4$ is not a perfect square trinomial, since:

$$2 \cdot \sqrt{25y^2} \cdot \sqrt{4} = 2(5y)(2) = 20y$$

$9x^2 - 24xy + 16y^2$ is a perfect square trinomial, since:

$$2 \cdot \sqrt{9x^2} \cdot \sqrt{16y^2} = 2(3x)(4y) = 24xy$$

Use the above test to decide whether each trinomial below is a perfect square or not.

a) $100t^2 - 20t + 1$ ________ b) $4p^2 + 10pq + 9q^2$ ________

Answers (previous frame):

a) Yes, since: $2 \cdot \sqrt{x^2} \cdot \sqrt{1} = 2x$

b) No, since: $2 \cdot \sqrt{y^2} \cdot \sqrt{9} = 6y$

c) No, since: $2 \cdot \sqrt{c^2} \cdot \sqrt{d^2} = 2cd$

d) Yes, since: $2 \cdot \sqrt{p^2} \cdot \sqrt{q^2} = 2pq$

---

66. Using the three tests, identify the perfect square trinomials below. ________

a) $t^2 + 6t - 9$ c) $p^2 + 2pq + q^2$

b) $x^2 - 12x + 36$ d) $-c^2 + 2cd + d^2$

Answers:

a) Yes, since: $2 \cdot \sqrt{100t^2} \cdot \sqrt{1} = 20t$

b) No, since: $2 \cdot \sqrt{4p^2} \cdot \sqrt{9q^2} = 12pq$

---

67. Which of the following are perfect square trinomials? ____________

a) $-4y^2 + 12y + 9$ c) $100R^2 - 20RT - T^2$

b) $25x^2 - 20x + 4$ d) $9a^2 + 9ab + b^2$

Answer: Only (b) and (c)

---

68. Any perfect square trinomial is the square of a binomial. <u>The terms of the binomial are the square roots of the first and last terms of the trinomial.</u> For example:

$$x^2 + 10x + 25 = (x + 5)^2$$

$$4p^2 + 12pq + 9q^2 = (2p + 3q)^2$$

Following the examples, complete these:

a) $m^2 + 16m + 64 = (\qquad)^2$

b) $25d^2 + 10dh + h^2 = (\qquad)^2$

Answer: Only (b)

---

69. When the middle term of the trinomial is negative, the binomial is a difference. For example:

$$y^2 - 6y + 9 = (y - 3)^2$$

$$25a^2 - 10ab + b^2 = (5a - b)^2$$

Following the examples, complete these.

a) $F^2 - 14F + 49 = (\qquad)^2$

b) $9x^2 - 12xy + 4y^2 = (\qquad)^2$

Answers:

a) $(m + 8)^2$

b) $(5d + h)^2$

70. Each trinomial below is also a perfect square.

$$x^6 + 2x^3 + 1 = (x^3 + 1)^2$$

$$p^2q^2 - 10pq + 25 = (pq - 5)^2$$

Following the examples, complete these:

a) $y^8 - 6y^4 + 9 = (\qquad)^2$

b) $4a^2b^2 + 4ab + 1 = (\qquad)^2$

a) $(F - 7)^2$

b) $(3x - 2y)^2$

---

a) $(y^4 - 3)^2$ b) $(2ab + 1)^2$

6-8 FACTORING THE SUM AND DIFFERENCE OF TWO CUBES

In this section, we will discuss the patterns used to factor the sum and difference of two cubes.

---

71. A cube root of a number N is a number whose cube is N. That is:

Since $3^3 = 27$, 3 is a cube root of 27.

Since $5^3 = 125$, 5 is a cube root of 125.

Instead of saying the cube root of, we use the symbol $\sqrt[3]{}$. That is

Since $2^3 = 8$, $\sqrt[3]{8} = 2$ Since $6^3 = 216$, $\sqrt[3]{216} =$ ______

---

72. Any whole number whose cube root is a whole number is a perfect cube. For example:

125 is a perfect cube, since $\sqrt[3]{125} = 5$

The perfect cubes with cube roots from 1 to 10 are given in the table below.

Perfect Cube	Cube Root	Perfect Cube	Cube Root
1	1	216	6
8	2	343	7
27	3	512	8
64	4	729	9
125	5	1000	10

Which of the following are perfect cubes? ____________

a) 1 b) 32 c) 64 d) 512 e) 800

6

73. When a power is cubed, the exponent is always a multiple of 3. For example:

$$(x)^3 = x^3 \qquad (y^2)^3 = y^6 \qquad (t^4)^3 = t^{12}$$

Therefore, any power whose exponent is a multiple of 3 is a perfect cube. To find the cube root of a power that is a perfect cube, we divide the exponent by 3. That is:

$$\sqrt[3]{m^3} = m^{\frac{3}{3}} = m^1 = m \qquad \sqrt[3]{b^9} = b^{\frac{9}{3}} = b^3$$

Find the cube root of each perfect cube.

a) $\sqrt[3]{t^3}$ = ______ b) $\sqrt[3]{d^6}$ = ______ c) $\sqrt[3]{p^{15}}$ = ______

(a), (c), and (d)

---

74. Both $8x^3$ and $64y^6$ are perfect cubes because their coefficients and powers are perfect cubes. To find the cube root of each, we find the cube root of the coefficient and the power. That is:

$$\sqrt[3]{8x^3} = 2x \qquad \sqrt[3]{64y^6} = 4y^2$$

Find the cube root of each perfect cube.

a) $\sqrt[3]{27t^3}$ = _____ b) $\sqrt[3]{125m^6}$ = _____ c) $\sqrt[3]{1,000q^{12}}$ = _____

a) t

b) d^2

c) p^5

---

75. We use the following pattern to factor the difference of two perfect cubes. Notice that x and y are the cube roots of x^3 and y^3.

$$x^3 - y^3 = (x - y)(x^2 + xy + y^2)$$

We used the pattern to factor $x^3 - 27$ below.

$$x^3 - 27 = (x - 3)(x^2 + 3x + 9)$$

Note:
1) x is the cube root of x^3.
2) 3 is the cube root of 27.
3) x^2 is the square of x.
4) $3x$ is the product of x and 3.
5) 9 is the square of 3.

Following the example, factor this expression.

$t^3 - 8$ = ______________________________

a) $3t$

b) $5m^2$

c) $10q^4$

$(t - 2)(t^2 + 2t + 4)$

76. We used the pattern to factor $8y^3 - 1$ below.

$$8y^3 - 1 = (2y - 1)(4y^2 + 2y + 1)$$

Note: 1) $2y$ is the cube root of $8y^3$.

2) "1" is the cube root of "1".

3) $4y^2$ is the square of $2y$.

4) $2y$ is the product of $2y$ and "1".

5) "1" is the square of "1".

Following the example, factor this expression.

$27d^3 - 1 =$ ______________________

Answer: $(3d - 1)(9d^2 + 3d + 1)$

---

77. We used the same pattern to factor $m^3 - 27p^3$ below.

$$m^3 - 27p^3 = (m - 3p)(m^2 + 3mp + 9p^2)$$

Note: 1) $\underline{m}$ is the cube root of m^3.

2) $3p$ is the cube root of $27p^3$.

3) m^2 is the square of $\underline{m}$.

4) $3mp$ is the product of $\underline{m}$ and $3p$.

5) $9p^2$ is the square of $3p$.

Following the example, factor each expression.

a) $64d^3 - t^3 =$ ______________________

b) $x^6 - y^3 =$ ______________________

Answers:

a) $(4d - t)(16d^2 + 4dt + t^2)$

b) $(x^2 - y)(x^4 + x^2y + y^2)$

---

78. We use the following pattern to factor the sum of two perfect cubes. Notice again that $\underline{x}$ and $\underline{y}$ are the cube roots of x^3 and y^3.

$$\boxed{x^3 + y^3 = (x + y)(x^2 - xy + y^2)}$$

We used the pattern above to factor each expression below.

$$x^3 + 125 = (x + 5)(x^2 - 5x + 25)$$

$$27y^3 + 1 = (3y + 1)(9y^2 - 3y + 1)$$

Use the same pattern to factor these:

a) $t^3 + 64 =$ ______________________

b) $125d^3 + 1 =$ ______________________

79. We used the same pattern to factor the expression below.

$$x^3 + 8y^3 = (x + 2y)(x^2 - 2xy + 4y^2)$$

Following the pattern, factor each expression.

a) $27m^3 + p^3$ = ______________________

b) $a^3 + b^6$ = ______________________

a) $(t+4)(t^2 - 4t + 16)$

b) $(5d+1)(25d^2 - 5d + 1)$

a) $(3m + p)(9m^2 - 3mp + p^2)$ b) $(a + b^2)(a^2 - ab^2 + b^4)$

SELF-TEST 19 (pages 247-259)

Factor each trinomial.

1. $5x^2 + 11x + 2$ = ______________

2. $6y^2 - 11y - 7$ = ______________

3. $3m^2 - 11mt + 10t^2$ = ______________

4. $4r^2s^2 + rs - 5$ = ______________

5. $d^2 - 14d + 49$ = ______________

6. $9x^2 + 30xy + 25y^2$ = ______________

Factor each binomial.

7. $36t^2 - 1$ = ______________

8. $25a^4 - 16b^2$ = ______________

9. $64d^3 - 1$ = ______________

10. $x^3 + 8y^3$ = ______________

ANSWERS:

1. $(5x + 1)(x + 2)$
2. $(3y - 7)(2y + 1)$
3. $(m - 2t)(3m - 5t)$
4. $(rs - 1)(4rs + 5)$
5. $(d - 7)^2$
6. $(3x + 5y)^2$
7. $(6t + 1)(6t - 1)$
8. $(5a^2 + 4b)(5a^2 - 4b)$
9. $(4d - 1)(16d^2 + 4d + 1)$
10. $(x + 2y)(x^2 - 2xy + 4y^2)$

6-9 FACTORING COMPLETELY

In this section, we will discuss what is meant by factoring completely.

---

80. When factoring a binomial, we always begin by looking for a common monomial factor. For example:

$$3x^2 - 12 = 3(x^2 - 4)$$

$$4y^3 + 8y = 4y(y^2 + 2)$$

To factor completely, we then look to see whether the binomial can be factored further. It can if it is the difference of two perfect squares. Therefore:

$3(x^2 - 4)$ can be factored to $3(x + 2)(x - 2)$

$4y(y^2 + 2)$ cannot be factored further.

Factor each of these completely.

a) $7a^2 - 7b^2$ = ____________________

b) $9pq^2 - 3pt^2$ = ____________________

---

81. When factoring a trinomial, we should also begin by looking for a common monomial factor. For example:

$$x^3 + 5x^2 + 6x = x(x^2 + 5x + 6)$$

$$4y^2 - 6y - 8 = 2(2y^2 - 3y - 4)$$

To factor completely, we then look to see whether the trinomial can be factored further. For example:

$x(x^2 + 5x + 6)$ can be factored to $x(x + 2)(x + 3)$

$2(2y^2 - 3y - 4)$ cannot be factored further.

Factor each of these completely.

a) $5b^2 - 10b + 15$ = ____________________

b) $ad^2 - 4ad - 5a$ = ____________________

Answers to 80:

a) $7(a + b)(a - b)$

b) $3p(3q^2 - t^2)$

---

82. After factoring out a common monomial from a trinomial, the remaining trinomial can be a perfect square. To factor completely, we must factor the perfect square. For example:

$$2x^2 - 12x + 18 = 2(x^2 - 6x + 9) = 2(x - 3)^2$$

Factor completely.

a) $6m^3 - 3m - 9$ = ____________________

b) $bx^2 + 2bxy + by^2$ = ____________________

Answers to 81:

a) $5(b^2 - 2b + 3)$

b) $a(d - 5)(d + 1)$

83. Factoring completely means this: Whenever you get a factor that can still be factored, factor it. Factor each of these completely.

a) $2xy^2 - 50x$ = ______________

b) $4bp^2 - 9bq^2$ = ______________

c) $3x^2 + 12x - 36$ = ______________

d) $9a^2t + 12abt + 4b^2t$ = ______________

Answers: a) $3(2m^3 - m - 3)$ b) $b(x + y)^2$

---

84. After factoring $(1 - 16t^4)$ below, we were able to factor $(1 - 4t^2)$ further.

$$1 - 16t^4 = (1 + 4t^2)(1 - 4t^2) = (1 + 4t^2)(1 + 2t)(1 - 2t)$$

Following the example, factor this one completely.

$x^4 - y^4$ = ______________

Answers: a) $2x(y + 5)(y - 5)$ b) $b(2p + 3q)(2p - 3q)$ c) $3(x + 6)(x - 2)$ d) $t(3a + 2b)^2$

---

85. After factoring the trinomial below, we were able to factor $(x^2 - 1)$ further.

$$x^4 - 3x^2 + 2 = (x^2 - 2)(x^2 - 1) = (x^2 - 2)(x + 1)(x - 1)$$

Following the example, factor this one completely.

$y^4 - 5y^2 + 4$ = ______________

Answer: $(x^2 + y^2)(x + y)(x - y)$

---

86. After factoring out the 2 below, we were able to factor $x^3 - y^3$ further.

$$2x^3 - 2y^3 = 2(x^3 - y^3) = 2(x - y)(x^2 + xy + y^2)$$

Following the example, factor this one completely.

$ap^3 + aq^3$ = ______________

Answer: $(y+2)(y-2)(y+1)(y-1)$

Answer: $a(p + q)(p^2 - pq + q^2)$

6-10 THE PRINCIPLE OF ZERO PRODUCTS

In this section, we will use the principle of zero products to solve equations that have two solutions or roots.

---

87. If one factor in a multiplication is 0, the product is 0. For example:

$$(5)(0) = 0 \qquad (0)(9) = 0$$

The principle of zero products is this: If the product in a multiplication is 0, one of the factors must be 0. That is:

If $ab = 0$, either $a = 0$ or $b = 0$

Using the above principle, complete this statement.

If $xy = 0$, either x or y must be ______.

Answer to 87: 0

88. The equation below involves a multiplication. The factors are $(x - 2)$ and $(x - 3)$.

$$(x - 2)(x - 3) = 0$$

Since the product is 0, the equation is true when either $(x - 2)$ or $(x - 3)$ is 0. Therefore, we can find the two solutions of the equation by solving the two equations below.

$$x - 2 = 0 \qquad x - 3 = 0$$

$$x = 2 \qquad x = 3$$

We checked 2 as one solution of the original equation. Check 3 as the second solution.

$$(x - 2)(x - 3) = 0 \qquad (x - 2)(x - 3) = 0$$

$$(2 - 2)(2 - 3) = 0$$

$$(0)(-1) = 0$$

$$0 = 0$$

Answer to 88:

$$(3 - 2)(3 - 3) = 0$$

$$(1)(0) = 0$$

$$0 = 0$$

89. The equation below is true if either $(y + 4)$ or $(y - 5)$ is 0.

$$(y + 4)(y - 5) = 0$$

Therefore, to find the two solutions, we solved the equations below.

$$y + 4 = 0 \qquad y - 5 = 0$$

$$y = -4 \qquad y = 5$$

The two solutions of the original equation are ______ and ______.

90. We solved the equation below by setting both factors equal to 0. Solve the other equation.

$(t + 2)(t + 8) = 0$		$(m + 10)(m - 10) = 0$
$t + 2 = 0$	$t + 8 = 0$	
$t = -2$	$t = -8$	

Answer: -4 and 5

---

91. The equation below is true if either $(3x - 2)$ or $(2x + 5)$ is 0.

$$(3x - 2)(2x + 5) = 0$$

Therefore to find the two solutions, we solved the equations below.

$3x - 2 = 0$	$2x + 5 = 0$
$3x = 2$	$2x = -5$
$x = \frac{2}{3}$	$x = -\frac{5}{2}$

The two solutions of the original equation are ______ and ______.

Answer: $m = -10$ and 10

---

92. We solved the equation below by setting both factors equal to 0. Solve the other equation.

$(3p + 1)(4p - 7) = 0$		$(6t + 1)(6t - 1) = 0$
$3p + 1 = 0$	$4p - 7 = 0$	
$3p = -1$	$4p = 7$	
$p = -\frac{1}{3}$	$p = \frac{7}{4}$	

Answer: $\frac{2}{3}$ and $-\frac{5}{2}$

---

93. The equation below is true if either (x) or $(x - 3)$ is 0.

$$x(x - 3) = 0$$

Therefore to find the two solutions, we solved the equations below.

$x = 0$	$x - 3 = 0$
	$x = 3$

The two solutions of the original equation are ______ and ______.

Answer: $t = -\frac{1}{6}$ and $\frac{1}{6}$

---

94. In the last frame, we saw that 0 is one of the solutions of $x(x - 3) = 0$. The 0 solution is frequently overlooked. Following the example, solve the other equation.

$m(m + 6) = 0$		$y(y - 12) = 0$
$m = 0$	$m + 6 = 0$	
	$m = -6$	

Answer: 0 and 3

95. We solved the equation below by setting both factors equal to 0. Solve the other equation.

$$b(2b - 1) = 0 \qquad\qquad x(5x + 6) = 0$$

$b = 0$ | $2b - 1 = 0$, $2b = 1$, $b = \frac{1}{2}$

Answer: y = 0 and 12

---

96. We solved one equation below. Solve the other equation.

$$4y(y + 3) = 0 \qquad\qquad 7t(4t - 3) = 0$$

$4y = 0$, $y = \frac{0}{4}$, $y = 0$ | $y + 3 = 0$, $y = -3$

Answer: $x = 0$ and $-\frac{6}{5}$

---

97. The principle of zero products applies only when the product is 0. Therefore, it could not be used with either equation below.

$$(x - 3)(x - 9) = 5 \qquad\qquad y(y + 8) = 9$$

In which equations below does the principle apply? ______________

a) $(3x - 1)(2x + 3) = 7$ b) $d(d - 1) = 0$ c) $p(p + 4) = 12$

Answer: $t = 0$ and $\frac{3}{4}$

Answer: Only (b)

6-11 SOLVING EQUATIONS BY FACTORING

In this section, we will solve equations by factoring and then using the principle of zero products.

---

98. To solve the equation below, we factored the left side and then used the principle of zero products.

$$x^2 + 3x - 10 = 0$$

$$(x - 2)(x + 5) = 0$$

$x - 2 = 0$, $x = 2$ | $x + 5 = 0$, $x = -5$

Continued on following page.

98. Continued

The two solutions are 2 and -5. We checked 2 as a solution of the original equation below. Check -5 as a solution.

$$x^2 + 3x - 10 = 0$$
$$(2)^2 + 3(2) - 10 = 0$$
$$4 + 6 - 10 = 0$$
$$10 - 10 = 0$$
$$0 = 0$$

$$x^2 + 3x - 10 = 0$$

$(-5)^2 + 3(-5) - 10 = 0$

$25 - 15 - 10 = 0$

$25 - 25 = 0$

$0 = 0$

99. We used the factoring method to solve one equation below. Solve the other equation.

$$y^2 + 7y + 6 = 0$$
$$(y + 1)(y + 6) = 0$$

$y + 1 = 0$	$y + 6 = 0$
$y = -1$	$y = -6$

$$m^2 - 8m - 9 = 0$$

---

m = -1 and 9, from:

$(m + 1)(m - 9) = 0$

100. Following the example, solve the other equation.

$$5d^2 + 3d - 2 = 0$$
$$(5d - 2)(d + 1) = 0$$

$5d - 2 = 0$	$d + 1 = 0$
$5d = 2$	$d = -1$
$d = \frac{2}{5}$	

$$2y^2 - 7y + 6 = 0$$

$y = \frac{3}{2}$ and 2, from:

$(2y - 3)(y - 2) = 0$

101. Following the example, solve the other equation.

$$9x^2 - 16 = 0$$
$$(3x + 4)(3x - 4) = 0$$

$3x + 4 = 0$	$3x - 4 = 0$
$3x = -4$	$3x = 4$
$x = -\frac{4}{3}$	$x = \frac{4}{3}$

$$t^2 - 49 = 0$$

---

t = -7 and 7, from:

$(t + 7)(t - 7) = 0$

102. We factored to solve the equation below. Notice that one root is 0. Solve the other equation.

$$x^2 + 8x = 0$$
$$x(x + 8) = 0$$

$x = 0$	$x + 8 = 0$
	$x = -8$

$$t^2 - t = 0$$

103. We used the factoring method to solve one equation below. Solve the other equation.

$$2x^2 - 3x = 0 \qquad\qquad 7y^2 + 5y = 0$$

$$x(2x - 3) = 0$$

$x = 0$	$2x - 3 = 0$
	$2x = 3$
	$x = \frac{3}{2}$

Answer: $t = 0$ and 1, from:

$t(t - 1) = 0$

---

104. When using the factoring method, we factor completely. An example is shown. Solve the other equation.

$$8x^2 + 12x = 0 \qquad\qquad 12y^2 - 6y = 0$$

$$4x(2x + 3) = 0$$

$4x = 0$	$2x + 3 = 0$
$x = 0$	$2x = -3$
	$x = -\frac{3}{2}$

Answer: $y = 0$ and $-\frac{5}{7}$, from:

$y(7y + 5) = 0$

$y = 0$	$7y + 5 = 0$
	$7y = -5$
	$y = -\frac{5}{7}$

---

105. Following the example, solve the other equation.

$$7x - x^2 = 0 \qquad\qquad 2y - 5y^2 = 0$$

$$x(7 - x) = 0$$

$x = 0$	$7 - x = 0$
	$x = 7$

Answer: $y = 0$ and $\frac{1}{2}$, from:

$6y(2y - 1) = 0$

---

106. To use the factoring method, all of the terms must be on one side with 0 on the other side. Therefore, we cannot use the factoring method with the equation below as it stands.

$$m^2 = 7m - 8$$

However, we can put it in the form needed for the factoring method by adding $-7m$ and 8 to both sides. We get:

$$m^2 - 7m + 8 = 7m - 7m - 8 + 8$$

$$m^2 - 7m + 8 = 0$$

Using the same steps, put each of these in the form needed for the factoring method.

a) $7x^2 = 3x + 4$ b) $2R^2 = 8R$

Answer: $y = 0$ and $\frac{2}{5}$, from:

$y(2 - 5y) = 0$

107. To solve the equation below, we began by putting it in the form needed for the factoring method. Solve the other equation.

$$y^2 = y + 6$$
$$y^2 - y - 6 = 0$$
$$(y + 2)(y - 3) = 0$$

$y + 2 = 0$	$y - 3 = 0$
$y = -2$	$y = 3$

$$3m^2 + 2 = 7m$$

Answers: a) $7x^2 - 3x - 4 = 0$ b) $2R^2 - 8R = 0$

108. To solve the equation below, we began by putting it in the form needed for the factoring method. Solve the other equation.

$$t^2 = 2t$$
$$t^2 - 2t = 0$$
$$t(t - 2) = 0$$

$t = 0$	$t - 2 = 0$
	$t = 2$

$$4d^2 = 1$$

Answer: $m = \frac{1}{3}$ and 2, from:

$3m^2 - 7m + 2 = 0$

$(3m - 1)(m - 2) = 0$

109. To solve the equation below, we began by putting it in the form needed for the factoring method. Solve the other equation.

$$3 + p(2p - 5) = 15$$
$$3 + 2p^2 - 5p = 15$$
$$2p^2 - 5p - 12 = 0$$
$$(2p + 3)(p - 4) = 0$$

$2p + 3 = 0$	$p - 4 = 0$
$2p = -3$	$p = 4$
$p = -\frac{3}{2}$	

$$4y(y - 1) = 3$$

Answer: $d = -\frac{1}{2}$ and $\frac{1}{2}$, from:

$4d^2 - 1 = 0$

$(2d + 1)(2d - 1) = 0$

110. To solve the equation below, we began by factoring out the largest common numerical factor. Use the same method to solve the other equation.

$$3x^2 + 12x + 9 = 0$$
$$3(x^2 + 4x + 3) = 0$$
$$\frac{1}{3}(3)(x^2 + 4x + 3) = \frac{1}{3}(0)$$
$$x^2 + 4x + 3 = 0$$
$$(x + 1)(x + 3) = 0$$

$x + 1 = 0$	$x + 3 = 0$
$x = -1$	$x = -3$

$$30t^2 - 80t + 50 = 0$$

Answer: $y = -\frac{1}{2}$ and $\frac{3}{2}$, from:

$4y^2 - 4y - 3 = 0$

Answer: $t = \frac{5}{3}$ and 1, from: $3t^2 - 8t + 5 = 0$

6-12 WORD PROBLEMS

In this section, we will solve some word problems that require the factoring method.

---

111. We translated the English sentence below to an equation.

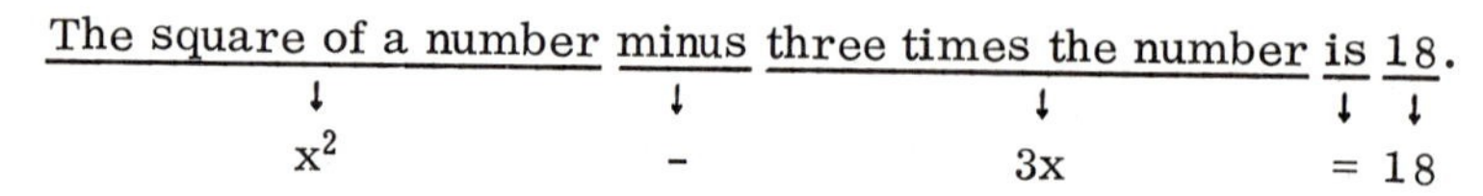

Following the example, translate each of these to an equation. Use x for the variable.

a) 10 more than the square of a number is seven times that number.

b) Two more than a number times one less than that number is 4.

---

a) $x^2 + 10 = 7x$

b) $(x + 2)(x - 1) = 4$

112. We used the four-step method introduced earlier to solve the problem below.

Problem: Six less than the square of a number is five times that number. Find the number.

1) Represent the unknown with a letter.

Let x equal the unknown number.

2) Translate the problem to an equation.

$$x^2 - 6 = 5x$$

3) Solve the equation.

$$x^2 - 6 = 5x$$

$$x^2 - 5x - 6 = 0$$

$$(x - 6)(x + 1) = 0$$

$x - 6 = 0$	$x + 1 = 0$
$x = 6$	$x = -1$

The two solutions are 6 and -1.

4) Check each solution in the original wording of the problem.

Checking 6. 6 squared is 36. 5 times 6 is 30. 6 less than 36 does equal 30.

Checking -1. -1 squared is 1. 5 times -1 is -5. 6 less than 1 does equal -5.

Continued on following page.

112. Continued

Following the example, solve these.

a) The square of a number minus that number is 12. Find the number.

b) If you subtract a number from three times its square, you get 2. Find the number.

---

113. Following the example, solve the other problem.

One more than a number times one less than that number is 15. Find the number.

$$(x + 1)(x - 1) = 15$$
$$x^2 - 1 = 15$$
$$x^2 - 16 = 0$$
$$(x + 4)(x - 4) = 0$$

$x + 4 = 0$	$x - 4 = 0$
$x = -4$	$x = 4$

Two solutions are -4 and 4.

Two more than a number times two less than that number is 45. Find the number.

Answers to 112:

a) 4 and -3, from: $x^2 - x = 12$

b) $-\frac{2}{3}$ and 1, from: $3x^2 - x = 2$

---

114. To solve the problem below, we let x equal the smaller integer and x + 1 equal the larger integer. Solve the other problem.

The product of two consecutive integers is 56. Find the integers.

$$x(x + 1) = 56$$
$$x^2 + x = 56$$
$$x^2 + x - 56 = 0$$
$$(x - 7)(x + 8) = 0$$

$x - 7 = 0$	$x + 8 = 0$
$x = 7$	$x = -8$

If $x = 7$, $x + 1 = 8$. The two consecutive integers are 7 and 8.

If $x = -8$, $x + 1 = -7$. The two consecutive integers are -8 and -7.

The product of two consecutive integers is 110. Find the integers.

Answer to 113:

-7 and 7, from: $(x + 2)(x - 2) = 45$

115. To solve the problem below, we let x equal the smaller integer and x + 2 equal the larger integer. Solve the other problem.

The product of two consecutive even integers is 48. Find the integers.

$$x(x + 2) = 48$$

$$x^2 + 2x = 48$$

$$x^2 + 2x - 48 = 0$$

$$(x - 6)(x + 8) = 0$$

$x - 6 = 0$, $x = 6$ | $x + 8 = 0$, $x = -8$

If $x = 6$, $x + 2 = 8$. The two consecutive even integers are 6 and 8.

If $x = -8$, $x + 2 = -6$. The two consecutive even integers are -8 and -6.

The product of two consecutive odd integers is 99. Find the integers.

10 and 11
or
-11 and -10

---

116. To solve the problem below, we let x equal the width and x + 5 equal the length. Solve the other problem.

The length of a rectangle is 5 ft more than the width. The area of the rectangle is 84 ft^2. Find the length and width.

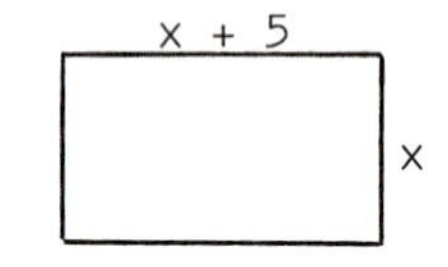

$$A = LW$$

$$84 = (x + 5)(x)$$

$$84 = x^2 + 5x$$

$$0 = x^2 + 5x - 84$$

$$0 = (x - 7)(x + 12)$$

$x - 7 = 0$, $x = 7$ | $x + 12 = 0$, $x = -12$

Though the solutions of the equation are 7 and -12, the width of a rectangle cannot have a negative value.

Therefore, W = 7ft and L = 12 ft.

The width of a rectangle is 7m less than the length. The area of the rectangle is $120m^2$. Find the length and width.

9 and 11
or
-11 and -9

L = 15m, W = 8m

117. Following the example, solve the other problem.

The height of a triangle is 10 cm more than the base. If the area of the triangle is 48 cm^2, find the height and base.

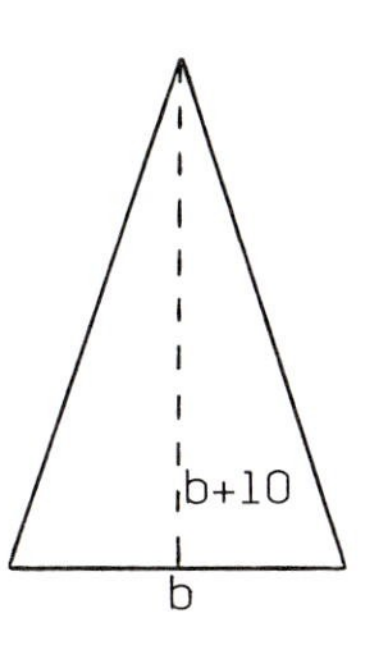

$$A = \frac{1}{2}bh$$

$$48 = \frac{1}{2}(b)(b + 10)$$

$$48 = \frac{1}{2}(b^2 + 10b)$$

$$96 = b^2 + 10b$$

$$0 = b^2 + 10b - 96$$

$$0 = (b - 6)(b + 16)$$

$b - 6 = 0$	$b + 16 = 0$
$b = 6$	$b = -16$

Though the solutions of the equation are 6 and -16, the base of a triangle cannot have a negative length.

Therefore, the base is 6 cm and the height is 16 cm.

The height of a triangle is 6m less than the base. If the area of the triangle is 36m^2, find the height and base.

---

118. If an object is dropped, the distance d it falls in t seconds (disregarding air resistance) is given by the formula below.

$$d = 16t^2$$

Continued on following page.

$b = 12m$, $h = 6m$

118. Continued

We used the formula to solve one problem below. Solve the other problem.

How long would it take an object to fall 64 ft?

$$d = 16t^2$$

$$64 = 16t^2$$

$$0 = 16t^2 - 64$$

$$0 = 16(t^2 - 4)$$

$$0 = t^2 - 4$$

$$0 = (t + 2)(t - 2)$$

$t + 2 = 0$	$t - 2 = 0$
$t = -2$	$t = 2$

Though the solutions of the equation are -2 and 2, a negative time interval does not make sense.

Therefore, it takes 2 seconds to drop 64 ft.

How long would it take an object to fall 400 ft?

---

119. If an object is projected straight up with an original velocity of v_0 feet per second, its height h after t seconds is given by the formula below.

t = 5 seconds

$$h = v_0t - 16t^2$$

We used the formula to solve one problem. To do so, we used the fact that the height h is 0 when the object hits the ground. Solve the other problem.

Suppose an object is projected upward with an original velocity of 80 feet per second. How long will it take the object to hit the ground?

$$h = v_0t - 16t^2$$

$$0 = 80t - 16t^2$$

$$0 = 16t(5 - t)$$

$16t = 0$	$5 - t = 0$
$t = 0$	$t = 5$

Though the two solutions of the equation are 0 and 5, a time of 0 seconds does not make sense for the rise and fall of the object.

Therefore, the object hits the ground at 5 seconds.

Suppose an object is projected upward with an original velocity of 192 feet per second. How long will it take the object to hit the ground?

---

t = 12 seconds

SELF-TEST 20 (pages 260-273)

Factor completely.

1. $3x^2 - 27$

2. $ab^2 - 2ab - 35a$

3. $4x^2 - 14x + 10$

4. $1 - 16y^4$

Solve by factoring.

5. $4d^2 - 5d = 0$

6. $k^2 - 7k - 18 = 0$

7. $5y^2 + 11 = 8(y + 1)$

8. $60t^2 + 10t = 20$

9. The product of two consecutive even integers is 80. Find the integers.

10. The length of a rectangle is 3m more than the width. The area of the rectangle is $88m^2$. Find the length and width.

ANSWERS:

1. $3(x + 3)(x - 3)$
2. $a(b + 5)(b - 7)$
3. $2(2x - 5)(x - 1)$
4. $(1 + 4y^2)(1 + 2y)(1 - 2y)$
5. $d = 0$ and $\frac{5}{4}$
6. $k = 9$ and -2
7. $y = \frac{3}{5}$ and 1
8. $t = -\frac{2}{3}$ and $\frac{1}{2}$
9. 8 and 10 or -10 and -8
10. L = 11m, W = 8m

SUPPLEMENTARY PROBLEMS - CHAPTER 6

Assignment 18

Write the missing factor in each blank.

1. $48 = (12)(\quad)$
2. $18x = (\quad)(3)$
3. $40y^2 = (\quad)(5y)$
4. $6t^2 = (t)(\quad)$
5. $7d^4 = (\quad)(d^2)$
6. $x^3y^3 = (xy^2)(\quad)$
7. $12a^2b^5 = (3a^2b^2)(\quad)$
8. $p^4q^5 = (\quad)(p^4q^5)$

Factor out the largest common monomial factor.

9. $8x + 20$
10. $18y - 12$
11. $4t + 4$
12. $25d - 5$
13. $9a^2 + 15$
14. $60V^2 - 36$
15. $3x^2 + 5x$
16. $4y^6 - y^4$
17. $12d^3 + 24d$
18. $40p^5 - 25p^3$
19. $x^3y + 2xy$
20. $10p^5q^5 - 5p^2q^2$
21. $12x^2 - 6x + 30$
22. $y^4 + 7y^2 - y$
23. $2t^3 - 6t^2 + 4t$
24. $9a^4b^3 + 6a^3b^2 - 3ab^2$

Factor by grouping.

25. $x^2 + 8x + 2x + 16$
26. $y^2 - 7y + 3y - 21$
27. $4t^2 + 8t + 5t + 10$
28. $6d^2 - 3d + 14d - 7$
29. $4a^2 + 12at + 3at + 9t^2$
30. $12p^2 - 8pq + 15pq - 10q^2$
31. $x^2 + 2x - 3x - 6$
32. $y^2 + 6y - 2y - 12$
33. $b^2 - 3b - 4b + 12$

Factor each trinomial.

34. $x^2 + 3x + 2$
35. $a^2 + 7a + 10$
36. $y^2 + 6y + 9$
37. $t^2 - 4t + 3$
38. $R^2 - 6R + 8$
39. $w^2 - w - 12$
40. $h^2 + 4h - 5$
41. $d^2 - 3d - 10$
42. $m^2 + 4m - 21$
43. $p^2 + 8pq + 15q^2$
44. $a^2 - ab - 6b^2$
45. $c^2d^2 - 6cd + 5$
46. $x^2y^2 + 5xy - 14$
47. $t^4 - 7t^2 + 12$
48. $p^6q^6 - 2p^3q^3 - 24$

Assignment 19

Factor each trinomial.

1. $2y^2 + 3y + 1$
2. $3x^2 + 7x + 2$
3. $3p^2 + 11p + 6$
4. $2r^2 - 5r + 2$
5. $5t^2 - 8t + 3$
6. $6w^2 - 11w + 4$
7. $4E^2 + 4E - 3$
8. $3d^2 + 2d - 8$
9. $7x^2 - 4x - 3$
10. $8m^2 - 6m - 9$
11. $2b^2 + 7bt + 3t^2$
12. $2r^2 + 5rs - 3s^2$
13. $5a^2 + 11ab + 2b^2$
14. $3t^2w^2 - 7tw - 6$
15. $4x^2 + 5xy - 6y^2$
16. $x^2 + 6x + 9$
17. $y^2 - 12y + 36$
18. $4b^2 - 4b + 1$
19. $9w^2 + 12w + 4$
20. $4a^2 + 20ab + 25b^2$
21. $16d^2h^2 - 8dh + 1$

Factor each binomial.

22. $x^2 - 49$
23. $y^2 - 100$
24. $25m^2 - 1$
25. $4t^2 - 81$
26. $a^4 - 36$
27. $9b^6 - 4$
28. $x^2 - y^2$
29. $c^2 - 16d^2$
30. $64p^2 - 9q^2$
31. $25x^4 - y^2$
32. $1 - b^2c^2$
33. $36a^2d^6 - 49$
34. $x^3 - 1$
35. $y^3 + 125$
36. $8b^3 - 27$
37. $125p^3 + 1$
38. $c^3 - d^3$
39. $x^3 + y^6$
40. $m^3 - 64t^3$
41. $27x^3 + y^3$

Assignment 20

Factor completely.

1. $5x^2 - 20$
2. $10ab^2 - 4ad^2$
3. $cp^2 - 8cp + 7c$
4. $2y^2 - 2y - 24$
5. $bp^2 + 2bpq + bq^2$
6. $81t^4 - 1$
7. $y^4 - 3y^2 - 4$
8. $8ax^3 + a$

Solve each equation by factoring.

9. $5x^2 - 3x = 0$
10. $3d^2 - d = 0$
11. $4y + 8y^2 = 0$
12. $t^2 = t$
13. $20w^2 = 5w$
14. $12R = 8R^2$
15. $x^2 - 3x - 10 = 0$
16. $k^2 + 6k - 7 = 0$
17. $r^2 - 2r + 1 = 0$
18. $t(t - 2) = 15$
19. $5F - 4 = F^2$
20. $20x^2 + 40x - 60 = 0$
21. $2t^2 + 3t - 2 = 0$
22. $3d^2 - 2d - 1 = 0$
23. $5p^2 - 11p + 2 = 0$
24. $9 - 7x = 6(x^2 + 1)$
25. $3h^2 = 2(h + 4)$
26. $10y^2 - 55y + 75 = 0$

Solve each problem.

27. The square of a number minus three times that number equals 10. Find the number.
28. If a number is subtracted from double its square, the result is 15. Find the number.
29. Three more than a number times three less than that number is 27. Find the number.
30. The product of two consecutive integers is 72. Find the integers.
31. The product of two consecutive odd integers is 63. Find the integers.
32. The product of two consecutive even integers is 120. Find the integers.
33. The length of a rectangle is 6 ft more than the width. The area of the rectangle is 40 ft^2. Find the length and width.
34. The height of a triangle is 3m less than the base. If the area of the triangle is 35m^2, find the height and base.
35. If an object is dropped, the distance d it falls in t seconds is given by the formula: $d = 16t^2$. How long would it take an object to fall 144 ft?
36. If an object is projected vertically upward with a velocity of 160 feet per second, it reaches a height of h feet in t seconds according to the formula: $h = 160t - 16t^2$? In how many seconds will it reach a height of 400 feet?

7 Rational Expressions

In this chapter, we will define rational expressions and discuss the basic operations with them. We will discuss a method for solving fractional equations and methods for solving various types of word problems involving fractional equations. Some evaluations and rearrangements of fractional formulas are included.

7-1 EQUIVALENT RATIONAL EXPRESSIONS

In this section, we will define rational expressions and discuss the procedures for obtaining equivalent rational expressions. The procedure for reducing rational expressions to lowest terms is emphasized.

1. A quotient of two polynomials is called a rational expression. Some examples of rational expressions are:

$$\frac{3x}{5} \qquad \frac{y}{y-2} \qquad \frac{t^2+5t+6}{t^2-9}$$

A rational expression indicates a division. That is:

$\frac{3x}{5}$ means $3x \div 5$ $\qquad$ $\frac{y}{y-2}$ means _____ ÷ _____

2. Some substitutions in rational expressions do not make sense because they lead to a division by 0, and division by 0 is not possible. For example:	$y \div (y - 2)$

If we substitute 0 for x below, we get a division by 0.

$$\frac{10}{x} = \frac{10}{0} \qquad \text{Note: } \frac{10}{0} \text{ is not possible.}$$

If we substitute 3 for x below, we get a division by 0.

$$\frac{x+5}{x-3} = \frac{3+5}{3-3} = \frac{8}{0} \qquad \text{Note: } \frac{8}{0} \text{ is not possible.}$$

To avoid the problem above, we will do the following in this text:

> We will avoid substitutions in rational expressions that do not make sense. That is, we will assume that a variable does not take on a value that leads to a division by 0.

---

3. To multiply two rational expressions, we multiply their numerators and multiply their denominators. For example:

$$\left(\frac{3}{x+2}\right)\left(\frac{x-1}{x-5}\right) = \frac{3(x-1)}{(x+2)(x-5)}$$

$$\left(\frac{y-9}{7}\right)\left(\frac{y+1}{y-3}\right) = \underline{\qquad\qquad}$$

Answer: $\frac{(y-9)(y+1)}{7(y-3)}$

4. To multiply a rational expression by a polynomial, we multiply the numerator by the polynomial. For example:

$$2x\left(\frac{x-1}{7}\right) = \frac{2x(x-1)}{7} \qquad \left(\frac{x+3}{x-2}\right)(x+4) = \frac{(x+3)(x+4)}{x-2}$$

Complete these:

a) $(y-3)\left(\frac{y-1}{y+1}\right) = \underline{\qquad\qquad}$ b) $\left(\frac{a-7}{5}\right)(3a) = \underline{\qquad\qquad}$

Answers: a) $\frac{(y-3)(y-1)}{y+1}$ b) $\frac{(a-7)(3a)}{5}$

5. The following two facts are used to get an expression that is equivalent to a rational expression.

1. If we multiply a rational expression by "1", the product is identical to the rational expression. That is:

$$(1)\left(\frac{3t}{7}\right) = \frac{3t}{7} \qquad \left(\frac{y+5}{y-1}\right)(1) = \frac{y+5}{y-1}$$

2. Any rational expression with the same numerator and denominator equals "1". That is:

$$\frac{3y}{3y} = 1 \qquad \frac{t-2}{t-2} = 1 \qquad \frac{7(x+5)}{7(x+5)} = 1$$

Continued on following page.

5. Continued

To get an expression that is equivalent to a rational expression, we multiply the rational expression by an expression that equals "1". For example:

$$\frac{y+5}{y-1} = \left(\frac{y+5}{y-1}\right)(1) = \left(\frac{y+5}{y-1}\right)\left(\frac{3y}{3y}\right) = \frac{(y+5)(3y)}{(y-1)(3y)}$$

$$\frac{3t}{7} = \left(\frac{3t}{7}\right)(1) = \left(\frac{3t}{7}\right)\left(\frac{t-2}{t-2}\right) = \underline{\qquad\qquad}$$

---

6. To reduce a rational expression to lowest terms, we simply reverse the procedure in the last frame. That is, we factor out an expression that equals "1". For example:

$$\frac{5x}{7x} = \left(\frac{5}{7}\right)\left(\frac{x}{x}\right) = \left(\frac{5}{7}\right)(1) = \frac{5}{7}$$

$$\frac{3y}{2y} = (\quad)(\quad) = (\quad)(\quad) = \underline{\qquad}$$

Answer: $\frac{3t(t-2)}{7(t-2)}$

---

7. We reduced each rational expression below to lowest terms.

$$\frac{4t}{t} = \frac{4t}{1t} = \left(\frac{4}{1}\right)\left(\frac{t}{t}\right) = (4)(1) = 4$$

$$\frac{cd}{9cd} = \frac{1cd}{9cd} = \left(\frac{1}{9}\right)\left(\frac{cd}{cd}\right) = \frac{1}{9}(1) = \frac{1}{9}$$

Following the examples, reduce these to lowest terms.

a) $\frac{7ab}{ab} =$ b) $\frac{y}{10y} =$

Answer: $\left(\frac{3}{2}\right)\left(\frac{y}{y}\right) = \left(\frac{3}{2}\right)(1) = \frac{3}{2}$

---

8. Following the example, reduce the other expression to lowest terms.

$$\frac{4x^2}{12x} = \left(\frac{4x}{4x}\right)\left(\frac{x}{3}\right) = (1)\left(\frac{x}{3}\right) = \frac{x}{3}$$

$$\frac{15y}{10y^2} =$$

Answer: a) 7 b) $\frac{1}{10}$

---

9. To reduce the expression below to lowest terms, we began by factoring out a common monomial factor from each binomial. Use the same method to reduce the other expression to lowest terms.

$$\frac{2m+8}{3m+12} = \frac{2(m+4)}{3(m+4)} = \left(\frac{2}{3}\right)\left(\frac{m+4}{m+4}\right) = \left(\frac{2}{3}\right)(1) = \frac{2}{3}$$

$$\frac{10t^2-5}{12t^2-6} =$$

Answer: $\frac{3}{2y}$, from $\left(\frac{5y}{5y}\right)\left(\frac{3}{2y}\right)$

---

10. Following the example, reduce the other expression to lowest terms.

$$\frac{3x^2+6x}{9x^2-3x} = \frac{3x(x+2)}{3x(3x-1)} = \left(\frac{3x}{3x}\right)\left(\frac{x+2}{3x-1}\right) = (1)\left(\frac{x+2}{3x-1}\right) = \frac{x+2}{3x-1}$$

$$\frac{8y^2-6y}{2y^2+4y} =$$

Answer: $\frac{5}{6}$, from $\frac{5(2t^2-1)}{6(2t^2-1)}$

11. To reduce the expression below to lowest terms, we began by factoring each trinomial. Use the same method to reduce the other expression to lowest terms.

$$\frac{x^2 + x - 6}{x^2 - 3x + 2} = \frac{(x + 3)(x - 2)}{(x - 1)(x - 2)} = \left(\frac{x + 3}{x - 1}\right)\left(\frac{x - 2}{x - 2}\right) = \left(\frac{x + 3}{x - 1}\right)(1) = \frac{x + 3}{x - 1}$$

$$\frac{y^2 + 5y + 4}{y^2 + 2y - 8} =$$

Answer: $\frac{4y - 3}{y + 2}$, from $\frac{2y(4y - 3)}{2y(y + 2)}$

---

12. Following the example, reduce the other expression to lowest terms.

$$\frac{t^2 + 6t + 5}{t^2 - 1} = \frac{(t + 5)(t + 1)}{(t + 1)(t - 1)} = \left(\frac{t + 1}{t + 1}\right)\left(\frac{t + 5}{t - 1}\right) = (1)\left(\frac{t + 5}{t - 1}\right) = \frac{t + 5}{t - 1}$$

$$\frac{m^2 - 1}{m^2 - 4m + 3} =$$

Answer: $\frac{y + 1}{y - 2}$, from $\left(\frac{y + 4}{y + 4}\right)\left(\frac{y + 1}{y - 2}\right)$

---

13. Following the example, reduce the other expression to lowest terms.

$$\frac{2x + 6}{8} = \frac{2(x + 3)}{2(4)} = \left(\frac{2}{2}\right)\left(\frac{x + 3}{4}\right) = (1)\left(\frac{x + 3}{4}\right) = \frac{x + 3}{4}$$

$$\frac{20y}{10y - 5} =$$

Answer: $\frac{m + 1}{m - 3}$

---

14. In $\frac{x - 3}{3 - x}$, the numerator and denominator are opposites. To reduce the expression to lowest terms, we use these facts.

1. $-(x - 3)$ can be substituted for $3 - x$ since:

$$-(x - 3) = -x + 3 = 3 - x$$

2. A negative sign can be moved from the denominator and become the sign of the fraction. That is:

$$\frac{x - 3}{-(x - 3)} = -\frac{x - 3}{x - 3}$$

Using the facts above, we reduced the expression to lowest terms below.

$$\frac{x - 3}{3 - x} = \frac{x - 3}{-(x - 3)} = -\left(\frac{x - 3}{x - 3}\right) = -1$$

Following the example, reduce these to lowest terms.

a) $\frac{a - b}{b - a} =$ ________ b) $\frac{2y - 5}{5 - 2y} =$ ________

Answer: $\frac{4y}{2y - 1}$

---

15. To reduce the expression below to lowest terms, we substituted $-(x - 2)$ for $(2 - x)$. Reduce the other expression to lowest terms.

$$\frac{x^2 - 4}{2 - x} = \frac{(x + 2)(x - 2)}{-(x - 2)} = -\left(\frac{x - 2}{x - 2}\right)(x + 2) = -1(x + 2) = -(x + 2)$$

$$\frac{p^2 - q^2}{q - p} =$$

Answer: a) -1 b) -1

Answer: $-(p + q)$

7-2 CANCELLING TO REDUCE TO LOWEST TERMS

Instead of reducing rational expressions to lowest terms by factoring out an expression that equals "1", we can use a shorter method called cancelling. We will discuss cancelling in this section.

16. The rational expression below contains x as a factor in both its numerator and its denominator. We reduced it to lowest terms by cancelling the x's. Use cancelling to reduce the other expressions to lowest terms.

$\frac{4\not{x}}{7\not{x}} = \frac{4}{7}$ a) $\frac{5y}{2y} =$ ______ b) $\frac{at^2}{bt^2} =$ ______

17. Before cancelling below, we substituted 1x for x and 1pq for pq.

$\frac{9x}{x} = \frac{9\not{x}}{1\not{x}} = 9$ $\frac{pq}{3pq} = \frac{1\not{pq}}{3\not{pq}} = \frac{1}{3}$

Following the examples, reduce these to lowest terms.

a) $\frac{6dt}{dt} =$ ______ b) $\frac{R}{bR} =$ ______

Answers to 16: a) $\frac{5\not{y}}{2\not{y}} = \frac{5}{2}$ b) $\frac{a\not{t^2}}{b\not{t^2}} = \frac{a}{b}$

18. To reduce the expression below to lowest terms, we cancelled the x in the denominator and one of the x's in the numerator. Reduce the other expressions to lowest terms.

$\frac{3x^{\not{2}\,1}}{5\not{x}} = \frac{3x}{5}$ a) $\frac{y^2}{2y} =$ ______ b) $\frac{7m}{4m^2} =$ ______

Answers to 17: a) $\frac{6\not{dt}}{1\not{dt}} = 6$ b) $\frac{1\not{R}}{b\not{R}} = \frac{1}{b}$

19. To reduce the expression below to lowest terms, we cancelled the x's and divided both the 6 and 8 by 2. Reduce the other expressions to lowest terms.

$\frac{\overset{3}{\not{6}}\not{x}}{\underset{4}{\not{8}}\not{x}} = \frac{3}{4}$ a) $\frac{12c}{8c} =$ ______ b) $\frac{7pq}{14pq} =$ ______

Answers to 18: a) $\frac{y^{\not{2}\,1}}{2\not{y}} = \frac{y}{2}$ b) $\frac{7\not{m}}{4m^{\not{2}\,1}} = \frac{7}{4m}$

20. Following the example, reduce the other rational expressions to lowest terms.

$\frac{\overset{1}{\not{3}}x^{\not{2}\,1}}{\underset{3}{\not{9}}\not{x}} = \frac{x}{3}$ a) $\frac{10y^2}{6y} =$ ______ b) $\frac{8t}{4t^2} =$ ______

Answers to 19: a) $\frac{\overset{3}{\not{12}}\not{c}}{\underset{2}{\not{8}}\not{c}} = \frac{3}{2}$ b) $\frac{\overset{1}{\not{7}}\not{pq}}{\underset{2}{\not{14}}\not{pq}} = \frac{1}{2}$

21. After factoring below, we got (x + 3) in both terms. We reduced to lowest terms by cancelling both binomials.

$$\frac{2x+6}{3x+9} = \frac{2(x+3)}{3(x+3)} = \frac{2}{3}$$

Following the example, reduce these to lowest terms.

a) $\frac{5x-5}{4x-4} =$ b) $\frac{12y+4}{18y+6} =$

Answers: a) $\frac{10y^2}{6y} = \frac{5y}{3}$ b) $\frac{8t}{4t^2} = \frac{2}{t}$

---

22. After factoring below, we cancelled the 4x in each term.

$$\frac{4x^2+4x}{4x^2-4x} = \frac{4x(x+1)}{4x(x-1)} = \frac{x+1}{x-1}$$

Following the example, reduce these to lowest terms.

a) $\frac{2y^2-6y}{4y^2+2y} =$ b) $\frac{3t^2-6t}{18t^2+12t} =$

Answers: a) $\frac{5(x-1)}{4(x-1)} = \frac{5}{4}$ b) $\frac{4(3y+1)}{6(3y+1)} = \frac{2}{3}$

---

23. Following the example, reduce the other expression to lowest terms.

$$\frac{x^2-4x-5}{x^2+3x+2} = \frac{(x-5)(x+1)}{(x+2)(x+1)} = \frac{x-5}{x+2}$$

$$\frac{y^2-7y+10}{y^2-2y-15} =$$

Answers: a) $\frac{2y(y-3)}{2y(2y+1)} = \frac{y-3}{2y+1}$ b) $\frac{3t(t-2)}{6t(3t+2)} = \frac{t-2}{2(3t+2)}$

---

24. Following the example, reduce the other expression to lowest terms.

$$\frac{x^2+5x+6}{x^2-4} = \frac{(x+2)(x+3)}{(x+2)(x-2)} = \frac{x+3}{x-2}$$

$$\frac{t^2-9}{t^2+t-12} =$$

Answer: $\frac{y-2}{y+3}$, from $\frac{(y-2)(y-5)}{(y+3)(y-5)}$

---

25. Following the example, reduce the other expression to lowest terms.

$$\frac{3x+6}{9} = \frac{3(x+2)}{3(3)} = \frac{x+2}{3} \qquad \frac{8y}{16y-4} =$$

Answer: $\frac{t+3}{t+4}$, from $\frac{(t+3)(t-3)}{(t+4)(t-3)}$

---

26. To reduce the expression below to lowest terms, we substituted $-(x-5)$ for $5-x$, made the negative sign the sign of the fraction, and then cancelled. reduce the other expression to lowest terms.

$$\frac{x-5}{5-x} = \frac{x-5}{-(x-5)} = -\frac{x-5}{x-5} = -1$$

$$\frac{p-q}{q-p} =$$

Answer: $\frac{2y}{4y-1}$, from $\frac{4(2y)}{4(4y-1)}$

27. Following the example, reduce the other expression to lowest terms.

$$\frac{y^2 - 9}{3 - y} = \frac{(y + 3)(y - 3)}{-(y - 3)} = -\frac{(y + 3)\cancel{(y - 3)}}{\cancel{y - 3}} = -(y + 3)$$

$$\frac{a^2 - b^2}{b - a} =$$

Answer: -1

---

28. When reducing fractions to lowest terms, we can only cancel common factors of the numerator and denominator. <u>A common error is to cancel common terms of the numerator and denominator</u>. For example:

$$\frac{\cancel{x} + 3}{\cancel{x}} = 3 \quad \text{(Common error)}$$

$$\frac{x + \cancel{5}}{y + \cancel{5}} = \frac{x}{y} \quad \text{(Common error)}$$

Reduce to lowest terms if possible.

a) $\frac{2x + 2}{2} =$

b) $\frac{x + 2}{2} =$

c) $\frac{x - 5}{y - 5} =$

d) $\frac{5x - 5}{5y - 5} =$

Answer: $-(a + b)$

Answers: a) $x + 1$ b) cannot be reduced c) cannot be reduced d) $\frac{x - 1}{y - 1}$

7-3 CANCELLING IN MULTIPLICATIONS

In multiplications of rational expressions, the product should always be reduced to lowest terms. To avoid products that are not in lowest terms, we can cancel before multiplying. We will discuss the method in this section.

---

29. When multiplying rational expressions, the product should always be reduced to lowest terms. For example:

$$\left(\frac{2x}{5}\right)\left(\frac{5}{3x}\right) = \frac{\overset{2}{\cancel{10x}}}{\underset{3}{\cancel{15x}}} = \frac{2}{3}$$

To avoid having to reduce the product above to lowest terms, we ordinarily cancel before multiplying. That is:

$$\left(\frac{2\cancel{x}}{\cancel{5}}\right)\left(\frac{\cancel{5}}{3\cancel{x}}\right) = \frac{2}{3}$$

Continued on following page.

29. Continued

Do these. Cancel before multiplying.

a) $\left(\frac{6}{7y}\right)\left(\frac{4y}{3}\right) =$ ____ b) $\left(\frac{8m}{5}\right)\left(\frac{1}{4m}\right) =$ ____

---

30. We cancelled before multiplying in the example below.

$$\left(\frac{2x^2}{3}\right)\left(\frac{5}{6x}\right) = \frac{5x}{9}$$

Do these by cancelling before multiplying.

a) $\left(\frac{3x^2}{10}\right)\left(\frac{5}{2x}\right) =$ ____ b) $\left(\frac{1}{ax^2}\right)\left(\frac{7ax}{8}\right) =$ ____

Answers: a) $\frac{8}{7}$, from $\left(\frac{6}{7y}\right)\left(\frac{4y}{3}\right)$ b) $\frac{2}{5}$, from $\left(\frac{8m}{5}\right)\left(\frac{1}{4m}\right)$

---

31. In the multiplication below, we cancelled after factoring. Complete the other multiplication.

$$\left(\frac{x^2 + 5x + 4}{x^2 - 1}\right)\left(\frac{x - 1}{x - 3}\right) = \left[\frac{(x + 1)(x + 4)}{(x + 1)(x - 1)}\right]\left(\frac{x - 1}{x - 3}\right) = \frac{x + 4}{x - 3}$$

$$\left(\frac{y - 5}{y + 2}\right)\left(\frac{y^2 - 4}{y^2 - 6y + 5}\right) =$$

Answers: a) $\frac{3x}{4}$, from $\left(\frac{3x^2}{10}\right)\left(\frac{5}{2x}\right)$ b) $\frac{7}{8x}$, from $\left(\frac{1}{ax^2}\right)\left(\frac{7ax}{8}\right)$

---

32. In the multiplication below, we also cancelled after factoring. Use the same method for the other multiplication.

$$\left(\frac{x^2 + 2x - 3}{8}\right)\left(\frac{4x}{2x^2 + 3x - 9}\right) = \left[\frac{(x + 3)(x - 1)}{8}\right]\left[\frac{4x}{(2x - 3)(x + 3)}\right] = \frac{x(x - 1)}{2(2x - 3)}$$

$$\left(\frac{6y^2}{y^2 - 5y - 14}\right)\left(\frac{y - 7}{3y}\right) =$$

Answer: $\frac{y - 2}{y - 1}$

---

33. In the multiplication below, $(x + 3)^2$ means $(x + 3)(x + 3)$. Notice how we cancelled one of those factors. Do the other multiplication.

$$\left[\frac{x^2 - 6x - 7}{(x + 3)^2}\right]\left(\frac{x + 3}{x - 7}\right) = \left[\frac{(x + 1)(x - 7)}{(x + 3)^2}\right]\left(\frac{x + 3}{x - 7}\right) = \frac{x + 1}{x + 3}$$

$$\left(\frac{y + 2}{y - 1}\right)\left[\frac{(y - 1)^2}{y^2 + 7y + 10}\right] =$$

Answer: $\frac{2y}{y + 2}$

Answer: $\frac{y - 1}{y + 5}$

34. Notice how we cancelled after factoring below. Do the other multiplication.

$$\left(\frac{x^2}{x^2 - 1}\right)\left(\frac{x^2 + 4x - 5}{x^2 + 5x}\right) = \left[\frac{\overset{1}{\cancel{x^2}}}{(x + 1)\cancel{(x - 1)}}\right]\left[\frac{\cancel{(x - 1)}\cancel{(x + 5)}}{\cancel{x}\cancel{(x + 5)}}\right] = \frac{x}{x + 1}$$

$$\left(\frac{y^2 - 3y}{y^2 + y - 12}\right)\left(\frac{y^2 - 16}{2y^2}\right) =$$

Answer: $\dfrac{y - 4}{2y}$

35. Following the example, do the other multiplications.

$$\left(\frac{x^2 - 5x}{x + 1}\right)\left(\frac{x + 1}{x}\right) = \left[\frac{\cancel{x}(x - 5)}{\cancel{x + 1}}\right]\left(\frac{\cancel{x + 1}}{\cancel{x}}\right) = x - 5$$

a) $\left(\dfrac{y + 4}{3y}\right)\left(\dfrac{3y^2 - 3y}{y - 1}\right) =$

b) $\left(\dfrac{cd - d^2}{5c}\right)\left(\dfrac{5c + 5d}{c^2d - d^3}\right) =$

Answers: a) $y + 4$ b) $\dfrac{1}{c}$

7-4 RECIPROCALS AND DIVISION

In this section, we will discuss reciprocals of rational expressions and then use that concept to perform divisions involving rational expressions.

36. Two quantities are reciprocals if their product is "1". For example:

Since $2x\left(\dfrac{1}{2x}\right) = \dfrac{2x}{2x} = 1$: the reciprocal of $2x$ is $\dfrac{1}{2x}$

the reciprocal of $\dfrac{1}{2x}$ is $2x$

Since $(y + 3)\left(\dfrac{1}{y + 3}\right) = \dfrac{y + 3}{y + 3} = 1$: the reciprocal of $y + 3$ is $\dfrac{1}{y + 3}$

the reciprocal of $\dfrac{1}{y + 3}$ is $y + 3$

Write the reciprocal of each quantity.

a) $3t$ _____ b) $\dfrac{1}{m}$ _____ c) $x - y$ _____ d) $\dfrac{1}{3b + 4}$ _____

Answers: a) $\dfrac{1}{3t}$ b) m c) $\dfrac{1}{x - y}$ d) $3b + 4$

37. Two rational expressions are reciprocals if their product is "1". For example:

Since $\left(\frac{x}{7}\right)\left(\frac{7}{x}\right) = \frac{7x}{7x} = 1$: the reciprocal of $\frac{x}{7}$ is $\frac{7}{x}$

the reciprocal of $\frac{7}{x}$ is $\frac{x}{7}$

Since $\left(\frac{y+3}{y-1}\right)\left(\frac{y-1}{y+3}\right) = 1$: the reciprocal of $\frac{y+3}{y-1}$ is $\frac{y-1}{y+3}$

the reciprocal of $\frac{y-1}{y+3}$ is $\frac{y+3}{y-1}$

Write the reciprocal of each expression.

a) $\frac{t}{9}$ ______ b) $\frac{c}{d}$ ______ c) $\frac{m+2}{m-5}$ ______ d) $\frac{p+q}{p-q}$ ______

Answers: a) $\frac{9}{t}$ b) $\frac{d}{c}$ c) $\frac{m-5}{m+2}$ d) $\frac{p-q}{p+q}$

38. Write the reciprocal of each expression.

a) $\frac{1}{3(x-5)}$ ____________ b) $\frac{y+1}{y^2 - 4y + 5}$ ____________

Answers: a) $3(x - 5)$ b) $\frac{y^2 - 4y + 5}{y+1}$

39. Any division involving a rational expression is written as a complex fraction in algebra. For example:

$\frac{x}{3} \div \frac{x}{5}$ is written $\frac{\frac{x}{3}}{\frac{x}{5}}$

To perform divisions like those above, we multiply the numerator by the reciprocal of the denominator. That is:

Numerator unchanged

$$\frac{\frac{x}{3}}{\frac{x}{5}} = \left(\frac{\cancel{x}}{3}\right)\left(\frac{5}{\cancel{x}}\right) = \frac{5}{3}$$

Reciprocal of denominator

Following the example, do these:

a) $\frac{\frac{2}{y}}{\frac{6}{y}} = (\quad)(\quad) =$ ______ b) $\frac{\frac{t}{2}}{\frac{3}{t}} = (\quad)(\quad) =$ ______

40. To perform the division below, we also multiplied the numerator by the reciprocal of the denominator. Do the other division.

$$\frac{\frac{x+1}{x-3}}{\frac{x+1}{x+2}} = \left(\frac{x+1}{x-3}\right)\left(\frac{x+2}{x+1}\right) = \frac{x+2}{x-3}$$

$$\frac{\frac{y+3}{y-2}}{\frac{y+5}{y-2}} = (\quad)(\quad) = \_\_\_\_\_\_$$

Answers: a) $\left(\frac{2}{y}\right)\left(\frac{y}{6}\right) = \frac{1}{3}$ b) $\left(\frac{t}{2}\right)\left(\frac{t}{3}\right) = \frac{t^2}{6}$

---

41. Complete each division.

a) $\frac{\frac{y-1}{10}}{\frac{y-1}{5}} =$ b) $\frac{\frac{c}{c+d}}{\frac{d}{c+d}} =$

Answer: $\left(\frac{y+3}{y-2}\right)\left(\frac{y-2}{y+5}\right) = \frac{y+3}{y+5}$

---

42. Notice how we factored to complete the division below. Do the other division.

$$\frac{\frac{3x+9}{x-1}}{\frac{x+3}{x^2-1}} = \left(\frac{3x+9}{x-1}\right)\left(\frac{x^2-1}{x+3}\right)$$

$$= \left[\frac{3(x+3)}{x-1}\right]\left[\frac{(x+1)(x-1)}{x+3}\right] = 3(x+1)$$

$$\frac{\frac{y^2-16}{6y+24}}{\frac{y-4}{8}} =$$

Answers: a) $\frac{1}{2}$ b) $\frac{c}{d}$

---

43. Notice how we cancelled in the division below. Do the other division.

$$\frac{\frac{x-y}{4y}}{\frac{(x-y)^2}{5y^2}} = \left(\frac{x-y}{4y}\right)\left[\frac{5y^2}{(x-y)^2}\right] = \frac{5y}{4(x-y)}$$

$$\frac{\frac{10a^2}{(a+b)^2}}{\frac{15a}{a+b}} =$$

Answer: $\frac{4}{3}$, from $\left[\frac{(y+4)(y-4)}{6(y+4)}\right]\left[\frac{8}{y-4}\right]$

Answer: $\frac{2a}{3(a+b)}$

44. Following the example, do the other division.

$$\frac{\frac{x^2 + 4x}{x^2 + 5x + 4}}{\frac{x^2 - 3x}{x^2 + 4x + 3}} = \left(\frac{x^2 + 4x}{x^2 + 5x + 4}\right)\left(\frac{x^2 + 4x + 3}{x^2 - 3x}\right)$$

$$= \left[\frac{\cancel{x}\cancel{(x + 4)}}{\cancel{(x + 1)}\cancel{(x + 4)}}\right]\left[\frac{(x + 3)\cancel{(x + 1)}}{\cancel{x}(x - 3)}\right] = \frac{x + 3}{x - 3}$$

$$\frac{\frac{y^2 - 7y + 10}{y^2 - 2y - 15}}{\frac{y^2 - 4y + 4}{y^2 + 6y + 9}} =$$

$\dfrac{y + 3}{y - 2}$

45. In the division below, the numerator of the complex fraction is a polynomial.

$$\frac{3(x + 1)}{\frac{x + 1}{x}} = 3\cancel{(x + 1)}\left(\frac{x}{\cancel{x + 1}}\right) = 3x$$

$$\frac{a^2 - b^2}{\frac{a + b}{b}} =$$

$b(a - b)$

46. In the division below, the denominator of the complex fraction is a polynomial. Do the other division.

$$\frac{\frac{(y + 3)^2}{5}}{y + 3} = \left[\frac{(y + 3)^{\cancel{2}\,1}}{5}\right]\left(\frac{1}{\cancel{y + 3}}\right) = \frac{y + 3}{5}$$

$$\frac{\frac{x^2 - 1}{x}}{x + 1} =$$

$\dfrac{x - 1}{x}$

SELF-TEST 21 (pages 276-288)

Reduce to lowest terms.

1. $\dfrac{3x^2}{9x} =$

2. $\dfrac{4y - 12}{8} =$

3. $\dfrac{a^2 - 25}{a^2 - 6a + 5} =$

Multiply. Report each product in lowest terms.

4. $12y^2\left(\dfrac{1}{8y}\right) =$

5. $\left(\dfrac{x^2 - 4x}{2x - 6}\right)\left(\dfrac{x - 3}{3x}\right) =$

6. $\left(\dfrac{p + 2}{p - 3}\right)\left[\dfrac{p^2 - 4p + 3}{(p + 2)^2}\right] =$

Divide. Report each quotient in lowest terms.

7. $\dfrac{\dfrac{m - 5}{8}}{\dfrac{m}{4}} =$

8. $\dfrac{\dfrac{a^2 - b^2}{a}}{a + b} =$

9. $\dfrac{\dfrac{x^2 + 9x + 18}{x^2 + 5x - 6}}{\dfrac{x^2 - 3x - 10}{x^2 - 6x + 5}} =$

ANSWERS:

1. $\dfrac{x}{3}$

2. $\dfrac{y - 3}{2}$

3. $\dfrac{a + 5}{a - 1}$

4. $\dfrac{3y}{2}$

5. $\dfrac{x - 4}{6}$

6. $\dfrac{p - 1}{p + 2}$

7. $\dfrac{m - 5}{2m}$

8. $\dfrac{a - b}{a}$

9. $\dfrac{x + 3}{x + 2}$

7-5 ADDITION AND SUBTRACTION WITH LIKE DENOMINATORS

In this section, we will discuss the procedure for adding and subtracting rational expressions with like or common denominators.

47. To add rational expressions with like denominators, we add their numerators and keep the same denominator. For example:

$$\frac{2x}{5} + \frac{1}{5} = \frac{2x + 1}{5} \qquad \frac{y}{y + 2} + \frac{3}{y + 2} = \frac{y + 3}{y + 2}$$

Following the examples, do these:

a) $\dfrac{3m}{7} + \dfrac{2}{7} =$ ________

b) $\dfrac{5t}{t - 1} + \dfrac{1}{t - 1} =$ ________

48. Notice how we combined like terms to simplify the numerator below.

$$\frac{2x}{3} + \frac{5x}{3} = \frac{2x + 5x}{3} = \frac{7x}{3}$$

Following the example, do these:

a) $\frac{4}{y} + \frac{1}{y} =$ ______ b) $\frac{m}{m + 4} + \frac{2m}{m + 4} =$ ______

Answers: a) $\frac{3m + 2}{7}$ b) $\frac{5t + 1}{t - 1}$

---

49. Notice how we combined like terms to simplify the numerator below. Do the other addition.

$$\frac{x + 5}{x^2 - 1} + \frac{x + 2}{x^2 - 1} = \frac{(x + 5) + (x + 2)}{x^2 - 1} = \frac{2x + 7}{x^2 - 1}$$

$$\frac{2y + 1}{y + 5} + \frac{y - 8}{y + 5} =$$

Answers: a) $\frac{5}{y}$ b) $\frac{3m}{m + 4}$

---

50. Notice how we reduced the sum below to lowest terms.

$$\frac{2x}{10} + \frac{2x}{10} = \frac{2x + 2x}{10} = \frac{\overset{2}{\cancel{4}}x}{\underset{5}{\cancel{10}}} = \frac{2x}{5}$$

Complete. Reduce to lowest terms if possible.

a) $\frac{2y}{9} + \frac{4y}{9} =$ ______ b) $\frac{7}{2t} + \frac{3}{2t} =$ ______

Answer: $\frac{3y - 7}{y + 5}$

---

51. Following the example, reduce the other sum to lowest terms.

$$\frac{x^2 + 6x}{x^2 - 3x} + \frac{x^2 - x}{x^2 - 3x} = \frac{(x^2 + 6x) + (x^2 - x)}{x^2 - 3x} = \frac{2x^2 + 5x}{x^2 - 3x} = \frac{\cancel{x}(2x + 5)}{\cancel{x}(x - 3)} = \frac{2x + 5}{x - 3}$$

$$\frac{y^2 - 3y}{y^2 + y} + \frac{y^2 - 4y}{y^2 + y} =$$

Answers: a) $\frac{2y}{3}$ b) $\frac{5}{t}$

---

52. To subtract rational expressions with like denominators, we subtract their numerators and keep the same denominator. For example:

$$\frac{3x}{8} - \frac{5}{8} = \frac{3x - 5}{8}$$

Following the example, do these:

a) $\frac{a}{c} - \frac{b}{c} =$ ______ b) $\frac{m}{m + 1} - \frac{1}{m + 1} =$ ______

Answer: $\frac{2y - 7}{y + 1}$

53. Notice how we simplified and reduced to lowest terms in this one.

$$\frac{5x}{6} - \frac{x}{6} = \frac{5x - x}{6} = \frac{\overset{2}{\cancel{4}}x}{\underset{3}{\cancel{6}}} = \frac{2x}{3}$$

Do these. Simplify and reduce to lowest terms if possible.

a) $\frac{3y}{10} - \frac{y}{10} =$ ________ b) $\frac{11}{8t} - \frac{5}{8t} =$ ________

a) $\frac{a - b}{c}$ b) $\frac{m - 1}{m + 1}$

---

54. In this one, notice how we wrote parentheses around $x + 3$ to make sure we subtracted that whole numerator.

$$\frac{5x}{x - 1} - \frac{x + 3}{x - 1} = \frac{5x - (x + 3)}{x - 1} = \frac{5x - x - 3}{x - 1} = \frac{4x - 3}{x - 1}$$

Following the example, do this one.

$$\frac{2y}{5} - \frac{y + 2}{5} =$$

a) $\frac{y}{5}$ b) $\frac{3}{4t}$

---

55. Following the example, complete the other subtraction.

$$\frac{3x - 1}{x} - \frac{x - 4}{x} = \frac{(3x - 1) - (x - 4)}{x} = \frac{3x - 1 - x + 4}{x} = \frac{2x + 3}{x}$$

$$\frac{4y + 7}{y + 3} - \frac{3y - 1}{y + 3} =$$

$\frac{y - 2}{5}$

---

56. Following the example, reduce the other answer to lowest terms.

$$\frac{5x}{8} - \frac{3x + 4}{8} = \frac{5x - (3x + 4)}{8} = \frac{5x - 3x - 4}{8} = \frac{2x - 4}{8} = \frac{\overset{1}{\cancel{2}}(x - 2)}{\underset{4}{\cancel{8}}} = \frac{x - 2}{4}$$

$$\frac{10}{3y} - \frac{1 - 3y}{3y} =$$

$\frac{y + 8}{y + 3}$

---

57. Following the example, reduce the other answer to lowest terms.

$$\frac{x}{x^2 - 1} - \frac{1}{x^2 - 1} = \frac{x - 1}{x^2 - 1} = \frac{\overset{1}{\cancel{x - 1}}}{(x + 1)\cancel{(x - 1)}} = \frac{1}{x + 1}$$

$$\frac{a^2}{a + b} - \frac{b^2}{a + b} =$$

$\frac{3 + y}{y}$

$a - b$

58. Following the example, reduce the other answer to lowest terms.

$$\frac{x^2 - 2}{x^2 - x - 12} - \frac{2x + 13}{x^2 - x - 12} = \frac{(x^2 - 2) - (2x + 13)}{x^2 - x - 12}$$

$$= \frac{x^2 - 2 - 2x - 13}{x^2 - x - 12}$$

$$= \frac{x^2 - 2x - 15}{x^2 - x - 12}$$

$$= \frac{\cancel{(x+3)}(x - 5)}{\cancel{(x+3)}(x - 4)} = \frac{x - 5}{x - 4}$$

$$\frac{5y - 4}{3y^2 + 5y - 2} - \frac{2y - 3}{3y^2 + 5y - 2} =$$

$\frac{1}{y + 2}$

7-6 FINDING LOWEST COMMON DENOMINATORS

To add rational expressions with unlike denominators, we use the lowest common denominator (LCD). We will discuss a method for finding lowest common denominators in this section.

59. To find the lowest common denominator (LCD) for unlike denominators, we begin by factoring the denominators into primes. For example:

$$\frac{7}{12} + \frac{11}{30} \qquad \begin{aligned} 12 &= (2)(2)(3) \\ 30 &= (2)(3)(5) \end{aligned}$$

Then to get the LCD, we use each factor the greatest number of times it appears in any denominator. Therefore, we get the following LCD for the addition above:

$$\text{LCD} = (2)(2)(3)(5) = 60$$

Note: Since 12 has 2 as a factor twice and 30 has 2 as a factor only once, the LCD has 2 as a factor twice (the greatest number of times it occurs).

We factored each denominator below. Use the method above to find the LCD.

$$\frac{4}{15} + \frac{7}{18} \qquad \begin{aligned} 15 &= (3)(5) \\ 18 &= (2)(3)(3) \end{aligned} \qquad \text{LCD} = \underline{\qquad\qquad}$$

60. The following principle is used after factoring each denominator.

> <u>To get the LCD, use each factor the greatest number of times it appears in any denominator.</u>

We used the principle above to find the LCD below.

$$\frac{5}{12}+\frac{1}{21} \qquad 12 = (2)(2)(3) \qquad 21 = (3)(7) \qquad \text{LCD} = (2)(2)(3)(7) = 84$$

For the addition below, factor each denominator into primes and then find the LCD.

$$\frac{7}{30}+\frac{1}{18} \qquad 30 = \_\_\_\_ \qquad 18 = \_\_\_\_ \qquad \text{LCD} = \_\_\_\_\_\_$$

Answer: (2)(3)(3)(5) = 90

---

61. We used the same method to find the LCD for the three denominators below.

$$\frac{5}{6}+\frac{4}{9}+\frac{17}{21} \qquad 6 = (2)(3) \qquad 9 = (3)(3) \qquad 21 = (3)(7) \qquad \text{LCD} = (2)(3)(3)(7) = 126$$

Following the example, find the LCD for this addition.

$$\frac{3}{4}+\frac{9}{10}+\frac{1}{12} \qquad 4 = \_\_\_\_ \qquad 10 = \_\_\_\_ \qquad 12 = \_\_\_\_ \qquad \text{LCD} = \_\_\_\_\_\_$$

Answer:
30 = (2)(3)(5)
18 = (2)(3)(3)
LCD = (2)(3)(3)(5) = 90

---

62. The same method is used when one or more denominators contains a variable. An example is shown.

$$\frac{5}{6x}+\frac{1}{10x} \qquad 6x = (2)(3)(x) \qquad 10x = (2)(5)(x) \qquad \text{LCD} = (2)(3)(5)(x) = 30x$$

Following the example, find the LCD below.

$$\frac{7}{12y}+\frac{3}{16y} \qquad 12y = \_\_\_\_ \qquad 16y = \_\_\_\_ \qquad \text{LCD} = \_\_\_\_\_\_$$

Answer:
4 = (2)(2)
10 = (2)(5)
12 = (2)(2)(3)
LCD = (2)(2)(3)(5) = 60

---

63. Another example of finding the LCD is shown below.

$$\frac{1}{2a}+\frac{5}{6b} \qquad 2a = (2)(a) \qquad 6b = (2)(3)(b) \qquad \text{LCD} = (2)(3)(a)(b) = 6ab$$

Find the LCD for each addition below.

a) $\frac{3}{x}+\frac{1}{2y}$

LCD = ______

b) $\frac{1}{10p}+\frac{1}{5q}$

LCD = ______

Answer:
12y = (2)(2)(3)(y)
16y = (2)(2)(2)(2)(y)
LCD = (2)(2)(2)(2)(3)(y) = 48y

64. We found the LCD for the addition below.

$$\frac{2}{3y^2} + \frac{5}{6y} \qquad \begin{aligned} 3y^2 &= (3)(y)(y) \\ 6y &= (2)(3)(y) \end{aligned} \qquad \text{LCD} = (2)(3)(y)(y) = 6y^2$$

Find the LCD for each addition.

a) $\frac{1}{x^2} + \frac{7}{3x}$ LCD = ________

b) $\frac{3}{4t} + \frac{9}{10t^2}$ LCD = ________

a) $2xy$ b) $10pq$

---

65. We found the LCD for the addition below.

$$\frac{1}{3x} + \frac{1}{x^2 - 5x} = \frac{1}{(3)(x)} + \frac{1}{(x)(x - 5)}$$

$$\text{LCD} = (3)(x)(x - 5) = 3x(x - 5)$$

Find the LCD for each addition.

a) $\frac{1}{2y} + \frac{1}{2y - 6}$ LCD = ________

b) $\frac{3}{t^2 + 2t} + \frac{7}{5t^2}$ LCD = ________

a) $3x^2$ b) $20t^2$

---

66. In the addition below, the denominators do not have a common factor. Therefore, the LCD is the product of the denominators.

$$\frac{4}{x + 3} + \frac{3}{x - 4} \qquad \text{LCD} = (x + 3)(x - 4)$$

Find the LCD for each addition.

a) $\frac{3}{2a} + \frac{7}{5b}$ LCD = ________

b) $\frac{6}{7t} + \frac{5}{t - 1}$ LCD = ________

a) $2y(y - 3)$

b) $5t^2(t + 2)$

---

67. We found the LCD for the addition below.

$$\frac{3}{x^2 - 2x} + \frac{5}{3x - 6} = \frac{3}{x(x - 2)} + \frac{5}{3(x - 2)}$$

$$\text{LCD} = 3x(x - 2)$$

Find the LCD for each addition.

a) $\frac{1}{2x + 2} + \frac{1}{3x + 3}$ LCD = ________

b) $\frac{3}{10y - 5} + \frac{7}{2y^2 - y}$ LCD = ________

a) $10ab$

b) $7t(t - 1)$

---

a) $6(x + 1)$ b) $5y(2y - 1)$

68. We found the LCD for the addition below.

$$\frac{x+1}{x^2-4} + \frac{x-1}{x^2+5x+6} = \frac{x+1}{(x+2)(x-2)} + \frac{x-1}{(x+2)(x+3)}$$

$$\text{LCD} = (x+2)(x-2)(x+3)$$

Find the LCD for each addition.

a) $\frac{y-3}{y+3} + \frac{2y+1}{y^2+8y+15}$

LCD = ________

b) $\frac{x-5}{x^2+3x-10} + \frac{3x+7}{x^2-10x+16}$

LCD = ________

---

a) $(y+3)(y+5)$

b) $(x-2)(x+5)(x-8)$

69. We found the LCD for the addition below.

$$\frac{3x}{(x-1)^2} + \frac{5x}{x^2-1} = \frac{3x}{(x-1)(x-1)} + \frac{5x}{(x+1)(x-1)}$$

$$\text{LCD} = (x-1)(x-1)(x+1)$$

$$= (x-1)^2(x+1)$$

Find the LCD for each addition.

a) $\frac{a}{(a+b)^2} + \frac{b}{a^2-b^2}$

LCD = ________

b) $\frac{y+7}{y^2-y-6} + \frac{y-2}{(y-3)^2}$

LCD = ________

---

a) $(a+b)^2(a-b)$

b) $(y-3)^2(y+2)$

70. We can use the same method to find the LCD when an addition contains three denominators. An example is shown.

$$\frac{x}{x+5} + \frac{3x}{x^2-25} + \frac{x-12}{x^2-3x-10}$$

$x + 5 = (x+5)$

$x^2 - 25 = (x+5)(x-5)$

$x^2 - 3x - 10 = (x+2)(x-5)$

$\text{LCD} = (x+5)(x-5)(x+2)$

Find the LCD for this addition.

$$\frac{3y}{y+2} + \frac{y+1}{(y-2)^2} + \frac{y-1}{y^2-4}$$

LCD = ____________

$(y-2)^2(y+2)$

7-7 ADDITION AND SUBTRACTION WITH UNLIKE DENOMINATORS

In this section, we will discuss the procedure for adding and subtracting rational expressions with unlike denominators.

---

71. To add rational expressions with unlike denominators, we use the following three steps.

    1. Find the lowest common denominator.
    2. Convert each rational expression to an equivalent expression with the lowest common denominator as the denominator.
    3. Then add in the usual way.

We used the previous steps for the addition below. The LCD is 6x. To convert the rational expressions to expressions with 6x as the denominator, we multiplied $\frac{1}{2x}$ by $\frac{3}{3}$ and $\frac{4}{3x}$ by $\frac{2}{2}$.

$$\frac{1}{2x} + \frac{4}{3x} = \left(\frac{1}{2x}\right)\left(\frac{3}{3}\right) + \left(\frac{4}{3x}\right)\left(\frac{2}{2}\right)$$

$$= \frac{3}{6x} + \frac{8}{6x}$$

$$= \frac{11}{6x}$$

Following the example, do these additions.

a) $\frac{5}{6y} + \frac{1}{4y} =$

b) $\frac{3}{t} + \frac{4}{t^2} =$

---

72. The LCD for the addition below is $(x + 1)(x - 2)$. To get the LCD for each expression, we multiplied $\frac{2}{x+1}$ by $\frac{x-2}{x-2}$ and $\frac{3}{x-2}$ by $\frac{x+1}{x+1}$. Notice how we simplified the numerator of the sum.

$$\frac{2}{x+1} + \frac{3}{x-2} = \left(\frac{2}{x+1}\right)\left(\frac{x-2}{x-2}\right) + \left(\frac{3}{x-2}\right)\left(\frac{x+1}{x+1}\right)$$

$$= \frac{2(x-2) + 3(x+1)}{(x+1)(x-2)}$$

$$= \frac{2x - 4 + 3x + 3}{(x+1)(x-2)}$$

$$= \frac{5x - 1}{(x+1)(x-2)}$$

Continued on following page.

Answers to 71:

a) $\frac{13}{12y}$, from: $\frac{10}{12y} + \frac{3}{12y}$

b) $\frac{3t + 4}{t^2}$, from: $\frac{3t}{t^2} + \frac{4}{t^2}$

72. Continued

Following the example, do this addition.

$$\frac{y}{y+3} + \frac{5}{2y} =$$

---

73. The LCD for the addition below is $5y(y - 1)$. Notice how we got common denominators and simplified the numerator of the sum.

$$\frac{y+2}{5y} + \frac{y-4}{y-1} = \left(\frac{y+2}{5y}\right)\left(\frac{y-1}{y-1}\right) + \left(\frac{y-4}{y-1}\right)\left(\frac{5y}{5y}\right)$$

$$= \frac{(y+2)(y-1) + (y-4)(5y)}{5y(y-1)}$$

$$= \frac{y^2 + y - 2 + 5y^2 - 20y}{5y(y-1)}$$

$$= \frac{6y^2 - 19y - 2}{5y(y-1)}$$

Following the example, do this addition.

$$\frac{x+1}{x-3} + \frac{x-5}{x+2} =$$

Answer: $\frac{2y^2 + 5y + 15}{2y(y+3)}$

---

74. To find the LCD below, we began by factoring the denominators. The LCD is $2x(x - 3)$.

$$\frac{5}{x^2 - 3x} + \frac{x}{2x - 6} = \frac{5}{x(x-3)} + \frac{x}{2(x-3)}$$

$$= \left[\frac{5}{x(x-3)}\right]\left(\frac{2}{2}\right) + \left[\frac{x}{2(x-3)}\right]\left(\frac{x}{x}\right)$$

$$= \frac{5(2) + x(x)}{2x(x-3)}$$

$$= \frac{x^2 + 10}{2x(x-3)}$$

Continued on following page.

Answer: $\frac{2x^2 - 5x + 17}{(x-3)(x+2)}$

74. Continued

Following the example, do this addition.

$$\frac{y}{6y - 3} + \frac{5}{2y^2 - y} =$$

Answer: $\frac{y^2 + 15}{3y(2y - 1)}$

---

75. To find the LCD below, we began by factoring the denominators. The LCD is $(x + 3)(x + 4)(x - 2)$.

$$\frac{2x}{x^2 + 7x + 12} + \frac{3}{x^2 + x - 6} = \frac{2x}{(x+3)(x+4)} + \frac{3}{(x+3)(x-2)}$$

$$= \left[\frac{2x}{(x+3)(x+4)}\right]\left(\frac{x-2}{x-2}\right) + \left[\frac{3}{(x+3)(x-2)}\right]\left(\frac{x+4}{x+4}\right)$$

$$= \frac{2x(x-2) + 3(x+4)}{(x+3)(x+4)(x-2)}$$

$$= \frac{2x^2 - 4x + 3x + 12}{(x+3)(x+4)(x-2)}$$

$$= \frac{2x^2 - x + 12}{(x+3)(x+4)(x-2)}$$

Following the example, do this addition.

$$\frac{y}{y^2 - 1} + \frac{5}{y^2 + 3y - 4} =$$

Answer: $\frac{y^2 + 9y + 5}{(y + 1)(y - 1)(y + 4)}$

---

76. To do the subtraction below, we had to get like denominators first. The LCD is 12x.

$$\frac{2}{3x} - \frac{1}{4x} = \left(\frac{2}{3x}\right)\left(\frac{4}{4}\right) - \left(\frac{1}{4x}\right)\left(\frac{3}{3}\right)$$

$$= \frac{8}{12x} - \frac{3}{12x}$$

$$= \frac{8 - 3}{12x} = \frac{5}{12x}$$

Continued on following page.

76. Continued

Do these subtractions.

a) $\frac{7}{8y} - \frac{5}{6y} =$

b) $\frac{5}{t^2} - \frac{1}{2t} =$

Answers: a) $\frac{1}{24y}$ b) $\frac{10 - t}{2t^2}$

---

77. The LCD for the subtraction below is $(x - 2)(x + 1)$. Notice how we wrote parentheses around $x^2 - 3x + 2$ to make sure we subtracted that whole numerator.

$$\frac{x + 3}{x - 2} - \frac{x - 1}{x + 1} = \left(\frac{x + 3}{x - 2}\right)\left(\frac{x + 1}{x + 1}\right) - \left(\frac{x - 1}{x + 1}\right)\left(\frac{x - 2}{x - 2}\right)$$

$$= \frac{(x + 3)(x + 1) - (x - 1)(x - 2)}{(x - 2)(x + 1)}$$

$$= \frac{x^2 + 4x + 3 - (x^2 - 3x + 2)}{(x - 2)(x + 1)}$$

$$= \frac{x^2 + 4x + 3 - x^2 + 3x - 2}{(x - 2)(x + 1)}$$

$$= \frac{7x + 1}{(x - 2)(x + 1)}$$

Following the example, do this subtraction.

$\frac{y - 1}{y + 3} - \frac{y + 4}{2y} =$

Answer: $\frac{y^2 - 9y - 12}{2y(y + 3)}$

78. To find the LCD below, we began by factoring the denominators. The LCD is $(x + 2)(x - 2)$. Notice how we wrote parentheses around $5x - 10$ to make sure we subtracted that whole numerator.

$$\frac{7}{x^2 - 4} - \frac{5}{x + 2} = \frac{7}{(x + 2)(x - 2)} - \frac{5}{x + 2}$$

$$= \frac{7}{(x + 2)(x - 2)} - \left(\frac{5}{x + 2}\right)\left(\frac{x - 2}{x - 2}\right)$$

$$= \frac{7 - 5(x - 2)}{(x + 2)(x - 2)}$$

$$= \frac{7 - (5x - 10)}{(x + 2)(x - 2)}$$

$$= \frac{7 - 5x + 10}{(x + 2)(x - 2)}$$

$$= \frac{17 - 5x}{(x + 2)(x - 2)}$$

Following the example, do this subtraction.

$$\frac{5a}{a^2 - b^2} - \frac{4}{a - b} =$$

Answer: $\frac{a - 4b}{(a + b)(a - b)}$

79. To find the LCD below, we began by factoring both denominators. The LCD is $(x - 3)(x - 3)(x + 3)$.

$$\frac{5x}{(x - 3)^2} - \frac{2}{x^2 - 9} = \frac{5x}{(x - 3)(x - 3)} - \frac{2}{(x + 3)(x - 3)}$$

$$= \left[\frac{5x}{(x - 3)(x - 3)}\right]\left(\frac{x + 3}{x + 3}\right) - \left[\frac{2}{(x + 3)(x - 3)}\right]\left(\frac{x - 3}{x - 3}\right)$$

$$= \frac{5x(x + 3) - 2(x - 3)}{(x - 3)(x - 3)(x + 3)}$$

$$= \frac{5x^2 + 15x - (2x - 6)}{(x - 3)(x - 3)(x + 3)}$$

$$= \frac{5x^2 + 15x - 2x + 6}{(x - 3)(x - 3)(x + 3)}$$

$$= \frac{5x^2 + 13x + 6}{(x - 3)(x - 3)(x + 3)}$$

Continued on following page.

79. Continued

Following the example, do this subtraction.

$$\frac{a}{a^2 - b^2} - \frac{b}{(a + b)^2} =$$

$$\frac{a^2 + b^2}{(a + b)(a + b)(a - b)}$$

7-8 COMPLEX FRACTIONS

In this section, we will discuss the procedure for simplifying complex fractions.

---

80. Any rational expression with a fraction in its numerator or its denominator or both is called a complex fraction. Some examples of complex fractions are:

$$\frac{\frac{x}{5}}{\frac{3}{4}} \qquad \frac{1 + \frac{4}{x}}{\frac{3}{2}} \qquad \frac{\frac{x}{x - y}}{\frac{1}{x} + \frac{1}{y}}$$

To simplify a complex fraction, we begin by adding or subtracting (if necessary) to get a single rational expression in both numerator and denominator. Then we divide by multiplying by the reciprocal of the denominator. An example is shown. Simplify the other complex fraction.

$$\frac{1 + \frac{4}{x}}{\frac{3}{2}} = \frac{1\left(\frac{x}{x}\right) + \frac{4}{x}}{\frac{3}{2}} \qquad\qquad \frac{\frac{y}{3} + 2}{\frac{7}{6}} =$$

$$= \frac{\frac{x + 4}{x}}{\frac{3}{2}}$$

$$= \left(\frac{x + 4}{x}\right)\left(\frac{2}{3}\right)$$

$$= \frac{2(x + 4)}{3x}$$

81. To simplify the complex fraction below, we began by doing the subtraction in the numerator and the addition in the denominator. Simplify the other complex fraction.

$$\frac{\frac{2}{y}-\frac{1}{3y}}{\frac{1}{2y}+\frac{5}{4y}} = \frac{\left(\frac{2}{y}\right)\left(\frac{3}{3}\right)-\frac{1}{3y}}{\left(\frac{1}{2y}\right)\left(\frac{2}{2}\right)+\frac{5}{4y}}$$

$$= \frac{\frac{6}{3y}-\frac{1}{3y}}{\frac{2}{4y}+\frac{5}{4y}}$$

$$= \frac{\frac{5}{3y}}{\frac{7}{4y}}$$

$$= \left(\frac{5}{3\cancel{y}}\right)\left(\frac{4\cancel{y}}{7}\right)$$

$$= \frac{20}{21}$$

$$\frac{\frac{3}{x}+\frac{1}{2x}}{1-\frac{4}{x}} =$$

Answer to previous frame: $\frac{2(y+6)}{7}$

---

82. Following the example, simplify the other complex fraction.

$$\frac{\frac{x+y}{x}}{\frac{1}{x}+\frac{1}{y}} = \frac{\frac{x+y}{x}}{\left(\frac{1}{x}\right)\left(\frac{y}{y}\right)+\left(\frac{1}{y}\right)\left(\frac{x}{x}\right)}$$

$$= \frac{\frac{x+y}{x}}{\frac{y+x}{xy}}$$

$$= \left(\frac{\cancel{x+y}}{\cancel{x}}\right)\left(\frac{\cancel{x}y}{\cancel{y+x}}\right)$$

$$= y$$

$$\frac{t-\frac{1}{t}}{t+\frac{1}{t}} =$$

Answer to frame 81: $\frac{7}{2(x-4)}$

Answer to frame 82: $\frac{t^2-1}{t^2+1}$

83. Notice how we factored $(x^2 - y^2)$ to simplify the expression below. Simplify the other expression.

$$\frac{\frac{x}{y} - \frac{y}{x}}{\frac{1}{y} + \frac{1}{x}} = \frac{\left(\frac{x}{y}\right)\left(\frac{x}{x}\right) - \left(\frac{y}{x}\right)\left(\frac{y}{y}\right)}{\left(\frac{1}{y}\right)\left(\frac{x}{x}\right) + \left(\frac{1}{x}\right)\left(\frac{y}{y}\right)}$$

$$= \frac{\frac{x^2}{xy} - \frac{y^2}{xy}}{\frac{x}{xy} + \frac{y}{xy}}$$

$$= \frac{\frac{x^2 - y^2}{xy}}{\frac{x + y}{xy}}$$

$$= \left(\frac{x^2 - y^2}{\cancel{xy}}\right)\left(\frac{\cancel{xy}}{x + y}\right)$$

$$= \frac{\cancel{(x + y)}(x - y)}{\cancel{x + y}}$$

$$= x - y$$

$$\frac{1 + \frac{1}{m}}{1 - \frac{1}{m^2}} =$$

---

84. In an earlier frame, we simplified the expression below and got $\frac{20}{21}$. We can simplify the same expression by multiplying both numerator and denominator by 12y, the LCD of all four denominators. The steps are shown. Use the same method to simplify the other expression. The LCD of all denominators is 6b.

Answer: $\frac{m}{m - 1}$

$$\frac{\frac{2}{y} - \frac{1}{3y}}{\frac{1}{2y} + \frac{5}{4y}} = \frac{12y\left(\frac{2}{y} - \frac{1}{3y}\right)}{12y\left(\frac{1}{2y} + \frac{5}{4y}\right)}$$

$$= \frac{12\cancel{y}\left(\frac{2}{\cancel{y}}\right) - \overset{4}{\cancel{12y}}\left(\frac{1}{\cancel{3y}}\right)}{\overset{6}{\cancel{12y}}\left(\frac{1}{\cancel{2y}}\right) + \overset{3}{\cancel{12y}}\left(\frac{5}{\cancel{4y}}\right)}$$

$$= \frac{24 - 4}{6 + 15}$$

$$= \frac{20}{21}$$

$$\frac{\frac{a}{b} - \frac{3}{2b}}{\frac{a}{3b} + \frac{1}{2b}} =$$

Answer: $\frac{3(2a - 3)}{2a + 3}$

SELF-TEST 22 (pages 288-303)

(pages 288-303)

Add or subtract. Report each answer in lowest terms.

1. $\frac{5}{8x} + \frac{1}{8x}$

2. $\frac{2y}{5} - \frac{y - 7}{5}$

3. $\frac{p}{p^2 - 1} - \frac{1}{p^2 - 1}$

4. $\frac{1}{6x} + \frac{3}{10x}$

5. $\frac{t}{t - 2} + \frac{4}{3t}$

6. $\frac{y + 1}{y - 2} - \frac{y - 1}{y + 3}$

7. $\frac{x}{x^2 - y^2} + \frac{3}{x + y}$

8. $\frac{x - 3}{x^2 - 2x - 8} - \frac{x}{(x + 2)^2}$

Simplify each expression.

9. $\dfrac{\frac{5}{y} + \frac{1}{3y}}{1 - \frac{2}{3y}}$

10. $\dfrac{\frac{1}{a} + \frac{1}{b}}{\frac{a}{b} - \frac{b}{a}}$

ANSWERS:

1. $\frac{3}{4x}$
2. $\frac{y + 7}{5}$
3. $\frac{1}{p + 1}$
4. $\frac{7}{15x}$
5. $\frac{3t^2 + 4t - 8}{3t(t - 2)}$
6. $\frac{7y + 1}{(y - 2)(y + 3)}$
7. $\frac{4x - 3y}{(x + y)(x - y)}$
8. $\frac{3(x - 2)}{(x + 2)(x + 2)(x - 4)}$
9. $\frac{16}{3y - 2}$
10. $\frac{1}{a - b}$

7-9 SOLVING FRACTIONAL EQUATIONS

To solve equations containing one or more rational expressions, we begin by clearing the fraction or fractions. We will discuss the method in this section.

---

85. In an earlier chapter, we solved equations containing one fraction. To do so, we cleared the fraction by multiplying both sides by the denominator of the fraction. An example is shown. Solve the other equation.

$$\frac{3x}{2} = 12$$

$$\not{2}\left(\frac{3x}{\not{2}}\right) = 2(12)$$

$$3x = 24$$

$$x = 8$$

$$\frac{5}{2y} = 4$$

---

86. To solve the equation below, we began by multiplying both sides by x. Solve the other equation.

Answer to 85: $y = \frac{5}{8}$

$$\frac{x-1}{x} = 4$$

$$\not{x}\left(\frac{x-1}{\not{x}}\right) = x(4)$$

$$x - 1 = 4x$$

$$-1 = 3x$$

$$x = -\frac{1}{3}$$

$$\frac{2y+7}{5} = 1$$

---

87. To solve the equation below, we began by multiplying both sides by (x + 4). Notice that we then had to multiply by the distributive principle on the right side. Solve the other equation.

Answer to 86: $y = -1$

$$\frac{9}{x+4} = 2$$

$$\cancel{(x+4)}\left(\frac{9}{\cancel{x+4}}\right) = 2(x+4)$$

$$9 = 2x + 8$$

$$1 = 2x$$

$$x = \frac{1}{2}$$

$$3 = \frac{y}{y-2}$$

Answer to 87: $y = 3$

88. A value that makes a denominator equal 0 cannot be the solution of a fractional equation. For example:

$x = 0$ cannot be the solution of $\frac{x - 3}{2x} = 1$

$y = 4$ cannot be the solution of $\frac{y}{y - 4} = 2$

Therefore, when you finish solving a fractional equation, always check to see whether the proposed solution makes a denominator equal 0. If so, it is not a solution.

---

89. The equation below has two terms on the left side. To clear the fraction, we multiplied both sides by 4. Notice that we multiplied by the distributive principle on the left side. Solve the other equation.

$$\frac{x}{4} + 2 = 3$$

$$4\left(\frac{x}{4} + 2\right) = 4(3)$$

$$\not{4}\left(\frac{x}{\not{4}}\right) + 4(2) = 12$$

$$x + 8 = 12$$

$$x = 4$$

$$1 = \frac{5y}{2} - 3$$

---

90. To clear the fraction below, we multiplied both sides by 3x. Solve the other equation.

$$2 - \frac{45}{3x} = 5$$

$$3x\left(2 - \frac{45}{3x}\right) = 3x(5)$$

$$3x(2) - \overset{1}{\cancel{3x}}\left(\frac{45}{\cancel{3x}}\right) = 15x$$

$$6x - 45 = 15x$$

$$-45 = 9x$$

$$x = -5$$

$$4 = \frac{3}{y} + 1$$

Answer to 89: $y = \frac{8}{5}$, from:

$2 = 5y - 6$

Answer to 90: $y = 1$, from:

$4y = 3 + y$

91. When an equation contains more than one fraction, we clear the fractions by multiplying both sides by the LCD of all denominators. For example, we cleared the fractions below by multiplying both sides by 24. Solve the other equation.

$$\frac{3x}{8} = \frac{1}{6}$$

$$\overset{3}{\cancel{24}}\left(\frac{3x}{\cancel{8}}\right) = \overset{4}{\cancel{24}}\left(\frac{1}{\cancel{6}}\right)$$

$$9x = 4$$

$$x = \frac{4}{9}$$

$$\frac{3}{2} = \frac{7y}{10}$$

Answer: $y = 1$, from: $4y = 3 + y$

---

92. Notice how we had to multiply by the distributive principle in the example below. Solve the other equation.

$$\frac{x + 4}{5} = \frac{x}{3}$$

$$\overset{3}{\cancel{15}}\left(\frac{x + 4}{\cancel{5}}\right) = \overset{5}{\cancel{15}}\left(\frac{x}{\cancel{3}}\right)$$

$$3(x + 4) = 5(x)$$

$$3x + 12 = 5x$$

$$12 = 2x$$

$$x = 6$$

$$\frac{2y - 1}{5} = \frac{1}{2}$$

Answer: $y = \frac{15}{7}$, from: $15 = 7y$

---

93. To clear the fractions below, we multiplied both sides by 5x. Solve the other equation.

$$\frac{x + 2}{x} = \frac{3}{5}$$

$$5\cancel{x}\left(\frac{x + 2}{\cancel{x}}\right) = \cancel{5}x\left(\frac{3}{\cancel{5}}\right)$$

$$5(x + 2) = x(3)$$

$$5x + 10 = 3x$$

$$2x = -10$$

$$x = \frac{-10}{2} = -5$$

$$\frac{t - 9}{t} = \frac{1}{2}$$

Answer: $y = \frac{7}{4}$, from: $4y - 2 = 5$

Answer: $t = 18$, from: $2t - 18 = t$

94. To clear the fractions below, we multiplied both sides by 5(x + 5). Solve the other equation.

$$\frac{x-3}{x+5} = \frac{3}{5}$$

$$\frac{y+4}{4y-1} = \frac{1}{2}$$

$$5\cancel{(x+5)}\left(\frac{x-3}{\cancel{x+5}}\right) = \cancel{5}(x+5)\left(\frac{3}{\cancel{5}}\right)$$

$$5(x-3) = (x+5)(3)$$

$$5x - 15 = 3x + 15$$

$$2x = 30$$

$$x = 15$$

$y = \frac{9}{2}$, from:

$2y + 8 = 4y - 1$

95. To clear the fractions below, we multiplied both sides by x(x − 3). Solve the other equation.

$$\frac{4}{x-3} = \frac{2}{x}$$

$$\frac{5}{y} = \frac{1}{3y+4}$$

$$x\cancel{(x-3)}\left(\frac{4}{\cancel{x-3}}\right) = \cancel{x}(x-3)\left(\frac{2}{\cancel{x}}\right)$$

$$x(4) = (x-3)2$$

$$4x = 2x - 6$$

$$2x = -6$$

$$x = -3$$

$y = -\frac{10}{7}$, from:

$15y + 20 = y$

96. To clear the fractions below, we multiplied both sides by (x + 1)(x − 4). Solve the other equation.

$$\frac{3}{x+1} = \frac{2}{x-4}$$

$$\frac{4}{y-1} = \frac{5}{y+3}$$

$$\cancel{(x+1)}(x-4)\left(\frac{3}{\cancel{x+1}}\right) = (x+1)\cancel{(x-4)}\left(\frac{2}{\cancel{x-4}}\right)$$

$$(x-4)(3) = (x+1)(2)$$

$$3x - 12 = 2x + 2$$

$$x = 14$$

$y = 17$, from:

$4y + 12 = 5y - 5$

97. Notice how the squared terms are eliminated in the solution below. Solve the other equation.

$$\frac{x-1}{x-2} = \frac{x-3}{x+2} \qquad\qquad \frac{y-3}{y-1} = \frac{y+4}{y+1}$$

$$\cancel{(x-2)}(x+2)\left(\frac{x-1}{\cancel{x-2}}\right) = (x-2)\cancel{(x+2)}\left(\frac{x-3}{\cancel{x+2}}\right)$$

$$(x+2)(x-1) = (x-2)(x-3)$$

$$x^2 + x - 2 = x^2 - 5x + 6$$

$$6x = 8$$

$$x = \frac{8}{6} = \frac{4}{3}$$

---

98. To clear the fractions below, we multiplied both sides by 12. Solve the other equation.

$y = \frac{1}{5}$, from:

$y^2 - 2y - 3 = y^2 + 3y - 4$

$$\frac{2x}{3} + \frac{x+1}{4} = 14 \qquad\qquad \frac{y}{3} + \frac{y+1}{5} = 5$$

$$12\left(\frac{2x}{3} + \frac{x+1}{4}\right) = 12(14)$$

$$\overset{4}{\cancel{12}}\left(\frac{2x}{\cancel{3}}\right) + \overset{3}{\cancel{12}}\left(\frac{x+1}{\cancel{4}}\right) = 168$$

$$4(2x) + 3(x+1) = 168$$

$$8x + 3x + 3 = 168$$

$$11x = 165$$

$$x = 15$$

---

99. To clear the fractions below, we multiplied both sides by 12, the LCD of all three denominators. Notice how we multiplied by the distributive principle on the left side. Solve the other equation.

$y = 9$, from:

$5y + 3y + 3 = 75$

$$\frac{x}{3} - \frac{1}{2} = \frac{3}{4} \qquad\qquad \frac{y}{6} + \frac{1}{3} = \frac{4}{9}$$

$$12\left(\frac{x}{3} - \frac{1}{2}\right) = \overset{3}{\cancel{12}}\left(\frac{3}{\cancel{4}}\right)$$

$$\overset{4}{\cancel{12}}\left(\frac{x}{\cancel{3}}\right) - \overset{6}{\cancel{12}}\left(\frac{1}{\cancel{2}}\right) = 9$$

$$4x - 6 = 9$$

$$4x = 15$$

$$x = \frac{15}{4}$$

100. To clear the fractions below, we multiplied both sides by 6x. Solve the other equation.

$$\frac{1}{2} - \frac{1}{3} = \frac{1}{x}$$

$$6x\left(\frac{1}{2} - \frac{1}{3}\right) = 6x\left(\frac{1}{x}\right)$$

$$6x\left(\frac{1}{2}\right) - 6x\left(\frac{1}{3}\right) = 6(1)$$

$$3x - 2x = 6$$

$$x = 6$$

$$\frac{1}{4} + \frac{1}{6} = \frac{1}{y}$$

$y = \frac{2}{3}$, from:

$3y + 6 = 8$

101. To clear the fractions below, we multiplied both sides by 2x. Solve the other equation.

$$\frac{3}{x} = \frac{4}{x} - \frac{1}{2}$$

$$2x\left(\frac{3}{x}\right) = 2x\left(\frac{4}{x} - \frac{1}{2}\right)$$

$$2(3) = 2x\left(\frac{4}{x}\right) - 2x\left(\frac{1}{2}\right)$$

$$6 = 8 - x$$

$$-2 = -x$$

$$x = 2$$

$$\frac{1}{3y} + \frac{5}{y} = 2$$

$y = \frac{12}{5}$, from

$3y + 2y = 12$

102. Since $x^2 - 4 = (x + 2)(x - 2)$, the LCD below is $(x + 2)(x - 2)$. Therefore, to clear the fractions, we multiplied both sides by $(x + 2)(x - 2)$.

$$\frac{3}{x + 2} + \frac{4x}{x^2 - 4} = \frac{1}{x - 2}$$

$$(x + 2)(x - 2)\left[\frac{3}{x + 2} + \frac{4x}{x^2 - 4}\right] = \left(\frac{1}{x - 2}\right)(x + 2)(x - 2)$$

$$(x + 2)(x - 2)\left(\frac{3}{x + 2}\right) + (x + 2)(x - 2)\left(\frac{4x}{x^2 - 4}\right) = 1(x + 2)$$

$$3x - 6 + 4x = x + 2$$

$$7x - 6 = x + 2$$

$$6x = 8$$

$$x = \frac{8}{6} = \frac{4}{3}$$

$y = \frac{8}{3}$, from:

$1 + 15 = 6y$

Continued on following page.

102. Continued

Using the same steps, solve this equation.

$$\frac{1}{x + 4} + \frac{1}{x - 4} = \frac{1}{x^2 - 16}$$

103. To clear the fraction below, we multiplied both sides by $(x - 2)$.

$$\frac{2}{x - 2} + 2 = \frac{x}{x - 2}$$

$$(x - 2)\left(\frac{2}{x - 2} + 2\right) = \cancel{(x - 2)}\left(\frac{x}{\cancel{x - 2}}\right)$$

$$\cancel{(x - 2)}\left(\frac{2}{\cancel{x - 2}}\right) + (x - 2)(2) = x$$

$$2 + 2x - 4 = x$$

$$2x - 2 = x$$

$$-2 = -x$$

$$x = 2$$

The proposed solution is $x = 2$. However, 2 is a value that makes both denominators equal 0. Therefore:

a) Is $x = 2$ the solution of the equation? ________

b) Does the equation have a solution? ________

Answer to 102: $x = \frac{1}{2}$, from:

$x - 4 + x + 4 = 1$

104. When clearing fractions, we can get an equation that has two solutions. An example is shown. Solve the other equation.

$$2x + \frac{3}{x} = 5 \qquad\qquad y + \frac{5}{y} = 6$$

$$x\left(2x + \frac{3}{x}\right) = 5(x)$$

$$x(2x) + \cancel{x}\left(\frac{3}{\cancel{x}}\right) = 5x$$

$$2x^2 + 3 = 5x$$

$$2x^2 - 5x + 3 = 0$$

$$(2x - 3)(x - 1) = 0$$

$2x - 3 = 0$	$x - 1 = 0$
$2x = 3$	$x = 1$
$x = \frac{3}{2}$	

Answers to 103:

a) No

b) No

Answer to 104: $y = 1$ and 5, from:

$y^2 - 6y + 5 = 0$

7-10 NUMBER PROBLEMS

In this section, we will solve some word problems involving an "unknown number". All of the problems involve a fractional equation.

---

105. We used the four-step method to solve the problem below.

Problem: One number is 4 more than another. The quotient of the larger divided by the smaller is 3. Find the two numbers.

1) Represent the unknowns with a letter or algebraic expression.

Let x equal the smaller number. Then the larger number is $x + 4$.

2) Translate the problem to an equation.

$$\frac{x+4}{x} = 3$$

3) Solve the equation.

$$\frac{x+4}{x} = 3$$

$$\not{x}\left(\frac{x+4}{\not{x}}\right) = x(3)$$

$$x + 4 = 3x$$

$$4 = 2x$$

$$x = 2 \quad (\text{and } x + 4 = 2 + 4 = 6)$$

4) Check the solution in the original words of the problem.

The two numbers are 6 and 2, since 6 is 4 more than 2 and 6 divided by 2 is 3.

Following the example, solve these.

a) One number is 12 more than another. The quotient of the larger divided by the smaller is 4. Find the two numbers.

b) One number is 3 more than another. The quotient of the larger divided by the smaller is $\frac{6}{5}$. Find the two numbers.

106. To solve the first problem we let $\underline{x}$ equal the original numerator and $\underline{x + 2}$ equal the original denominator. Solve the other problem.

The denominator of a fraction is 2 more than the numerator. If 3 is subtracted from both the numerator and denominator, the resulting fraction is $\frac{2}{3}$. Find the original fraction.

$$\frac{x - 3}{(x + 2) - 3} = \frac{2}{3}$$

$$\frac{x - 3}{x - 1} = \frac{2}{3}$$

$$3(\cancel{x - 1})\left(\frac{x - 3}{\cancel{x - 1}}\right) = \cancel{3}(x - 1)\left(\frac{2}{\cancel{3}}\right)$$

$$3(x - 3) = (x - 1)(2)$$

$$3x - 9 = 2x - 2$$

$$x = 7$$

The original fraction is:

$$\frac{x}{x + 2} = \frac{7}{9}$$

In a certain fraction, the denominator is 7 larger than the numerator. If 4 is added to both the numerator and denominator, the result is $\frac{1}{2}$. Find the original fraction.

a) Smaller is 4; larger is 16.

$$\frac{x + 12}{x} = 4$$

b) Smaller is 15; larger is 18.

$$\frac{x + 3}{x} = \frac{6}{5}$$

---

107. To solve the first problem below, we let $\underline{x}$ equal the original numerator and $\underline{2x}$ equal the original denominator. Solve the other problem.

The denominator of a fraction is twice the numerator. If one is added to the numerator and subtracted from the denominator, the result equals $\frac{2}{3}$. Find the original fraction.

$$\frac{x + 1}{2x - 1} = \frac{2}{3}$$

$$3(\cancel{2x - 1})\left(\frac{x + 1}{\cancel{2x - 1}}\right) = \cancel{3}(2x - 1)\left(\frac{2}{\cancel{3}}\right)$$

$$3(x + 1) = (2x - 1)(2)$$

$$3x + 3 = 4x - 2$$

$$x = 5$$

The original fraction is:

$$\frac{x}{2x} = \frac{5}{10}$$

The denominator of a fraction is 4 times the numerator. If 6 is added to the numerator and subtracted from the denominator, the result equals "1". Find the original fraction.

The original fraction is $\frac{3}{10}$, from:

$$\frac{x + 4}{(x + 7) + 4} = \frac{1}{2}$$

108. Following the example, solve the other problem.

One half of a number is 2 more than one third of the same number. Find the number.	One fourth of a number is 5 more than one eighth of the same number. Find the number.

$$\frac{1}{2}(x) = \frac{1}{3}(x) + 2$$

$$\frac{x}{2} = \frac{x}{3} + 2$$

$$\overset{3}{\not{6}}\left(\frac{x}{\not{2}}\right) = 6\left(\frac{x}{3} + 2\right)$$

$$3(x) = \overset{2}{\not{6}}\left(\frac{x}{\not{3}}\right) + 6(2)$$

$$3x = 2x + 12$$

$$x = 12$$

The number is 12.

Answer: $\frac{4}{16}$, from: $\frac{x + 6}{4x - 6} = 1$

---

109. We know that the reciprocal of a number is "1" divided by that number. For example:

The reciprocal of 8 is $\frac{1}{8}$.

Following the pattern above, we used the letter x to translate these phrases to algebraic expressions.

The reciprocal of a number is $\frac{1}{x}$.

The reciprocal of 3 times a number is $\frac{1}{3x}$.

The reciprocal of 2 more than a number is $\frac{1}{x + 2}$.

Translate each phrase to an algebraic expression. Use x as the variable.

a) The reciprocal of twice a number is ________.

b) The reciprocal of 5 less than a number is ________.

Answer: 40, from: $\frac{x}{4} = \frac{x}{8} + 5$

Answers:

a) $\frac{1}{2x}$

b) $\frac{1}{x - 5}$

110. Following the example, solve the other problem.

The reciprocal of 3 plus the reciprocal of 5 is the reciprocal of what number?

$$\frac{1}{3} + \frac{1}{5} = \frac{1}{x}$$

$$15x\left(\frac{1}{3} + \frac{1}{5}\right) = 15\cancel{x}\left(\frac{1}{\cancel{x}}\right)$$

$$\overset{5}{\cancel{15}}x\left(\frac{1}{\cancel{3}}\right) + \overset{3}{\cancel{15}}x\left(\frac{1}{\cancel{5}}\right) = 15(1)$$

$$5x + 3x = 15$$

$$8x = 15$$

$$x = \frac{15}{8}$$

The number is $\frac{15}{8}$.

The reciprocal of 4 plus the reciprocal of 6 is the reciprocal of what number?

---

111. Following the example, solve the other problem.

$\frac{12}{5}$, from:

$$\frac{1}{4} + \frac{1}{6} = \frac{1}{x}$$

The reciprocal of 4 less than a number is 3 times the reciprocal of the number itself. Find the number.

$$\frac{1}{x - 4} = 3\left(\frac{1}{x}\right)$$

$$\frac{1}{x - 4} = \frac{3}{x}$$

$$x\cancel{(x - 4)}\left(\frac{1}{\cancel{x - 4}}\right) = \cancel{x}(x - 4)\left(\frac{3}{\cancel{x}}\right)$$

$$x(1) = (x - 4)(3)$$

$$x = 3x - 12$$

$$-2x = -12$$

$$x = \frac{-12}{-2} = 6$$

The number is 6.

The reciprocal of 2 more than a number is twice the reciprocal of the number itself. Find the number.

-4, from:

$$\frac{1}{x + 2} = \frac{2}{x}$$

112. Following the example, solve the other problem.

Find two consecutive integers such that one-half of the smaller plus one-third of the larger is equal to 7.

$$\frac{1}{2}(x) + \frac{1}{3}(x + 1) = 7$$

$$\frac{x}{2} + \frac{x + 1}{3} = 7$$

$$6\left(\frac{x}{2} + \frac{x + 1}{3}\right) = 6(7)$$

$$\overset{3}{\not{6}}\left(\frac{x}{\not{2}}\right) + \overset{2}{\not{6}}\left(\frac{x + 1}{\not{3}}\right) = 42$$

$$3x + 2x + 2 = 42$$

$$5x = 40$$

$$x = 8$$

The two consecutive integers are 8 and 9.

Find two consecutive integers such that two-fifths of the smaller plus one-third of the larger is equal to 15.

The two consecutive integers are 20 and 21, from:

$$\frac{2x}{5} + \frac{x + 1}{3} = 15$$

---

113. Notice that we got two solutions for the problem below. Solve the other problem.

If 3 times the reciprocal of a number is subtracted from the number, we get 2. Find the number.

$$x - 3\left(\frac{1}{x}\right) = 2$$

$$x - \frac{3}{x} = 2$$

$$x\left(x - \frac{3}{x}\right) = x(2)$$

$$x(x) - \not{x}\left(\frac{3}{\not{x}}\right) = 2x$$

$$x^2 - 3 = 2x$$

$$x^2 - 2x - 3 = 0$$

$$(x - 3)(x + 1) = 0$$

$$x - 3 = 0 \quad | \quad x + 1 = 0$$

$$x = 3 \quad | \quad x = -1$$

The number is either 3 or -1.

If twice a number is added to 5 times its reciprocal, the answer is 7. Find the number.

$\frac{5}{2}$ or 1, from:

$$2x + \frac{5}{x} = 7$$

7-11 WORK PROBLEMS

In this section, we will solve some problems involving the time needed for two people, working together, to complete a job. Each problem involves a fractional equation.

---

114. If a job can be done in x hours (or some other unit of time), then $\frac{1}{x}$ of the job can be done in 1 hour. For example:

If a job can be done in 6 hours, $\frac{1}{6}$ of the job can be done in 1 hour.

If a job can be done in 4 days, $\frac{1}{4}$ of the job can be done in 1 day.

If a job can be done in x weeks, ____ of the job can be done in 1 week.

Answer: $\frac{1}{x}$

115. The relationship in the last frame is used to solve the problem below.

Problem: John can paint a room in 4 hours. Mike can paint the same room in 6 hours. How long would it take them, working together, to paint the room?

To solve the problem, we will set up a relationship based on the fractional part of the job that can be done in 1 hour. We know these facts:

1. Working alone, John can paint $\frac{1}{4}$ of the room in 1 hour and Mike can paint $\frac{1}{6}$ of the room in 1 hour.
2. Working together, they can paint $\frac{1}{4} + \frac{1}{6}$ of the room in 1 hour.
3. If together they can paint the room in x hours, they can paint $\frac{1}{x}$ of the room in 1 hour.

Using the two expressions for the fractional part of the job that can be done in 1 hour, we can set up the following equation:

$$\frac{1}{4} + \frac{1}{6} = \frac{1}{x}$$

To clear the fractions, we multiply both sides by 12x. We get:

$$12x\left(\frac{1}{4} + \frac{1}{6}\right) = 12\cancel{x}\left(\frac{1}{\cancel{x}}\right)$$

$$\overset{3}{\cancel{12}}x\left(\frac{1}{\cancel{4}}\right) + \overset{2}{\cancel{12}}x\left(\frac{1}{\cancel{6}}\right) = 12(1)$$

$$3x + 2x = 12$$

$$5x = 12$$

$$x = \frac{12}{5} = 2\frac{2}{5} \text{ hours}$$

Therefore, working together, John and Mike can paint the room in

________ hours.

116. Following the example, solve the other problem.

A man can plant his garden in 5 hours. His wife can plant the same garden in 4 hours. How long would it take them if they worked together?

$$\frac{1}{5} + \frac{1}{4} = \frac{1}{x}$$

$$20x\left(\frac{1}{5} + \frac{1}{4}\right) = 20\cancel{x}\left(\frac{1}{\cancel{x}}\right)$$

$$\overset{4}{\cancel{20}}x\left(\frac{1}{\cancel{5}}\right) + \overset{5}{\cancel{20}}x\left(\frac{1}{\cancel{4}}\right) = 20(1)$$

$$4x + 5x = 20$$

$$9x = 20$$

$$x = \frac{20}{9} = 2\frac{2}{9}$$

Together, they plant the garden in $2\frac{2}{9}$ hours.

A swimming pool can be filled in 8 hours by pipe A alone and in 6 hours by pipe B alone. How long would it take to fill the tank if both pipes were used together?

$2\frac{2}{5}$ hours

---

117. Use the same method for these.

a) Sue can do a job in 4 days, but Mary needs only 3 days. How long would it take them working together?

b) Pete can dig a trench in 6 hours, but his son needs 9 hours. How long would it take them working together?

$3\frac{3}{7}$ hours, from:

$$\frac{1}{6} + \frac{1}{8} = \frac{1}{x}$$

118. To translate the problem below to an equation, we used x for the number of hours Joe needs and x + 12 for the number of hours Tom needs.

Problem: Working together, Joe and Tom can complete a job in 8 hours. It would take Tom 12 more hours working alone than it would take Joe. How long would it take each of them working alone?

$$\frac{1}{x} + \frac{1}{x+12} = \frac{1}{8}$$

$$8x(x+12)\left(\frac{1}{x} + \frac{1}{x+12}\right) = \cancel{8}x(x+12)\left(\frac{1}{\cancel{8}}\right)$$

$$8\cancel{x}(x+12)\left(\frac{1}{\cancel{x}}\right) + 8x\cancel{(x+12)}\left(\frac{1}{\cancel{x+12}}\right) = x(x+12)$$

$$8(x+12) + 8x(1) = x(x+12)$$

$$8x + 96 + 8x = x^2 + 12x$$

$$x^2 - 4x - 96 = 0$$

$$(x-12)(x+8) = 0$$

$$x = 12 \text{ and } -8$$

Since -8 does not make sense for the number of hours, the answer is 12 hours for Joe and 12 + 12 = 24 hours for Tom.

Using the same method, solve this one.

Problem: Working together, Peggy and Joan can complete a job in 4 hours. Joan alone can do the job in 6 hours less than Peggy alone. How long would it take each of them working alone?

a) $\frac{12}{7} = 1\frac{5}{7}$ days

b) $\frac{18}{5} = 3\frac{3}{5}$ hours

Peggy alone takes 12 hours and Joan alone takes 6 hours.

SELF-TEST 23 (pages 304-319)

Solve each equation.

1. $\frac{x - 5}{x} = \frac{3}{2}$

2. $\frac{1}{y - 1} = \frac{3}{y + 3}$

3. $\frac{1}{2y} - 3 = \frac{5}{y}$

4. $3x - \frac{2}{x} = 5$

Solve each problem.

5. One number is 12 more than another. If the larger is divided by the smaller, we get 3. Find the two numbers.

6. The reciprocal of 8 less than a number is twice the reciprocal of the number itself. Find the number.

7. Find two consecutive integers such that two-thirds of the smaller plus one-fifth of the larger is equal to 8.

8. Mary can type a report in 5 hours, but Lois needs 6 hours. How long would it take them if they worked together?

ANSWERS:

1. $x = -10$
2. $y = 3$
3. $y = -\frac{3}{2}$
4. $x = -\frac{1}{3}$ and 2
5. 6 and 18
6. 16
7. 9 and 10
8. $2\frac{8}{11}$ hours

7-12 MOTION PROBLEMS

In this section, we will solve some motion problems based on the formula $d = rt$. Some of the problems involve a fractional equation.

---

119. The following formula relates three variables: distance traveled, rate (or speed), and time traveled.

$$d = rt$$

where: d = distance traveled
r = rate (or speed)
t = time traveled

The formula can be used to solve many applied problems. Here is an example:

Problem: A car and a bus leave a city at the same time traveling in opposite directions. The car travels 50 mph and the bus travels 60 mph. In how many hours will they be 275 miles apart?

To solve a problem of this type, it is helpful to make a diagram as we have done below.

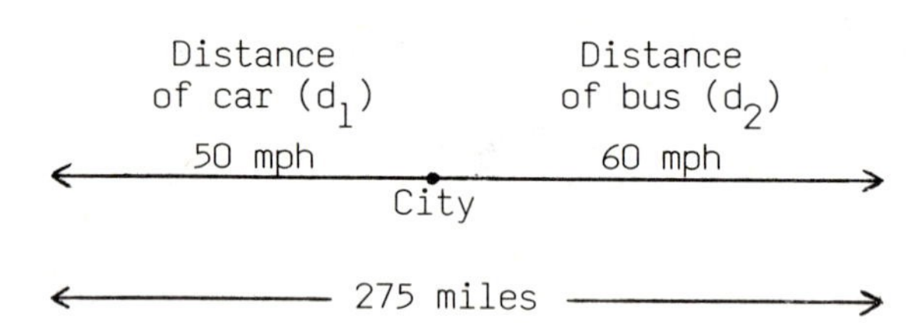

We can see that the two distances (d_1 and d_2) are not equal because the car and bus traveled at different rates for the same period of time. However, the time t is equal for both. We can organize what we know in the chart below.

	Distance	Rate	Time
Car	$d_1 = 50t$	50	t
Bus	$d_2 = 60t$	60	t

From the diagram, we know that the sum of the distance traveled by the car and bus equals 275 miles. That is:

$$d_1 + d_2 = 275$$

But based on $d = rt$, $d_1 = 50t$ and $d_2 = 60t$, as we can see from the chart. Substituting, we get:

$$50t + 60t = 275$$

Continued on following page.

119. Continued

Now solving the equation, we get:

$$50t + 60t = 275$$

$$110t = 275$$

$$t = \frac{275}{110} = \frac{5}{2} = 2\frac{1}{2} \text{ hours}$$

The solution checks in the original problem because $2\frac{1}{2} \times 50 = 125$, $2\frac{1}{2} \times 60 = 150$, and $125 + 150 = 275$.

To solve the problem below, draw a diagram and make up a chart like those in the example problem.

Problem: A car and a bus leave a city at the same time traveling in opposite directions. The car travels 45 mph and the bus travels 57 mph. In how many hours will they be 170 miles apart?

$\frac{5}{3} = 1\frac{2}{3}$ hours, from:

$45x + 57x = 170$

120. The following problem involves the same formula.

Problem: On a two-day trip, Mel drove 510 miles in a total of 10 hours. On the first day, he averaged 55 mph. On the second day, he averaged 45 mph. How long and how far did he travel each day?

As a help, we diagrammed the problem. We have no reason to believe that d_1 and d_2 are equal.

d_1 55 mph → d_2 45 mph →

← 510 miles →

Continued on following page.

120. Continued

The data of the problem is organized in a chart below. Since we let t be the time for Day 1, the time for Day 2 must be 10 - t. We used $d = rt$ to get 55t and 45(10 - t).

	Distance	Rate	Time
Day 1	$d_1 = 55t$	55	t
Day 2	$d_2 = 45(10 - t)$	45	10 - t

From the diagram, we know that the sum of the distance traveled on the two days is 510 miles. That is:

$$d_1 + d_2 = 510$$

Substituting 55t for d_1 and 45(10 - t) for d_2 and solving, we get:

$$55t + 45(10 - t) = 510$$
$$55t + 450 - 45t = 510$$
$$10t = 60$$
$$t = 6 \text{ hours}$$

Therefore, he travels 6 hours on Day 1 and 10 - 6 = 4 hours on Day 2. On Day 1, he traveled 6(55) = 330 miles. On Day 2, he traveled 4(45) = 180 miles. This checks since 330 + 180 = 510.

Following the example, draw a diagram and make up a chart to solve this one.

Problem: A woman walks and jogs for 2 hours and covers 11 miles. If she walks at 4 mph and jogs at 8 mph, how long and how far did she walk and jog?

121. We used the same general method for the problem below. In this problem, the distance traveled is the same for both cars.

Problem: One car leaves a motel traveling 42 mph. A second car leaves the same motel one hour later traveling 56 mph on the same highway. How long does it take the second car to catch up to the first car?

We diagrammed the problem below. The two distances are equal. Since we want to find the time traveled by the second car, we used t for the time of the second car.

Car 1

42 mph (t + 1) hour →

Car 2

56 mph t hours →

We organized the data in the chart below. We used $d = rt$ to get $42(t + 1)$ and $56t$.

	Distance	Rate	Time
Car 1	$d = 42(t + 1)$	42	$t + 1$
Car 2	$d = 56t$	56	t

From the diagram, we can see that the two distances are equal. Therefore, we can set up and solve the following equation.

$$42(t + 1) = 56t$$

$$42t + 42 = 56t$$

$$42 = 14t$$

$$t = 3 \text{ hours (and } t + 1 = 4 \text{ hours)}$$

Therefore, it takes the second car 3 hours to catch up to the first car (which travels 4 hours). This checks since $4(42) = 168$ and $3(56) = 168$.

Solve this one. Use a diagram and a chart.

Problem: A small plane leaves an airport and flies due north at 180 km/h. A jet leaves the same airport two hours later and flies due north at 900 km/h. How long will it take the jet to catch up to the small plane?

Walks 5 miles in $1\frac{1}{4}$ hours and jogs 6 miles in $\frac{3}{4}$ hours.

From:

$4t + 8(2 - t) = 11$

122. In the problem below, the two distances are also equal.

Problem: A boat travels 5 hours downstream with a current of 4 mph. The return trip takes 10 hours. Find the speed of the boat in still water.

We diagrammed the problem below. The two distances are equal. x is the speed of the boat in still water.

Downstream

5 hours (x + 4) mph →

Upstream

← 10 hours (x - 4) mph

The data is summarized in the chart below.

	Distance	Rate	Time
Downstream	$d = 5(x + 4)$	$x + 4$	5
Upstream	$d = 10(x - 4)$	$x - 4$	10

Since the two distances are equal, we can set up and solve the following equation.

$$5(x + 4) = 10(x - 4)$$

$$5x + 20 = 10x - 40$$

$$60 = 5x$$

$$x = 12 \text{ mph}$$

Therefore, the speed of the boat in still water is 12 mph. This checks since $5(12 + 4) = 5(16) = 80$ and $10(12 - 4) = 10(8) = 80$.

Using a diagram and a chart, solve this one.

Problem: A small plane flew 3 hours with a 15 km/h tail wind. The return flight into the same wind took 4 hours. Find the speed of the plane in still air.

$\frac{1}{2}$ hour, from:

$180(t + 2) = 900t$

123. In this problem, the two times are equal. Therefore, we have to solve a fractional equation to solve the problem.

Problem: One car travels 12 miles an hour faster than another. While one car travels 200 miles, the other travels 260 miles. Find the speed of each car.

We diagrammed the problem below.

Car 1

200 miles ——— x mph →

Car 2

260 miles ——— (x + 12) mph →

The data is summarized in the chart below. The time t for each is equal. We used $t = \frac{d}{r}$ to get $\frac{200}{x}$ and $\frac{260}{x + 12}$.

	Distance	Rate	Time
Car 1	200	x	$t = \frac{200}{x}$
Car 2	260	x + 12	$t = \frac{260}{x + 12}$

Since the two times are equal, we can set up and solve the following fractional equation.

$$\frac{200}{x} = \frac{260}{x + 12}$$

$$\cancel{x}(x + 12)\left(\frac{200}{\cancel{x}}\right) = x\cancel{(x + 12)}\left(\frac{260}{\cancel{x + 12}}\right)$$

$$200x + 2400 = 260x$$

$$2400 = 60x$$

$$x = 40 \text{ mph}$$

Therefore, the speed of Car 1 is 40 mph and the speed of Car 2 is 40 + 12 = 52 mph. These values check since $\frac{200}{40} = 5$ and $\frac{260}{52} = 5$. Therefore, they both travel 5 hours at those speeds.

Using a diagram and a chart, do this one:

Problem: A freight train travels 30 mph slower than a passenger train. While the freight train travels 220 miles, the passenger train travels 385 miles. Find the speed of each train.

105 km/h, from:

$3(x + 15) = 4(x - 15)$

124. Since the two times are equal, we use a fractional equation to solve the problem below.

Problem: A river has a current of 2 mph. A boat takes as long to go 18 miles downstream as to go 12 miles upstream. What is the speed of the boat in still water?

We diagrammed the problem below. The two distances are not equal.

Downstream

18 miles (x + 2) mph →

Upstream

← 12 miles (x - 2) mph

We organized the data in the chart below. The time t for each trip is equal. We used $t = \frac{d}{r}$ to get $\frac{18}{x+2}$ and $\frac{12}{x-2}$.

	Distance	Rate	Time
Downstream	18	x + 2	$t = \frac{18}{x+2}$
Upstream	12	x - 2	$t = \frac{12}{x-2}$

Since the two times are equal, we can set up and solve the equation below.

$$\frac{18}{x+2} = \frac{12}{x-2}$$

$$\cancel{(x+2)}(x-2)\left(\frac{18}{\cancel{x+2}}\right) = (x+2)\cancel{(x-2)}\left(\frac{12}{\cancel{x-2}}\right)$$

$$18x - 36 = 12x + 24$$

$$6x = 60$$

$$x = 10 \text{ mph}$$

Therefore, the speed of the boat is 10 mph in still water. This checks since $\frac{18}{12} = \frac{3}{2}$ and $\frac{12}{8} = \frac{3}{2}$. Therefore, both trips take $1\frac{1}{2}$ hours.

Using a diagram and chart, solve this one.

Problem: An airplane takes as long to fly 425 miles with the wind as it does to fly 375 miles against the wind. If the wind is blowing at 10 mph, what is the speed of the plane without a wind?

The speed of the freight train is 40 mph; the speed of the passenger train is 70 mph.

From either:

$$\frac{220}{x} = \frac{385}{x+30}$$

or $$\frac{385}{x} = \frac{220}{x-30}$$

125. We also use a fractional equation to solve the problem below.

Problem: A jet flies at an average speed of 500 mph without a wind. If it takes as long to fly 910 miles with the wind as it does to fly 840 miles against the wind, how strong is the wind?

We diagrammed the problem below. The two distances are not equal.

With the wind

910 miles (500 + x) mph →

Against the wind

← 840 miles (500 - x) mph

We organized the data in the chart below.

	Distance	Rate	Time
With the wind	910	500 + x	$t = \frac{910}{500 + x}$
Against the wind	840	500 - x	$t = \frac{840}{500 - x}$

Since the two times are equal, we can set up and solve this equation where x is the speed of the wind.

$$\frac{910}{500 + x} = \frac{840}{500 - x}$$

$$\cancel{(500 + x)}(500 - x)\left(\frac{910}{\cancel{500 + x}}\right) = (500 + x)\cancel{(500 - x)}\left(\frac{840}{\cancel{500 - x}}\right)$$

$$455,000 - 910x = 420,000 + 840x$$

$$35,000 = 1750x$$

$$x = 20 \text{ mph}$$

Therefore, the wind is blowing at 20 mph. This checks since $\frac{910}{520} = \frac{7}{4}$ and $\frac{840}{480} = \frac{7}{4}$. Therefore, both flights take $1\frac{3}{4}$ hours.

Using a diagram and chart, solve this one:

Problem: Joe can row 4 miles per hour in still water. It takes as long to row 15 miles upstream as it does to row 25 miles downstream. How fast is the current?

160 mph, from:

$$\frac{425}{x + 10} = \frac{375}{x - 10}$$

1 mph, from:

$$\frac{15}{4 - x} = \frac{25}{4 + x}$$

7-13 RATIO AND PROPORTION

In this section, we will discuss ratios and proportions. Some applied problems are included.

---

126. A ratio is a comparison of two quantities by division. A ratio can be expressed in three ways: with a fraction, with the word to, or with a colon. For example, if there are 11 boys and 14 girls in a class, the ratio of boys to girls can be expressed in one of the three ways below:

$$\frac{11}{14} \qquad 11 \text{ to } 14 \qquad 11:14$$

A team wins 5 games and loses 4 games. Express the ratio of wins to losses in three different ways.

________ ________ ________

Answer: $\frac{5}{4}$ 5 to 4 5:4

127. When ratio language is used, the ratio is ordinarily reduced to lowest terms. For example:

If there are 10 smokers and 20 non-smokers in a group, the ratio of smokers to non-smokers is $\frac{10}{20} = \frac{1}{2}$ or 1 to 2.

When a ratio is expressed as a fraction that reduces to a whole number, we ordinarily write "1" as the denominator to make it clear that we are comparing two quantities. For example:

If an alloy contains 24 grams of tin and 2 grams of silver, the ratio of tin to silver is $\frac{24}{2} = \frac{12}{1}$ or _____ to _____.

Answer: 12 to 1

128. Ratios can be used to express rates or costs. For example:

If a woman drives 318 miles in 6 hours, the ratio of miles to hours is $\frac{318 \text{ miles}}{6 \text{ hours}}$ or $\frac{53}{1}$.

If a man pays 90 cents for 5 pears, the ratio of cents to pears is $\frac{90 \text{ cents}}{5 \text{ pears}} =$ ________ .

Answer: $\frac{18}{1}$

129. A <u>proportion</u> is an equation stating that two ratios are equal. For example:

$$\frac{2}{3} = \frac{12}{18}$$

One of the numbers is unknown in the proportion below. To solve the proportion, we multiplied both sides by (30)(42) rather than trying to find the LCD. Solve the other proportion.

$$\frac{10}{30} = \frac{x}{42} \qquad\qquad \frac{18}{15} = \frac{x}{50}$$

$$(\cancel{30})(42)\left(\frac{10}{\cancel{30}}\right) = \left(\frac{x}{\cancel{42}}\right)(30)(\cancel{42})$$

$$420 = 30x$$

$$x = \frac{420}{30}$$

$$x = 14$$

---

130. To clear the fraction below, we multiplied both sides by 44x. Solve the other proportion.

Answer to 129: x = 60

$$\frac{8}{44} = \frac{10}{x} \qquad\qquad \frac{11}{12} = \frac{77}{x}$$

$$\cancel{44}x\left(\frac{8}{\cancel{44}}\right) = \left(\frac{10}{\cancel{x}}\right)(44\cancel{x})$$

$$8x = 440$$

$$x = 55$$

---

131. To solve the problem below, we set up a proportion <u>with</u> units, solved the same proportion <u>without</u> units, and then stated the solution in terms of the original wording of the problem.

Answer to 130: x = 84

A car travels 90 miles in 2 hours. At that rate, how many miles would it travel in 7 hours?

$$\frac{90 \text{ miles}}{2 \text{ hours}} = \frac{x \text{ miles}}{7 \text{ hours}} \qquad\qquad \frac{90}{2} = \frac{x}{7}$$

$$(\cancel{2})(7)\left(\frac{90}{\cancel{2}}\right) = \left(\frac{x}{\cancel{7}}\right)(2)(\cancel{7})$$

$$630 = 2x$$

$$x = 315$$

Therefore, the car would travel ________ miles in 7 hours.

Answer to 131: 315

132. When using a proportion to solve a problem, the following steps should be used to set up the proportion correctly.

1. Set up the known ratio on the left side.
2. Then make sure that the units in the ratio on the right side correspond to the units on the left side.

For example, we set up a proportion correctly below. Notice that the ratio is $\boxed{\frac{\text{gallons}}{\text{miles}}}$ on both sides. Set up the other proportions.

If a car uses 2 gallons of gasoline to travel 30 miles, how many gallons will it need to travel 75 miles?

$$\frac{2 \text{ gallons}}{30 \text{ miles}} = \frac{x \text{ gallons}}{75 \text{ miles}}$$

a) If 3 grams of sodium chloride are needed to make 1 liter of a solution, how many grams are needed to make 12 liters of the same solution?

b) If 2 inches on a map represent 80 miles, what is the distance between two cities that are 5 inches apart?

---

a) $\frac{3 \text{ grams}}{1 \text{ liter}} = \frac{x \text{ grams}}{12 \text{ liters}}$

b) $\frac{2 \text{ inches}}{80 \text{ miles}} = \frac{5 \text{ inches}}{x \text{ miles}}$

133. Use a proportion to solve each problem.

a) A baseball player gets 40 hits in 125 at bats. At that rate, how many hits would he get in 400 at bats?

b) A woman saved $75 in 5 weeks. At that rate, how long would it take her to save $180?

134. Proportions can be used for conversions of units. An example is shown.

There are 36 inches in 1 yard. How many inches are there in 3.5 yards?

$$\frac{36 \text{ inches}}{1 \text{ yard}} = \frac{x \text{ inches}}{3.5 \text{ yards}}$$

$$\frac{36}{1} = \frac{x}{3.5}$$

$$3.5\left(\frac{36}{1}\right) = \overset{1}{\cancel{3.5}}\left(\frac{x}{\cancel{3.5}}\right)$$

$$126 = x$$

Therefore, there are 126 inches in 3.5 yards.

Using the same steps, solve these.

a) There are 28 grams in 1 ounce. How many grams are there in 5.5 ounces?

b) There are 15 milliliters in 1 tablespoon. Therefore, 108 milliliters equals how many tablespoons?

a) 128 hits, from: $\frac{40}{125} = \frac{x}{400}$

b) 12 weeks, from: $\frac{75}{5} = \frac{180}{x}$

---

135. Proportions can be used to solve problems containing ratio language. An example is shown.

If the ratio of girls to boys in a class is 5 to 4, how many girls are there if there are 36 boys?

$$\frac{5 \text{ girls}}{4 \text{ boys}} = \frac{x \text{ girls}}{36 \text{ boys}}$$

$$\frac{5}{4} = \frac{x}{36}$$

$$\overset{9}{\cancel{36}}\left(\frac{5}{\cancel{4}}\right) = \cancel{36}\left(\frac{x}{\cancel{36}}\right)$$

$$45 = x$$

Therefore, there are 45 girls if there are 36 boys.

a) In an alloy, the ratio of copper to iron is 3:10. How much iron is there if there are 12 pounds of copper?

b) The ratio of the weight of an object on the moon to its weight on earth is 0.16 to 1. How much would a 175 lb man weigh on the moon?

a) 154 grams, from: $\frac{28}{1} = \frac{x}{5.5}$

b) 7.2 tablespoons, from: $\frac{15}{1} = \frac{108}{x}$

136. We can sometimes set up a ratio even when specific numbers are not given for the terms. An example is shown.

There are 25 students in a class. Some are male; some are female. To set up the ratio of males to females, we use these steps.

1. Let x equal the number of males.
2. Then 25 - x equals the number of females.
3. Therefore, the ratio of males to females is:

$$\frac{x \text{ males}}{(25 - x)\text{females}}$$

We used the same method to set up a ratio in the proportion below.

Two cities will send a total of 42 delegates to a convention. The ratio of the population of the two cities is 3 : 4. How many delegates should be sent from the smaller city?

$$\frac{3}{4} = \frac{x \text{ delegates}}{(42 - x)\text{delegates}}$$

$$\frac{3}{4} = \frac{x}{42 - x}$$

$$\cancel{4}(42 - x)\left(\frac{3}{\cancel{4}}\right) = \left(\frac{x}{\cancel{42 - x}}\right)(4)\cancel{(42 - x)}$$

$$126 - 3x = 4x$$

$$126 = 7x$$

$$x = 18$$

Therefore, the smaller city should send 18 delegates.

Using the same steps, solve these:

a) The ratio of girls to boys in a graduating class is 3 : 2. If there are 80 in the class, how many girls are there?

b) The ratio of copper to silver in dimes is 1 : 9. How much copper is there in 50 pounds of dimes?

---

a) 40 pounds, from: $\frac{3}{10} = \frac{12}{x}$

b) 28 pounds, from: $\frac{0.16}{1} = \frac{x}{175}$

---

a) 48 girls, from: $\frac{3}{2} = \frac{x}{80 - x}$

b) 5 pounds, from: $\frac{1}{9} = \frac{x}{50 - x}$

7-14 FORMULAS

In this section, we will do some evaluations and rearrangements with fractional formulas. Some rearrangements that require multiplying or factoring by the distributive principle are also included.

---

137. To do the evaluation below, we had to solve a proportion. Do the other evaluation.

In the formula below, find d_1 when $F_1 = 12$, $F_2 = 20$, and $d_2 = 40$.

$$\frac{F_1}{F_2} = \frac{d_1}{d_2}$$

$$\frac{12}{20} = \frac{d_1}{40}$$

$$\overset{2}{\cancel{40}}\left(\frac{12}{\cancel{20}}\right) = \cancel{40}\left(\frac{d_1}{\cancel{40}}\right)$$

$$24 = d_1$$

In the formula below, find P_1 when $P_2 = 12$, $V_2 = 12$, and $V_1 = 18$.

$$\frac{P_1}{P_2} = \frac{V_2}{V_1}$$

---

138. To do the evaluation below, we had to solve a fractional equation. Do the other evaluation.

$P_1 = 8$

In the formula below, find R_1 when $R_t = 10$ and $R_2 = 30$.

$$\frac{1}{R_t} = \frac{1}{R_1} + \frac{1}{R_2}$$

$$\frac{1}{10} = \frac{1}{R_1} + \frac{1}{30}$$

$$\overset{3}{\cancel{30}}R_1\left(\frac{1}{\cancel{10}}\right) = 30R_1\left(\frac{1}{R_1} + \frac{1}{30}\right)$$

$$3R_1 = 30\cancel{R_1}\left(\frac{1}{\cancel{R_1}}\right) + \cancel{30}R_1\left(\frac{1}{\cancel{30}}\right)$$

$$3R_1 = 30 + R_1$$

$$2R_1 = 30$$

$$R_1 = 15$$

In the formula below, find d when $D = 90$ and $f = 30$.

$$\frac{1}{D} + \frac{1}{d} = \frac{1}{f}$$

$d = 45$

139. To solve for P_1 below, we began by multiplying both sides by P_2V_1 to clear the fractions. Solve for T_1 in the other formula.

$$\frac{P_1}{P_2} = \frac{V_2}{V_1} \qquad\qquad \frac{V_1}{V_2} = \frac{T_1}{T_2}$$

$$\not{P_2}V_1\left(\frac{P_1}{\not{P_2}}\right) = P_2\not{V_1}\left(\frac{V_2}{\not{V_1}}\right)$$

$$P_1V_1 = P_2V_2$$

$$P_1 = \frac{P_2V_2}{V_1}$$

---

140. To solve for T_2 below, we began by multiplying both sides by T_1T_2. Solve for $\underline{d}$ in the other formula.

$$T_1 = \frac{V_1T_2}{V_2}$$

$$\frac{P_1V_1}{T_1} = \frac{P_2V_2}{T_2} \qquad\qquad \frac{D}{d} = \frac{F}{f}$$

$$\not{T_1}T_2\left(\frac{P_1V_1}{\not{T_1}}\right) = T_1\not{T_2}\left(\frac{P_2V_2}{\not{T_2}}\right)$$

$$P_1T_2V_1 = P_2T_1V_2$$

$$T_2 = \frac{P_2T_1V_2}{P_1V_1}$$

---

141. Sometimes we have to factor by the distributive principle when rearranging a formula. Two factorings of that type are shown.

$$d = \frac{Df}{F}$$

$$ab + ac = a(b + c)$$

$$CV - CT = C(V - T)$$

In $E = IR_1 + IR_2$, the variable I appears in both terms on the right side. To solve for I, we begin by factoring by the distributive principle as we have done below. Use the same method to solve for S in the other formula.

$$E = IR_1 + IR_2 \qquad\qquad MS - QS = R$$

$$E = I(R_1 + R_2)$$

$$I = \frac{E}{R_1 + R_2}$$

$$S = \frac{R}{M - Q}$$

142. To solve for S below, we isolated both S terms on one side before factoring. Solve for P_2 in the other formula.

$$MS = R - QS$$

$$MS + QS = R$$

$$S(M + Q) = R$$

$$S = \frac{R}{M + Q}$$

$$P_1P_2 = I - RP_2$$

$$P_2 = \frac{I}{P_1 + R}$$

143. To solve for $\underline{t}$ below, we isolated both $\underline{t}$ terms on one side before factoring. Solve for T in the other formula.

$$bt + cm = dt$$

$$cm = dt - bt$$

$$cm = t(d - b)$$

$$t = \frac{cm}{d - b}$$

$$RV + ST = QT$$

$$T = \frac{RV}{Q - S}$$

144. To solve for $\underline{c}$ below, we multiplied both sides by $\underline{abc}$ to clear the fractions and then factored by the distributive principle. Solve for $\underline{f}$ in the other formula.

$$\frac{1}{a} + \frac{1}{b} = \frac{1}{c}$$

$$abc\left(\frac{1}{a} + \frac{1}{b}\right) = abc\left(\frac{1}{c}\right)$$

$$\cancel{a}bc\left(\frac{1}{\cancel{a}}\right) + a\cancel{b}c\left(\frac{1}{\cancel{b}}\right) = ab\cancel{c}\left(\frac{1}{\cancel{c}}\right)$$

$$bc + ac = ab$$

$$c(b + a) = ab$$

$$c = \frac{ab}{b + a}$$

$$\frac{1}{D} + \frac{1}{d} = \frac{1}{f}$$

$$f = \frac{Dd}{d + D}$$

145. To solve for d below, we had to isolate the d terms before factoring. Solve for C_t in the other formula.

$$\frac{1}{D} + \frac{1}{d} = \frac{1}{f}$$

$$Ddf\left(\frac{1}{D} + \frac{1}{d}\right) = Ddf\left(\frac{1}{f}\right)$$

$$\cancel{D}df\left(\frac{1}{\cancel{D}}\right) + D\cancel{d}f\left(\frac{1}{\cancel{d}}\right) = Dd\cancel{f}\left(\frac{1}{\cancel{f}}\right)$$

$$df + Df = Dd$$

$$Df = Dd - df$$

$$Df = d(D - f)$$

$$d = \frac{Df}{D - f}$$

$$\frac{1}{C_1} = \frac{1}{C_t} - \frac{1}{C_2}$$

Answer: $C_t = \dfrac{C_1C_2}{C_1 + C_2}$

146. To solve for B below, we had to isolate the B terms before factoring B - AB. Solve for D in the other formula.

$$A = \frac{B}{B + 1}$$

$$A(B + 1) = \cancel{(B + 1)}\left(\frac{B}{\cancel{B + 1}}\right)$$

$$AB + A = B$$

$$A = B - AB$$

$$A = B(1 - A)$$

$$B = \frac{A}{1 - A}$$

$$C = \frac{DF}{D + F}$$

Answer: $D = \dfrac{CF}{F - C}$

SELF-TEST 24 (pages 320-337)

1. A plane leaves an airport traveling 160 mph. A second plane leaves the same airport two hours later traveling 400 mph in the same direction. How long will it take the second plane to catch up to the first plane?

2. One car travels 8 miles an hour faster than another. While one car travels 240 miles, the other travels 280 miles. Find the speed of each car.

3. A small plane flies 340 miles in two hours. At that rate, how many miles would it fly in 5 hours?

4. There are 162 students in a graduation class. The ratio of females to males is 5:4. How many females and males are there?

5. In the formula below, find P_2 when $P_1 = 36$, $V_2 = 40$ and $V_1 = 30$.

$$\frac{P_1}{P_2} = \frac{V_2}{V_1}$$

6. Solve for a:

$$\frac{1}{a} + \frac{1}{b} = \frac{1}{c}$$

ANSWERS:

1. $1\frac{1}{3}$ hours
2. 48 mph and 56 mph
3. 850 miles
4. 90 females, 72 males
5. $P_2 = 27$
6. $a = \dfrac{bc}{b - c}$

SUPPLEMENTARY PROBLEMS - CHAPTER 7

Assignment 21

Reduce to lowest terms.

1. $\frac{7x}{8x}$
2. $\frac{3ab}{ab}$
3. $\frac{m}{5m^2}$
4. $\frac{8pq}{20pq}$
5. $\frac{9d^2}{6d}$
6. $\frac{4x - 2}{8x - 4}$
7. $\frac{4x^2 - 4x}{4x^2 + 6x}$
8. $\frac{t^2 - 6t + 8}{t^2 - t - 12}$
9. $\frac{y^2 + 7y + 10}{y^2 - 4}$
10. $\frac{10a}{5a - 15}$
11. $\frac{y - 7}{7 - y}$
12. $\frac{x^2 - y^2}{y - x}$

Multiply. Report each product in lowest terms.

13. $\left(\frac{x}{3}\right)\left(\frac{2}{y}\right)$
14. $\left(\frac{4}{5}\right)\left(\frac{x + 1}{x - 1}\right)$
15. $\left(\frac{8}{5y}\right)\left(\frac{3y}{4}\right)$
16. $10y\left(\frac{1}{5y}\right)$
17. $\left(\frac{5x^2}{6}\right)\left(\frac{3}{2x}\right)$
18. $\left(\frac{1}{ay^2}\right)\left(\frac{5ay}{6}\right)$
19. $\left(\frac{y - 3}{y + 1}\right)\left(\frac{y^2 - 1}{y^2 - 5y + 6}\right)$
20. $\left(\frac{x^2 + 4x - 5}{10}\right)\left(\frac{5x}{2x^2 + x - 3}\right)$
21. $\left(\frac{t + 1}{t - 2}\right)\left[\frac{t^2 - 5t + 6}{(t + 1)^2}\right]$
22. $\left(\frac{p^2 - 3p}{p^2 + 2p - 15}\right)\left(\frac{p^2 - 25}{3p}\right)$
23. $\left(\frac{x - 3}{2x}\right)\left(\frac{2x^2 - 2x}{x - 1}\right)$
24. $\left(\frac{ab - b^2}{7a}\right)\left(\frac{7a + 7b}{a^2b - b^3}\right)$

Divide. Report each quotient in lowest terms.

25. $\dfrac{\frac{x}{7}}{\frac{x}{2}}$
26. $\dfrac{\frac{y - 1}{9}}{\frac{y}{3}}$
27. $\dfrac{\frac{a}{a + b}}{\frac{b}{a + b}}$
28. $\dfrac{\frac{x^2 - 36}{3x + 9}}{\frac{x + 6}{6}}$
29. $\dfrac{\frac{12y^2}{(x + y)^2}}{\frac{16y}{x + y}}$
30. $\dfrac{\frac{x^2 - 5x + 4}{x^2 - x - 12}}{\frac{x^2 - 2x + 1}{x^2 + 6x + 9}}$
31. $\dfrac{x^2 - y^2}{\frac{x + y}{x}}$
32. $\dfrac{\frac{(a + 5)^2}{7}}{a + 5}$

Assignment 22

Add or subtract. Report each answer in lowest terms.

1. $\frac{2x}{3} + \frac{1}{3}$
2. $\frac{y}{y + 2} + \frac{3y}{y + 2}$
3. $\frac{3t + 2}{t - 1} + \frac{t - 7}{t - 1}$
4. $\frac{3x}{10} + \frac{3x}{10}$
5. $\frac{y^2 + y}{y^2 - y} + \frac{y^2 - 6y}{y^2 - y}$
6. $\frac{2p}{p + 3} - \frac{7}{p + 3}$
7. $\frac{5}{6x} - \frac{1}{6x}$
8. $\frac{3y}{7} - \frac{y + 5}{7}$
9. $\frac{3d - 5}{d - 3} - \frac{2d - 7}{d - 3}$
10. $\frac{7}{2x} - \frac{1 - 4x}{2x}$
11. $\frac{y}{y^2 - 4} - \frac{2}{y^2 - 4}$
12. $\frac{3t + 5}{t^2 - t - 6} - \frac{2t + 3}{t^2 - t - 6}$

Find the LCD only. (Do not add or subtract.)

13. $\frac{7}{24} + \frac{1}{18}$

14. $\frac{11}{12} + \frac{1}{6} + \frac{8}{15}$

15. $\frac{5}{8x} - \frac{1}{6x}$

16. $\frac{3}{4x} - \frac{1}{2y}$

17. $\frac{5}{6t^2} + \frac{3}{10t}$

18. $\frac{2}{3y} - \frac{1}{3y - 9}$

19. $\frac{x}{4x + 2} + \frac{1}{2x^2 + x}$

20. $\frac{y - 1}{y^2 - 1} - \frac{y + 1}{y^2 + 3y - 4}$

21. $\frac{m - 5}{(m + 2)^2} + \frac{m + 4}{m^2 - 3m - 10}$

Add or subtract. Find the LCD first.

22. $\frac{3}{4x} + \frac{1}{10x}$

23. $\frac{m}{m - 1} + \frac{3}{2m}$

24. $\frac{y + 1}{y - 2} + \frac{y - 7}{y + 3}$

25. $\frac{x}{3x - 6} + \frac{5}{x^2 - 2x}$

26. $\frac{2y}{y^2 - y - 6} + \frac{1}{y^2 - 9}$

27. $\frac{3}{2x} - \frac{2}{3x}$

28. $\frac{p + 3}{p - 2} - \frac{p + 1}{p - 3}$

29. $\frac{5x}{x^2 - 16} - \frac{2}{x - 4}$

30. $\frac{x}{(x - y)^2} - \frac{y}{x^2 - y^2}$

Simplify each expression.

31. $\dfrac{1 + \frac{3}{x}}{\frac{2}{5}}$

32. $\dfrac{\frac{y}{4} + 3}{\frac{7}{8}}$

33. $\dfrac{\frac{3}{y} - \frac{1}{2y}}{\frac{1}{y} + \frac{4}{3y}}$

34. $\dfrac{1 - \frac{1}{t}}{1 + \frac{1}{t}}$

35. $\dfrac{\frac{a}{b} - \frac{b}{a}}{\frac{1}{a} + \frac{1}{b}}$

36. $\dfrac{\frac{x}{y} - \frac{1}{3y}}{\frac{x}{3y} + \frac{1}{6y}}$

Assignment 23

Solve each equation.

1. $\frac{7}{3x} = 4$

2. $\frac{y - 1}{y} = 2$

3. $4 = \frac{1}{x - 2}$

4. $\frac{2y}{3} - 1 = 5$

5. $6 = \frac{10}{x} - 1$

6. $\frac{t - 3}{t} = \frac{2}{3}$

7. $\frac{m + 5}{3m - 1} = \frac{1}{2}$

8. $\frac{2}{y} = \frac{5}{y - 4}$

9. $\frac{2}{x + 1} = \frac{4}{x - 4}$

10. $\frac{p + 1}{p + 2} = \frac{p - 2}{p - 3}$

11. $\frac{3d}{2} + \frac{d + 1}{3} = 15$

12. $\frac{y}{8} + \frac{1}{2} = \frac{3}{4}$

13. $\frac{1}{8} + \frac{1}{6} = \frac{1}{x}$

14. $\frac{1}{y} = \frac{5}{2y} - \frac{1}{2}$

15. $m - \frac{20}{m} = 1$

16. $\frac{1}{x + 3} + \frac{1}{x^2 - 9} = \frac{2}{x - 3}$

17. $\frac{1}{x + 2} + \frac{1}{x - 4} = \frac{1}{x^2 - 2x - 8}$

Solve each problem.

18. One number is 10 more than another. The quotient of the larger divided by the smaller is 6. Find the two numbers.

19. One number is 4 less than another. The quotient of the smaller divided by the larger is $\frac{4}{5}$. Find the two numbers.

20. The denominator of a fraction is 3 more than the numerator. If 5 is subtracted from both the numerator and denominator, the resulting fraction is $\frac{1}{2}$. Find the original fraction.

21. The denominator of a fraction is 3 times the numerator. If 5 is added to the numerator and subtracted from the denominator, the result equals $\frac{3}{5}$. Find the original fraction.

22. One-fourth of a number is 2 more than one-fifth of the same number. Find the number.

23. The reciprocal of 2 plus the reciprocal of 5 is the reciprocal of what number?

24. The reciprocal of 5 less than a number is twice the reciprocal of the number itself. Find the number.

25. Find two consecutive integers such that four-fifths of the smaller plus one-half of the larger is equal to 20.

26. If 5 times the reciprocal of a number is subtracted from the number, we get 4. Find the number.

27. A man can cut 2 acres of grass in 4 hours on a riding mower. His son can cut the 2 acres of grass in 6 hours with a walking mower. How long would it take them if they worked together.

28. Peggy can type a report in 4 hours, but Joan needs 5 hours. How long would it take them working together?

29. Dan can lay some sod in 6 hours, but his son needs 8 hours. How long would it take them working together?

30. Working together, Mike and Joe can complete a job in 6 hours. It would take Mike 9 more hours working alone than it would take Joe. How long would it take each of them working alone?

Assignment 24

Solve each problem.

1. A car and a bus leave a city at the same time traveling in opposite directions. The car travels 49 mph and the bus travels 55 mph. In how many hours will they be 234 miles apart?

2. On a two-day trip, Mary drove 582 miles in 12 hours. On the first day, she averaged 46 mph. On the second day, she averaged 52 mph. How long and how far did she travel each day?

3. Two buses leave a city at the same time going in the same direction. One travels 48 mph and the other travels 57 mph. In how many hours will they be 72 miles apart?

4. A small plane leaves an airport and flies due south at 240 km/h. A jet leaves the same airport two hours later and flies due south at 960 km/h. How long will it take the jet to catch up to the small plane?

5. A freight train and a passenger train travel from Crystal City to Cloville. The freight train travels at 60 km/h while the passenger train travels at 100 km/h. The passenger train takes 2 hours less for the trip. How far is it from Crystal City to Cloville?

6. A motor boat travels 3 hours downstream with a current of 3 mph. The return trip takes 4 hours. Find the speed of the boat in still water.

7. One car travels 8 miles an hour faster than another. While one car travels 230 miles, the other travels 270 miles. Find the speed of each car.

8. An airplane takes as long to fly 380 miles with the wind as it does to fly 320 miles against the wind. If the wind is blowing at 12 mph, what is the speed of the plane without a wind?

9. Sue can row 3 miles per hour in still water. It takes as long to row 6 miles upstream as it does to row 18 miles downstream. How fast is the current?

Solve each problem.

10. Express the ratio of 10 wins to 6 losses.

11. Express the ratio of 240 miles to 10 gallons.

12. A car travels 135 miles in 3 hours. At that rate, how long would it take to travel 225 miles?

13. If 2 inches on a map represent 150 miles, what is the distance between two cities that are 5 inches apart?

14. A hospital charges a patient $6.60 for 12 pills. At that rate, what would the charge be for 30 pills?

15. If a machine can drill 20 holes in 4 minutes, how long would it take to drill 150 holes?

16. There are 36 inches in 1 yard. How many inches are there in 2.75 yards?

17. There are 28 grams in 1 ounce. Therefore, 126 grams equal how many ounces?

18. The ratio of the weight of an object on Mars to its weight on earth is 0.4 to 1. How much would a 180 lb man weigh on Mars?

19. If the ratio of women to men in a club is 3 to 2, how many men are there if there are 42 women?

20. The winner of an election won by a vote of 5 to 4. If 144 people voted, how many votes did the loser get?

Complete each evaluation.

21. In $\frac{F_1}{F_2} = \frac{d_1}{d_2}$, find F_1 when $F_2 = 60$, $d_1 = 54$, and $d_2 = 90$.

22. In $\frac{P_1}{P_2} = \frac{V_2}{V_1}$, find V_1 when $P_1 = 20$, $P_2 = 15$, and $V_2 = 48$.

23. In $\frac{1}{R_t} = \frac{1}{R_1} + \frac{1}{R_2}$, find R_2 when $R_t = 20$ and $R_1 = 60$.

24. In $\frac{1}{D} + \frac{1}{d} = \frac{1}{f}$, find D when $d = 90$ and $f = 60$.

Complete each rearrangement.

25. Solve for P_1.

$$\frac{P_1V_1}{T_1} = \frac{P_2V_2}{T_2}$$

26. Solve for F.

$$\frac{D}{d} = \frac{F}{f}$$

27. Solve for P_2.

$$\frac{P_1}{P_2} = \frac{V_2}{V_1}$$

28. Solve for T_2.

$$\frac{V_1}{V_2} = \frac{T_1}{T_2}$$

29. Solve for A.

$$B = AC + AD$$

30. Solve for $\underline{t}$.

$$\frac{1}{m} + \frac{1}{p} = \frac{1}{t}$$

31. Solve for R.

$$\frac{1}{F} + \frac{1}{R} = \frac{1}{S}$$

32. Solve for D.

$$C = \frac{D}{D + 1}$$

33. Solve for T.

$$H = \frac{PT}{P + T}$$

8 Graphing Linear Equations

A linear equation is an equation whose graph is a straight line. In this chapter, we will introduce the coordinate system and discuss the procedure for graphing linear equations. Intercepts, slope, slope-intercept form, and point-slope form are discussed. Graphs of formulas and the meaning of direct variation are also discussed.

8-1 LINEAR EQUATIONS IN TWO VARIABLES

In this section, we will discuss solutions of linear equations in two variables.

1. A linear equation in two variables is an equation that either is in the form or can be put in the form:

$$\boxed{Ax + By = C}$$

where A, B, and C can be any number, but A and B cannot both be 0.

Continued on following page.

1. Continued

The following equations are linear equations because they are in the form on the preceding page.

$2x + 3y = 6$ $\qquad$ $5x - y = 10$

$x + 4y = 8$ $\qquad$ $x - y = 0$

The following equations are linear equations because they can be put in the form on the preceding page.

$y = 3x + 1$ $\qquad$ $y = 7 - x$

$y = 6x$ $\qquad$ $2y - 3x = 6$

In a linear equation, each term containing a variable is a first-degree term. The following equations are not linear equations because they contain a second-degree term.

$y = x^2$ $\qquad$ $y = 2x^2 - 1$

Which of the following are linear equations? ______________

a) $y = 2x^2$ b) $y = 2x$ c) $y = 5x + 1$ d) $y = 5x^2 + 1$

(b) and (c)

2. A solution of a linear equation is a pair of values that satisfies the equation. For example:

$x = 3$ and $y = 7$ is a solution of $y = 2x + 1$, since:

$$y = 2x + 1$$
$$7 = 2(3) + 1$$
$$7 = 6 + 1$$
$$7 = 7$$

$x = 7$ and $y = 5$ is not a solution of $2x - y = 7$, since:

$$2x - y = 7$$
$$2(7) - 5 \neq 7$$
$$14 - 5 \neq 7$$
$$9 \neq 7$$

a) Is $x = 2$, $y = 8$ a solution of $y = 3x$? __________

b) Is $x = 5$, $y = 7$ a solution of $x + y = 12$? __________

---

a) No, since: $8 \neq 3(2)$

b) Yes, since: $5 + 7 = 12$

3. $x = 4$ and $y = 8$ is a solution of $y = 2x$. Usually, the phrase "$x = 4$ and $y = 8$" is abbreviated to:

$$(4, 8)$$

In the abbreviation, the two numbers are separated by a comma and enclosed in parentheses. Since the x-value is always written first, the order in which the numbers are written makes a difference. That is:

(4, 8) means: $x = 4$, $y = 8$

(8, 4) means: $x = 8$, $y = 4$

Continued on following page.

3. Continued

Since the order in which the numbers are written makes a difference, two numbers written in this form are called an ordered pair.

a) Write $x = 1$, $y = 7$ as an ordered pair. ________

b) Write $y = 4$, $x = 9$ as an ordered pair. ________

---

4. To decide whether an ordered pair is a solution of a linear equation, we substitute the values and see whether they satisfy the equation. For example:

$(4, 3)$ is a solution of $2x - y = 5$, since $2(4) - 3 = 5$

$(5, 8)$ is not a solution of $y = 2x$, since $8 \neq 2(5)$

a) Is $(10, 4)$ a solution of $x - y = 3$? ________

b) Is $(2, 1)$ a solution of $3x + 2y = 8$? ________

a) $(1, 7)$

b) $(9, 4)$

---

5. The solution of a linear equation can contain one or two negative values. For example:

$(-2, -8)$ is a solution of $y = 4x$, since $-8 = 4(-2)$

a) Is $(-4, 1)$ a solution of $y = x + 5$? ________

b) Is $(-3, -6)$ a solution of $2x - y = 4$? ________

a) No, since:

$10 - 4 \neq 3$

b) Yes, since:

$3(2) + 2(1) = 8$

---

6. A solution can contain one or two fractions. For example:

$\left(\frac{5}{4}, -\frac{1}{4}\right)$ is a solution of $x + y = 1$, since $\frac{5}{4} + \left(-\frac{1}{4}\right) = 1$

a) Is $\left(\frac{2}{3}, 8\right)$ a solution of $y = 6x$? ________

b) Is $\left(\frac{3}{2}, -\frac{1}{2}\right)$ a solution of $x - y = 2$? ________

a) Yes, since:

$1 = -4 + 5$

b) No, since:

$2(-3) - (-6) \neq 4$

---

7. A solution can contain one or two decimal numbers. For example:

$(2.5, 1.5)$ is a solution of $x - y = 1$, since $2.5 - 1.5 = 1$

a) Is $(3.4, 2.6)$ a solution of $x + y = 5$? ________

b) Is $(2.5, 10)$ a solution of $y = 4x$? ________

a) No, since:

$8 \neq 6\left(\frac{2}{3}\right)$

b) Yes, since:

$\frac{3}{2} - \left(-\frac{1}{2}\right) = 2$

---

a) No, since:

$3.4 + 2.6 \neq 5$

b) Yes, since:

$10 = 4(2.5)$

8. Any linear equation has many solutions. For example, each ordered pair below is a solution of $y = 2x$.

(5, 10)	(-1, -2)
(1.5, 3)	(-2.5, -5)
$\left(\frac{1}{2}, 1\right)$	$\left(-\frac{7}{2}, -7\right)$
(0, 0)	(-40, -80)

How many more solutions are there for $y = 2x$? __________

Answer: An infinite number

9. We completed the solutions (4,) and (, 7) for $y = 2x + 3$ below.

To complete (4,), we substituted 4 for x and found the corresponding value of y by evaluating.

$$y = 2x + 3$$
$$y = 2(4) + 3$$
$$y = 8 + 3$$
$$y = 11$$

The ordered pair is (4, 11).

To complete (, 7), we substituted 7 for y and found the corresponding value of x by solving an equation.

$$y = 2x + 3$$
$$7 = 2x + 3$$
$$4 = 2x$$
$$x = 2$$

The ordered pair is (2, 7).

Following the examples, complete the solutions (-2,) and (, -5) for $y = 2x + 3$ below.

a) (-2,)

b) (, -5)

Answers: a) (-2, -1) b) (-4, -5)

10. We completed the solutions for (-3,) and (, 0) for $2x + 3y = 6$ below.

To complete (-3,), we substituted -3 for x and got:

$$2x + 3y = 6$$
$$2(-3) + 3y = 6$$
$$-6 + 3y = 6$$
$$3y = 12$$
$$y = 4$$

The ordered pair is (-3, 4).

To complete (, 0), we substituted 0 for y and got:

$$2x + 3y = 6$$
$$2x + 3(0) = 6$$
$$2x = 6$$
$$x = 3$$

The ordered pair is (3, 0).

Continued on following page.

10. Continued

Complete (1,) and (,-2) for $2x + 3y = 6$ below.

a) (1,)

b) (,-2)

Answers: a) $\left(1, \frac{4}{3}\right)$ b) (6,-2)

11. We completed the solutions (2,) and (,-8) for $4x - y = 12$ below.

To complete (2,), we substituted 2 for x and got:

$$\begin{aligned} 4x - y &= 12 \\ 4(2) - y &= 12 \\ 8 - y &= 12 \\ -y &= 4 \\ y &= -4 \end{aligned}$$

The ordered pair is (2,-4).

To complete (,-8), we substituted -8 for y and got:

$$\begin{aligned} 4x - y &= 12 \\ 4x - (-8) &= 12 \\ 4x + 8 &= 12 \\ 4x &= 4 \\ x &= 1 \end{aligned}$$

The ordered pair is (1,-8).

Complete (0,) and (,8) for $4x - y = 12$ below.

a) (0,)

b) (,8)

Answers: a) (0,-12) b) (5,8)

8-2 THE COORDINATE SYSTEM

In this section, we will introduce the rectangular (or Cartesian) coordinate system that is used to graph equations. We will read points on the coordinate system and plot points on the coordinate system.

---

12. The rectangular coordinate system is shown at the right. It consists of a horizontal number line used for x-values and a vertical number line used for y-values.

The horizontal number line is called the horizontal axis or x-axis.

The vertical number line is called the vertical axis or y-axis.

The two number lines together are called coordinate axes.

The two coordinate axes intersect at 0 on each. The point is called the ________________.

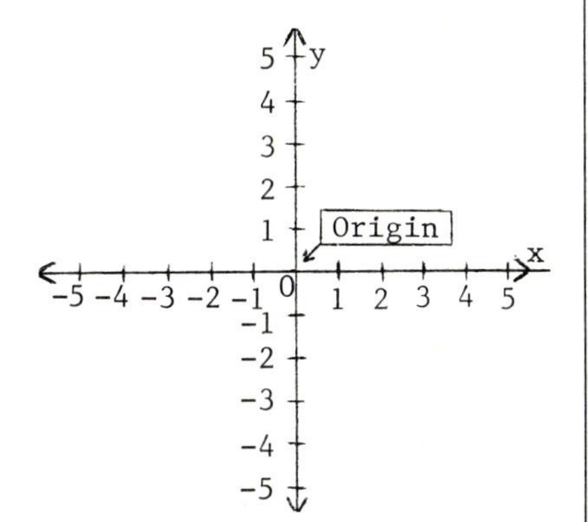

---

13. Each point on the coordinate system represents an ordered pair. To find the ordered pair for point A:

1. We drew an arrow down to the horizontal axis. The arrow points to x = 3.
2. We drew an arrow over to the vertical axis. The arrow points to y = 4.
3. Therefore, point A represents (3,4).

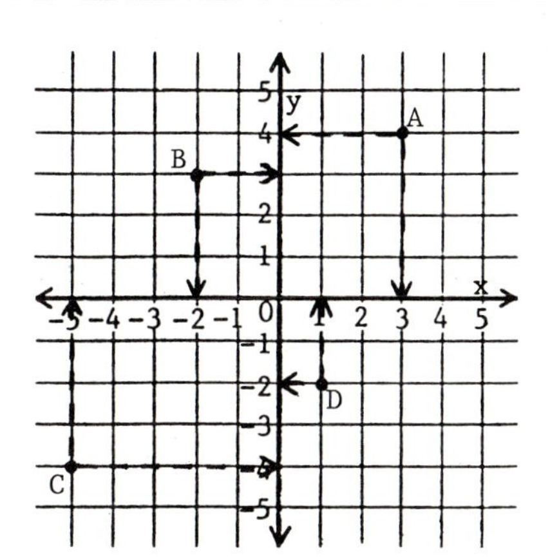

Similarly: Point B represents (,).

Point C represents (,).

Point D represents (,).

Answer: origin

Answers: B (-2,3)
C (-5,-4)
D (1,-2)

14. Point A at the right represents (4,2).

4 and 2 are called the coordinates of point A.

4 is called the <u>x-coordinate</u>.

2 is called the <u>y-coordinate</u>.

Write the coordinates of the other three points below.

B ____________

C ____________

D ____________

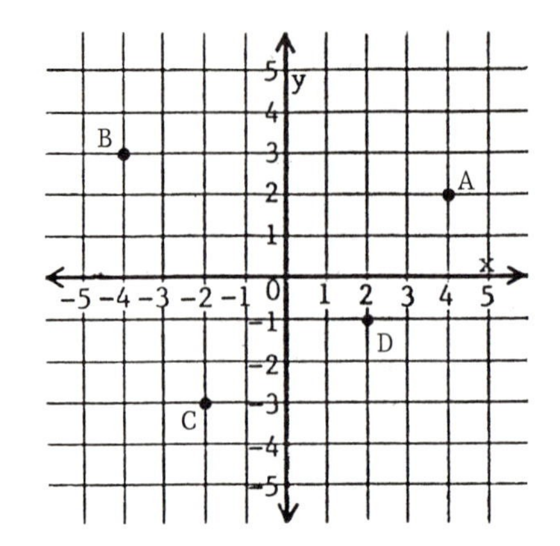

---

15. Write the coordinates of the six points below.

A ____________

B ____________

C ____________

D ____________

E ____________

F ____________

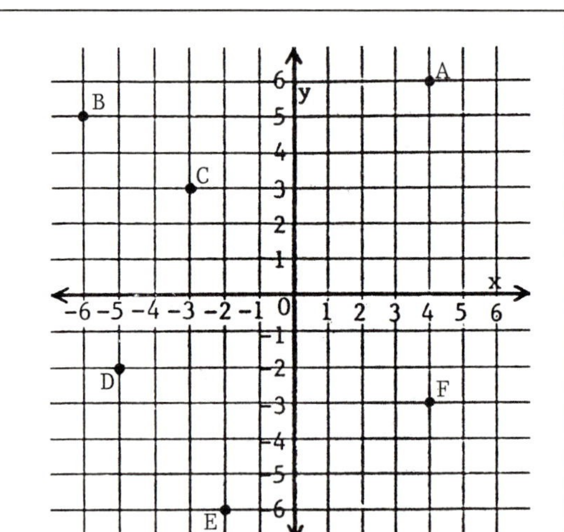

B (-4, 3)

C (-2, -3)

D (2, -1)

---

16. When scaling the axes at the right, we counted by 5's on the x-axis and by 10's on the y-axis. Write the coordinates of the four points below.

A ____________

B ____________

C ____________

D ____________

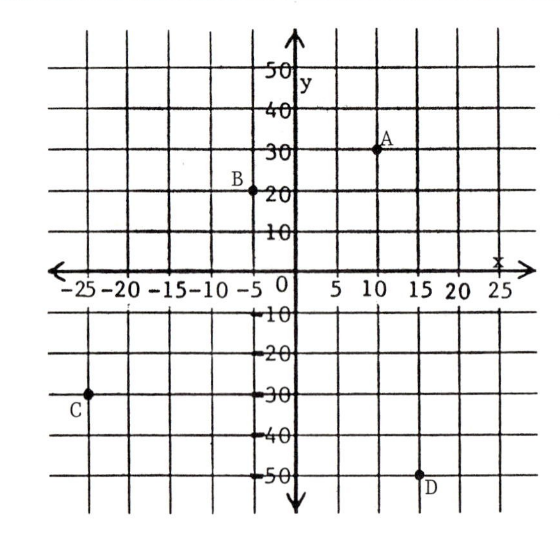

A (4, 6)

B (-6, 5)

C (-3, 3)

D (-5, -2)

E (-2, -6)

F (4, -3)

17. Points A, B, C, and D lie on the horizontal axis. The y-coordinate for all points of that type is 0. Using that fact, write the coordinates of the four points below.

A ____________

B ____________

C ____________

D ____________

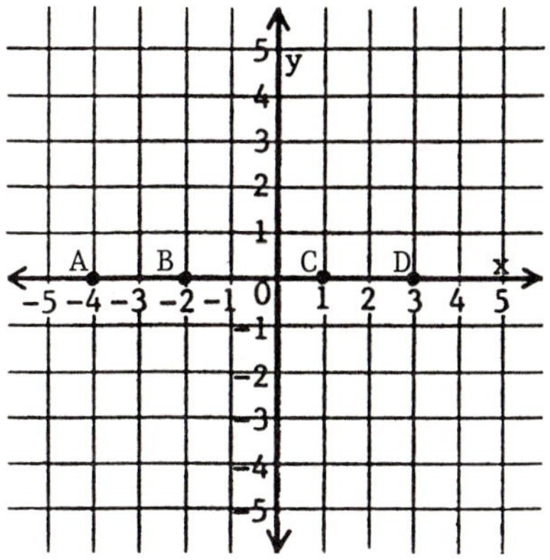

A (10, 30)

B (-5, 20)

C (-25, -30)

D (15, -50)

18. Points A, B, C, and D lie on the vertical axis. The x-coordinate for all points of that type is 0. Using that fact, write the coordinates of the four points below.

A ____________

B ____________

C ____________

D ____________

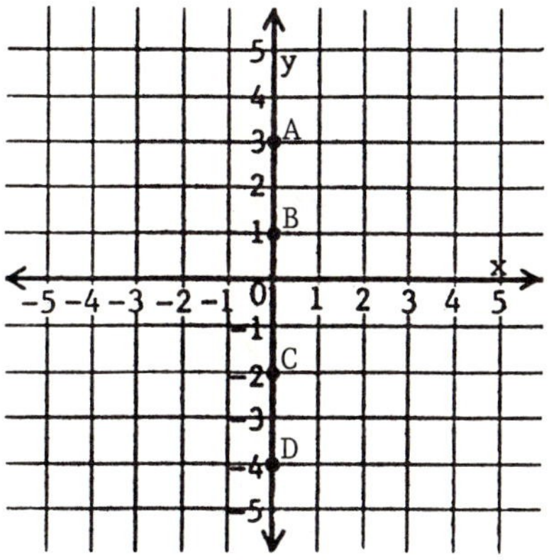

A (-4, 0)

B (-2, 0)

C (1, 0)

D (3, 0)

19. All four points at the right are on an axis. Write the coordinates of each:

A ____________

B ____________

C ____________

D ____________

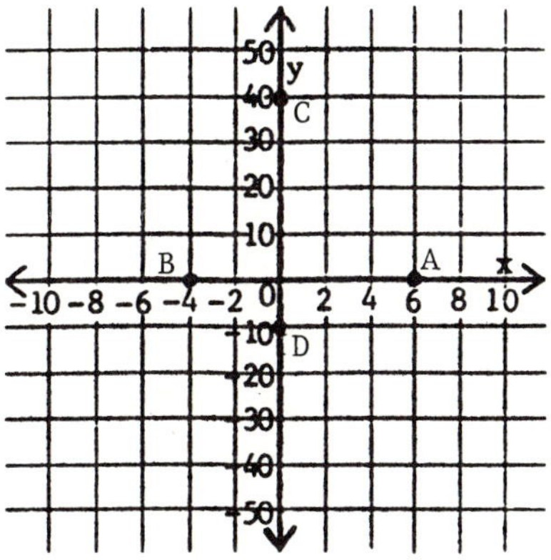

A (0, 3)

B (0, 1)

C (0, -2)

D (0, -4)

A (6, 0)

B (-4, 0)

C (0, 40)

D (0, -10)

20. Point A at the right does not lie at an intersection of coordinate lines. Therefore, we have to estimate its coordinates. We get:

$$\left(2\frac{1}{2}, 7\right)$$

Estimate the coordinates of these points.

B ______________

C ______________

D ______________

E ______________

F ______________

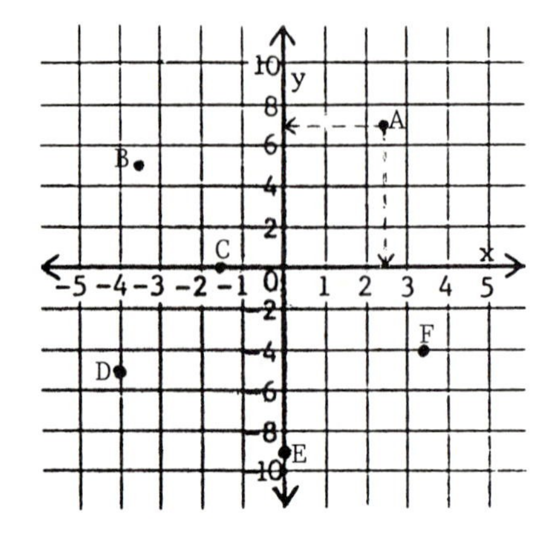

21. Points A and B at the right do not lie at an intersection of coordinate lines. Estimating their coordinates, we get:

A (25,45)

B (-32,18)

Estimate the coordinates of these points.

C ______________

D ______________

E ______________

F ______________

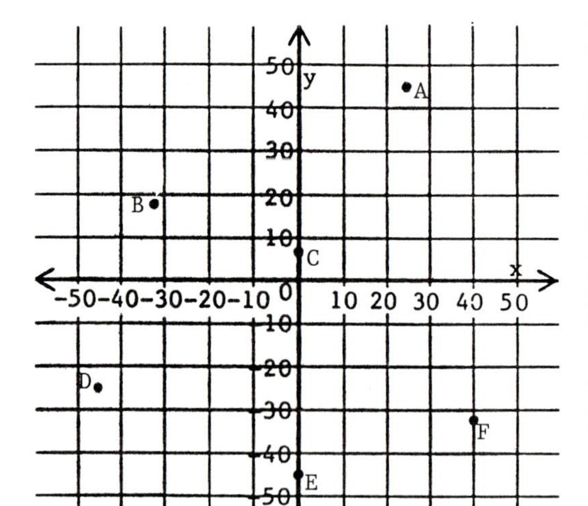

Note: Your answers should be close to these:

B $\left(-3\frac{1}{2}, 5\right)$

C $\left(-1\frac{1}{2}, 0\right)$

D (-4,-5)

E (0,-9)

F $\left(3\frac{1}{2}, -4\right)$

22. The coordinate axes divide the coordinate system into four parts called quadrants. We labeled the four quadrants in Figure 1 below. Notice that they are numbered in a counter-clockwise direction, beginning with the upper right quadrant.

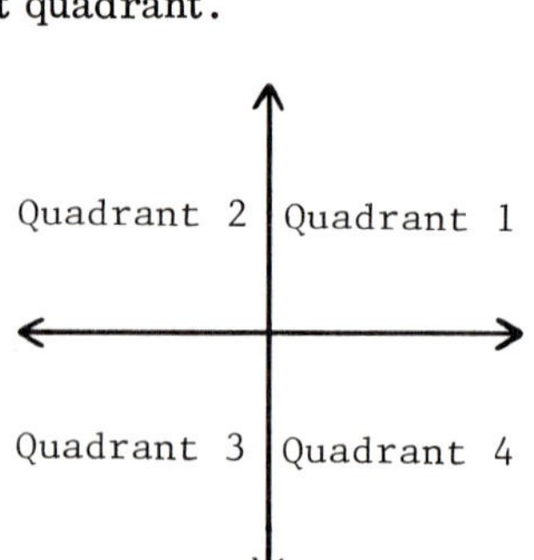

Figure 1

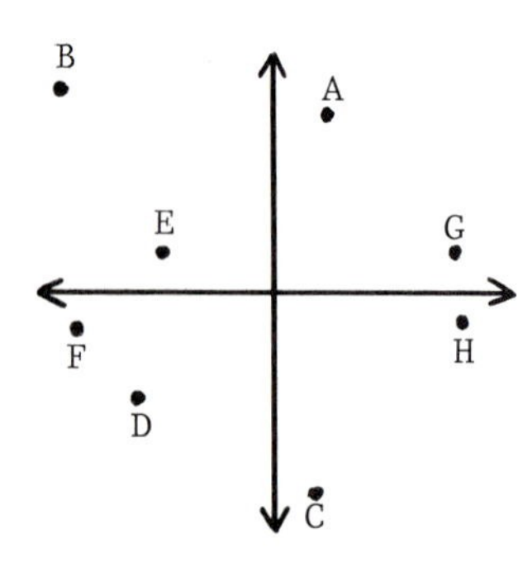

Figure 2

On the coordinate system in Figure 2 above, there are two points labeled in each quadrant.

a) Points B and E lie in Quadrant ______.

b) Points C and H lie in Quadrant ______.

Note: Your answers should be close to these:

C (0,7)

D (-46, -25)

E (0,-45)

F (40,-32)

a) Quadrant 2

b) Quadrant 4

23. To locate or <u>plot</u> a point on the coordinate system, we simply reverse the procedure for reading the coordinates of a point. For example, we have plotted points A and B at the right by drawing arrows from the axes.

A (2, 4)

B (-4, -3)

Plot and label the four points below on the same coordinate system.

C (3, 1) E (-2, 5)

D (3, -2) F (-2, -4)

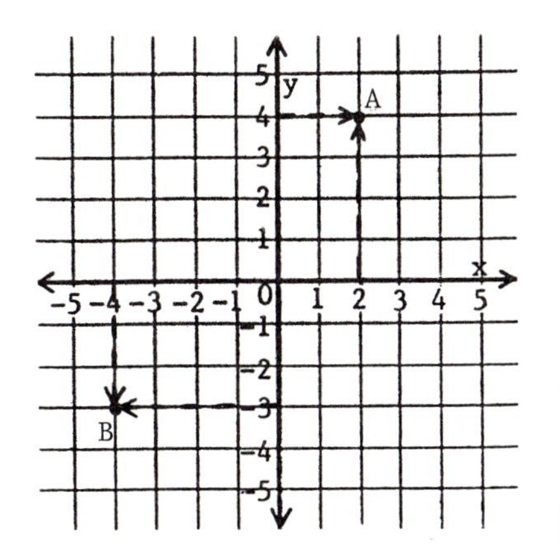

See graph at left for answer.

---

24. The correct plotting of the points from the last frame is shown at the right.

Point C is in quadrant _____.

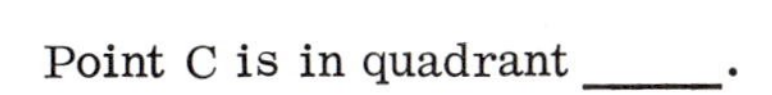

Point D is in quadrant _____.

Point E is in quadrant _____.

Point F is in quadrant _____.

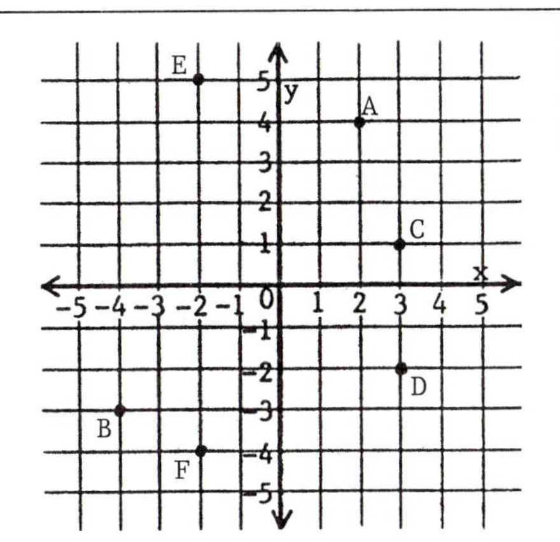

C (quadrant 1)

D (quadrant 4)

E (quadrant 2)

F (quadrant 3)

---

25. When the scales on the axes are different, be careful when plotting points. For example, we plotted points A and B on the coordinate system at the right.

A (-10, 20)

B (2, -30)

Plot and label the four points below on the same coordinate system.

C (-6, -10) E (2, 50)

D (-4, 40) F (10, -20)

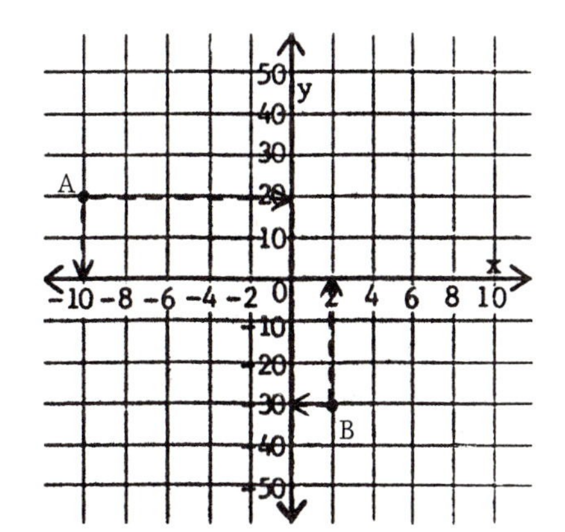

See next frame for answer.

26. The correct plotting of the points from the last frame is shown at the right.

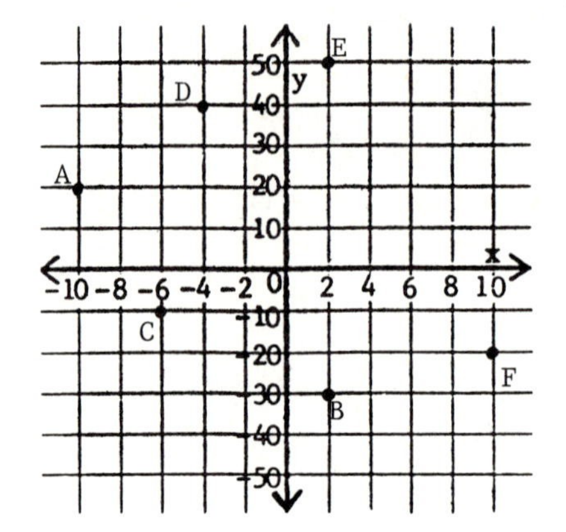

Two other names are used for the coordinates of a point. That is:

The x-coordinate is called the abscissa.

The y-coordinate is called the ordinate.

Since the coordinates of A are (-10, 20):

its abscissa is -10.

its ordinate is 20.

Since the coordinates of B are (2, -30):

a) its abscissa is __________.

b) its ordinate is __________.

---

27. Notice the points summarized in the box below.

The labels "horizontal", "x", and "abscissa" go together.
The labels "vertical", "y", and "ordinate" go together.
To help you remember what goes together, notice that alphabetically:
horizontal comes before vertical
x comes before y
abscissa comes before ordinate

Write the coordinates of a point:

a) if its abscissa is 5 and its ordinate is 3. __________

b) if its ordinate is -1 and its abscissa is -4. __________

Answers to 26: a) 2 b) -30

---

28. All four points below lie on an axis. Plot them at the right.

A (4, 0)

B (0, 4)

C (-3, 0)

D (0, -3)

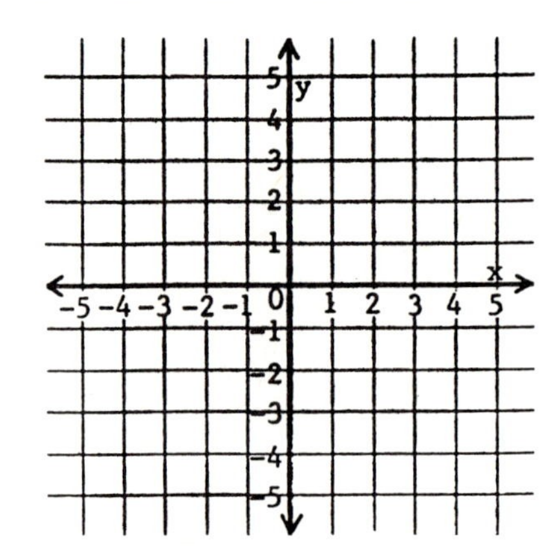

Answers to 27: a) (5, 3) b) (-4, -1)

Answer to 28: See next frame for answer.

29. The correct plotting of the points from the last frame is shown at the right.

As you can see from the graph, the point where the two axes intersect is called the <u>origin</u>. Since the origin lies on both axes:

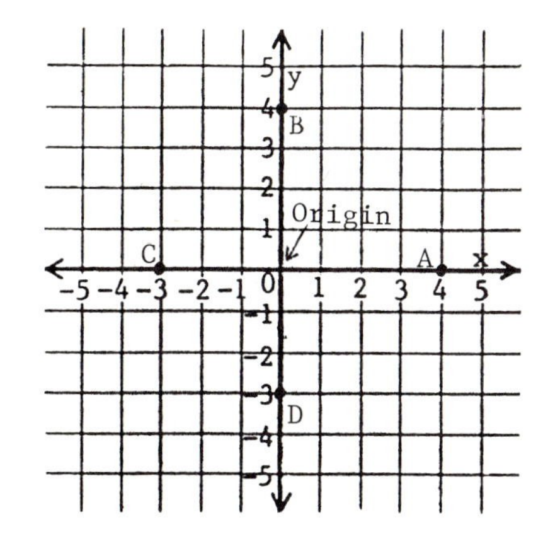

a) its x-coordinate is _____.

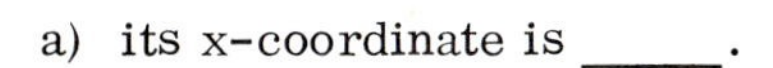

b) its y-coordinate is _____.

c) its coordinates are _____.

---

a) 0

b) 0

c) (0, 0)

---

30. To plot the following points at the right, we had to estimate their positions.

A (50, 90)

B (-75, 45)

Plot and label these points.

C (0, 32)

D (-47, -85)

E (30, -68)

F (73, 0)

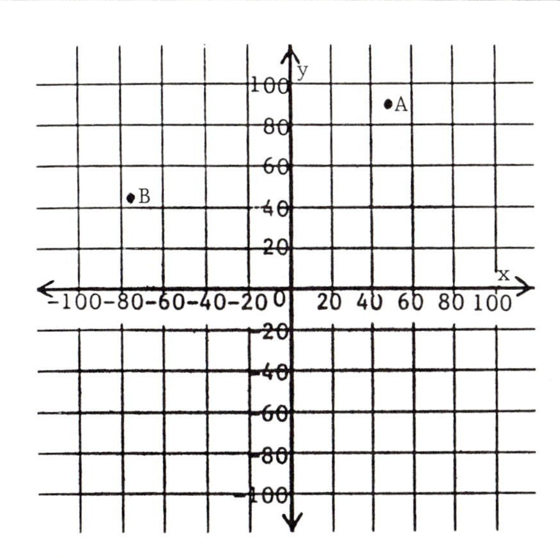

<u>Answer to Frame 30:</u>

Your plotted points should be approximately like those below.

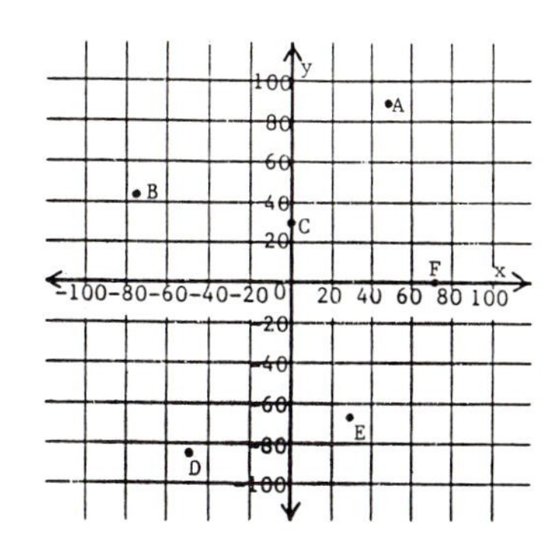

8-3 GRAPHING LINEAR EQUATIONS

The graph of any linear equation is a straight line. We will discuss the method for graphing linear equations in this section.

---

31. A graph of a linear equation is a drawing of its solution on the coordinate system. Though a linear equation has an infinite number of solutions, we only need some solutions to graph the equation. Suppose we want to graph the equation below.

$$y = x + 3$$

To find some solutions to use for graphing, we substitute values for x and find the corresponding values for y. We substitute a few positive values, 0, and a few negative values for x. For example, we substituted 5, 3, 0, -3, and -5 below.

When $x = 5$, $y = 5 + 3 = 8$

When $x = 3$, $y = 3 + 3 = 6$

When $x = 0$, $y = 0 + 3 = 3$

When $x = -3$, $y = -3 + 3 = 0$

When $x = -5$, $y = -5 + 3 = -2$

The solutions we found above are listed as ordered pairs below. They are also listed in a solution-table.

(5, 8)
(3, 6)
(0, 3)
(-3, 0)
(-5, -2)

x	y
5	8
3	6
0	3
-3	0
-5	-2

Are all possible solutions for $y = x + 3$ listed in the table? ______

---

32. The three steps needed to graph the following equation are discussed below.

$$y = x + 1$$

Step 1: Make up a solution-table.

	x	y
A	4	5
B	2	3
C	0	1
D	-2	-1
E	-4	-3

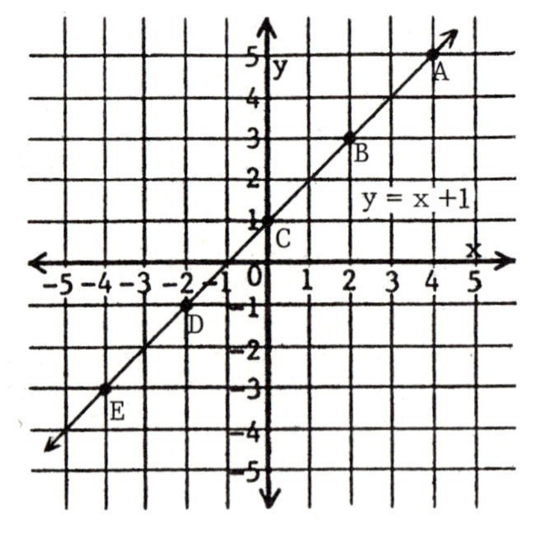

Step 2: Plot the points in the solution-table.

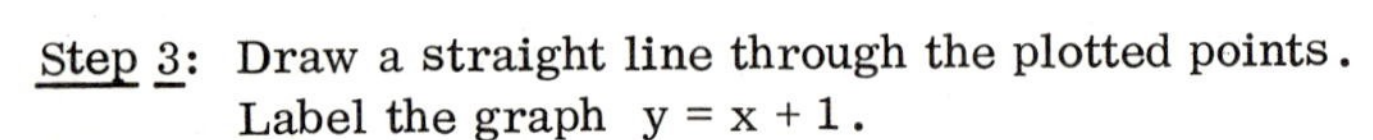

Step 3: Draw a straight line through the plotted points. Label the graph $y = x + 1$.

The straight line shown is the graph of $y = x + 1$.

a) Does the straight line (or graph) pass through the origin? ________

b) The straight line (or graph) passes through what three quadrants?

____

No. There are an infinite number of possible solutions.

33. We graphed $y = 2x$ below. When making up a solution-table, we avoid values of $\underline{x}$ that give points that are off the graph. For example, we did not use $x = 4$ below because (4, 8) is off the graph.

$y = 2x$

	x	y
A	3	6
B	1	2
C	0	0
D	-1	-2
E	-3	-6

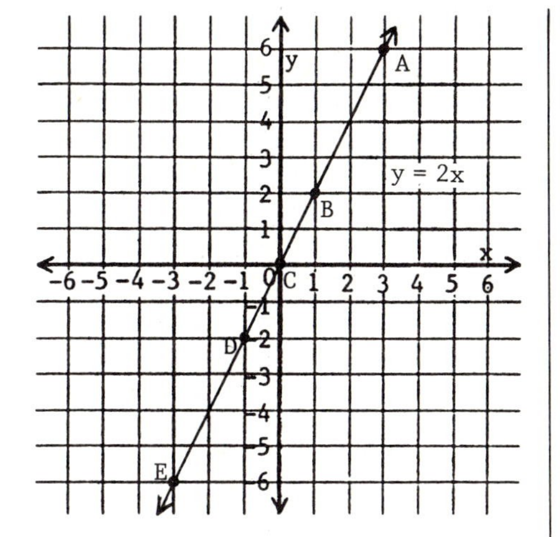

a) Does the graphed line pass through the origin? ________

b) The graphed line passed through what quadrants? ________

a) No

b) 1, 2, 3

---

34. We used the solution-table shown to graph $x + y = 1$ in two different ways below.

a) Yes

b) 1 and 3

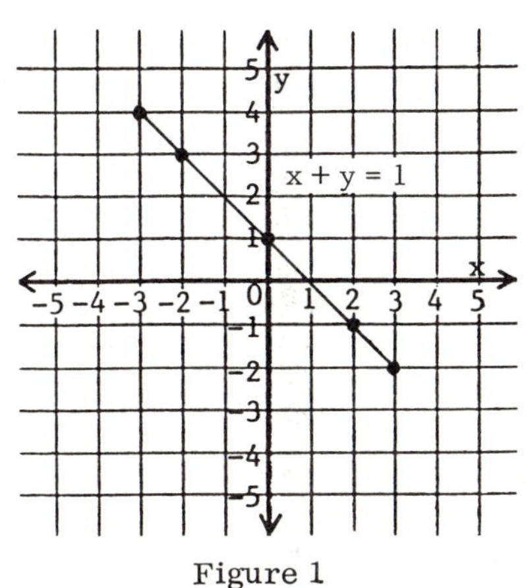

Figure 1

x	y
3	-2
2	-1
0	1
-2	3
-3	4

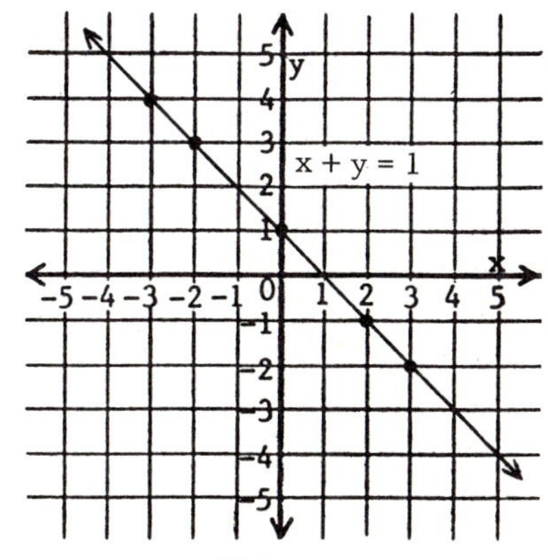

Figure 2

Note: In Figure 1, we stopped the graphed line at the last plotted point on each end.

In Figure 2, we extended the graphed line to the edge of the coordinate system shown and put arrowheads at each end of it.

Figure 2 is the correct graphing of $x + y = 1$ for these reasons:

1) The line should be extended to the edge of the coordinate system to show that there are other solutions, like (5, -4) and (-4, 5), beyond (-3, 4) and (3, -2).

2) Arrowheads should be put at each end of the graphed line to show that there are solutions, like (10, -9) and (-20, 21), beyond the edge of the coordinate system shown.

35. Complete the tables provided and then graph each equation. Be sure to extend each line to the edge of the coordinate system and put arrowheads on each end of each line.

$y = 2x + 1$

x	y
2	
1	
0	
-1	
-2	

$x + y = 2$

x	y
4	
2	
0	
-2	
-4	

36. Since two points determine a straight line, we only need to plot two points to graph a line. However, we always plot a third point as a check. Complete the table below and graph the equation.

$y = -3x$

x	y
2	
0	
-2	

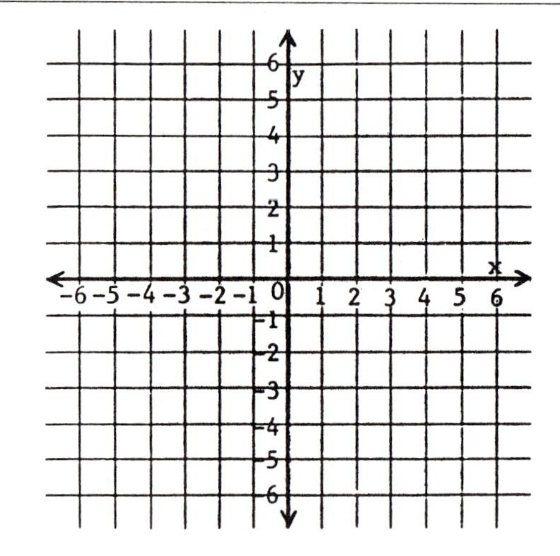

Answer to Frame 35:

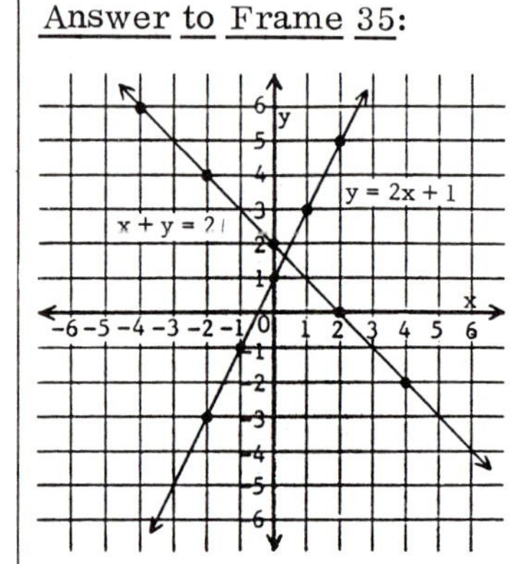

37. In each equation below, the numerical coefficient of x is a fraction. To avoid fractions for y, we substitute multiples of the denominator for x. That is, we substitute multiples of 3 at the left and multiples of 2 at the right. Complete the tables and graph each equation.

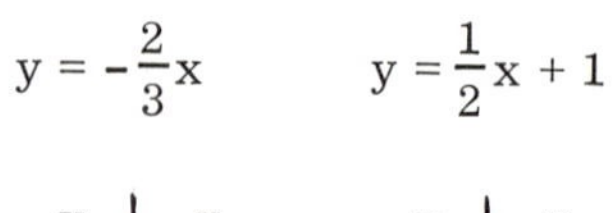

$y = -\frac{2}{3}x$ $\qquad$ $y = \frac{1}{2}x + 1$

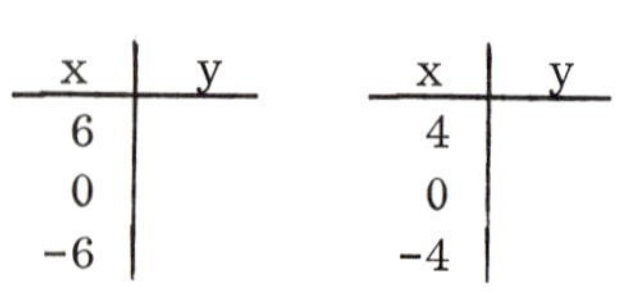

x	y
6	
0	
-6	

x	y
4	
0	
-4	

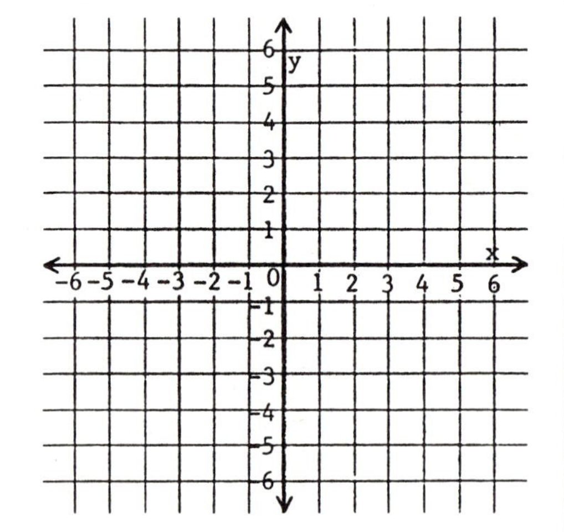

Answer to Frame 36:

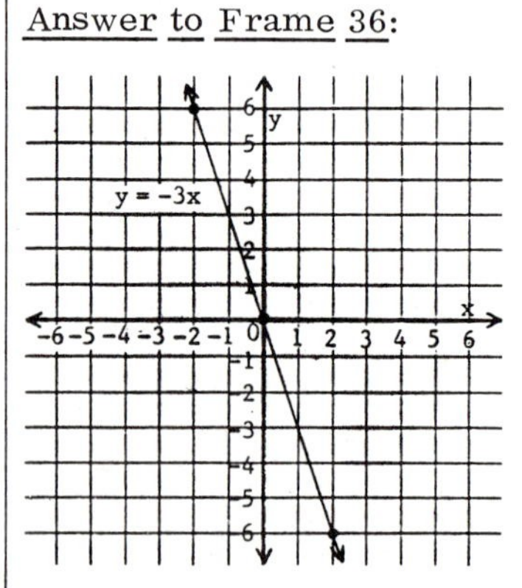

38. When graphing a straight line, pick three points that are far enough apart so that it is easy to draw the line accurately. Pick your own three points to graph the equations below.

$y = 3x$ $\qquad$ $2x + y = 1$

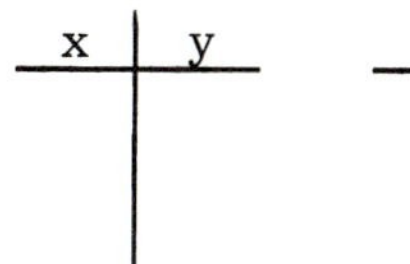

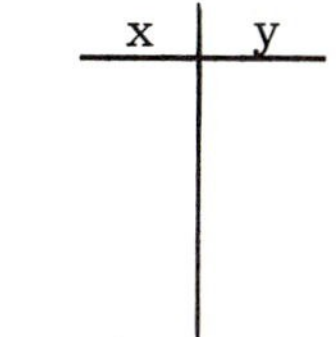

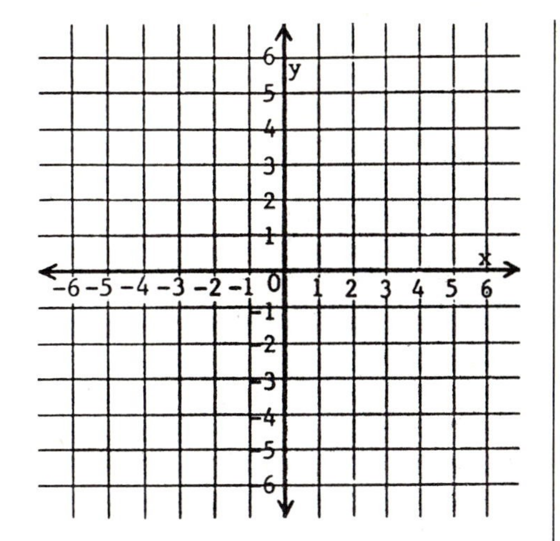

Answer to Frame 37:

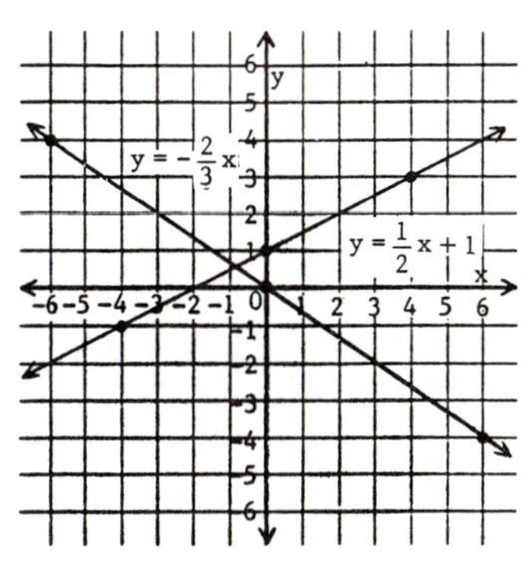

39. Pick your own three points to graph the equations below.

$y = 2x - 5$

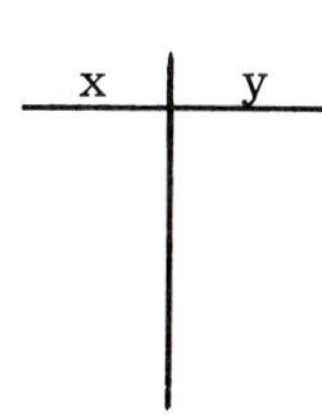

$y = -\frac{1}{2}x + 2$

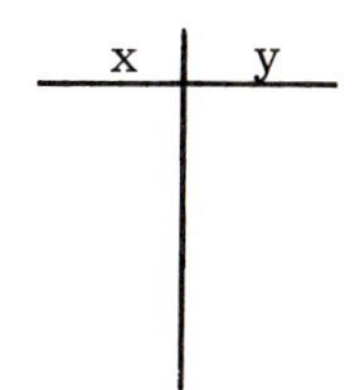

Answer to Frame 38:

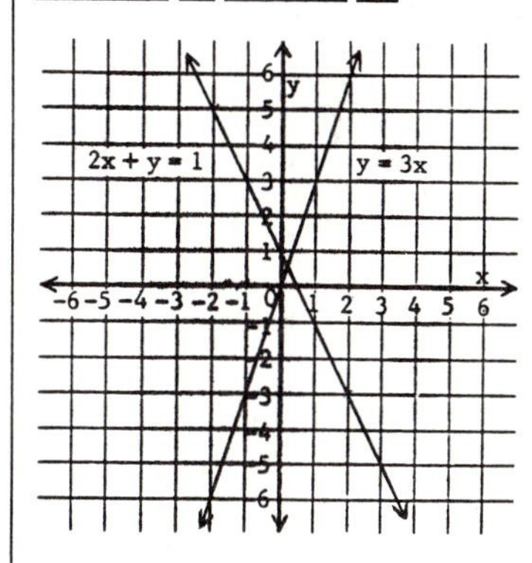

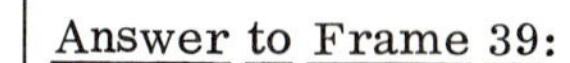

Answer to Frame 39:

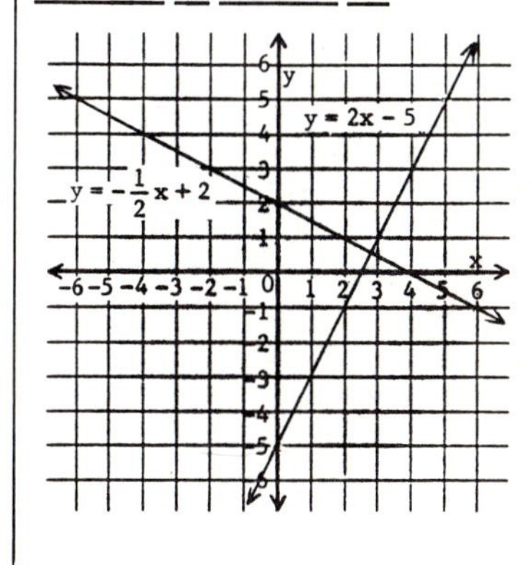

SELF-TEST 25 (pages 342-358)

Decide whether the given ordered pair is a solution of the given equation.

1. $(4, -3)$ for $x - 2y = 8$
2. $(4, -4)$ for $y = \frac{1}{4}x - 5$

Complete the solution for each equation.

3. $(5, \quad)$ for $x - y = 10$
4. $(\quad, -7)$ for $y = 4x - 3$

5. Write the coordinates of these points.

A ________ D ________

B ________ E ________

C ________ F ________

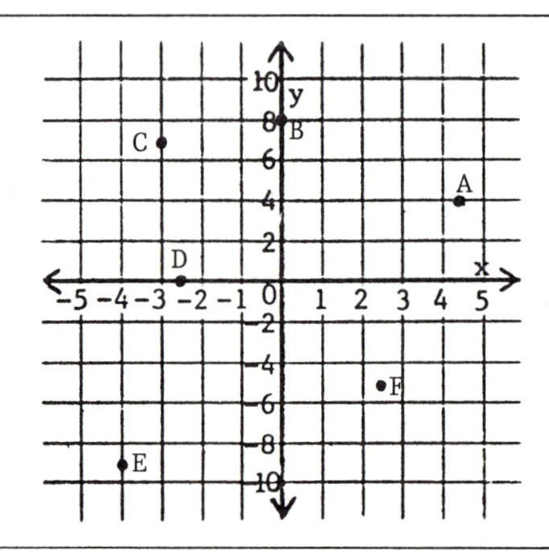

6. The point $(4, -1)$ lies in quadrant ______.
7. The point $(0, -3)$ lies on which axis? ________
8. The coordinates of the origin are ________.

Graph each equation.

9. $x - y = 2$

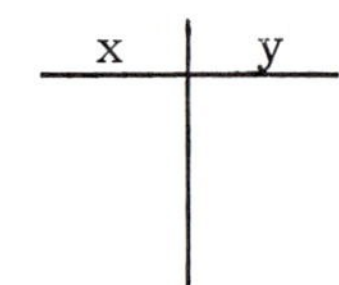

10. $y = -\frac{1}{2}x$

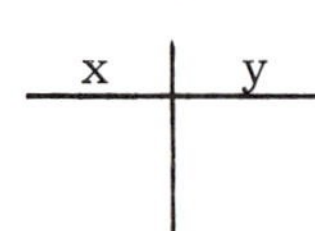

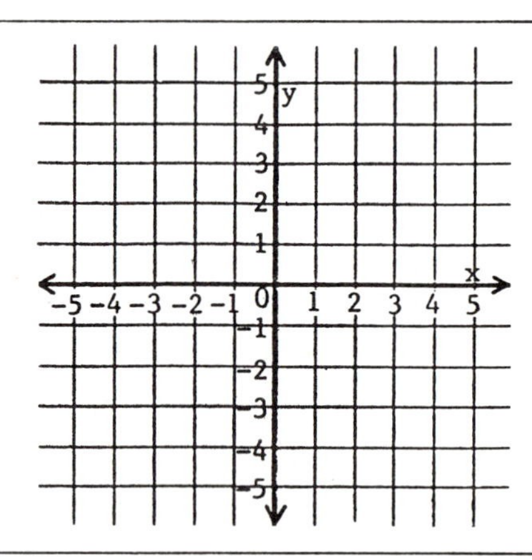

ANSWERS:

1. No
2. Yes
3. $(5, -5)$
4. $(-1, -7)$
5. A $\left(4\frac{1}{2}, 4\right)$
   B $(0, 8)$
   C $(-3, 7)$
   D $\left(-2\frac{1}{2}, 0\right)$
   E $(-4, -9)$
   F $\left(2\frac{1}{2}, -5\right)$
6. 4
7. y-axis
8. $(0, 0)$

9-10.

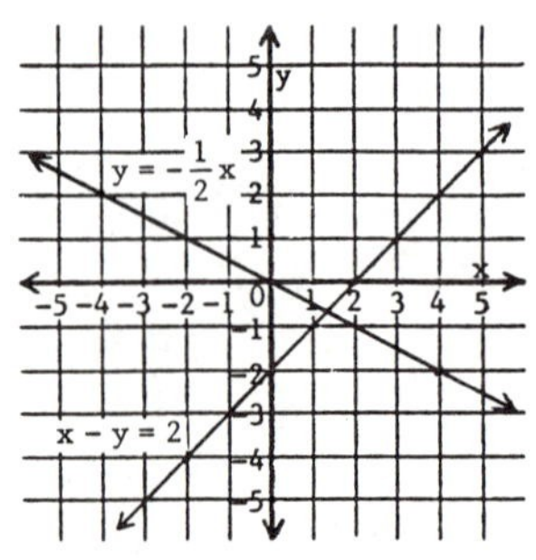

8-4 GRAPHING USING INTERCEPTS

In this section, we will discuss intercepts and show how they can be used to graph linear equations.

---

40. We graphed the linear equation below at the right.

$$3x - 2y = 6$$

The points where the line crosses the axes are called intercepts.

The x-intercept is (2, 0).

The y-intercept is (0, -3).

We can get the coordinates of the intercepts directly from the equation.

1) Finding the x-intercept

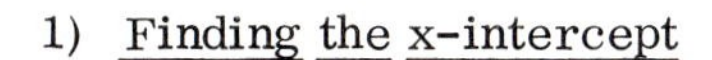

Since the y-coordinate is 0, we substitute 0 for y and solve for x.

$$3x - 2(0) = 6$$
$$3x = 6$$
$$x = 2$$

The x-intercept is (2, 0)

2) Finding the y-intercept

Since the x-coordinate is 0, we substitute 0 for x and solve for y.

$$3(0) - 2y = 6$$
$$-2y = 6$$
$$y = -3$$

The y-intercept is (0, -3).

---

41. To find the coordinates of the intercepts, we do the following:

The coordinates of the x-intercept are (a, 0). To find a, let $y = 0$.
The coordinates of the y-intercept are (0, b). To find b, let $x = 0$.

Find the coordinates of both intercepts for the equations below.

a) $2x + 5y = 10$

x-intercept (,)

y-intercept (,)

b) $4y - 3x = 12$

x-intercept (,)

y-intercept (,)

42. Find the coordinates of both intercepts for the equations below.

a) $y = 2x - 3$

x-intercept (,)

y-intercept (,)

b) $x - 3y = 1$

x-intercept (,)

y-intercept (,)

---

a) x-intercept (5, 0)
y-intercept (0, 2)

b) x-intercept (-4, 0)
y-intercept (0, 3)

---

43. When graphing a linear equation, <u>we frequently use the intercepts</u> as the two points needed to determine the line.

To graph $x + 2y = 4$, we plotted the two intercepts. Their coordinates are:

(0, 2) (4, 0)

We also plotted (-4, 4) as a check.

a) Does (-4, 4) lie on the line? ______

b) Is the graph probably correct? ______

---

a) x-intercept $\left(\frac{3}{2}, 0\right)$
y-intercept (0, -3)

b) x-intercept (1, 0)
y-intercept $\left(0, -\frac{1}{3}\right)$

---

44. Find the intercepts for each equation below and use them to graph the lines. <u>Plot a third point as a check.</u>

$4x - 2y = 8$ $2y = 5 - x$

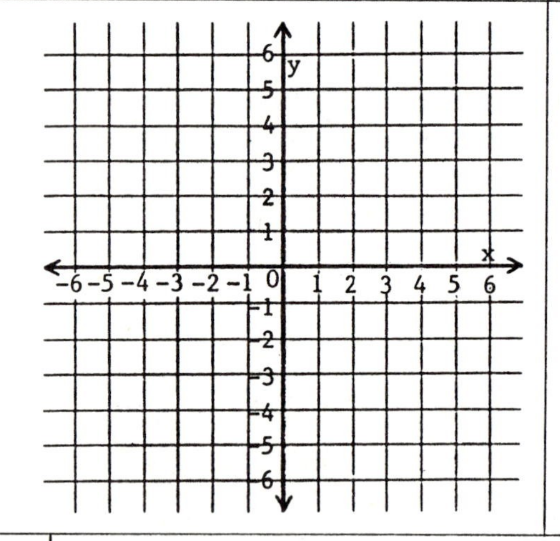

---

a) Yes

b) Yes

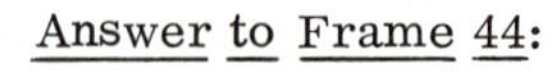

Answer to Frame 44:

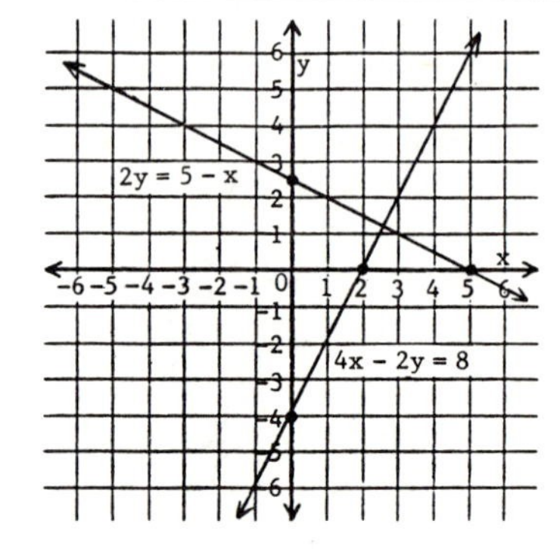

45. If we substitute 0 for x or y in the equation below, we get 0 for the other variable. Therefore, both intercepts are at the origin which is (0, 0). To graph the equation, we need one more point and a third point as a check. Graph the equation at the right.

$$y = -2x$$

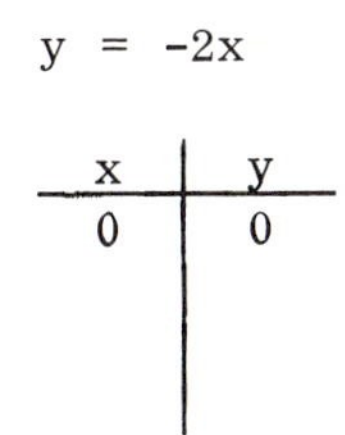

x	y
0	0

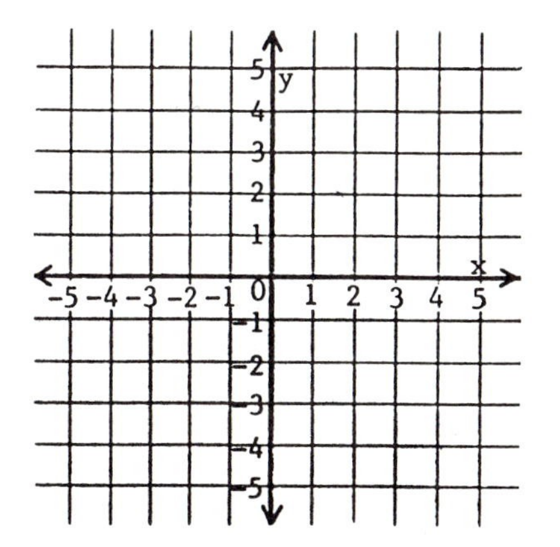

Answer to Frame 45: y = -2x

8-5 HORIZONTAL AND VERTICAL LINES

In this section, we will discuss linear equations whose graphs are either horizontal or vertical lines.

46. Two horizontal lines are shown at the right.

For line A, the y-coordinate is 2 for every value of x. Therefore, the equation of the line is $y = 2$.

$y = 2$ means: For every x-value, $y = 2$.

For line B, the y-coordinate is -3 for every value of x. Therefore, the equation of the line is $y = -3$.

$y = -3$ means: For every x-value, y = __________.

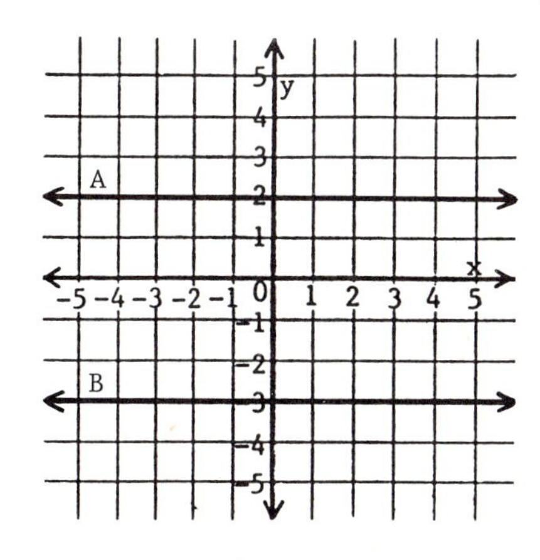

y = -3

47. Four horizontal lines are drawn at the right. Following the examples from the last frame, write the equation of each line.

A: ________

B: ________

C: ________

D: ________

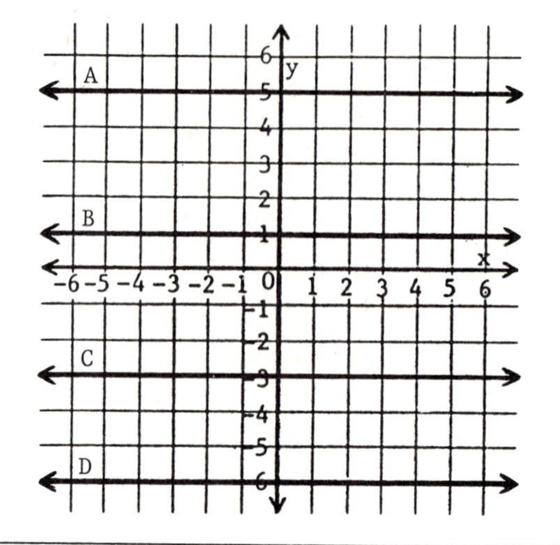

---

48. The x-axis is also a horizontal line on the coordinate system.

a) On the x-axis, for every x-value, $y =$ ______.

b) Therefore, the equation of the x-axis is ______.

A: $y = 5$ C: $y = -3$

B: $y = 1$ D: $y = -6$

---

49. Two vertical lines are shown at the right.

For line A, the x-coordinate is 4 for every value of y. Therefore, the equation of the line is $x = 4$.

$x = 4$ means: For every y-value, $x = 4$.

For line B, the x-coordinate is -2 for every value of y. Therefore, the equation of the line is $x = -2$.

$x = -2$ means: For every y-value, $x =$ ______.

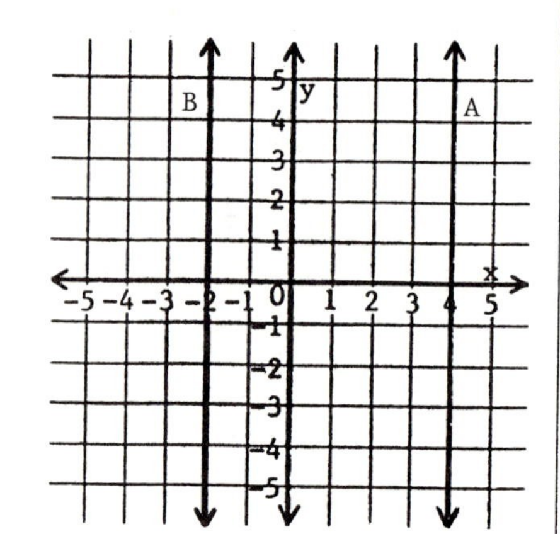

a) 0

b) $y = 0$

---

50. Four vertical lines are drawn at the right. Following the examples from the last frame, write the equation of each line.

A: ________

B: ________

C: ________

D: ________

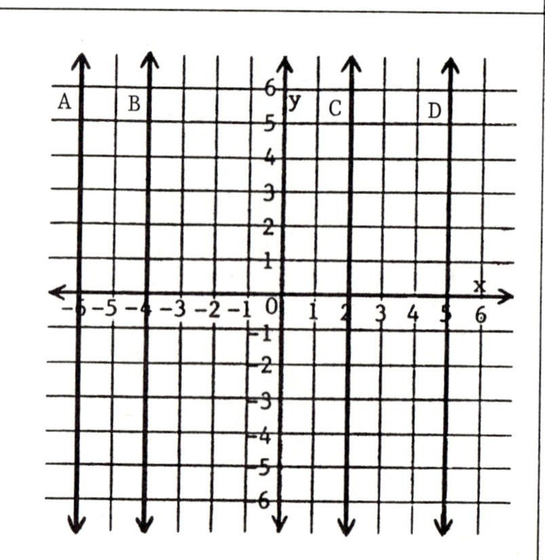

$x = -2$

51. The y-axis is also a vertical line on the coordinate system.

a) On the y-axis, for every y-value, x = ______.

b) Therefore, the equation of the y-axis is __________.

A: x = -6	C: x = 2
B: x = -4	D: x = 5

---

52. Four lines are drawn at the right. Write the equation of each line.

A: __________

B: __________

C: __________

D: __________

a) 0

b) x = 0

---

53. a) The graph of x = 0 is the ____________ (x-axis, y-axis).

b) The graph of y = 0 is the ____________ (x-axis, y-axis).

A: y = 1	C: x = 3
B: x = -1	D: y = -4

---

a) y-axis b) x-axis

8-6 SLOPE OF A LINE

The slope of a graphed line is a measure of the steepness of the rise or fall of the line from left to right. We will discuss the slope of a line in this section.

---

54. Points A (1, 2) and B (3, 5) are plotted on the line at the right. The changes in x and y from A to B are shown by arrows.

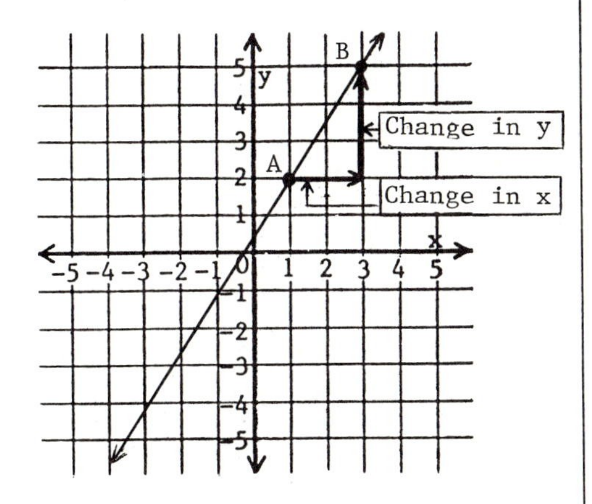

The horizontal arrow shows an increase in x from 1 to 3. The change in x can be represented by the signed number 2.

The vertical arrow shows an increase in y from 2 to 5. What signed number represents the change? ______

3

55. The symbol Δ (pronounced delta) is used as an abbreviation for the phrase change in. Therefore:

Δx means change in x.

Δy means change in y.

Points C (-2, 3) and D (3, -1) are plotted on the line at the right. Δx and Δy are the changes in x and y from C to D.

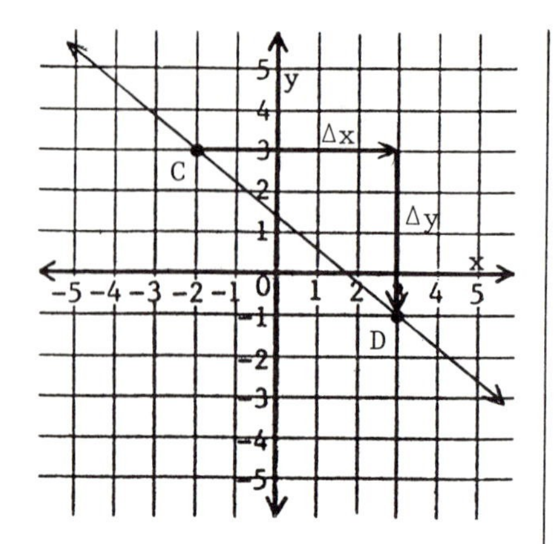

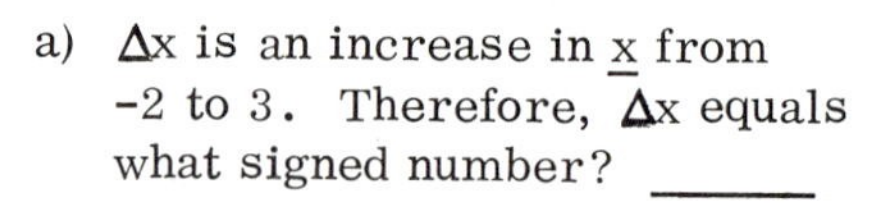

a) Δx is an increase in x from -2 to 3. Therefore, Δx equals what signed number? _______

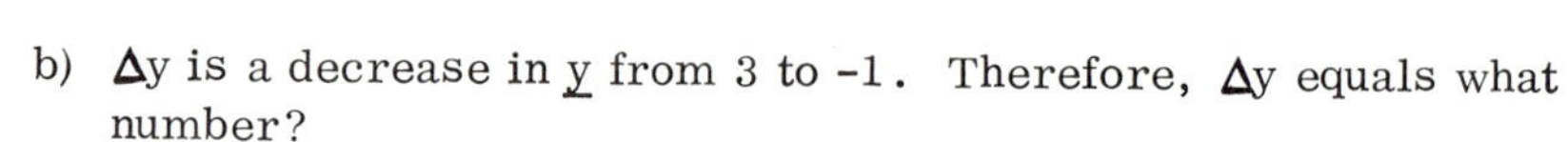

b) Δy is a decrease in y from 3 to -1. Therefore, Δy equals what number?

---

a) 5

b) -4

56. The slope of a line is a ratio of the change in y to the change in x from one point to another point on the line. That is:

$$\text{Slope} = \frac{\Delta y}{\Delta x} = \frac{\text{increase or decrease in } y}{\text{increase in } x}$$

Let's use the changes from P to Q to compute the slope of the line at the right.

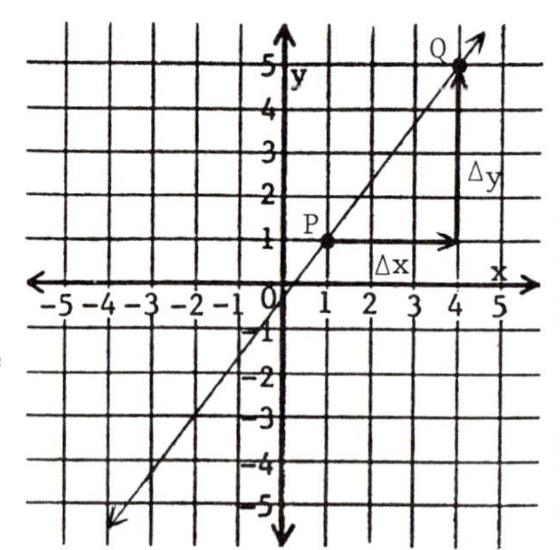

a) Since x increases from 1 to 4, Δx = _______.

b) Since y increases from 1 to 5, Δy = _______.

c) Therefore, the slope $= \dfrac{\Delta y}{\Delta x} =$ _______

---

a) 3

b) 4

c) $\frac{4}{3}$

57. Let's use the changes from S to T to compute the slope of the line at the right.

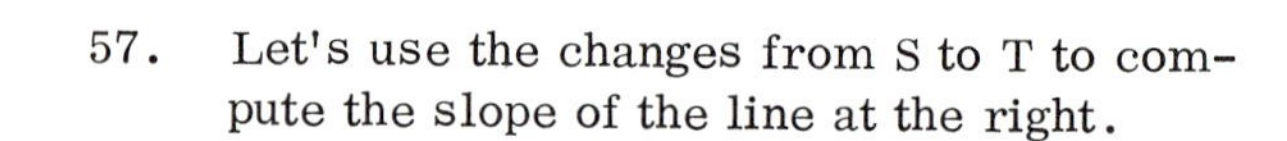

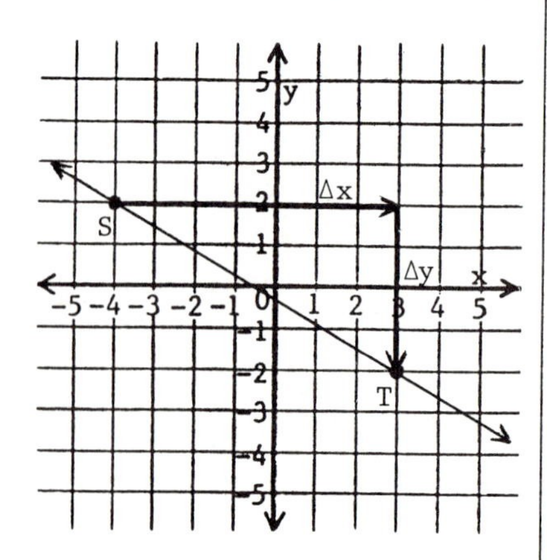

a) Since x increases from -4 to 3, Δx = _______.

b) Since y decreases from 2 to -2, Δy = _______.

c) Therefore, the slope $= \dfrac{\Delta y}{\Delta x} =$ _______

58. Slope is a ratio or fraction. When computing a slope, <u>the ratio should always be reduced to lowest terms</u>.

a) If $\Delta x = 6$ and $\Delta y = 4$, the slope is ______.

b) If $\Delta x = 8$ and $\Delta y = -10$, the slope is ______.

a) 7

b) -4

c) $-\frac{4}{7}$

---

59. When computing a slope, sometimes the ratio reduces to a whole number.

a) If $\Delta x = 4$ and $\Delta y = 12$, the slope is ______.

b) If $\Delta x = 1$ and $\Delta y = -2$, the slope is ______.

a) $\frac{2}{3}$

b) $-\frac{5}{4}$

---

60. No matter which pair of points we choose to compute the slope of a line, we always get the same value for the slope. As an example, we graphed the changes from A to B and from C to D on the line at the right.

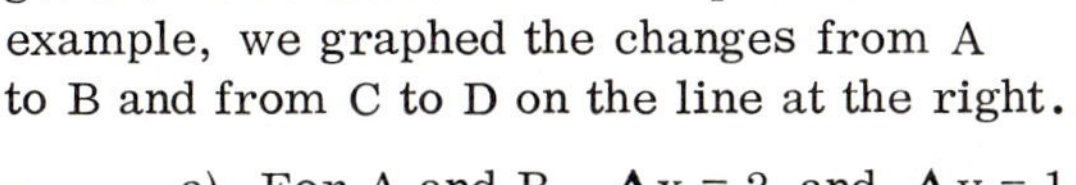

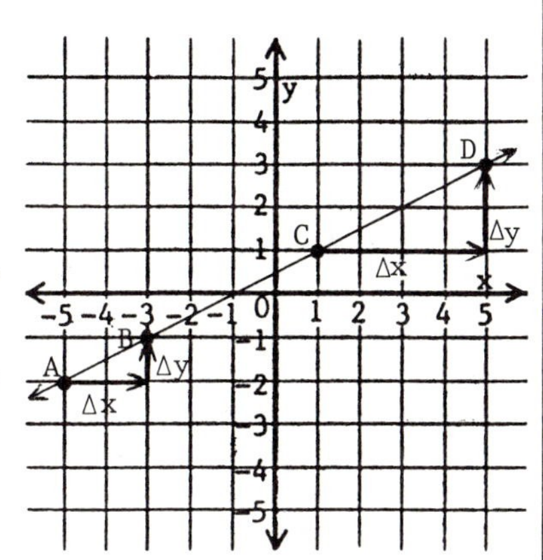

a) For A and B, $\Delta x = 2$ and $\Delta y = 1$. Therefore, the slope = ______.

b) For C and D, $\Delta x = 4$ and $\Delta y = 2$. Therefore, the slope = ______.

c) Did we get the same value for the slope with each pair? ______

a) 3

b) -2

---

61. The <u>sign</u> of the slope tells us whether a line rises or falls from left to right.

If its slope is <u>positive</u>, the line <u>rises</u>.

If its slope is <u>negative</u>, the line <u>falls</u>.

On the graph at the right, we have drawn four lines and labeled them #1, #2, #3, and #4.

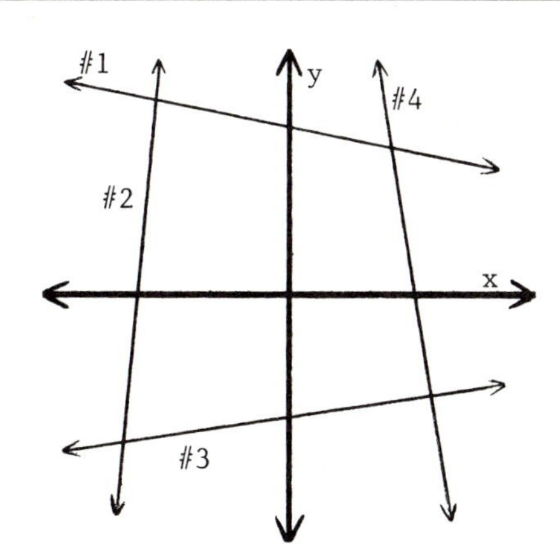

a) Which lines have a positive slope? ____________

b) Which lines have a negative slope? ____________

a) $\frac{1}{2}$

b) $\frac{1}{2}$ (from $\frac{2}{4}$)

c) Yes

---

a) lines #2 and #3

b) lines #1 and #4

62. The absolute value of the slope tells us how steep the rise or fall is.

We graphed two lines at the right. The rise of line #1 is steeper than the rise of line #2.

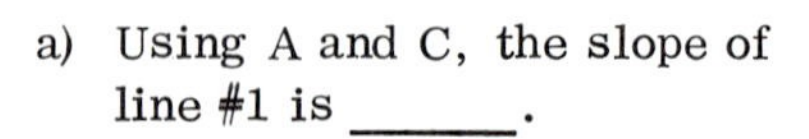

a) Using A and C, the slope of line #1 is ______.

b) Using A and B, the slope of line #2 is ______.

c) The rise of line #1 is steeper. Does line #1 have a slope with a larger absolute value? ______

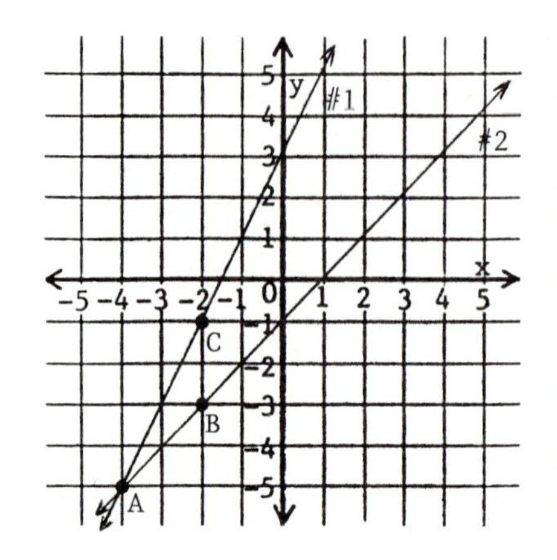

---

63. We graphed two lines at the right. The fall of line #1 is steeper than the fall of line #2.

a) Using P and R, the slope of line #1 is ______.

b) Using P and Q, the slope of line #2 is ______.

c) The fall of line #1 is steeper. Does line #1 have a slope with a larger absolute value? ______

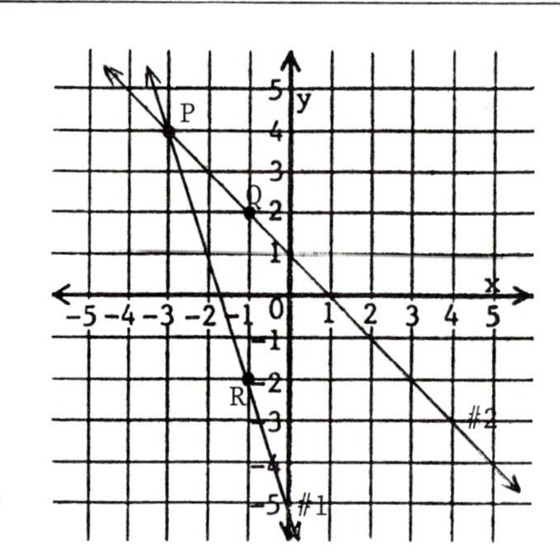

a) 2 (from $\frac{4}{2}$)

b) 1 (from $\frac{2}{2}$)

c) Yes

---

64. The slope of a line represents a ratio of changes in x and y.

A slope of $\frac{6}{5}$ means: For any 5-unit increase in x, there is a 6-unit increase in y.

A slope of $-\frac{1}{2}$ means: For any 2-unit increase in x, there is a ___-unit decrease in y.

a) -3 (from $\frac{-6}{2}$)

b) -1 (from $\frac{-2}{2}$)

c) Yes

---

65. When the slope is a whole number or decimal number, the slope is still a ratio. For example:

A slope of 3 means $\frac{3}{1}$. A slope of -1.5 means $\frac{-1.5}{1}$.

Therefore, a whole-number or decimal-number slope also states a ratio of changes in x and y.

a) A slope of 3 or $\frac{3}{1}$ means: For any 1-unit increase in x, there is a _____-unit increase in y.

b) A slope of -1.5 or $\frac{-1.5}{1}$ means: For any 1-unit increase in x, there is a _____-unit decrease in y.

1-unit

66. a) A slope of "1" or $\frac{1}{1}$ means: For any 1-unit increase in x, there is a _____-unit __________ (increase/decrease) in y.

b) A slope of -1 or $\frac{-1}{1}$ means: For any 1-unit increase in x, there is a _____-unit __________ (increase/decrease) in y.

a) 3-unit increase

b) 1.5-unit decrease

---

67. The equation of the horizontal line at the right is y = 2. Since a horizontal line does not rise or fall, it seems that its slope should be 0. Let's use points C and D to confirm that fact.

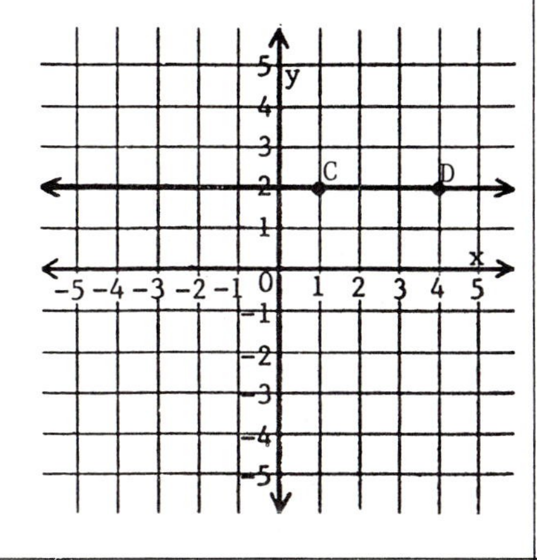

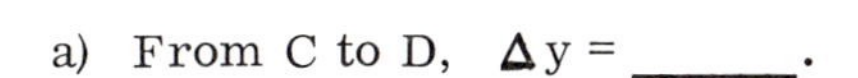

a) From C to D, Δy = ______.

b) From C to D, Δx = ______.

c) Therefore:

the slope $= \frac{\Delta y}{\Delta x} =$ ______

a) 1-unit increase

b) 1-unit decrease

---

68. The equation of the vertical line at the right is x = 3. Let's examine the slope of the line. From point P to point Q:

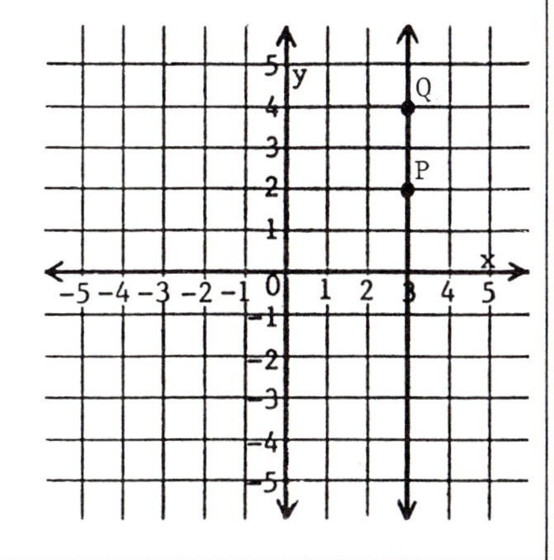

$\Delta y = 2$

$\Delta x = 0$

$\text{slope} = \frac{\Delta y}{\Delta x} = \frac{2}{0}$

But division by 0 is undefined. Therefore, the slope of the line is undefined.

a) 0

b) 3

c) $\frac{0}{3} = 0$

---

69. In the last two frames, we saw these facts:

1) The slope of any horizontal line is 0.

2) The slope of any vertical line is undefined.

What is the slope of each line below?

a) y = 8 slope = ________

c) y = -5 slope = ________

b) x = -9 slope = ________

d) x = 15 slope = ________

70. Since the x-axis is a horizontal line:

a) its slope is ________ b) its equation is ________

Since the y-axis is a vertical line:

c) its slope is ________ d) its equation is ________

a) 0
b) undefined
c) 0
d) undefined

---

a) 0	b) $y = 0$
c) undefined	d) $x = 0$

8-7 THE TWO-POINT FORMULA FOR SLOPE

If we know the coordinates of two points on a line, there is a formula we can use to compute its slope. We will discuss that formula in this section.

---

71. On the line at the right, $P_1(x_1, y_1)$ and $P_2(x_2, y_2)$ represent any two points. We can use their coordinates to find Δy and Δx. That is:

$$\Delta y = y_2 - y_1$$

$$\Delta x = x_2 - x_1$$

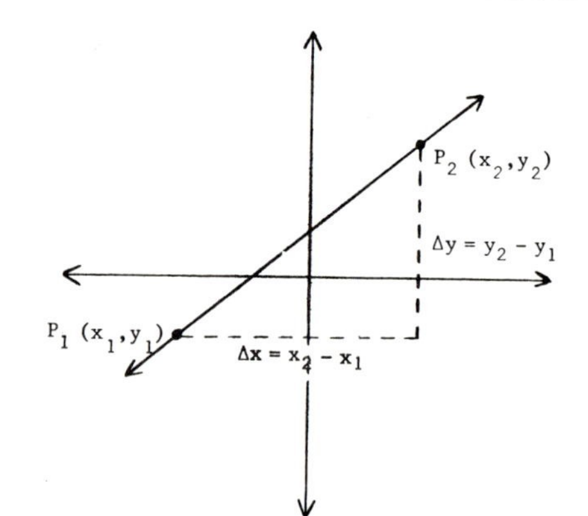

Using the letter m for slope, we can write the following formula for the slope of a line.

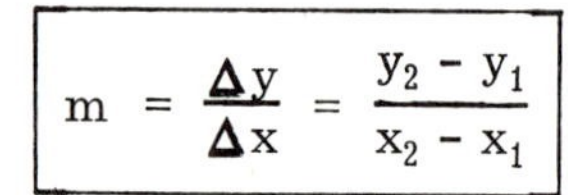

$$m = \frac{\Delta y}{\Delta x} = \frac{y_2 - y_1}{x_2 - x_1}$$

The line at the right passes through the points (2,-1) and (4,3). From the graph, you can see this fact:

$$m = \frac{\Delta y}{\Delta x} = \frac{4}{2} = 2$$

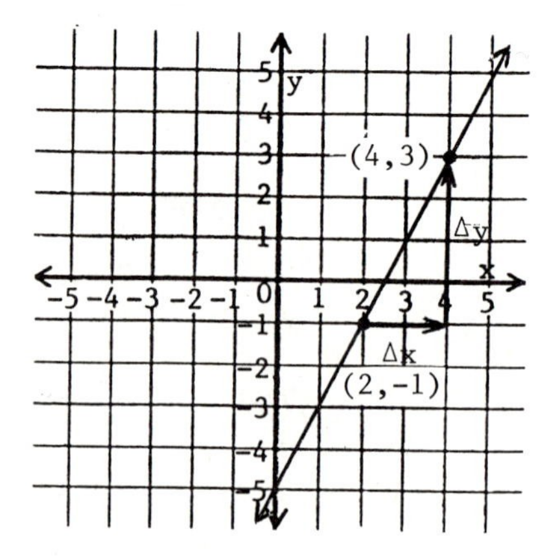

Let's use the two-point formula above to compute the slope. We can use either (2,-1) or (4,3) for (x_2, y_2).

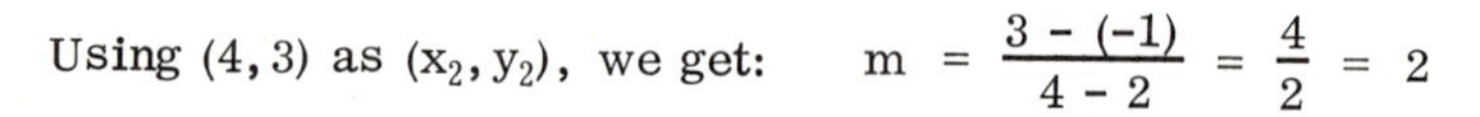

Using (4,3) as (x_2, y_2), we get: $m = \frac{3-(-1)}{4-2} = \frac{4}{2} = 2$

Using (2,-1) as (x_2, y_2), we get: $m = \frac{(-1)-3}{2-4} = \frac{-4}{-2} = 2$

Did we get the correct value for the slope using both methods? ________

Yes

72. Let's use the same formula to find the slope of the line through (-2,1) and (3,-3). To show that we can use either point as (x_2, y_2), we will do it both ways.

a) Using (3,-3) as (x_2, y_2), we get:

$$m = \frac{y_2 - y_1}{x_2 - x_1}$$

$$= \frac{(-3) - 1}{3 - (-2)} = \underline{\hspace{3cm}}$$

b) Using (-2,1) as (x_2, y_2), we get:

$$m = \frac{y_2 - y_1}{x_2 - x_1}$$

$$= \frac{1 - (-3)}{-2 - 3} = \underline{\hspace{3cm}}$$

---

a) $\frac{-4}{5} = -\frac{4}{5}$

b) $\frac{4}{-5} = -\frac{4}{5}$

---

73. Let's use the same formula to find the slope of the line through (-4,0) and (0,2).

Note: When using the formula, it is helpful to sketch the two points as we have done to avoid gross errors.

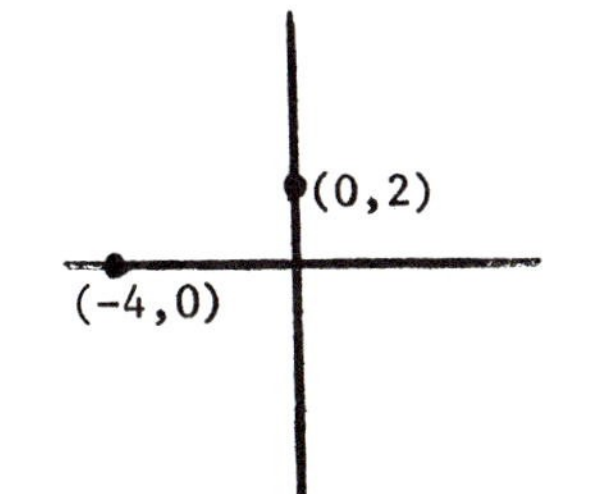

Using either (-4,0) or (0,2) as (x_2, y_2):

$$m = \frac{y_2 - y_1}{x_2 - x_1} = \underline{\hspace{3cm}}$$

$m = \frac{1}{2}$, from:

$$\frac{2 - 0}{0 - (-4)} = \frac{2}{4}$$

or

$$\frac{0 - 2}{-4 - 0} = \frac{-2}{-4}$$

---

74. Use the formula to find the slope of the line through each pair of points below. (Sketch the points first.)

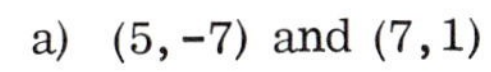

a) (5,-7) and (7,1)

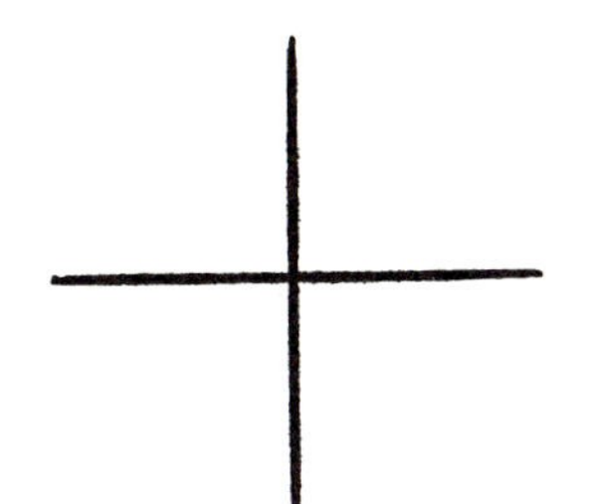

$m = \underline{\hspace{3cm}}$

b) (1,0) and (-5,4)

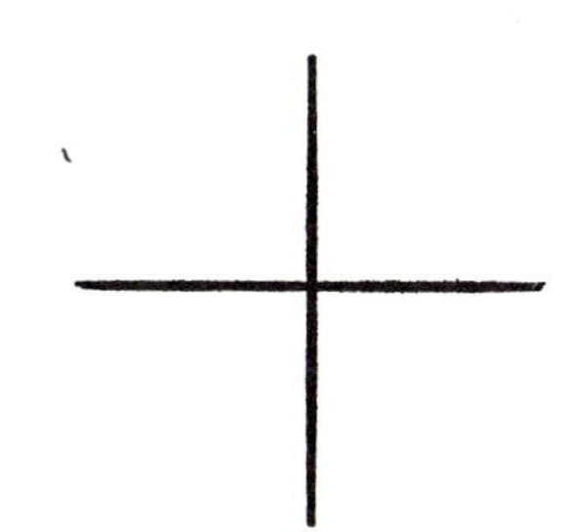

$m = \underline{\hspace{3cm}}$

---

a) $m = 4$

b) $m = -\frac{2}{3}$

75. Find the slope of the line through each pair of points. (Sketch the points first.)

a) (-8, 0) and (0, 20)

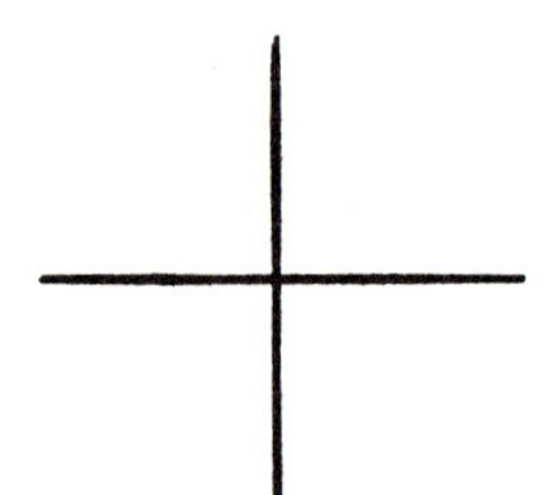

m = ____________

b) (17, -11) and (-12, 18)

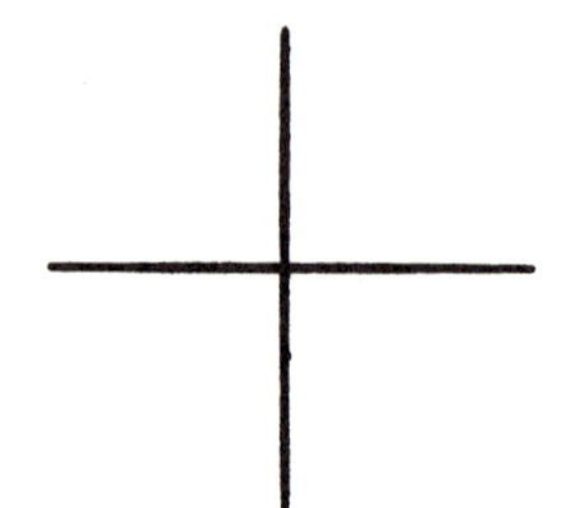

m = ____________

---

76. We know that the line at the right passes through (0, 0) and (2, 10). Let's use the two-point formula to find its slope.

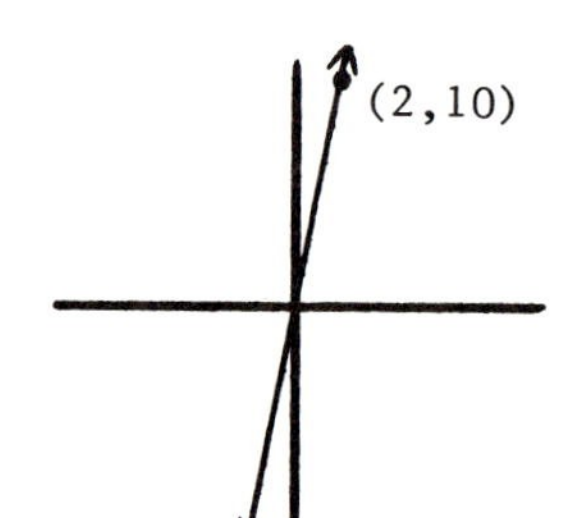

$$m = \frac{y_2 - y_1}{x_2 - x_1} = \frac{10 - 0}{2 - 0} = \frac{10}{2} = 5$$

or

$$m = \frac{y_2 - y_1}{x_2 - x_1} = \frac{0 - 10}{0 - 2} = \_\_\_\_\_\_$$

Answers to 75:

a) $m = \frac{5}{2}$

b) $m = -1$

---

77. Use the two-point formula to find the slope of a line passing through the origin and each point below. (Make a sketch.)

a) (-10, 8)

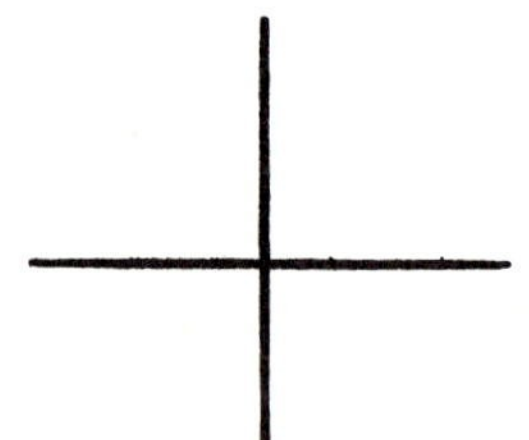

m = ________

b) (-6, -7)

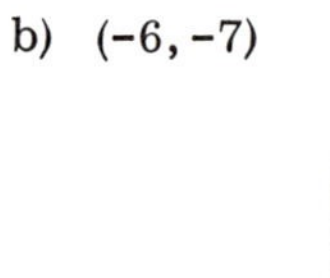

m = ________

Answer to 76:

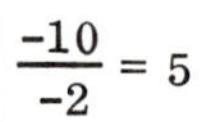

$\frac{-10}{-2} = 5$

Answers to 77:

a) $m = -\frac{4}{5}$

b) $m = \frac{7}{6}$

SELF-TEST 26 (pages 359-371)

Use the intercept method to graph each equation.

1. $5x - 4y = 20$

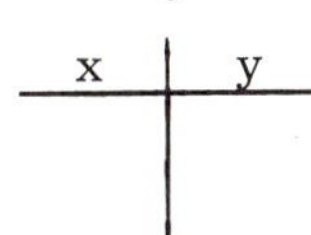

2. $3x + y = 3$

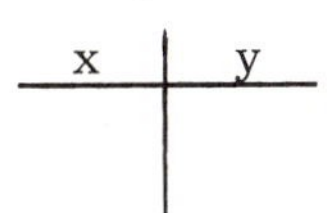

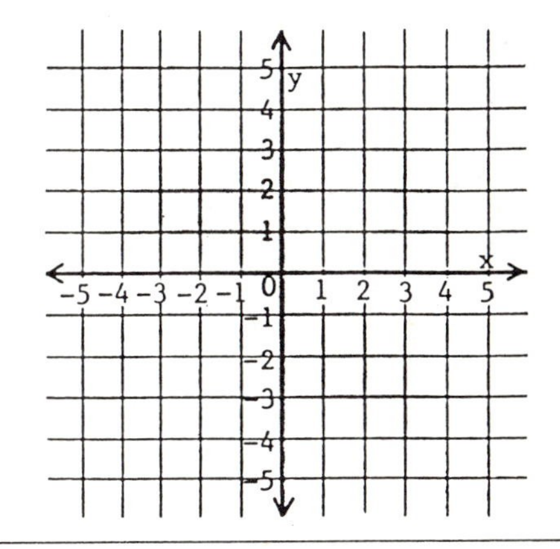

Write the equation of:

3. line A ______ 4. line B ______

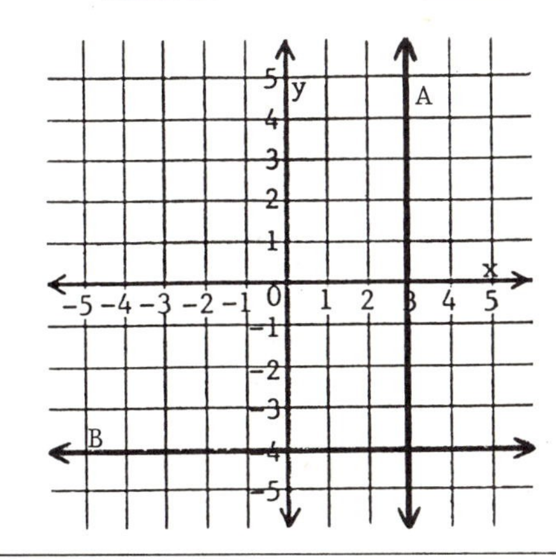

Find the slope of:

5. line A ______ 6. line B ______

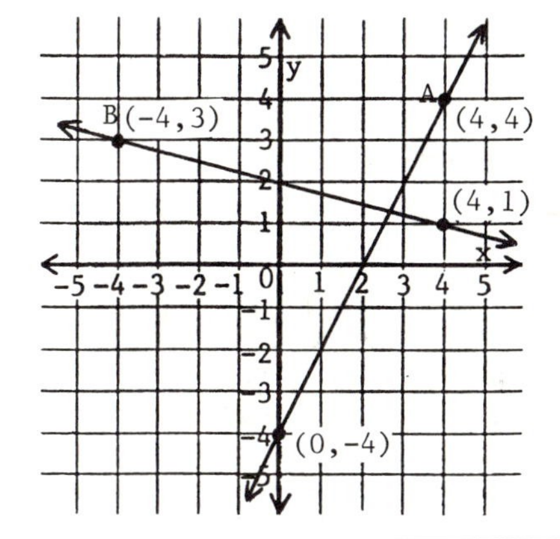

7. What is the equation of the y-axis? ________

8. What is the slope of the x-axis? m = ______

Find the slope of the line through each pair of points.

9. (-2,2) and (1,-7)

10. (0,0) and (-10,-5)

ANSWERS: 1-2.

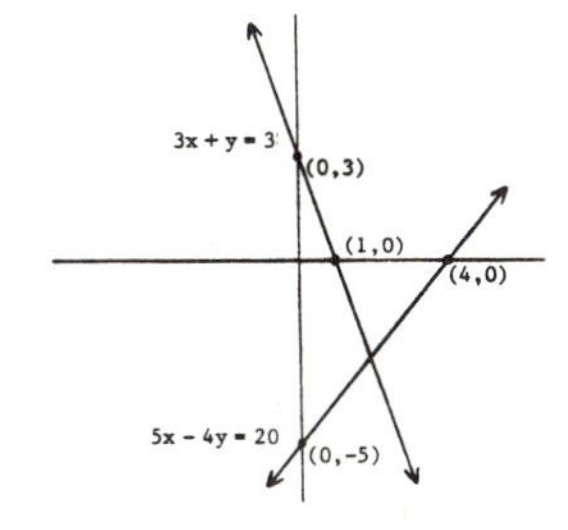

3. $x = 3$
4. $y = -4$
5. $m = 2$
6. $m = -\frac{1}{4}$
7. $x = 0$
8. $m = 0$
9. $m = -3$
10. $m = \frac{1}{2}$

8-8 SLOPE-INTERCEPT FORM

In this section, we will discuss the slope-intercept form of linear equations. We will show how linear equations in other forms can be rearranged to slope-intercept form.

---

78. The following form of a linear equation is called slope-intercept form.

$$\boxed{y = mx + b}$$ where: m is the slope of the line. b is the y-intercept of the line.

As an example, we graphed $y = 2x - 3$ below. The coordinates of points A and B are given.

1. In the equation, $m = 2$. To show that 2 is the slope of the line, we can use the changes in y and x from A to B.

$$m = \frac{\Delta y}{\Delta x} = \frac{4}{2} = 2$$

2. In the equation, $b = -3$. A is the y-intercept. Its coordinates are $(0,-3)$. -3 is the y-coordinate of A.

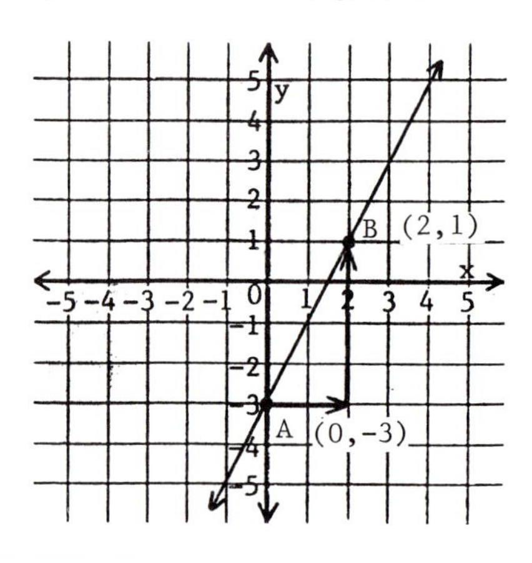

---

79. The equation $y = -\frac{2}{3}x + 3$ is graphed at the right. The coordinates of points C and D are given.

a) In the equation, $m = -\frac{2}{3}$. Use the changes in y and x from C to D to show that $-\frac{2}{3}$ is the slope of the line.

$$m = \frac{\Delta y}{\Delta x} = \_\_\_\_\_\_\_\_$$

b) In the equation, $b = 3$, we can see that 3 is the y-coordinate of C, the y-intercept.

The coordinates of C are ________.

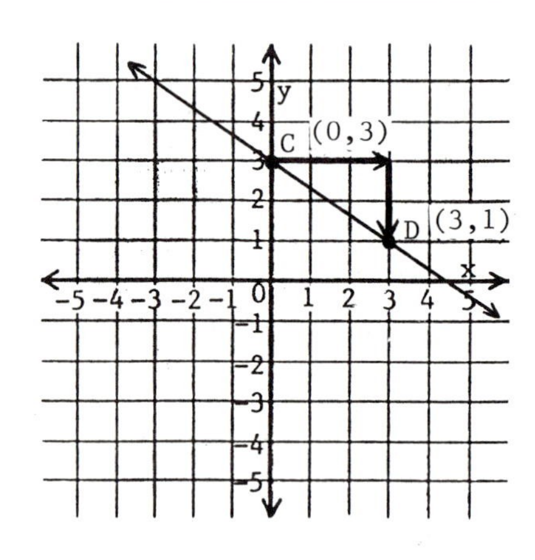

---

80. In the equations below, m is not explicitly written.

$$y = x - 2 \qquad\qquad y = -x + 5$$

However, since $x = 1x$ and $-x = -1x$:

a) In $y = x - 2$, $m =$ _____ b) In $y = -x + 5$, $m =$ _____

Answers to 79: a) $-\frac{2}{3}$ b) $(0, 3)$

81. In $y = mx + b$, b is the y-coordinate of the y-intercept. Therefore:

For $y = x - 2$, the coordinates of the y-intercept are $(0, -2)$.

For $y = -3x + 9$, the coordinates of the y-intercept are (,).

Answers: a) 1 b) -1

---

82. Following the example, write the slope-intercept form of the linear equations with the following slopes and y-intercepts.

	Slope	y-intercept	Equation
	4	$(0, -1)$	$y = 4x - 1$
a)	$-\frac{5}{2}$	$(0, 4)$	________
b)	-1	$(0, -\frac{1}{2})$	________

Answer: $(0, 9)$

---

83. To put the equation below in slope-intercept form, we solved for y. Notice how we wrote the x-term first on the right side. Put the other equation in slope-intercept form.

$$y - 2x = 7$$
$$y - 2x + 2x = 2x + 7$$
$$y = 2x + 7$$

$$x + y = 10$$

Answers: a) $y = -\frac{5}{2}x + 4$ b) $y = -x - \frac{1}{2}$

---

84. To get the opposite of a binomial, we replace each term by its opposite. That is:

The opposite of $-9x + 5$ is $9x - 5$.

The opposite of $x - 1$ is $-x + 1$.

To put each equation below in slope-intercept form, we used the oppositing principle. That is, we replaced each side by its opposite.

$$-y = -2x + 3 \qquad -y = x - 9$$
$$y = 2x - 3 \qquad y = -x + 9$$

Use the oppositing principle to put each equation below in slope-intercept form.

a) $-y = -8x + 1$ b) $-y = 4x - 3$ c) $-y = -x + 7$

________ ________ ________

Answer: $y = -x + 10$

Answers: a) $y = 8x - 1$ b) $y = -4x + 3$ c) $y = x - 7$

85. Notice how we used the oppositing principle to put each equation below in slope-intercept form.

$$3x - y = 10$$
$$(-3x) + 3x - y = (-3x) + 10$$
$$-y = -3x + 10$$
$$y = 3x - 10$$

$$x = 5 - y$$
$$x + (-5) = (-5) + 5 - y$$
$$x - 5 = -y$$
$$y = -x + 5$$

Following the examples, put each equation below in slope-intercept form.

a) $x - y = 4$

b) $8 - y = 7x$

---

86. To put the equation below in slope-intercept form, we divided $x + 3$ by 5. Put the other equation in slope-intercept form.

Answers to 85: a) $y = x - 4$ b) $y = -7x + 8$

$$5y = x + 3$$
$$y = \frac{x + 3}{5}$$
$$y = \frac{1}{5}x + \frac{3}{5}$$

$$4y = -5x + 1$$

---

87. Notice how we reduced to lowest terms below. Put the other equation in slope-intercept form.

Answer to 86: $y = -\frac{5}{4}x + \frac{1}{4}$

$$3y = x + 6$$
$$y = \frac{x + 6}{3}$$
$$y = \frac{x}{3} + \frac{6}{3}$$
$$y = \frac{1}{3}x + 2$$

$$2y = -6x + 1$$

---

88. Following the example, put the other equation in slope-intercept form.

Answer to 87: $y = -3x + \frac{1}{2}$

$$4x + 3y = 12$$
$$3y = -4x + 12$$
$$y = \frac{-4x + 12}{3}$$
$$y = -\frac{4}{3}x + 4$$

$$x + 5y = 3$$

89. Notice how we used the oppositing principle below. Put the other equation in slope-intercept form.

$$x - 4y = 7$$
$$-4y = 7 - x$$
$$4y = x - 7$$
$$y = \frac{x-7}{4}$$
$$y = \frac{1}{4}x - \frac{7}{4}$$

$$5x - 3y = 15$$

Answer: $y = -\frac{1}{5}x + \frac{3}{5}$

---

90. After putting an equation in slope-intercept form, we can easily identify its slope and y-intercept. For example:

Since $2x + y = 5$ is equivalent to $y = -2x + 5$:

Its slope is -2. Its y-intercept is $(0, 5)$.

Since $5x - 2y = 6$ is equivalent to $y = \frac{5}{2}x - 3$:

a) Its slope is ______. b) Its y-intercept is ______.

Answer: $y = \frac{5}{3}x - 5$

---

91. The general slope-intercept form of linear equations is $y = mx + b$. However, the y-intercept of all lines through the origin is $(0,0)$. Since $b = 0$ for all lines through the origin, their slope-intercept form is:

$$\boxed{y = mx}$$

If we know the slope of a line through the origin, we can write its equation. That is:

If $m = \frac{3}{4}$, $y = \frac{3}{4}x$ If $m = -\frac{5}{2}$, $y =$ ______

Answer: a) $\frac{5}{2}$ b) $(0,-3)$

---

92. We can find the slope of $y - 5x = 0$ by putting it in slope-intercept form. We get: $y = 5x$. Therefore, $m = 5$.

Find the slope of each line below by putting it in slope-intercept form.

a) $4x + 3y = 0$ $m =$ ______ b) $x - y = 0$ $m =$ ______

Answer: $y = -\frac{5}{2}x$

Answer: a) $m = -\frac{4}{3}$ b) $m = 1$

8-9 POINT-SLOPE FORM

In this section, we will discuss the point-slope form of linear equations.

---

93. For the line at the right, we know the slope m and the coordinates of point $P_1(x_1, y_1)$. Let $P(x, y)$ be any other point on the line. Then using the two-point formula for slope, we get:

$$\frac{y - y_1}{x - x_1} = m$$

Clearing the fraction, we get the point-slope form of a linear equation. It is:

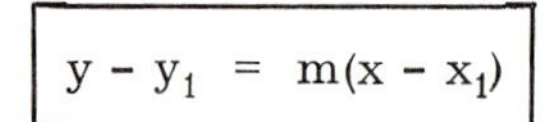

$$y - y_1 = m(x - x_1)$$

where: m is the slope of the line. $P_1(x_1, y_1)$ is a point whose coordinates are known.

Let's use the form above to get the equation of a line that passes through (4, 5) and has a slope of 2.

$$y - y_1 = m(x - x_1)$$
$$y - 5 = 2(x - 4)$$
$$y - 5 = 2x - 8$$

Write the equation above in slope-intercept form. ________________

---

94. We used the point-slope form to find the equation of one line below. We wrote it in slope-intercept form. Do the same for the other line.

Answer to 93: $y = 2x - 3$

Find the line with a slope of 4 that passes through (2, -3).

$$y - y_1 = m(x - x_1)$$
$$y - (-3) = 4(x - 2)$$
$$y + 3 = 4x - 8$$
$$y = 4x - 11$$

Find the line with a slope of $-\frac{1}{2}$ that passes through (4, -1).

---

95. We found the equation of one line below. Find the equation of the other line.

Answer to 94: $y = -\frac{1}{2}x + 1$

Find the line with a slope of -1 that passes through (-2, 4).

$$y - y_1 = m(x - x_1)$$
$$y - 4 = -1[x - (-2)]$$
$$y - 4 = -1(x + 2)$$
$$y - 4 = -x - 2$$
$$y = -x + 2$$

Find the line with a slope of $\frac{2}{3}$ that passes through (-6, -3).

Answer to 95: $y = \frac{2}{3}x + 1$

8-10 GRAPHS OF FORMULAS

In this section, we will graph some linear formulas and show how the graphs can be used to solve applied problems.

---

96. The formula below shows the relationship between degrees-Celsius (C) and degrees-Fahrenheit (F). We used three points to graph the formula. Notice that we plotted C on the horizontal axis and F on the vertical axis. This choice is arbitrary.

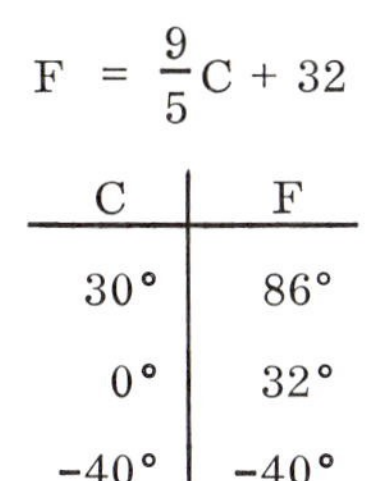

$$F = \frac{9}{5}C + 32$$

C	F
30°	86°
0°	32°
-40°	-40°

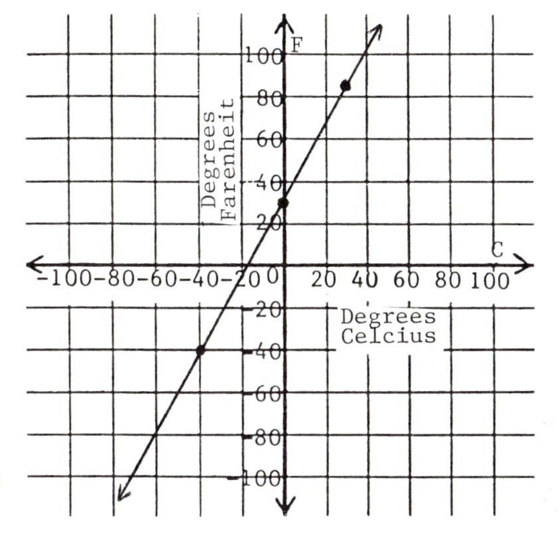

The formula above is in slope-intercept form. Therefore:

a) The slope of the line is __________.

b) The coordinates of the F-intercept are __________.

a) 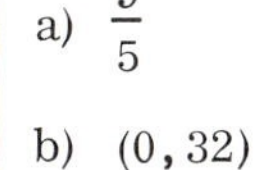 $\frac{9}{5}$

b) (0,32)

---

97. Though temperature has negative values, most quantities (like distances, time, weight, pressure, and so on) do not have negative values. Therefore, <u>only quadrant 1 is used when graphing most formulas</u>. An example is discussed below.

The formula below shows the relationship between distance traveled (d) and time (t) for an object traveling at a constant velocity (or speed) of 50 miles per hour. We used three points to graph the formula.

$$d = 50t$$

t	d
0	0
2	100
5	250

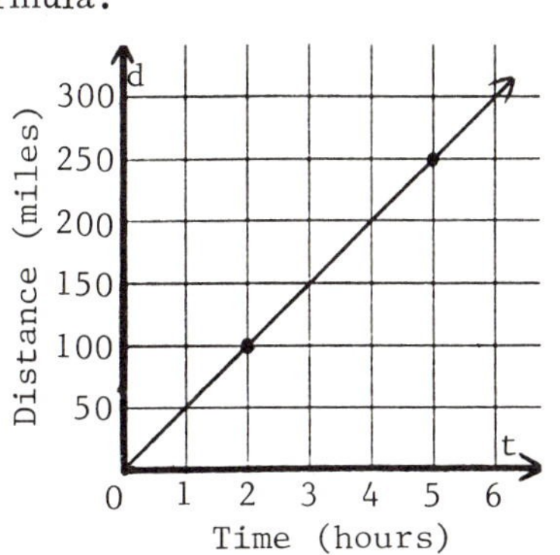

The formula is in slope-intercept form for a line through the origin. Therefore:

a) The slope of the line is __________.

b) The coordinates of the d-intercept are __________.

a) 50

b) (0,0)

98. The formula below is related to the concept of load line in transistor electronics. The two variables are voltage (E) and current (I).

$$E + 8I = 24$$

Use the intercept method to graph the formula at the right.

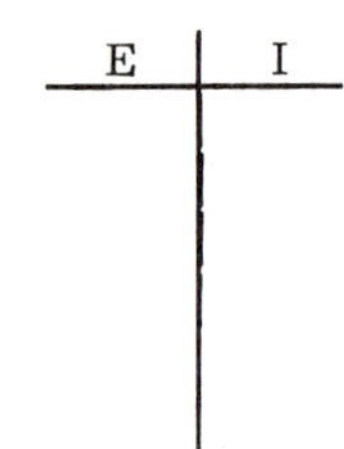

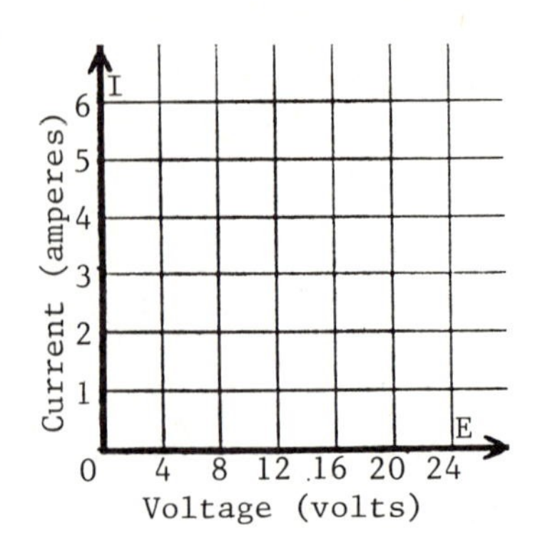

---

99. The graph below shows the relationship between distance traveled (d) and time (t) at an average velocity (or speed) of 100 kilometers per hour.

We can use the graph to solve this problem.

How far will a train travel in 3 hours at an average velocity of 100 kilometers per hour?

To solve the problem, we use these steps:

1) Draw an arrow from 3 on the time axis to the graphed line.

2) Then draw an arrow from that point to the distance axis.

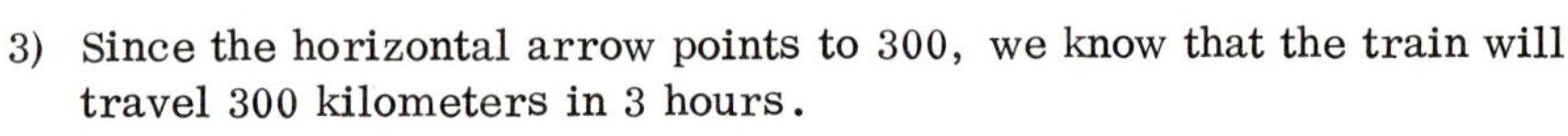

3) Since the horizontal arrow points to 300, we know that the train will travel 300 kilometers in 3 hours.

Use the graph for these:

a) How far would the train travel in 2 hours? ________

b) How far would the train travel in 5 hours? ________

---

a) 200 kilometers

b) 500 kilometers

100. The graph below shows the relationship between distance traveled (d) and time (t) at an average velocity (or speed) of 50 miles per hour.

We can use the graph to solve this problem.

How long would it take a car to drive 150 miles at an average speed of 50 miles per hour?

To solve the problem, we use these steps:

1) Draw an arrow from 150 on the distance axis to the graphed line.

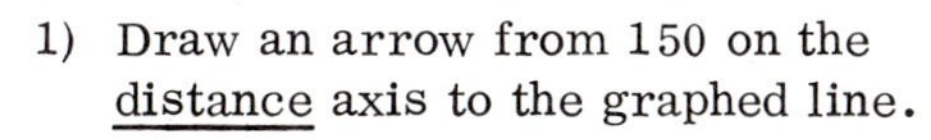

2) Then draw an arrow from that point to the time axis.

3) Since the vertical arrow points to 3, we know that it would take the car 3 hours to drive 150 miles at that speed.

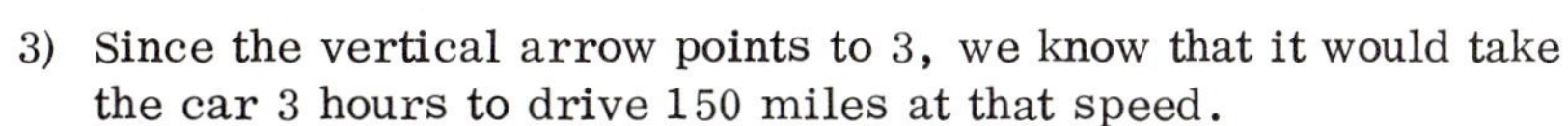

Use the graph to answer these:

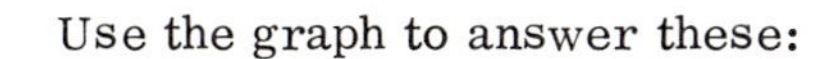

a) How long would it take the car to drive 100 miles? ________

b) How long would it take the car to drive 250 miles? ________

---

a) 2 hours

b) 5 hours

101. The graph below is a specific example of Hooke's Law. That is, it shows how far a spring will stretch when various amounts of force are applied to it. Use it to complete these:

a) If a force of 150 grams is applied, the spring will stretch approximately _____ centimeters.

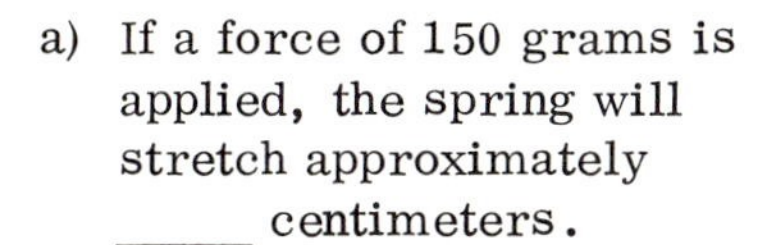

b) If a force of 75 grams is applied, the spring will stretch approximately _____ centimeters.

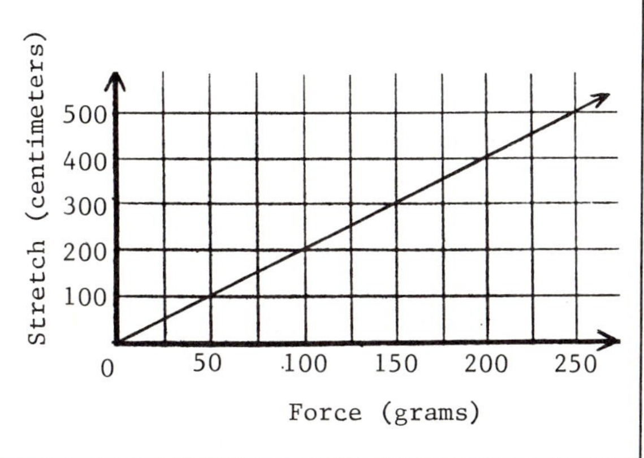

c) To stretch the spring 400 centimeters; we need a force of ______ grams.

---

a) 300 centimeters

b) 150 centimeters

c) 200 grams

102. The graph below is a specific example of Ohm's Law. That is, it shows how much current we get in an electric circuit when various voltages are applied with a constant resistance. Use it to complete these:

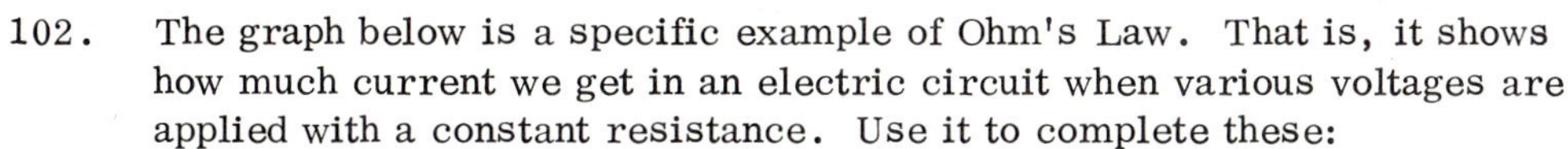

a) If a voltage of 40 volts is applied, we get a current of ______ amperes.

b) To get a current of 6 amperes, we must apply a voltage of ______ volts.

c) To get a current of 13 amperes, we must apply a voltage of ______ volts.

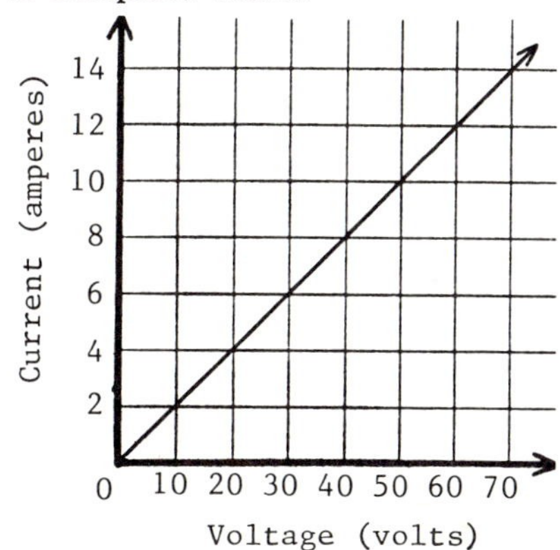

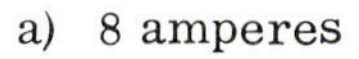

a) 8 amperes

b) 30 volts

c) 65 volts

8-11 DIRECT VARIATION

Any relationship of the form $y = kx$ is called direct variation. We will discuss direct variation in this section.

---

103. When two variables are related so that their ratio is always constant, we say that they are directly proportional or that one of the variables varies directly as the other. A relationship of that type is called direct variation. It is expressed by an equation or formula of the form:

$$y = kx$$

where k is called the constant of variation or the constant of proportionality

Some examples of direct variation are:

$$y = 4x \qquad d = 100t$$

a) In $y = 4x$, the constant of variation is ______.

b) In $d = 100t$, the constant of proportionality is ______.

---

104. The following language is used to state a direct variation.

For $y = 4x$, we say: y varies directly as x.
y is directly proportional to x.

a) For $d = 100t$, we say: ______ varies directly as ______.

b) For $E = 6R$, we say: ______ is directly proportional to ______.

Answers to 103: a) 4 b) 100

---

105. Using the given constant of variation, we wrote each direct variation below as an equation or formula.

y varies directly as x, and the constant of variation is k.	$y = kx$
P is directly proportional to H, and the constant of proportionality is 15.	$P = 15H$

Write each direct variation below as an equation or formula.

a) m varies directly as t, and the constant of proportionality is 50. ____________

b) R is directly proportional to S, and the constant of variation is k. ____________

Answers to 104: a) d ... t b) E ... R

106. The graph of any direct variation is a straight line through the origin or starting at the origin. As examples, we graphed two direct variations below.

$y = 2x$ $\qquad\qquad$ $d = 50t$

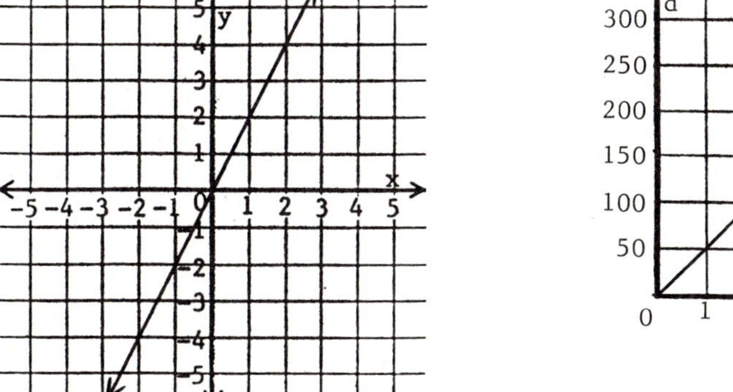

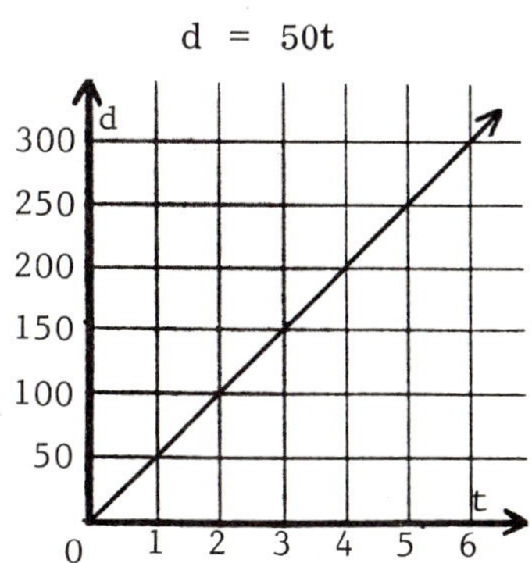

Only straight lines through the origin represent direct variations. Other straight lines that do not pass through the origin, like $y = 3x + 2$ or $P = 5V - 1$, are not direct variations.

Which of the following are direct variations? ______________

a) $y = 10x$ $\quad$ b) $R = 4t + 3$ $\quad$ c) $y = x + 9$ $\quad$ d) $E = 12I$

Answers: a) $m = 50t$ b) $R = kS$

---

107. In a direct variation, the solved-for variable is called the dependent variable because its value depends on the value substituted for the other variable. The other variable is called the independent variable. For example:

In $y = 4x$: $\quad$ y is the dependent variable.
$\qquad\qquad\quad$ x is the independent variable.

In $E = 8I$: $\quad$ a) the dependent variable is ______.

$\qquad\qquad\quad$ b) the independent variable is ______.

Answer: Only (a) and (d)

---

108. In a direct variation, whatever happens to the independent variable also happens to the dependent variable. Some specific changes are:

Independent Variable	Dependent Variable
Doubled	Doubled
Tripled	Tripled
Cut in half	Cut in half

As an example, some pairs of values for $d = 50t$ are given at the right.

d	t
50	1
100	2
150	3
200	4
250	5
300	6

If t is doubled from 1 to 2,
d is doubled from 50 to 100.

If t is tripled from 2 to 6,
d is tripled from 100 to 300.

If t is cut in half from 4 to 2,
d is cut in half from ______ to ______.

Answers: a) E b) I

109. If we are given one specific pair of values in a direct variation, we can find k by substitution. For example, if y varies directly as x and $y = 10$ when $x = 2$, we get:

$$y = kx$$
$$10 = k(2)$$
$$k = \frac{10}{2} = 5$$

Since $k = 5$, the direct variation is $y = 5x$. Using that equation, we can find y for other values of x. For example, we found y when $x = 3$ below. Find y when $x = 10$.

$y = 5x$	$y = 5x$
$y = 5(3)$	
$y = 15$	________

Answer: 200 to 100

---

110. Using the two steps from the last frame, we solved the problem below.

If E is directly proportional to R and $E = 30$ when $R = 3$, find E when $R = 5$.

$E = kR$	$E = 10R$
$30 = k(3)$	$E = 10(5)$
$k = \frac{30}{3} = 10$	$E = 50$

Using the same steps, solve this one.

If d varies directly as t and $d = 600$ when $t = 10$, find d when $t = 7$.

Answer: $y = 50$

---

111. Using the same two steps, solve each problem.

a) When the resistance is constant, the amount of current (I) in an electric circuit varies directly as the applied voltage (E). If $I = 30$ amperes when $E = 6$ volts, find I when $E = 24$ volts.

b) The amount of stretch (s) in a spring is directly proportional to the force (F) applied to it. If $s = 6$ centimeters when $F = 40$ kilograms, find s when $F = 90$ kilograms.

Answer: $d = 420$, from: $d = 60t$

Answers:

a) 120 amperes, from: $I = 5E$

b) 13.5 centimeters, from: $s = 0.15F$

SELF-TEST 27 (pages 372-383)

For each equation: (a) put it in slope-intercept form; (b) identify its slope; (c) identify the coordinates of its y-intercept.

1. $x + y = 7$

2. $3x - 5y = 10$

3. Find the slope-intercept form of the equation of the line whose slope is 4 and which passes through (3, -5).

4. Find the slope-intercept form of the equation of the line whose slope is $-\frac{3}{2}$ and which passes through (-2, 7).

The graph at the right shows the relationship between distance traveled (in miles) and time (in hours) for a small airplane.

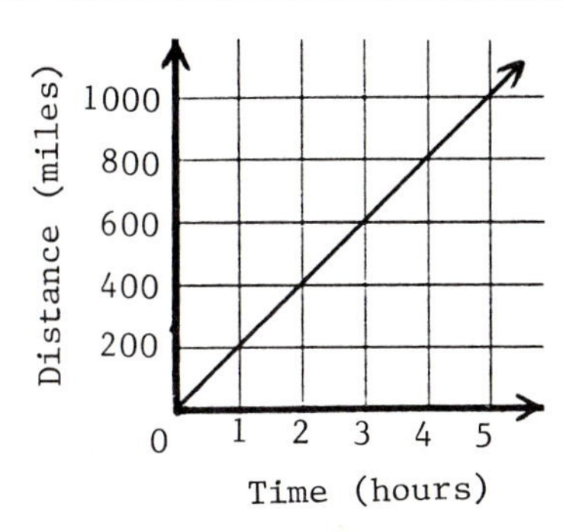

How far would the plane travel:

5. in 2 hours? ________

6. in $3\frac{1}{2}$ hours? ________

How long would it take the plane to travel:

7. 800 miles? ________

8. 300 miles? ________

9. If y is directly proportional to x and $y = 24$ when $x = 8$, find y when $x = 5$.

10. If velocity v is directly proportional to time t and $v = 20$ m/sec when $t = 8$ sec, find v when $t = 50$ sec.

ANSWERS:

1. a) $y = -x + 7$
   b) $m = -1$
   c) (0, 7)

2. a) $y = \frac{3}{5}x - 2$
   b) $m = \frac{3}{5}$
   c) (0, -2)

3. $y = 4x - 17$

4. $y = -\frac{3}{2}x + 4$

5. 400 miles

6. 700 miles

7. 4 hours

8. $1\frac{1}{2}$ hours

9. $y = 15$

10. $v = 125$ m/sec

SUPPLEMENTARY PROBLEMS - CHAPTER 8

Assignment 25

Decide whether the given ordered pair is a solution of the given equation.

1. (2,10) for $y = 5x$
2. (-1,-4) for $y = 3x - 1$
3. (-2,12) for $2x + y = 9$
4. (0,-3) for $x - y = 3$
5. (-10,-5) for $y = -\frac{1}{2}x$
6. (-6,-5) for $y = -\frac{2}{3}x - 1$

Complete the solutions below for $y = 3x + 6$.

7. (-1,) 8. (,-6) 9. (0,) 10. (,0)

Complete the solutions below for $4x + y = 8$.

11. (3,) 12. (,4) 13. (0,) 14. (,0)

15. Write the coordinates of each point.

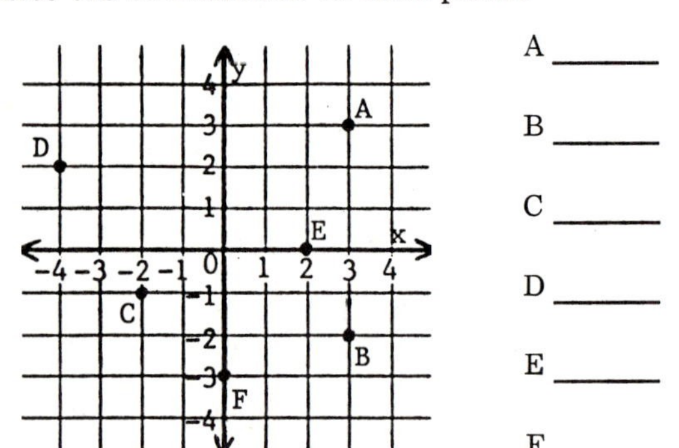

A ______
B ______
C ______
D ______
E ______
F ______

16. Estimate the coordinates of each point.

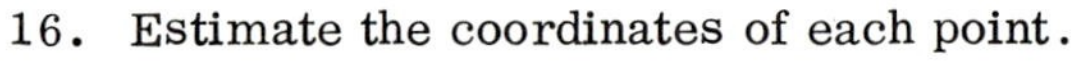

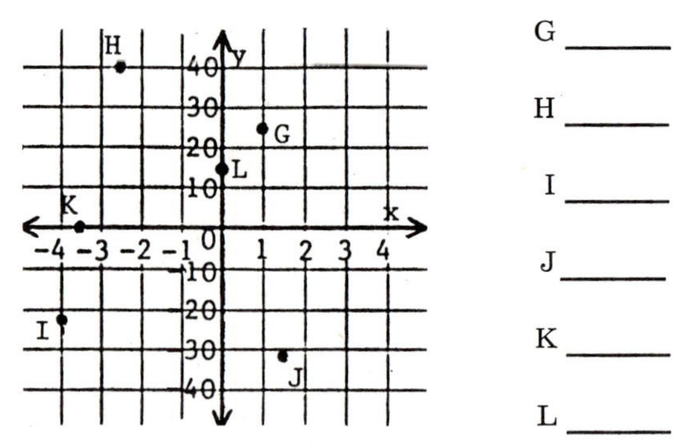

G ______
H ______
I ______
J ______
K ______
L ______

17. State the number of the quadrant in which each point lies.

a) (-8,1) b) (20,-5) c) (2,13) d) (-1,-3) e) (-32,50)

18. Which of the following points lie on the x-axis?

a) (0,-7) b) (-7,0) c) (0,0) d) (30,0) e) (0,1)

19. Which of the following points lie on the y-axis?

a) (0,4) b) (1,-1) c) (-3,0) d) (0,0) e) (0,-50)

20. What name is given to the point (0,0)?

21. The abscissa of (9,-4) is ______.

22. The ordinate of (-1,-10) is ______.

23. Write the coordintates of the point whose ordinate is -9 and whose abscissa is 17.

Graph each equation.

24. $y = x$

x	y

25. $y = -2x - 1$

x	y

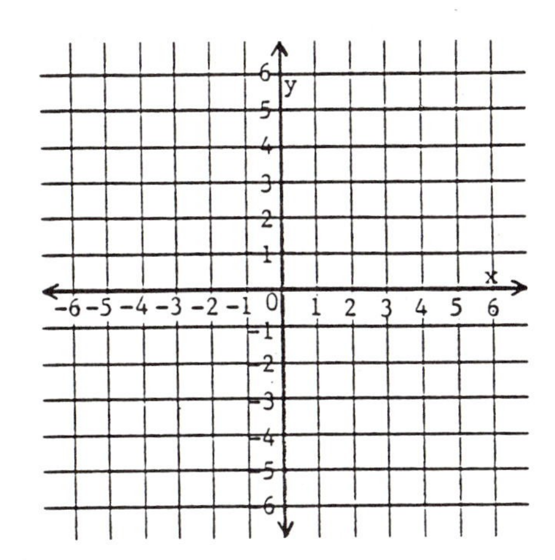

Graph each equation.

26. $x + y = 2$

x	y

27. $y = \frac{1}{2}x - 3$

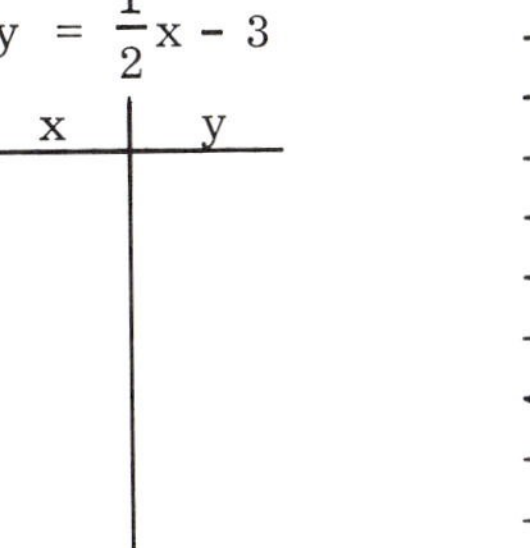

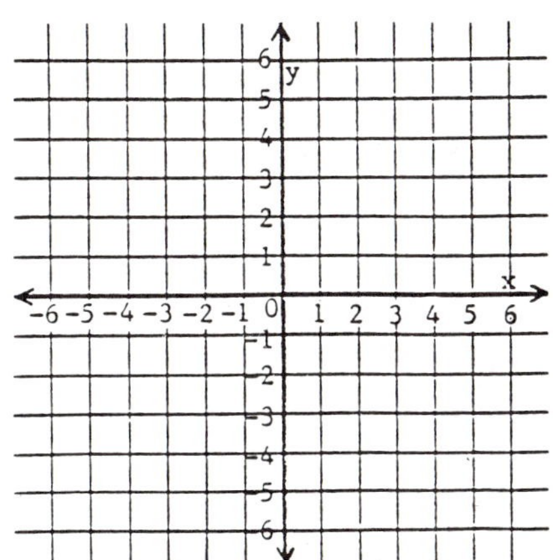

Assignment 26

Find the coordinates of the x-intercept of each equation.

1. $y = 2x + 6$
2. $3x + 5y = 15$
3. $4x - 3y = 2$
4. $2y = 5x + 8$

Find the coordinates of the y-intercept of each equation.

5. $y = 3x - 9$
6. $2x + 5y = 10$
7. $3y = 4x + 1$
8. $6x - 2y = 5$

Find the intercepts of each equation and use them to graph the equation.

9. $y = x + 2$

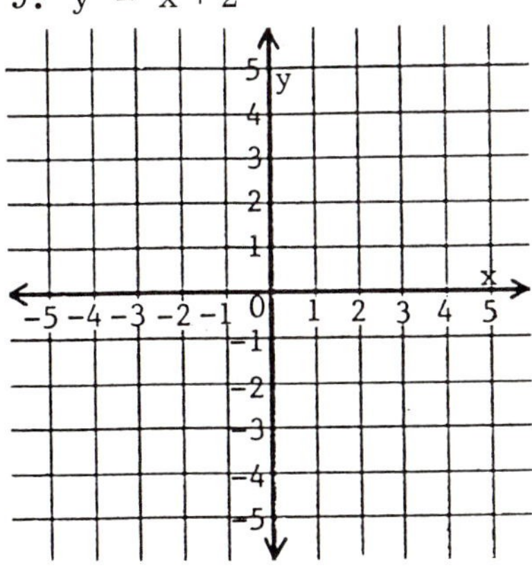

10. $x + 3y = 3$

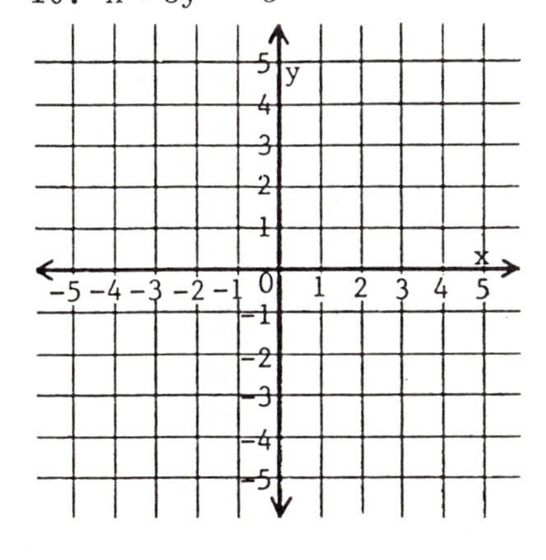

11. $3x - 2y = 6$

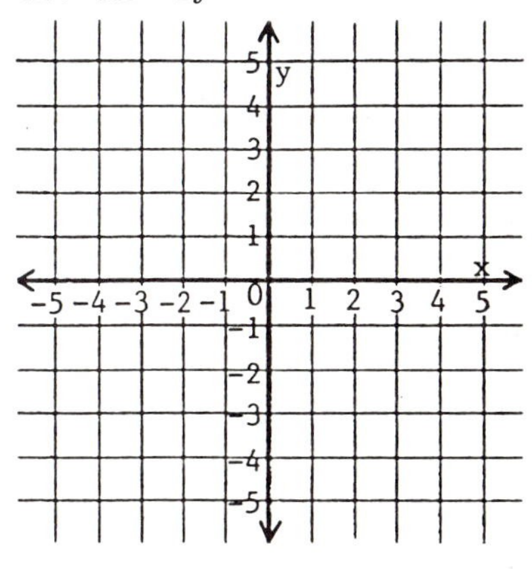

Write the equation of each line.

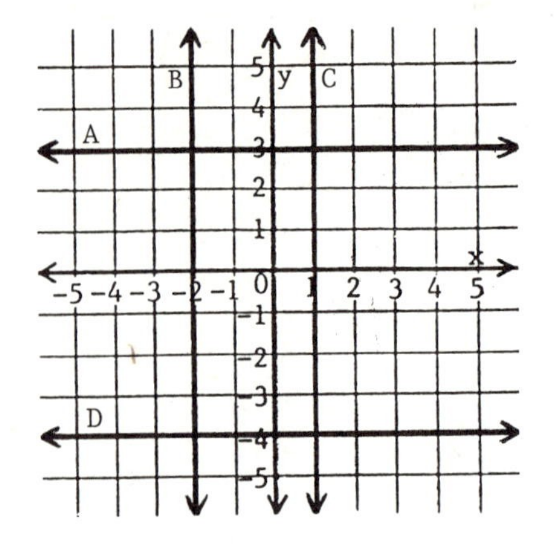

12. line A: ____________

13. line B: ____________

14. line C: ____________

15. line D: ____________

16. the x-axis: ____________

17. the y-axis: ____________

Find the slope of each line graphed below.

18. Line A

19. Line B

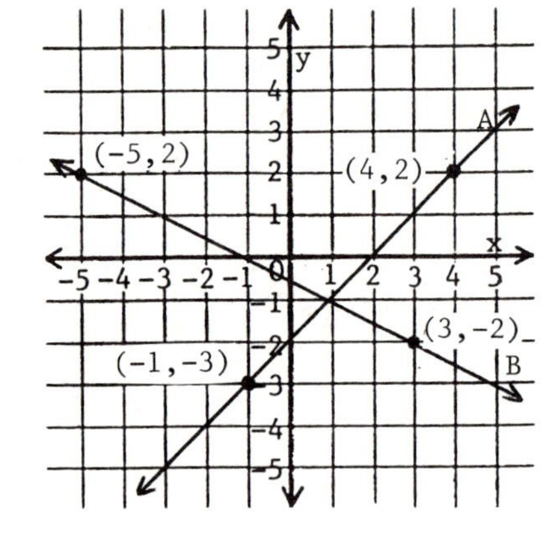

20. Line C

21. Line D

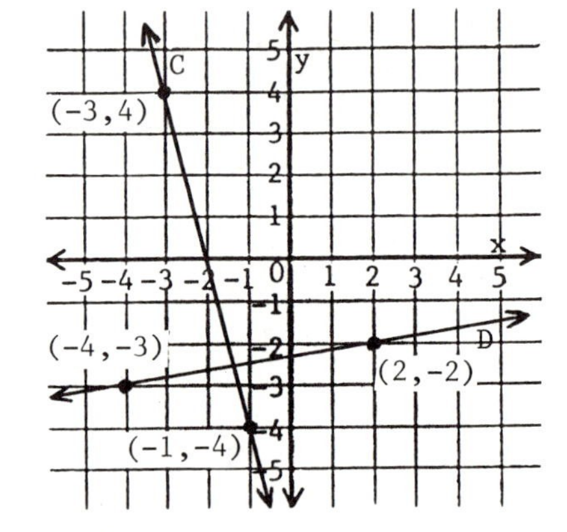

Five lines have these slopes: Line A: 4 Line B: -1 Line C: -5 Line D: $\frac{1}{2}$ Line E: -3

22. Which lines rise from left to right?

23. Which line has the greatest rise?

24. Which lines fall from left to right?

25. Which line has the greatest fall?

26. Which line is steepest?

What is the slope of: 27. Any horizontal line? 28. Any vertical line?

Find the slope of the line through each pair of points.

29. (2,3) and (-1,-3)

30. (6,8) and (10,2)

31. (-4,4) and (0,8)

32. (-4,2) and (2,-4)

33. (-3,-4) and (5,-2)

34. (0,6) and (10,0)

35. (-20,10) and (30,35)

36. (0,0) and (-6,-12)

37. (0,0) and (2,-8)

Assignment 27

For each equation, identify the slope and the coordinates of the y-intercept.

1. $y = 5x + 2$

2. $y = x - 3$

3. $y = -\frac{2}{5}x + \frac{3}{2}$

4. $y = \frac{1}{2}x$

Write the slope-intercept form of the linear equation whose:

5. Slope is -1 and y-intercept is (0, 2)

6. Slope is $\frac{1}{2}$ and y-intercept is (0, -5)

7. Slope is $\frac{1}{3}$ and y-intercept is (0, -1)

8. Slope is $\frac{8}{3}$ and y-intercept is (0, 0)

Put each equation in slope-intercept form.

9. $y = 6 - 4x$

10. $y - 3x = 2$

11. $x - y = 5$

12. $4y = x - 8$

13. $3x + 6y = 10$

14. $x - 5y = 0$

Find the slope-intercept form of the equation of the line whose:

15. Slope is 3 and passes through (2, 0)

16. Slope is -1 and passes through (3, -1)

17. Slope is $\frac{1}{3}$ and passes through (-6, 3)

18. Slope is -4 and passes through (1, 2)

19. Slope is $-\frac{1}{2}$ and passes through (-4, 5)

20. Slope is 5 and passes through (0, 0)

The graph at the right shows how much current we get in a particular electric circuit when various voltages are applied with a constant resistance.

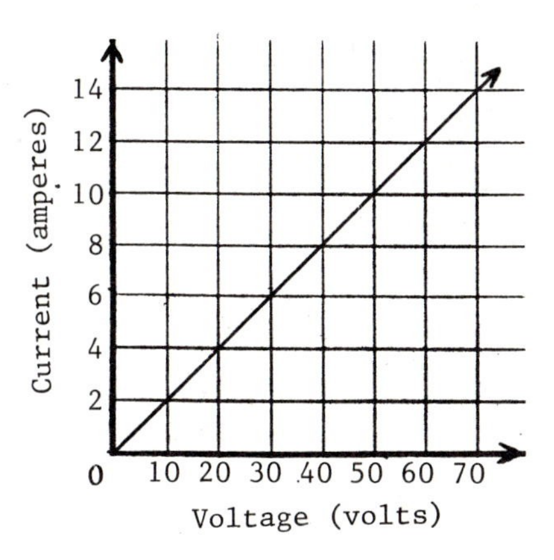

How much current do we get if we apply the following voltages?

21. 20 volts __________ amperes

22. 55 volts __________ amperes

How much voltage must we apply to get the following currents?

23. 12 amperes __________ volts

24. 7 amperes __________ volts

25. If $\underline{y}$ is directly proportional to $\underline{x}$, and if $y = 6$ when $x = 2$, find:

a) The constant of variation $\underline{k}$.

b) The equation relating $\underline{y}$ and $\underline{x}$.

26. Which of the following are examples of direct variation?

a) $y = 4x$ b) $y = 7$ c) $y = 2x - 1$ d) $p = 3t + 2$ e) $b = 50a$

27. If G is directly proportional to R, and if $G = 40$ when $R = 8$, find G when $R = 20$.

28. If $\underline{w}$ is directly proportional to $\underline{s}$, and if $w = 16$ when $s = 24$, find $\underline{w}$ when $s = 60$.

29. The distance $\underline{d}$ traveled by a car is directly proportional to time $\underline{t}$. If $d = 150$ kilometers when $t = 2$ hours, find $\underline{d}$ when $t = 5$ hours.

30. The amount of stretch (s) in a spring is directly proportional to the force (F) applied to it. If $s = 10$ centimeters when $F = 40$ kilograms, find $\underline{s}$ when $F = 140$ kilograms.

9 Systems of Equations

In this chapter, we will define a system of two equations and solve systems of that type by the graphing method, the addition method, and the substitution method. Various types of word problems involving systems of equations are included.

9-1 THE GRAPHING METHOD

In this section, we will define a system of two equations and discuss the graphing method for solving systems of equations.

---

1. It is sometimes easier to solve a problem if we use two variables and two equations. An example is shown.

   The sum of two numbers is 21. The difference between the larger number and the smaller number is 5. Find the two numbers.

   Using $\underline{x}$ for the larger number and $\underline{y}$ for the smaller number, we can set up two different equations. That is:

   The sum of two numbers is 21.

   $$x + y = 21$$

Continued on following page.

1. Continued

The difference between the larger and smaller is 5.

$$x - y = 5$$

Therefore, the problem has been translated to the following <u>pair</u> of equations which is called a <u>system of equations</u>.

$$x + y = 21$$
$$x - y = 5$$

The solution of a system of equations is an ordered pair that satisfies <u>both</u> equations. For example, for the system above:

(15, 6) <u>is not a solution</u> because it only satisfies the top equation.

(10, 5) <u>is not a solution because</u> it only satisfies the bottom equation.

(13, 8) <u>is the solution</u> because it satisfies <u>both</u> equations.

Show that (13, 8) satisfies both equations in the system.

$x + y = 21$ $\qquad\qquad$ $x - y = 5$

$$x + y = 21$$
$$13 + 8 = 21$$
$$21 = 21$$

$$x - y = 5$$
$$13 - 8 = 5$$
$$5 = 5$$

---

2. Remember that an ordered pair is a solution of a system of equations <u>only if it satisfies both equations</u>.

Which ordered pair below is the solution of the system at the right? ________

$$y = x + 3$$
$$x + y = 5$$

(2, 5) $\qquad$ (3, 2) $\qquad$ (1, 4) $\qquad$ (0, 3)

(1, 4)

---

3. We graphed both equations in the system below at the right.

$$x - y = 1$$
$$x + y = 5$$

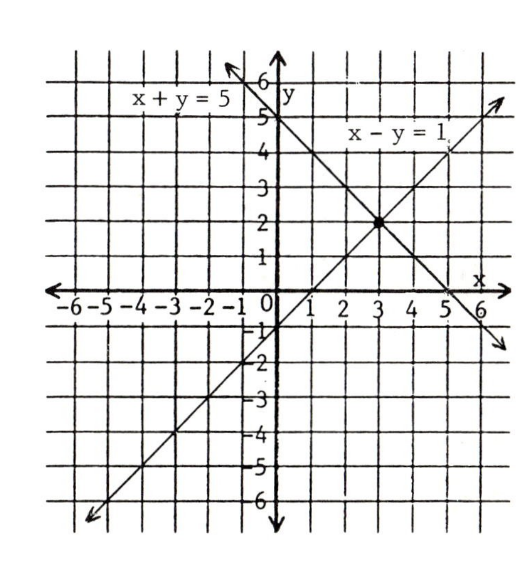

The point where the two graphed lines cross is called the <u>point of intersection</u>. Since that point lies on both lines, its coordinates satisfy both equations. Therefore, <u>its coordinates are the solution of the system</u>. Let's confirm that fact by answering the questions below.

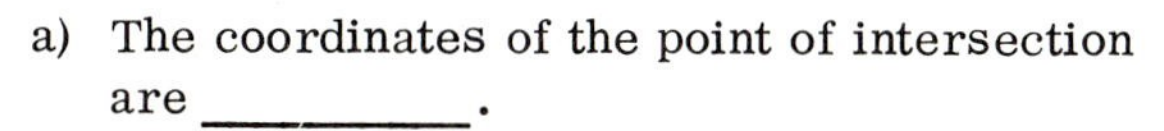

a) The coordinates of the point of intersection are ________.

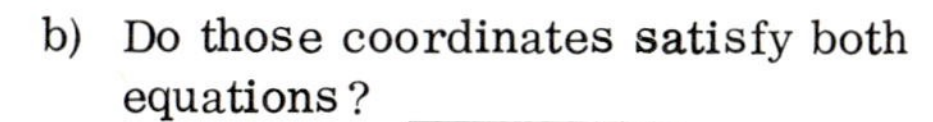

b) Do those coordinates satisfy both equations? ________

c) Therefore, the solution of the system is ________.

4. When the graphs of both equations are straight lines and the lines intersect, there is only one point of intersection. Therefore, there is only one solution of the system of equations.

We graphed the system of equations below at the right.

$$3x + y = 0$$
$$x - 2y = 7$$

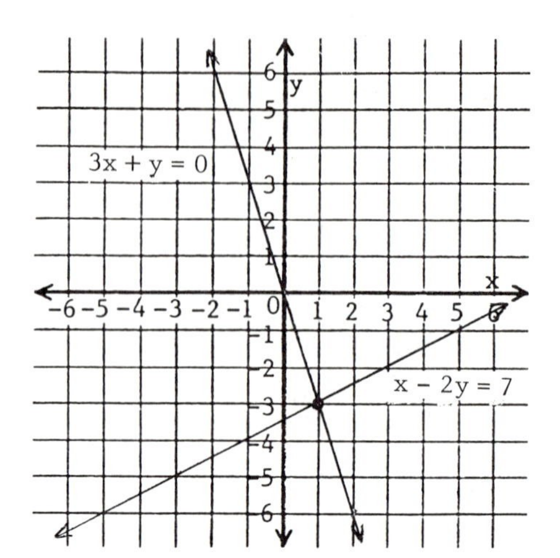

a) Using the coordinates of the point of intersection, write the solution of the system. ________

b) Show that the solution satisfies both equations.

$3x + y = 0$ $\qquad\qquad$ $x - 2y = 7$

a) (3, 2)

b) Yes

c) (3, 2)

---

5. Let's use the graphing method to solve the system below.

$$y = 2x$$
$$y - x = 2$$

a) Using the tables provided, find some solutions and graph the equations.

$y = 2x$ $\qquad$ $y - x = 2$

x	y

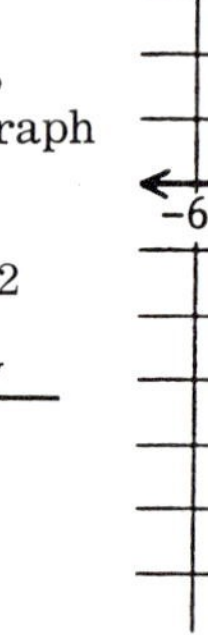

x	y

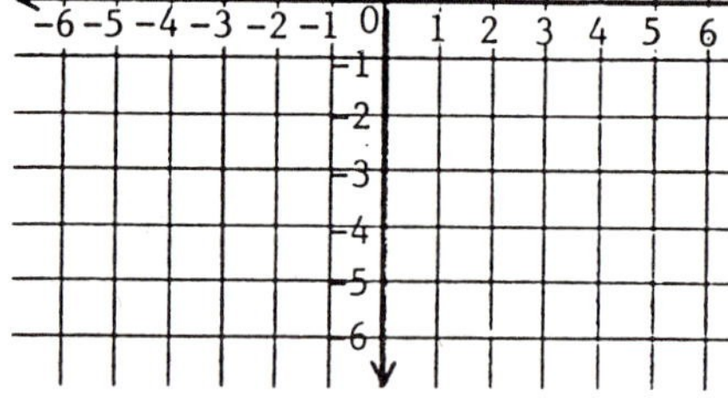

b) The solution of the system is: ________

c) Check the solution in both equations.

$y = 2x$ $\qquad\qquad$ $y - x = 2$

a) (1, −3)

b)
$$3x + y = 0$$
$$3(1) + (-3) = 0$$
$$3 + (-3) = 0$$
$$0 = 0$$

$$x - 2y = 7$$
$$1 - 2(-3) = 7$$
$$1 + 6 = 7$$
$$7 = 7$$

6. Graph the following system.

$$2x - y = 3$$
$$x + y = 0$$

$2x - y = 3$ $\qquad\qquad$ $x + y = 0$

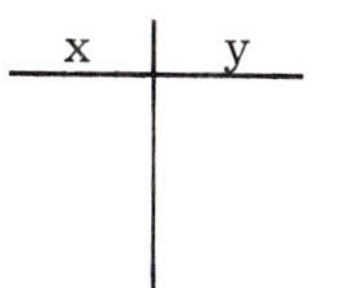

x	y

x	y

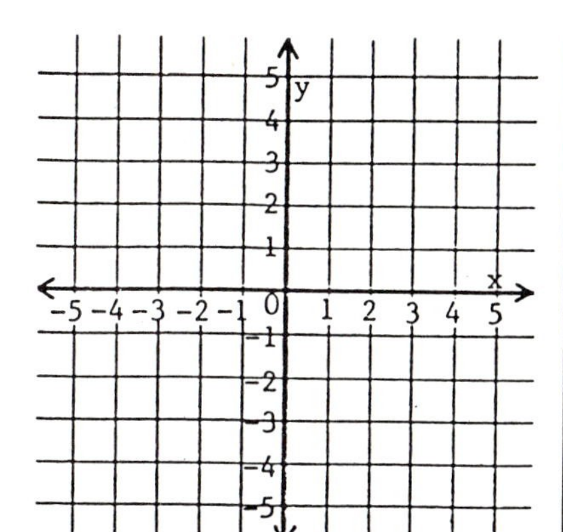

The solution of the system is: ________

a)

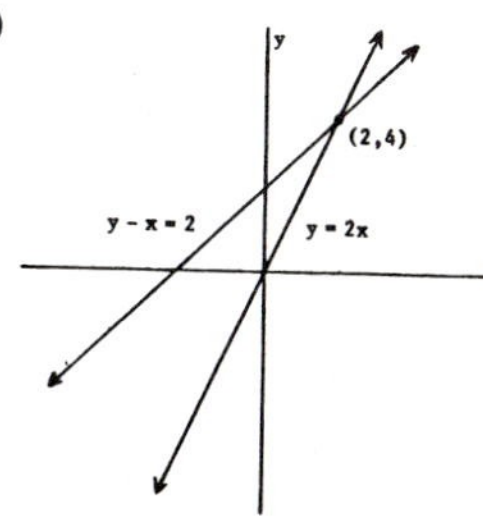

b) (2, 4)

c) $y = 2x$ $\qquad$ $y - x = 2$
$4 = 2(2)$ $\qquad$ $4 - 2 = 2$
$4 = 4$ $\qquad$ $2 = 2$

---

7. A system has one solution if its graph is two straight lines that intersect at one point. A system can have no solution or an infinite number of solutions. An example of each type is discussed below.

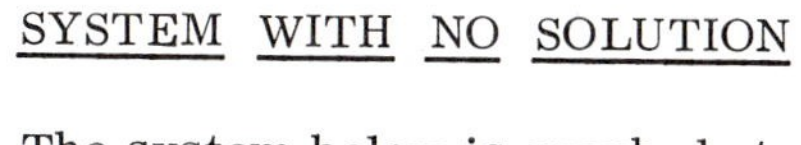

SYSTEM WITH NO SOLUTION

The system below is graphed at the right. Notice that the straight lines are parallel. They have the same slope, $m = 3$.

$$y = 3x + 2$$
$$y = 3x - 1$$

Since the parallel lines do not intersect, the equations have no common solution. Therefore, the system has no solution.

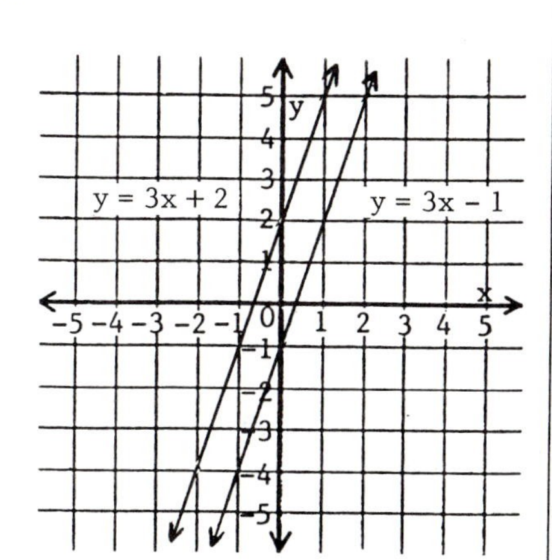

SYSTEM WITH AN INFINITE NUMBER OF SOLUTIONS

In the system below, the bottom equation can be obtained by multiplying the top equation by 2. The system is graphed at the right. Notice that each equation has the same graph.

$$x + y = 2$$
$$2x + 2y = 4$$

Since the lines are identical, the equations have an infinite number of common solutions. Therefore, the system has an infinite number of solutions.

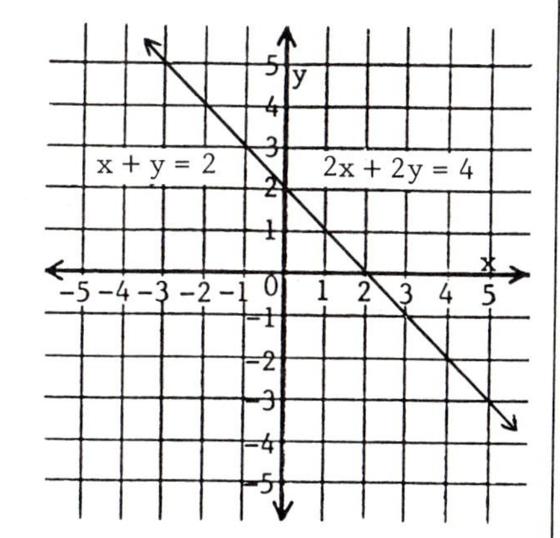

(1, -1)

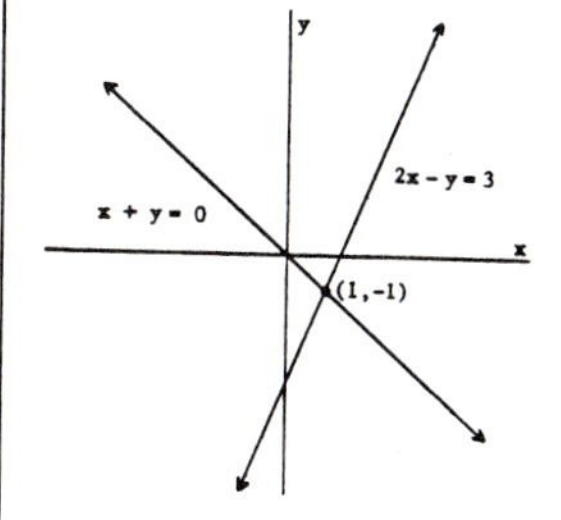

9-2 THE ADDITION METHOD

Because the graphing method for solving systems of equations is time-consuming and only approximate when the solutions are not whole numbers, an algebraic method is ordinarily used. In this section, we will discuss an algebraic method called the addition method for solving systems.

---

8. The addition method for solving systems is based on the following property of addition.

$$\begin{aligned} \text{If } A &= B \\ \text{and } C &= D, \\ \text{then } A + C &= B + D. \end{aligned}$$

The property above can be used to add the two equations in a system. For example, we added the two equations below and got $4x + 3y = 11$. Add the other two equations.

$$\begin{array}{rcl} x + 5y &=& 7 \\ 3x - 2y &=& 4 \\ \hline 4x + 3y &=& 11 \end{array} \qquad \begin{array}{rcl} x - 3y &=& 4 \\ 5x + y &=& 0 \\ \hline \end{array}$$

---

9. In the addition method for solving a system, we add the equations in order to eliminate a variable. For example, we eliminated y by adding the equations below. Add the other two equations.

$$\begin{array}{rcl} x + 2y &=& 3 \\ x - 2y &=& 5 \\ \hline 2x &=& 8 \end{array} \qquad \begin{array}{rcl} a + 4b &=& 0 \\ 2a - 4b &=& 7 \\ \hline \end{array}$$

Answer (8): $6x - 2y = 4$

---

10. Let's use the addition method to solve the system at the right. The two steps are described.

$$\begin{aligned} x + y &= 5 \\ x - y &= 1 \end{aligned}$$

1. Finding the value of x. By adding the two equations, we can eliminate y and solve for x.

$$\begin{array}{rcl} x + y &=& 5 \\ x - y &=& 1 \\ \hline 2x &=& 6 \\ x &=& 3 \end{array}$$

2. Finding the value of y. We can now find the corresponding value of y by substituting 3 for x in either of the original equations.

$$\begin{aligned} x + y &= 5 \\ 3 + y &= 5 \\ y &= 2 \end{aligned} \qquad \begin{aligned} x - y &= 1 \\ 3 - y &= 1 \\ y &= 2 \end{aligned}$$

Answer (9): $3a = 7$

Continued on following page.

10. Continued

The obtained solution is (3,2) or $x = 3$, $y = 2$. Check that solution in each original equation below.

a) $x + y = 5$

b) $x - y = 1$

---

11. Let's use the addition method to solve the system at the right.

$$s + 2t = 5$$
$$3s - 2t = 7$$

1. Finding the value of s.

By adding the equations, we can eliminate t and solve for s.

$$\begin{aligned} s + 2t &= 5 \\ 3s - 2t &= 7 \\ \hline 4s &= 12 \\ s &= 3 \end{aligned}$$

2. Finding the value of t.

To find the corresponding value of t, we substituted 3 for s in the top equation.

$$\begin{aligned} s + 2t &= 5 \\ 3 + 2t &= 5 \\ 2t &= 2 \\ t &= 1 \end{aligned}$$

Note: We could also have substituted 3 for s in the bottom equation.

The obtained solution is $s = 3$, $t = 1$. Show that the solution satisfies each of the original equations.

a) $s + 2t = 5$

b) $3s - 2t = 7$

Answers to 10:

a) $3 + 2 = 5$
$5 = 5$

b) $3 - 2 = 1$
$1 = 1$

---

12. Use the addition method to solve each system below.

a) $2x + y = 10$
$2x - y = 6$

b) $p - 4q = 16$
$p + 4q = 0$

Answers to 11:

a) $3 + 2(1) = 5$
$3 + 2 = 5$
$5 = 5$

b) $3(3) - 2(1) = 7$
$9 - 2 = 7$
$7 = 7$

Answers to 12:

a) $x = 4$, $y = 2$ b) $p = 8$, $q = -2$

9-3 MULTIPLICATION IN THE ADDITION METHOD

When using the addition method to solve a system, we sometimes have to multiply one or both equations by a number before adding the equations. We still discuss solutions of that type in this section.

---

13. When using the addition method to solve a system, we sometimes have to multiply one or both equations by a number. When doing so, we must multiply both sides by the number. Two examples are shown.

We multiplied $x - 2y = 4$ by 3 below.

$$3(x - 2y) = 3(4)$$
$$3x - 6y = 12$$

We multiplied $p - 2q = 3$ by -1 below.

$$-1(p - 2q) = -1(3)$$
$$-p + 2q = -3$$

Following the examples, do these:

a) Multiply this equation by 5.

$$4x - y = 1$$

b) Multiply this equation by -2.

$$2a + 3b = 4$$

---

a) $20x - 5y = 5$

b) $-4a - 6b = -8$

14. If we add the equations in the system at the right, we get $3x + y = 11$. Neither variable is eliminated.

$$x + 2y = 7$$
$$2x - y = 4$$

However, if the -y in the bottom equation were a -2y, we could eliminate y by adding. To get a -2y at the right, we multiply both sides of the bottom equation by 2. Then we add and solve for x.

$$x + 2y = 7$$
$$4x - 2y = 8 \quad \text{(Multiplied by 2)}$$
$$5x = 15$$
$$x = 3$$

Substituting 3 for x in one of the original equations, we can find the corresponding value of y.

$$x + 2y = 7$$
$$3 + 2y = 7$$
$$2y = 4$$
$$y = 2$$

We get (3,2) as a solution. Check it in each original equation below.

$$x + 2y = 7 \qquad\qquad 2x - y = 4$$

15. We could eliminate y in the system at the right if the 3y were -3y in either equation.

$$\begin{aligned} 2x + 3y &= 8 \\ x + 3y &= 7 \end{aligned}$$

To get a -3y in the bottom equation, we multiply it by -1 at the right and then solve for x.

$$\begin{aligned} 2x + 3y &= 8 \\ -x - 3y &= -7 \quad \text{(Multiplied by -1)} \\ \hline x &= 1 \end{aligned}$$

Substituting "1" for x in the original bottom equation, we find the corresponding value of y at the right.

$$\begin{aligned} x + 3y &= 7 \\ 1 + 3y &= 7 \\ 3y &= 6 \\ y &= 2 \end{aligned}$$

Check (1, 2) in each original equation below.

a) $2x + 3y = 8$

b) $x + 3y = 7$

Answers:

$$\begin{aligned} 3 + 2(2) &= 7 \\ 3 + 4 &= 7 \\ 7 &= 7 \end{aligned}$$

$$\begin{aligned} 2(3) - 2 &= 4 \\ 6 - 2 &= 4 \\ 4 &= 4 \end{aligned}$$

---

16. Let's solve the system at the right by eliminating x. To do so, we must multiply the bottom equation by -2.

$$\begin{aligned} 4x + y &= 1 \\ 2x + 3y &= 13 \end{aligned}$$

a) Write the new system obtained if the bottom equation is multiplied by -2.

b) Solve the system.

Answers:

a)
$$\begin{aligned} 2(1) + 3(2) &= 8 \\ 2 + 6 &= 8 \\ 8 &= 8 \end{aligned}$$

b)
$$\begin{aligned} 1 + 3(2) &= 7 \\ 1 + 6 &= 7 \\ 7 &= 7 \end{aligned}$$

---

17. Let's solve the system at the right by eliminating p. To do so, we can multiply either equation by -1. Let's multiply the bottom equation by -1.

$$\begin{aligned} p + q &= 7 \\ p + 3q &= 15 \end{aligned}$$

a) Write the new system obtained if the bottom equation is multiplied by -1.

b) Solve the system.

Answers:

a)
$$\begin{aligned} 4x + y &= 1 \\ -4x - 6y &= -26 \end{aligned}$$

b) (-1, 5)

18. We multiply an equation by a number to get a pair of opposites so that a variable is eliminated. When doing so, it is important that you know what number to use. For example:

a) To eliminate t in the system at the right, we multiply the bottom equation by ______.

$$\begin{aligned} s + 5t &= 9 \\ 2s - t &= 0 \end{aligned}$$

b) To eliminate y in the system at the right, we multiply the top equation by ______.

$$\begin{aligned} x + 2y &= 0 \\ 3x - 4y &= 5 \end{aligned}$$

c) To eliminate a in the system at the right, we multiply the bottom equation by ______.

$$\begin{aligned} 3a + 2b &= 5 \\ a + 3b &= 9 \end{aligned}$$

d) To eliminate p in the system at the right, we can multiply either the top equation or the bottom equation by ______.

$$\begin{aligned} 4p + 3q &= 7 \\ 4p + q &= 1 \end{aligned}$$

Answers:

a) $$\begin{aligned} p + q &= 7 \\ -p - 3q &= -15 \end{aligned}$$

b) $p = 3$, $q = 4$

---

19. By eliminating either letter, solve each system below.

a) $$\begin{aligned} x + 4y &= 10 \\ 3x - 2y &= 16 \end{aligned}$$

b) $$\begin{aligned} 2k + m &= 0 \\ 2k + 3m &= 8 \end{aligned}$$

Answers:

a) 5

b) 2

c) -3

d) -1

---

20. Sometimes we have to multiply both equations by a number before adding the equations. An example is discussed below.

We could eliminate y in the system at the right if the 3y were a 15y and the 5y were a -15y.

$$\begin{aligned} 5x + 3y &= 2 \\ 3x + 5y &= -2 \end{aligned}$$

To get a 15y in the top equation, we multiply it by 5 at the right. To get a -15y in the bottom equation, we multiply it by -3 at the right. Then we add to eliminate y and solve for x.

$$\begin{aligned} 25x + 15y &= 10 \quad \text{(Multiplied by 5)} \\ -9x - 15y &= 6 \quad \text{(Multiplied by -3)} \\ \hline 16x &= 16 \\ x &= 1 \end{aligned}$$

Substituting "1" for x in the original top equation, we find the corresponding value of y at the right.

$$\begin{aligned} 5x + 3y &= 2 \\ 5(1) + 3y &= 2 \\ 5 + 3y &= 2 \\ 3y &= -3 \\ y &= -1 \end{aligned}$$

Continued on following page.

Answers:

a) $x = 6$, $y = 1$

b) $k = -2$, $m = 4$

20. Continued

Check (1,-1) in each original equation.

$5x + 3y = 2$ $\qquad$ $3x + 5y = -2$

---

21. Let's solve the system at the right by eliminating t. We can do so if we get a -12t in the top equation and a 12t in the bottom equation.

$$\begin{aligned} 2v - 3t &= -1 \\ 3v + 4t &= 24 \end{aligned}$$

a) Write the new system obtained if we multiply the top equation by 4 and the bottom equation by 3.

b) Solve the system.

Answer to 20:

$$\begin{aligned} 5(1) + 3(-1) &= 2 \\ 5 + (-3) &= 2 \\ 2 &= 2 \end{aligned}$$

$$\begin{aligned} 3(1) + 5(-1) &= -2 \\ 3 + (-5) &= -2 \\ -2 &= -2 \end{aligned}$$

---

22. Let's solve the system at the right by eliminating a. We can do so if we get a 10a in the top equation and a -10a in the bottom equation.

$$\begin{aligned} 5a - 2b &= 0 \\ 2a - 3b &= -11 \end{aligned}$$

a) Write the new system obtained if the top equation is multiplied by 2 and the bottom equation is multiplied by -5.

b) Solve the system

Answer to 21:

a) $$\begin{aligned} 8v - 12t &= -4 \\ 9v + 12t &= 72 \end{aligned}$$

b) $v = 4$, $t = 3$

23. Sometimes we have to multiply both equations by a number to get a pair of opposites. When doing so, it is important that you know what numbers to use. For example:

To eliminate y at the right, we can multiply the top equation by 7 and the bottom equation by 2.

$$\begin{aligned} 3x + 2y &= 7 \\ 2x - 7y &= 0 \end{aligned}$$

a) To eliminate q at the right, we can multiply the top equation by ______ and the bottom equation by ______.

$$\begin{aligned} 5p - 4q &= 9 \\ 2p + 6q &= 1 \end{aligned}$$

b) To eliminate v at the right, we can multiply the top equation by ______ and the bottom equation by ______.

$$\begin{aligned} 2v + 5t &= 10 \\ 5v + 2t &= 15 \end{aligned}$$

Answers:

a) $10a - 4b = 0$
$-10a + 15b = 55$

b) $a = 2$, $b = 5$

---

24. By eliminating either letter, solve each system below.

a) $2p + 5q = 9$
$3p - 2q = 4$

b) $3x + 4y = 6$
$2x + 3y = 5$

Answers:

a) top by 3, bottom by 2

b) top by 5, bottom by -2
or
top by -5, bottom by 2

---

25. Earlier we saw that a system of equations could have no solution or an infinite number of solutions. Let's see what happens with the addition method for these two special cases.

SYSTEM WITH NO SOLUTION

If we graphed the equations at the right, the lines would be parallel. Therefore, the system has no solution.

$$\begin{aligned} y - 2x &= 5 \\ y - 2x &= 3 \end{aligned}$$

We tried the addition method at the right by multiplying the bottom equation by -1. We got the false equation $0 = 2$.

$$\begin{aligned} y - 2x &= 5 \\ \underline{-y + 2x} &= \underline{-3} \\ 0 &= 2 \end{aligned}$$

If you try the addition method and get a false equation like $0 = 2$, the system has no solution.

Continued on following page.

Answers:

a) $p = 2$, $q = 1$

b) $x = -2$, $y = 3$

25. Continued

SYSTEM WITH AN INFINITE NUMBER OF SOLUTIONS

If we graphed the system at the right, the lines would be identical. Therefore, the system has an infinite number of solutions.

$$x - 2y = 3$$
$$2x - 4y = 6$$

We tried the addition method at the right by multiplying the top equation by -2. We got the equation $0 = 0$.

$$-2x + 4y = -6$$
$$\underline{2x - 4y = 6}$$
$$0 = 0$$

If you try the addition method and get the equation $0 = 0$, the system has an infinite number of solutions.

SELF-TEST 28 (pages 388-399)

1. Use the graphing method to solve this system of equations.

$$x - 2y = 6$$
$$2x + y = 2$$

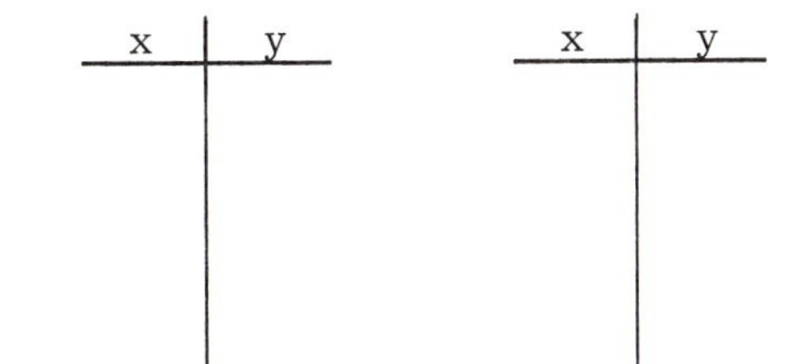

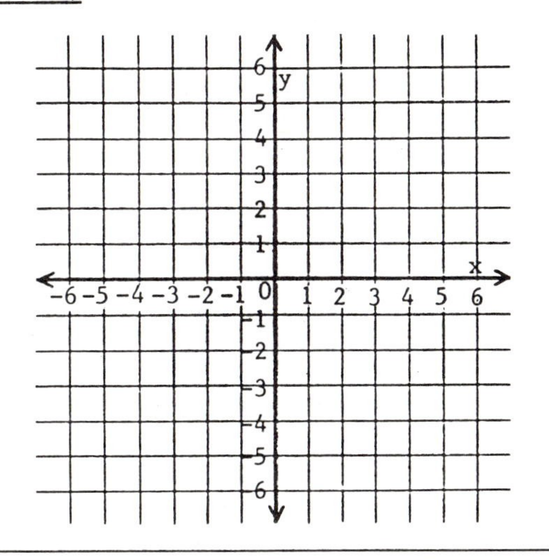

Use the addition method to solve each system of equations.

2. $3x + y = 13$
   $2x - 3y = 5$

3. $2t - 3w = 7$
   $5t + 2w = 8$

ANSWERS: 1. $x = 2$, $y = -2$

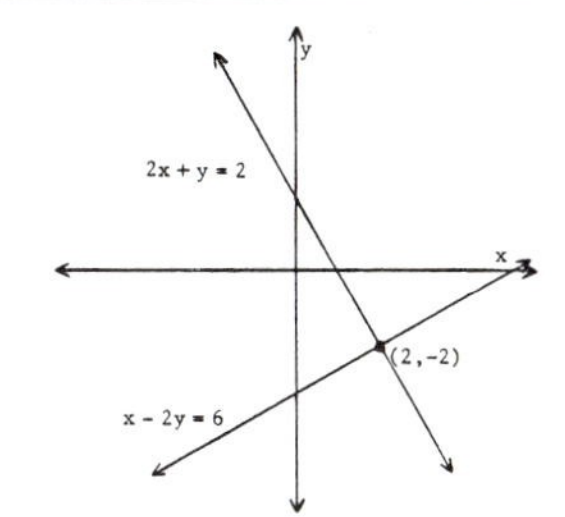

2. $x = 4$, $y = 1$

3. $t = 2$, $w = -1$

9-4 CONVERTING TO STANDARD FORM

To solve a system by the addition method, both equations must be in standard form. Sometimes we have to convert an equation to standard form before using the addition method. We will discuss solutions of that type in this section.

---

26. In the two systems below, the equations are in standard form $Ax + By = C$. That is, the variables are on the left side and the number is on the right side.

$$\begin{aligned} x + y &= 5 \\ x - y &= 0 \end{aligned} \qquad \begin{aligned} 3a + b &= 7 \\ a + 5b &= 9 \end{aligned}$$

An equation in a system is not in standard form if it has a variable on the right. For example, in the systems below, $y = x - 8$ and $s = 2t$ are not in standard form.

$$\begin{aligned} y &= x - 8 \\ 5x + y &= 4 \end{aligned} \qquad \begin{aligned} 3s - t &= 4 \\ s &= 2t \end{aligned}$$

In each system below, circle the equation that is not in standard form.

a) $x + y = 10$
   $y = 1 - 3x$

b) $3p + 2q = 5$
   $4q = 6p - 1$

---

a) $y = 1 - 3x$

b) $4q = 6p - 1$

27. When an equation in a system is not in standard form, we must convert it to standard form before using the addition method. Two examples are shown. Notice how we lined up the variables on the left side.

Original System

$$\begin{aligned} x + y &= 10 \\ y &= 2x + 3 \end{aligned}$$

Converting 2nd Equation

$$\begin{aligned} y &= 2x + 3 \\ (-2x) + y &= 2x + (-2x) + 3 \\ -2x + y &= 3 \end{aligned}$$

Standard System

$$\begin{aligned} x + y &= 10 \\ -2x + y &= 3 \end{aligned}$$

Original System

$$\begin{aligned} a &= b + 7 \\ 2a + b &= 5 \end{aligned}$$

Converting 1st Equation

$$\begin{aligned} a &= b + 7 \\ a + (-b) &= b + (-b) + 7 \\ a - b &= 7 \end{aligned}$$

Standard System

$$\begin{aligned} a - b &= 7 \\ 2a + b &= 5 \end{aligned}$$

Convert each system below to standard form.

a) $2p + q = 9$
   $q = 4p + 5$

b) $t = 2d - 1$
   $t + d = 10$

28. Two more examples of conversions to standard form are shown below. Notice in each that we lined up the variables on the left side.

Original System	Original System
$p - q = 7$	$x = 5 - y$
$q = 2p$	$2x + y = 3$

Converting 2nd Equation	Converting 1st Equation
$q = 2p$	$x = 5 - y$
$(-2p) + q = 2p + (-2p)$	$x + y = 5 - y + y$
$-2p + q = 0$	$x + y = 5$

Standard System	Standard System
$p - q = 7$	$x + y = 5$
$-2p + q = 0$	$2x + y = 3$

Convert each system below to standard form.

a) $b = a$
$2a - 3b = 8$

b) $t + 6m = 1$
$t = 4 - 3m$

Answers:
a) $2p + q = 9$
$-4p + q = 5$
b) $t - 2d = -1$
$t + d = 10$

---

29. In the system at the right, $y = 3x - 8$ is not in standard form.

$x - y = 2$
$y = 3x - 8$

To solve the system, we begin by converting $y = 3x - 8$ to standard form.

$-3x + y = 3x + (-3x) - 8$
$-3x + y = -8$

Then we can use the addition method to solve the standard system at the right.

$x - y = 2$
$-3x + y = -8$

a) Solve the standard system.

b) Check your solution in each original equation.

$x - y = 2$ $\qquad$ $y = 3x - 8$

Answers:
a) $-a + b = 0$
$2a - 3b = 8$
b) $t + 6m = 1$
$t + 3m = 4$

30. Following the steps in the last frame, solve each system.

a) $2p + q = 30$
$p = 2q$

b) $2a = 7 - 3b$
$3a - 2b = 4$

a) (3,1)

b) $3 - 1 = 2$
$2 = 2$

$1 = 3(3) - 8$
$1 = 9 - 8$
$1 = 1$

a) $p = 12,\ q = 6$ b) $a = 2,\ b = 1$

9-5 DECIMAL COEFFICIENTS

When solving word problems, we sometimes have to solve systems with decimal coefficients. With systems of that type, it is easier to clear the decimals before using the addition method. We will discuss systems of that type in this section.

31. To clear the decimal coefficients in the equation below, we multiplied both sides by 10. Notice how we multiplied by the distributive principle on the left side.

$$.3x + .9y = 4$$
$$10(.3x + .9y) = 10(4)$$
$$10(.3x) + 10(.9y) = 40$$
$$3x + 9y = 40$$

Multiply both sides by 10 to clear the decimals in these.

a) $.5x + .1y = 70$ b) $.8b - .7d = 1.5$

32. To clear the decimal coefficients below, we multiplied both sides by 100.

$$.45x + .8y = 20$$
$$100(.45x + .8y) = 100(20)$$
$$100(.45x) + 100(.8y) = 2{,}000$$
$$45x + 80y = 2{,}000$$

Multiply both sides by 100 to clear the decimals in these.

a) $.2x - .65y = 7.5$ b) $.95s + .35t = 45$

a) $5x + y = 700$

b) $8b - 7d = 15$

33. When multiplying by a power of ten to clear the decimals, the power of ten is determined by the number with the most decimal places. If its last digit is:

in the tenths place, we multiply by 10.

in the hundredths place, we multiply by 100.

Using the facts above, clear the decimals in these.

a) $.7x + .15y = 6.5$ b) $.4a - .1b = 200$

a) $20x - 65y = 750$

b) $95s + 35t = 4,500$

---

34. Since it is easier to solve a system without decimal coefficients, we usually clear any decimal coefficients before using the addition method. An example is shown.

Original System

$$.2x + .5y = 1.6$$
$$3x + 2y = 12$$

Equivalent System

$$2x + 5y = 16$$
$$3x + 2y = 12$$

Write the equivalent system obtained after clearing the decimals in these.

a) $.1x - .25y = 40$
$5x + 4y = 1,200$

b) $1.4p + .9q = 65$
$.75p + .2q = 1.5$

a) $70x + 15y = 650$

b) $4a - b = 2,000$

---

35. Let's solve the system at the right.

$$x + y = 100$$
$$.3x + .7y = 46$$

a) Clear the decimals and write the equivalent system.

b) Solve the equivalent system.

c) Show that the solution satisfies each original equation.

$x + y = 100$ $.3x + .7y = 46$

a) $10x - 25y = 4,000$
$5x + 4y = 1,200$

b) $14p + 9q = 650$
$75p + 20q = 150$

Answer to Frame 35:

a) $x + y = 100$
$3x + 7y = 460$

b) (60, 40)

c) $60 + 40 = 100$
$100 = 100$

$.3(60) + .7(40) = 46$
$18 + 28 = 46$
$46 = 46$

9-6 THE SUBSTITUTION METHOD

The substitution method is a second algebraic method for solving systems of equations. We will discuss the substitution method in this section.

---

36. Instead of adding equations to eliminate a variable, we can substitute to eliminate a variable. An example is discussed below.

In the system at the right, y is solved for in the top equation.

$$y = 2x$$
$$x + 4y = 45$$

Since $y = 2x$, we can eliminate y by substituting 2x for y in the bottom equation. We get:

$$x + 4y = 45$$
$$x + 4(2x) = 45$$
$$x + 8x = 45$$
$$9x = 45$$
$$x = 5$$

Substituting 5 for x in the top equation, we can solve for y. We get:

$$y = 2x$$
$$y = 2(5)$$
$$y = 10$$

Check (5, 10) in each original equation below.

$y = 2x$ $\qquad\qquad$ $x + 4y = 45$

$10 = 2(5)$
$10 = 10$

$5 + 4(10) = 45$
$5 + 40 = 45$
$45 = 45$

37. Another example of the substitution method is given below.

In the system at the right, b is solved for in the bottom equation.

$$3a - 2b = 4$$
$$b = a + 2$$

We can eliminate b by substituting $a + 2$ for b in the top equation. We substituted, simplified, and solved for a at the right.

$$3a - 2b = 4$$
$$3a - 2(a + 2) = 4$$
$$3a - (2a + 4) = 4$$
$$3a - 2a - 4 = 4$$
$$a = 8$$

Substituting 8 for a in the bottom equation, we can solve for b. We get:

$$b = a + 2$$
$$b = 8 + 2$$
$$b = 10$$

Check $(a = 8, b = 10)$ in each original equation below.

$3a - 2b = 4$ $\qquad\qquad$ $b = a + 2$

38. Use the substitution method to solve each system.

a)
$$\begin{aligned} y &= 3x \\ 5x - y &= 4 \end{aligned}$$

b)
$$\begin{aligned} 4x + y &= 7 \\ y &= x - 8 \end{aligned}$$

$$\begin{aligned} 3(8) - 2(10) &= 4 \\ 24 - 20 &= 4 \\ 4 &= 4 \end{aligned}$$

$$\begin{aligned} 10 &= 8 + 2 \\ 10 &= 10 \end{aligned}$$

---

39. Neither variable is solved for in the system below.

$$\begin{aligned} x - 3y &= 0 \\ 2x + y &= 4 \end{aligned}$$

To use the substitution method, we must solve for one variable in one equation.

If we solve for $\underline{x}$ in the top equation or $\underline{y}$ in the bottom equation, we get non-fractional solutions.

$$\begin{aligned} x - 3y &= 0 \\ x &= 3y \end{aligned} \qquad \begin{aligned} 2x + y &= 4 \\ y &= 4 - 2x \end{aligned}$$

If we solve for $\underline{y}$ in the top equation or $\underline{x}$ in the bottom equation, we get fractional solutions.

$$\begin{aligned} x - 3y &= 0 \\ x &= 3y \\ y &= \frac{x}{3} \end{aligned} \qquad \begin{aligned} 2x + y &= 4 \\ 2x &= 4 - y \\ x &= \frac{4 - y}{2} \end{aligned}$$

Since it is easier to substitute non-fractional solutions, we would use either $x = 3y$ or $y = 4 - 2x$ for the substitution. Notice that we got the non-fractional solutions by solving for a variable whose coefficient is "1".

Which variable in this system would you solve for to get a non-fractional solution?

$$\begin{aligned} 4a - 5b &= 0 \\ a - 2b &= 8 \end{aligned}$$

____

a) (2,6)

b) (3,-5)

---

40. Let's use the substitution method to solve this system.

$$\begin{aligned} a + 4b &= 6 \\ 5a - 3b &= 7 \end{aligned}$$

a) Solve for $\underline{a}$ in the top equation to get a non-fractional solution.

b) Substitute that solution for $\underline{a}$ in the bottom equation and then find the value of $\underline{b}$.

c) Substitute in one of the original equations to find the value of $\underline{a}$.

d) The solution is: a = ______, b = ______

$\underline{a}$ in the bottom equation

41. Let's use the substitution method to solve this system.

$$3x - 4y = 0$$
$$5y - 2x = 7$$

1. Solve for x in the top equation. (Note: We can't get a non-fractional solution because no variable has "1" as a coefficient.)

$$3x = 4y$$
$$x = \frac{4y}{3}$$

2. Substitute that solution for x in the bottom equation and then find the value of y.

$$5y - 2\left(\frac{4y}{3}\right) = 7$$
$$5y - \frac{8y}{3} = 7$$
$$3\left(5y - \frac{8y}{3}\right) = 3(7)$$
$$3(5y) - \cancel{3}\left(\frac{8y}{\cancel{3}}\right) = 21$$
$$15y - 8y = 21$$
$$7y = 21$$
$$y = 3$$

3. Substitute 3 for y in the top equation to find the value of x.

$$3x - 4y = 0$$
$$3x - 4(3) = 0$$
$$3x = 12$$
$$x = 4$$

The solution of the system is ________.

a) $a = 6 - 4b$

b) $5(6 - 4b) - 3b = 7$
$30 - 20b - 3b = 7$
$30 - 23b = 7$
$-23b = -23$
$b = 1$

c) $a = 2$

d) $a = 2,\ b = 1$

---

42. When using the substitution method, we sometimes get a fraction as a solution for a variable. Another example is discussed.

Let's use the substitution method to solve this system.

$$2x - 3y = 9$$
$$4x + 3y = 9$$

1. Solve for x in the top equation.

$$2x = 3y + 9$$
$$x = \frac{3y + 9}{2}$$

2. Substitute that solution for x in the bottom equation and then find the value of y.

$$\overset{2}{\cancel{4}}\left(\frac{3y + 9}{\cancel{2}}\right) + 3y = 9$$
$$6y + 18 + 3y = 9$$
$$9y + 18 = 9$$
$$9y = -9$$
$$y = -1$$

3. Substitute -1 for y in the top equation and find the value of x.

$$2x - 3y = 9$$
$$2x - 3(-1) = 9$$
$$2x + 3 = 9$$
$$2x = 6$$
$$x = 3$$

The solution of the system is ________.

(4, 3)

(3, -1)

43. When deciding whether to use the addition method or the substitution method, the following suggestions can be used.

> 1. When a variable is solved for in one equation, the substitution method is probably better.
>
> 2. When neither variable is solved for but one has a coefficient of "1" (so that the solution for it is non-fractional), either method can be used.
>
> 3. Otherwise, use the addition method.
>
> 4. When in doubt, use the addition method.

Identify the method you would use to solve each of these.

a) $3x - y = 10$
$x + 3y = 10$

b) $a + b = 6$
$a = 2b$

c) $2p + 5q = 9$
$3p - 2q = 4$

---

44. To solve the system at the right, we begin by clearing the fractions in each equation. We did so below.

$$\frac{x}{2} - \frac{y}{3} = \frac{5}{6}$$

$$\frac{x}{5} - \frac{y}{4} = \frac{1}{10}$$

For the top equation, the LCD is 6. We get:

$$6\left(\frac{x}{2} - \frac{y}{3}\right) = 6\left(\frac{5}{6}\right)$$

$$\overset{3}{\cancel{6}}\left(\frac{x}{\cancel{2}}\right) - \overset{2}{\cancel{6}}\left(\frac{y}{\cancel{3}}\right) = \overset{1}{\cancel{6}}\left(\frac{5}{\cancel{6}}\right)$$

$$3x - 2y = 5$$

For the bottom equation, the LCD is 20. We get:

$$20\left(\frac{x}{5} - \frac{y}{4}\right) = 20\left(\frac{1}{10}\right)$$

$$\overset{4}{\cancel{20}}\left(\frac{x}{\cancel{5}}\right) - \overset{5}{\cancel{20}}\left(\frac{y}{\cancel{4}}\right) = \overset{2}{\cancel{20}}\left(\frac{1}{\cancel{10}}\right)$$

$$4x - 5y = 2$$

The resulting system with non-fractional equations is shown at the right.

$$3x - 2y = 5$$
$$4x - 5y = 2$$

a) Which method would you use to solve the system? ______________

b) Solve the system.

Answers to 43:

a) Either method

b) The substitution method

c) The addition method

Answers to 44:

a) Addition Method

b) (3,2)

SELF-TEST 29 (pages 400-408)

Use the substitution method to solve each system.

1. $$\begin{aligned} 3x + 4y &= 10 \\ x &= 2y \end{aligned}$$

2. $$\begin{aligned} 8t - 3p &= 1 \\ p &= 2 - 2t \end{aligned}$$

Use any method to solve each system.

3. $$\begin{aligned} \frac{r}{2} + \frac{s}{3} &= 6 \\ \frac{r}{4} - 1 &= \frac{s}{6} \end{aligned}$$

4. $$\begin{aligned} 1.5t + 2w &= 6.5 \\ t - 2.5 &= 0.5w \end{aligned}$$

ANSWERS: 1. $x = 2,\ y = 1$ 2. $p = 1,\ t = \frac{1}{2}$ 3. $r = 8,\ s = 6$ 4. $t = 3,\ w = 1$

9-7 WORD PROBLEMS

Systems of equations can be used to solve various word problems. We will discuss problems of that type in this section.

---

45. The following problem can be solved by setting up and solving a system of equations.

    The sum of two numbers is 50. The difference between the larger and smaller numbers is 14. Find the two numbers.

    Using x for the larger number and y for the smaller number, we can set up two equations.

    The sum of two numbers is 50.

    $$x + y = 50$$

    The difference between the larger and smaller numbers is 14.

    $$x - y = 14$$

    Therefore, the problem can be solved by solving the system at the right.

    $$x + y = 50$$
    $$x - y = 14$$

    Adding the two equations, we can find the value of x.

    $$2x = 64$$
    $$x = 32$$

    Substituting 32 for x in the top equation, we can find the value of y.

    $$32 + y = 50$$
    $$y = 18$$

    Therefore, the larger number is _____ and the smaller number is _____.

---

46. Here is another problem that can be solved by setting up and solving a system of equations.

    A portable radio costs twice as much as a digital clock. If the total cost of the radio and clock is $96, how much does each cost?

    The two statements in the problem lead to two equations involving the cost of the radio (r) and the cost of the clock (c). That is:

    A radio costs twice as much as a clock.

    $$r = 2c$$

    The total cost of the radio and clock is $96.

    $$r + c = \$96$$

    Therefore, the problem can be solved by solving the system at the right.

    $$r = 2c$$
    $$r + c = 96$$

    Substituting 2c for r in the bottom equation, we can find the value of c.

    $$2c + c = 96$$
    $$3c = 96$$
    $$c = 32$$

    Substituting 32 for c in the top equation, we can find the value of r.

    $$r = 2(32)$$
    $$r = 64$$

    Therefore, the radio costs ______ and the clock costs ______.

Answers (45): larger is 32

smaller is 18

47. Following the examples in the last two frames, solve these:

a) The sum of two numbers is 64. The larger is 18 more than the smaller. Find the two numbers.

b) There are 65 adults at a party. If there are 4 times as many non-smokers as smokers, how many non-smokers and smokers are there?

radio costs $64

clock costs $32

---

48. A system of equations can be used to solve the problem below.

A 10-foot board is cut into two parts. The larger part is 2 feet longer than the smaller. How long is each part?

Using $\underline{x}$ for the larger part and $\underline{y}$ for the smaller part, we can set up two equations.

The sum of the two parts is 10 feet.

$$x + y = 10$$

The larger part is 2 feet longer than the smaller part.

$$x = y + 2$$

Therefore, we can solve the problem by solving the system at the right.

$$\begin{aligned} x + y &= 10 \\ x &= y + 2 \end{aligned}$$

Substituting $y + 2$ for $\underline{x}$ in the top equation, we can find the value of $\underline{y}$.

$$\begin{aligned} (y + 2) + y &= 10 \\ 2y + 2 &= 10 \\ 2y &= 8 \\ y &= 4 \end{aligned}$$

Substituting 4 for $\underline{y}$ in the bottom equation, we can find the value of $\underline{x}$.

$$\begin{aligned} x &= y + 2 \\ x &= 4 + 2 \\ x &= 6 \end{aligned}$$

Therefore, the larger part is _____ feet and the smaller part is _____ feet.

a) larger is 41
smaller is 23

b) 52 non-smokers
13 smokers

---

larger is 6 feet

smaller is 4 feet

49. The geometric problem below can be solved by means of a system of equations.

> The perimeter of a rectangle is 180 centimeters. If its length is 20 centimeters longer than its width, what are its length and width?

Using L for length and W for width, we can set up two equations.

The perimeter of a rectangle is 180 centimeters.

$$2L + 2W = 180$$

The length is 20 centimeters longer than the width.

$$L = W + 20$$

Therefore, we can solve the problem by solving the system at the right.

$$\begin{aligned} 2L + 2W &= 180 \\ L &= W + 20 \end{aligned}$$

Putting the bottom equation in standard form, we get:

$$\begin{aligned} 2L + 2W &= 180 \\ L - W &= 20 \end{aligned}$$

Multiplying the bottom equation by 2 and adding, we can eliminate W and solve for L.

$$\begin{aligned} 2L + 2W &= 180 \\ \underline{2L - 2W} &= \underline{40} \\ 4L &= 220 \\ L &= 55 \end{aligned}$$

Substituting 55 for L in the top equation, we get:

$$\begin{aligned} 2(55) + 2W &= 180 \\ 110 + 2W &= 180 \\ 2W &= 70 \\ W &= 35 \end{aligned}$$

Therefore, its length is ______ centimeters and its width is ______ centimeters.

---

50. Following the examples in the last two frames, solve these.

a) A piece of wire 54 cm long is cut into two parts. If the larger part is 12 cm longer than the smaller part, how long is each part?

b) The perimeter of a rectangular lot is 144 meters. If the length is twice the width, find the length and width.

length is 55 cm

width is 35 cm

51. We can use a system of equations to solve this problem.

A small plane flies 120 mph with the wind and 90 mph into the wind. Find the speed of the wind and the speed of the plane in still air.

Letting x equal the speed of the plane in still air and y equal the speed of the wind, we can set up the system at the right.

$$x + y = 120$$
$$x - y = 90$$

Using the addition method, we solved the system at the right.

$$2x = 210$$
$$x = 105$$

$$x + y = 120$$
$$105 + y = 120$$
$$y = 15$$

Therefore, the speed of the plane in still air is ______ mph and the speed of the wind is ______ mph.

a) larger is 33 cm
smaller is 21 cm

b) L = 48 meters
W = 24 meters

---

52. We can also use a system of equations to solve this problem.

It takes a boat $2\frac{1}{2}$ hours to go 20 miles downstream and 5 hours to return. Find the speed of the current and the speed of the boat in still water.

Letting x be the speed of the boat in still water and y be the speed of the current, we organized the data in the chart below.

	Distance	Rate	Time
Downstream	20	x + y	$2\frac{1}{2}$
Upstream	20	x - y	5

Using the relationship $r = \frac{d}{t}$ from d = rt, we can set up the following two equations.

$$x + y = \frac{20}{2\frac{1}{2}}$$

$$x + y = \frac{20}{\frac{5}{2}}$$

$$x + y = \overset{4}{\cancel{20}}\left(\frac{2}{\cancel{5}}\right)$$

$$x + y = 8$$

$$x - y = \frac{20}{5}$$

$$x - y = 4$$

We can now set up and solve the system of equations at the right.

$$x + y = 8$$
$$x - y = 4$$

$$2x = 12$$
$$x = 6$$

$$x + y = 8$$
$$6 + y = 8$$
$$y = 2$$

Therefore, the speed of the boat in still water is ______ mph and the speed of the current is ______ mph.

speed of plane in still air = 105 mph

speed of wind = 15 mph

53. Following the examples in the last two frames, solve these.

a) If a plane can fly 575 mph with the wind and 435 mph into the wind, find the speed of the wind and the speed of the plane in still air.

b) A plane took 2 hours to fly 800 km against a headwind. The return trip with the wind took $1\frac{2}{3}$ hours. Find the speed of the plane in still air.

speed in still water = 6 mph

speed of current = 2 mph

---

a) speed in still air = 505 mph
speed of wind = 70 mph

b) speed in still air = 440 km/h

9-8 MIXTURE PROBLEMS

Systems of equations can be used to solve various mixture problems. We will discuss problems of that type in this section.

---

54. A system of equations can be used to solve the mixture problem below.

A clerk wants to mix nuts worth $6 per pound with nuts worth $3 per pound to get 60 pounds of a mixture worth $4 per pound. How many pounds of the $6 and $3 nuts will he need?

Using $\underline{a}$ for the number of pounds of the $6 nuts and $\underline{b}$ for the number of pounds of the $3 nuts, we can set up two equations.

The sum of the weights of the two types of nuts must equal the weight of the mixture. That is:

$$a + b = 60$$

The sum of the costs of the two types of nuts must equal the cost of the mixture. That is:

$$6a + 3b = 4(60)$$

Continued on following page.

54. Continued

Therefore, the problem can be solved by solving the system at the right.

$$\begin{aligned} a + b &= 60 \\ 6a + 3b &= 240 \end{aligned}$$

Multiplying the top equation by -3 and adding, we can eliminate b and solve for a.

$$\begin{aligned} -3a - 3b &= -180 \\ 6a + 3b &= 240 \\ \hline 3a &= 60 \\ a &= 20 \end{aligned}$$

Substituting 20 for a in the top equation, we can solve for b.

$$\begin{aligned} a + b &= 60 \\ 20 + b &= 60 \\ b &= 40 \end{aligned}$$

Therefore, he would need ______ pounds of the $6 nuts and ______ pounds of the $3 nuts.

---

55. Here is another problem that can be solved with a system of equations.

> A chemist has one solution that is 70% acid and a second solution that is 20% acid. She wants to mix them to get 100 liters of a solution that is 50% acid. How many liters of each solution should she use?

Letting x represent the number of liters of the 70% acid and y represent the number of liters of the 20% acid, we can summarize the information given in the table below.

Liters of solution	Percent	Liters of acid
x	70	.7x
y	20	.2y
100	50	.5(100) = 50

She must mix x liters of the first solution and y liters of the second solution to get 100 liters. Therefore:

$$x + y = 100$$

The amount of acid in the new solution must be 50% of 100 liters, which is .5(100) = 50 liters. This amount will be made up by the acid in the two solutions mixed. These amounts of acid are 70% of x or .7x and 20% of y or .2y. Therefore:

$$\begin{aligned} 70\%x + 20\%y &= 50\%(100) \\ .7x + .2y &= 50 \end{aligned}$$

Multiplying by 10 to clear the decimals, we get:

$$\begin{aligned} 10(.7x + .2y) &= 10(50) \\ 10(.7x) + 10(.2y) &= 10(50) \\ 7x + 2y &= 500 \end{aligned}$$

Continued on following page.

20 pounds of the $6 nuts

40 pounds of the $3 nuts

55. Continued

Therefore, the problem can be solved by solving the system at the right.

$$x + y = 100$$
$$7x + 2y = 500$$

Multiplying the top equation by -2 and adding, we can eliminate y and solve for x.

$$-2x - 2y = -200$$
$$7x + 2y = 500$$
$$5x = 300$$
$$x = 60$$

Substituting 60 for x in the top equation, we can solve for y.

$$x + y = 100$$
$$60 + y = 100$$
$$y = 40$$

Therefore, she should use ______ liters of the 70% acid and ______ liters of the 20% acid.

Answer: 60 liters of the 70% acid

40 liters of the 20% acid

56. Following the examples in the last two frames, solve these:

a) The total cost of a 50 pound mixture of two different grades of coffee is $160. If grade A costs $4 per pound and grade B costs $2 per pound, how many pounds of each grade were used?

b) Solution A is 30% alcohol and solution B is 80% alcohol. How much of each is needed to make 100 milliliters of a solution that is 70% alcohol?

Answers:

a) 30 pounds of grade A

20 pounds of grade B

b) 20 milliliters of solution A

80 milliliters of solution B

57. For the problem below, we get a system in which one equation contains only one variable. Therefore, we don't need the addition method or substitution method to eliminate a variable.

Brine is a solution of salt and water. We want to mix pure water and a solution of brine that is 30% salt to get 100 gallons of a brine solution that is 24% salt. How many gallons of each should be mixed?

Continued on following page.

57. Continued

Gallons of solution	Percent	Gallons of salt
x	30	.3x
y	0	0y = 0
100	24	.24(100) = 24

We must mix x gallons of 30% brine and y gallons of water to get 100 gallons. Therefore:

$$x + y = 100$$

The amount of salt in the new solution must be 24% of 100 gallons, which is .24(100) = 24 gallons. This amount is made up of the salt in the 30% brine solution only, because there is no salt in pure water. Therefore:

$$30\%x \quad 0\%y = 24\%(100)$$
$$.3x = 24$$

Multiplying by 10 to clear the decimal, we get:

$$10(.3x) = 10(24)$$
$$3x = 240$$

Therefore, we can solve the problem by solving the system at the right.

$$x + y = 100$$
$$3x = 240$$

Solving for x immediately in the bottom equation, we get:

$$x = \frac{240}{3} = 80$$

Substituting 80 for x in the top equation, we can solve for y.

$$x + y = 100$$
$$80 + y = 100$$
$$y = 20$$

Therefore, we should mix ______ gallons of the 30% brine and ______ gallons of water.

---

80 gallons of the 30% brine

20 gallons of water

58. Following the example in the last frame, solve this one.

We want to mix some milk containing 6% butterfat with skimmed milk (containing 0% butterfat) to get 100 gallons of milk containing 4.5% butterfat. How many gallons of the 6% milk and the skimmed milk should we use?

Answer to Frame 58:

75 gallons of the 6% milk and

25 gallons of the skimmed milk

SELF-TEST 30 (pages 409-417)

Solve each problem.

1. The sum of two numbers is 131. The larger is 15 greater than the smaller. Find the two numbers.

2. The perimeter of a rectangular lot is 360 meters. If the length is three times the width, find the length and width.

3. A plane flies 567 mph with the wind and 513 mph into the wind. Find the speed of the wind and the speed of the plane in still air.

4. We want to add water to a 60% solution of radiator coolant to get 15 liters of a 40% solution of radiator coolant. How many liters of each should be mixed?

ANSWERS:

1. 73 and 58
2. L = 135m, W = 45m
3. wind is 27 mph, speed in still air is 540 mph
4. 10 liters of 60% coolant, 5 liters of water

SUPPLEMENTARY PROBLEMS - CHAPTER 9

Assignment 28

Use the graphing method to solve each system.

1. $x + y = 4$
   $x - y = 2$

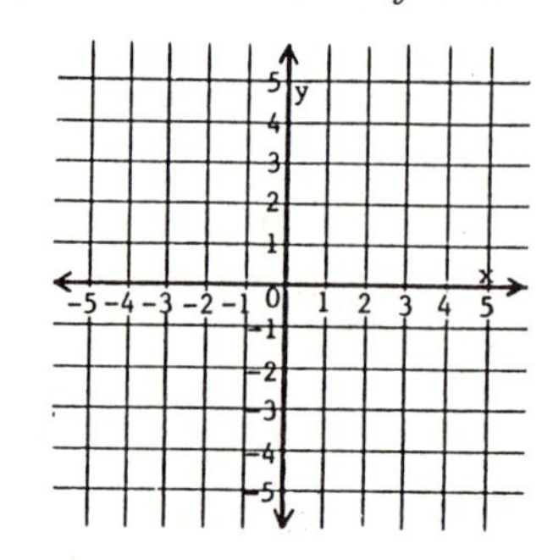

2. $2x - y = -4$
   $x + y = 1$

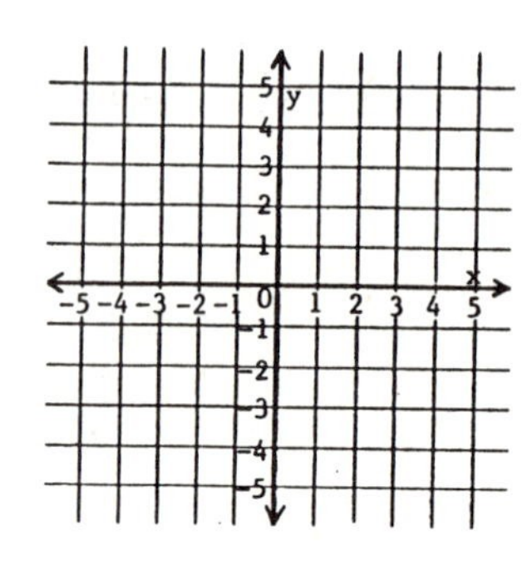

Use the addition method to solve each system of equations.

3. $x - y = 5$
   $x + y = 3$
4. $3h + 2m = 24$
   $h - 2m = 0$
5. $5x - 4y = 23$
   $3x + 4y = 1$
6. $7r - 10v = 40$
   $3r + 10v = 60$
7. $5h - 2k = 35$
   $h + 2k = 7$
8. $3b - m = 20$
   $2b + m = 0$
9. $p + 9r = 65$
   $4p - 9r = 35$
10. $x - y = 2.8$
    $x + y = 5.2$

Use the addition method to solve each system of equations.

11. $x - 3y = 2$
    $2x + y = 11$
12. $4x + y = 3$
    $7x + 3y = 4$
13. $3h + 2k = 28$
    $2h - k = 7$
14. $5r + 2s = 3$
    $3r + 6s = 21$
15. $3c - 4d = 11$
    $7c - 2d = 11$
16. $6v - w = 19$
    $2v - w = 7$
17. $7x + 3y = 26$
    $9x - 2y = 10$
18. $9p - 2r = 11$
    $7p + 5r = 2$
19. $5h + 7k = 32$
    $4h + 5k = 25$
20. $8x - 3y = 10$
    $5x - 2y = 7$
21. $11m - 6n = 8$
    $12m - 5n = 1$
22. $10b + 3d = 15$
    $9b + 5d = 2$

Assignment 29

Convert to standard form and then solve by the addition method.

1. $2x + y = 13$
   $y = x - 2$
2. $b = 3a$
   $2a = b - 3$
3. $t - w = 5$
   $3w + 9 = 2t$
4. $r + 2v = 2$
   $r = v + 5$
5. $h = 4m$
   $2m + 50 = 3h$
6. $s = p + 3$
   $2s - 5 = 3p$
7. $2d = 6 - 5h$
   $2h = 4d - 36$
8. $3k + 13 = 4m$
   $5m - 11 = 2k$

Clear the decimal coefficients and then solve by the addition method.

9. $.3r - .1t = 1.1$
   $.2r + .1t = .9$
10. $.21p + .08w = .5$
    $.17p + .08w = .42$
11. $1.5x - y = 1$
    $3x - 1.5y = 3$
12. $2d = 1.25p + 1$
    $d = 4.4 - .35p$

Use the substitution method to solve each system of equations.

13. $d = 3t$
    $2d + t = 14$
14. $3E - 8R = 12$
    $E = 4R$
15. $5x - y = 8$
    $y = 2x - 5$
16. $v = 2p - 3$
    $7p - 3v = 8$
17. $2b = 5m$
    $3m - b = 2$
18. $4y - 5x = 2$
    $4x = 3y$
19. $3F_1 - 2F_2 = 40$
    $4F_1 - 3F_2 = 50$
20. $7h + 3k = 12$
    $5h - 14 = 6k$

Use any method to solve each system of equations.

21. $7x - 3y = 34$
    $5x + 3y = 14$

22. $p + 4r = 5$
    $p + 5r = 7$

23. $s + 2w = 15$
    $s = 4w$

24. $5a - 4b = 17$
    $2a - 5b = 17$

25. $\frac{b}{2} + \frac{d}{3} = 4$
    $\frac{3b}{4} + \frac{d}{6} = 4$

26. $2x + \frac{y}{5} = 20$
    $5x - y = 5$

27. $.1b + .2d = 3$
    $.5b - .4d = 8$

28. $2.4r + 3.2 = w$
    $3w - 12 = 4r$

Assignment 30

Use a system of equations to solve each problem.

1. The sum of two numbers is 218 and their difference is 82. Find the two numbers.
2. The sum of two numbers is 64. The first number is three times the second number. Find the two numbers.
3. The total cost of a camera and calculator is $275. If the price of the camera is four times the price of the calculator, find the price of each.
4. There are 285 students in a graduation class. If there are 25 more females than males, find the number of males and females.
5. A metal rod 100 centimeters long is cut into two pieces. One piece is 10 centimeters longer than the other. Find the length of each piece.
6. A wire 24.6 meters long is cut into two parts so that the longer part is twice the shorter part. How long is each part?
7. We want to construct a rectangle whose perimeter is 60 centimeters and whose width is 6 centimeters less than its length. Find the length and width.
8. The perimeter of a rectangular room is 66 feet. If the length is twice the width, find the length and width.
9. A plane averages 156 mph with the wind and 130 mph into the wind. Find the speed of the wind and the speed of the plane in still air.
10. It takes a boat 2 hours to go 14 miles downstream and $3\frac{1}{2}$ hours to return. Find the speed of the current and the speed of the boat in still water.
11. Solution A is 30% acid and solution B is 60% acid. How much of each solution should be mixed to get 60 liters of a solution that is 40% acid?
12. How much of a 10% solution of alcohol and a 50% solution of alcohol should be mixed to make 200 milliliters of a 35% solution?
13. Alloy A contains 8% chromium and alloy B contains 5% chromium. How much of each alloy should be mixed to get 300 pounds of an alloy containing 6% chromium?
14. How much pure zinc (100% zinc) and how much of an alloy containing 40% zinc must be melted together to get 15 kilograms of an alloy containing 60% zinc?
15. We want to mix some milk containing 6% butterfat with skimmed milk (0% butterfat) to get 200 gallons of milk containing 4.8% butterfat. How many gallons of the 6% milk and the skimmed milk should we use?

10 Radical Expressions

In this chapter, we will discuss the basic operations with square root radicals. We will also discuss a method for solving radical equations. Some evaluations and rearrangements with radical formulas are included.

10-1 RADICAL EXPRESSIONS

In this section, we will discuss radical expressions involving square roots.

---

1. A square root of a number N is a number whose square is N. Any positive number has two square roots, one positive and one negative. For example:

   4 is a square root of 16, since $(4)^2 = 16$.

   -4 is a square root of 16, since $(-4)^2 = 16$.

   Complete: a) The positive square root of 36 is ________.

   b) The negative square root of 100 is ________.

   c) The two square roots of 49 are ________ and ________.

2. Though negative square roots are called negative square roots, positive square roots are called principal square roots.

The symbol $\sqrt{\quad}$, called a radical sign, is used for principal square roots. That is:

$\sqrt{25} = 5$ $\sqrt{81} = 9$ $\sqrt{400} = 20$

The symbol $-\sqrt{\quad}$ is used for negative square roots. That is:

$-\sqrt{9} = -3$ $-\sqrt{1} = -1$ $-\sqrt{225} = -15$

Complete these:

a) $\sqrt{1}$ = ____ b) $-\sqrt{64}$ = ____ c) $-\sqrt{4}$ = ____

a) 6
b) -10
c) 7 and -7

---

3. Any number whose square roots are integers is a perfect square. The perfect squares up to 400 are given in the table below. Only principal square roots are shown.

N	$\sqrt{N}$	N	$\sqrt{N}$
1	1	121	11
4	2	144	12
9	3	169	13
16	4	196	14
25	5	225	15
36	6	256	16
49	7	289	17
64	8	324	18
81	9	361	19
100	10	400	20

Using the table, complete these:

a) $-\sqrt{144}$ = ____ b) $\sqrt{256}$ = ____ c) $-\sqrt{361}$ = ____

a) 1
b) -8
c) -2

---

4. The number 0 has only one square root. It is 0. Therefore:

a) $\sqrt{0}$ = ____ b) $-\sqrt{169}$ = ____ c) $\sqrt{324}$ = ____

a) -12
b) 16
c) -19

---

5. Except for perfect squares, the square roots of numbers are non-terminating decimal numbers. We can use a calculator to find square roots. Using a calculator, we get:

$\sqrt{19} = 4.3588989\ldots$

$\sqrt{84} = 9.1651514\ldots$

Continued on following page.

a) 0
b) -13
c) 18

5. Continued

A table of square roots for numbers from 1 to 100 is given on the inside of the back cover. In the table, square roots are rounded to three decimal places. For example, the table has these entries which are rounded from the values above.

N	$\sqrt{N}$
19	4.359
84	9.165

Though the entries above are not exact, we will treat them as if they were exact. That is:

We will say: $\sqrt{19} = 4.359$ $\sqrt{84} = 9.165$

Using the table, complete these:

a) $\sqrt{28}$ = ______ b) $-\sqrt{41}$ = ______ c) $\sqrt{98}$ = ______

---

6. Each expression below is called a radical expression.

$\sqrt{17}$ $\sqrt{x}$ $\sqrt{5y^2}$ $\sqrt{\frac{x-3}{4}}$

The expression under the radical is called the radicand. For example:

In $\sqrt{17}$, the radicand is 17.

In $\sqrt{5y^2}$, the radicand is $5y^2$.

In $\sqrt{\frac{x-3}{4}}$, the radicand is ______.

Answers to 5: a) 5.292 b) -6.403 c) 9.899

---

7. To evaluate the radical expression below when $x = 12$, we substituted, simplified, and found the square root.

$$\sqrt{x+4} = \sqrt{12+4} = \sqrt{16} = 4$$

Evaluate each expression below when $x = 3$.

a) $\sqrt{4x^2}$ = b) $\sqrt{70-2x}$ =

Answer to 6: $\frac{x-3}{4}$

---

8. To evaluate the radical expression below when $x = 7$, we used the square-root table.

$$\sqrt{\frac{x^2-9}{2}} = \sqrt{\frac{7^2-9}{2}} = \sqrt{\frac{49-9}{2}} = \sqrt{\frac{40}{2}} = \sqrt{20} = 4.472$$

Evaluate this expression when $x = 8$. Use the table.

$\sqrt{\frac{x+8}{x}}$ =

Answers to 7: a) 6 b) 8

Answer to 8: 1.414

10-2 IRRATIONAL NUMBERS AND REAL NUMBERS

In this section, we will define irrational numbers and real numbers.

---

9. As we saw in an earlier chapter, a rational number is a number that can be expressed in the form $\frac{a}{b}$, where a and b are integers and b is not 0. Some examples are shown. The numbers could also be negative.

1) $\frac{2}{5}$	All fractions are rational numbers.
2) $6 = \frac{6}{1}$	All integers are rational numbers.
3) $.75 = \frac{75}{100}$	All terminating decimals are rational numbers.
4) $.666\ldots = \frac{2}{3}$	All non-terminating decimals with a repeating pattern are rational numbers.
5) $1\frac{1}{2} = \frac{3}{2}$	All mixed numbers are rational numbers.

An irrational number is a number that cannot be expressed in the form $\frac{a}{b}$, where a and b are integers and b is not 0. Any non-terminating decimal without a repeating pattern of digits is an irrational number. Two examples are shown.

1.454554555455554... (Pattern does not repeat)

0.0168168816888... (Pattern does not repeat)

State whether each number is rational or irrational.

a) 0.125 (terminating)

b) 3.424242... (pattern repeats)

c) 4.030030030003... (pattern does not repeat)

---

10. The number π is an irrational number because it is a non-terminating, non-repeating decimal. That is:

$$\pi = 3.1415926535\ldots$$

When 3.14 is used for π, it is only an approximation.

Since 3.14 is a terminating decimal, it is a ______________ (rational/irrational) approximation for π.

Answers to 9: a) rational b) rational c) irrational

Answer to 10: rational

11. When a number is a perfect square, its square root is rational because it is an integer. For example:

$\sqrt{36}$ is rational, since $\sqrt{36} = 6$

$\sqrt{81}$ is rational, since $\sqrt{81} = 9$

When a number is not a perfect square, its square root is irrational because it is a non-repeating, non-terminating number. For example:

$\sqrt{7}$ is irrational, since $\sqrt{7} = 2.6457513...$

$\sqrt{61}$ is irrational, since $\sqrt{61} = 7.8102497...$

State whether each number is rational or irrational.

a) $\sqrt{3}$ b) $\sqrt{16}$ c) $\sqrt{49}$ d) $\sqrt{88}$

---

12. Any number that is either rational or irrational is called a <u>real number</u>. There is a <u>real number</u> for each point on the number line. For example, all of the numbers shown below are real numbers.

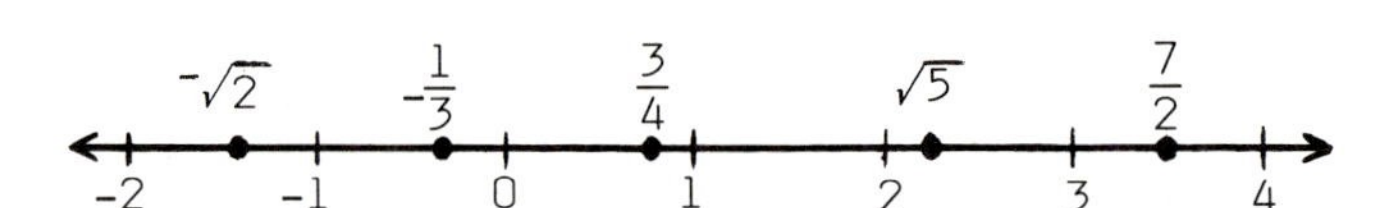

The order of real numbers is shown on the number line. That is:

Since $-\sqrt{2} < -\frac{1}{3}$, $-\sqrt{2}$ is to the left of $-\frac{1}{3}$ on the number line.

Since $\frac{7}{2} > \sqrt{5}$, $\frac{7}{2}$ is to the __________ of $\sqrt{5}$ on the number line.

Answers to 11: a) irrational b) rational c) rational d) irrational

---

13. Negative numbers do not have real number square roots because the square of any number is positive. For example:

$\sqrt{-16} \neq 4$, because $(4)^2 = 16$.

$\sqrt{-16} \neq -4$, because $(-4)^2 = 16$.

Therefore, the following expressions with negative radicands are meaningless in terms of real numbers.

$\sqrt{-81}$ $\sqrt{-25}$ $\sqrt{-7}$

Also, any substitution that makes a radicand negative does not make sense. For example, if we substitute 2 for <u>x</u> below, we get an expression that is meaningless in terms of real numbers.

$$\sqrt{x - 5} = \sqrt{2 - 5} = \sqrt{-3}$$

<u>Note:</u> <u>To avoid meaningless expressions in this text, we will assume that all radicands are either positive or 0.</u>

Answer to 12: right

10-3 MULTIPLYING RADICALS

In this section, we will discuss the procedure for multiplying radicals when their radicands are not negative.

---

14. To multiply radicals, we multiply their radicands. That is:

$$\sqrt{x} \cdot \sqrt{y} = \sqrt{xy}$$

To show that the definition above makes sense, a multiplication with perfect-square radicands is shown below.

$$\sqrt{9} \cdot \sqrt{4} = \sqrt{(9)(4)}$$
$$3 \cdot 2 = \sqrt{36}$$
$$6 = 6$$

When multiplying radicals, we simplify the radicand of the product. For example:

$$\sqrt{2} \cdot \sqrt{3} = \sqrt{(2)(3)} = \sqrt{6}$$
$$\sqrt{7} \cdot \sqrt{2x} = \sqrt{(7)(2x)} = \sqrt{14x}$$
$$\sqrt{5x} \cdot \sqrt{6y} = \sqrt{(5x)(6y)} = \_\_\_\_\_\_$$

---

15. Sometimes the law of exponents for multiplication is used to find the radicand of the product. For example:

Answer: $\sqrt{30xy}$

$$\sqrt{x^4} \cdot \sqrt{x} = \sqrt{(x^4)(x)} = \sqrt{x^5}$$
$$\sqrt{3y} \cdot \sqrt{5y} = \sqrt{(3y)(5y)} = \sqrt{15y^2}$$
$$\sqrt{at^3} \cdot \sqrt{a^2t^2} = \sqrt{(at^3)(a^2t^2)} = \_\_\_\_\_\_$$

---

16. Complete these:

Answer: $\sqrt{a^3t^5}$

a) $\sqrt{7} \cdot \sqrt{11} = \_\_\_\_\_\_$

b) $\sqrt{3a} \cdot \sqrt{8b} = \_\_\_\_\_\_$

c) $\sqrt{m^3} \cdot \sqrt{m^3} = \_\_\_\_\_\_$

d) $\sqrt{2x} \cdot \sqrt{3x^2y} = \_\_\_\_\_\_$

---

17. We use the same procedure when the radicands are fractions. For example:

Answers: a) $\sqrt{77}$ b) $\sqrt{24ab}$ c) $\sqrt{m^6}$ d) $\sqrt{6x^3y}$

$$\sqrt{\frac{1}{2}} \cdot \sqrt{\frac{3}{4}} = \sqrt{\left(\frac{1}{2}\right)\left(\frac{3}{4}\right)} = \sqrt{\frac{3}{8}}$$
$$\sqrt{\frac{1}{3}} \cdot \sqrt{\frac{5}{2}} = \sqrt{\left(\frac{1}{3}\right)\left(\frac{5}{2}\right)} = \_\_\_\_\_\_$$

Answer: $\sqrt{\frac{5}{6}}$

18. We use the same procedure when one or both radicands is a binomial. For example:

$$\sqrt{5} \cdot \sqrt{x + 3} = \sqrt{5(x + 3)} = \sqrt{5x + 15}$$
$$\sqrt{y + 2} \cdot \sqrt{y + 1} = \sqrt{(y + 2)(y + 1)} = \sqrt{y^2 + 3y + 2}$$

Following the examples, complete these:

a) $\sqrt{7} \cdot \sqrt{a - b} =$ ______________________

b) $\sqrt{x - 3} \cdot \sqrt{x - 2} =$ ______________________

Answers to 18: a) $\sqrt{7a - 7b}$ b) $\sqrt{x^2 - 5x + 6}$

19. To indicate a multiplication of a non-radical and a radical, we write the non-radical in front of the radical. That is:

$2\sqrt{5}$ means: multiply 2 and $\sqrt{5}$.

To simplify the expressions below, we multiplied the non-radical factors.

$2 \cdot 4\sqrt{7} = 8\sqrt{7}$ $\qquad$ $y \cdot 6\sqrt{y^3} = 6y\sqrt{y^3}$

Following the examples, simplify these:

a) $3x \cdot 5\sqrt{x} =$ __________ $\qquad$ b) $t^2 \cdot t^3\sqrt{7} =$ __________

Answers to 19: a) $15x\sqrt{x}$ b) $t^5\sqrt{7}$

20. To simplify the expressions below, we multiplied the radical factors.

$7\sqrt{3} \cdot \sqrt{2} = 7\sqrt{6}$ $\qquad$ $a\sqrt{x} \cdot \sqrt{2x} = a\sqrt{2x^2}$

Following the examples, simplify these:

a) $9\sqrt{y^3} \cdot \sqrt{y^3} =$ __________ $\qquad$ b) $at\sqrt{t} \cdot \sqrt{a^2t^2} =$ __________

Answers to 20: a) $9\sqrt{y^6}$ b) $at\sqrt{a^2t^3}$

21. To simplify the expressions below, we multiplied both the non-radical and the radical factors.

$$2\sqrt{3} \cdot 5\sqrt{7} = 2 \cdot 5 \cdot \sqrt{3} \cdot \sqrt{7} = 10\sqrt{21}$$
$$a\sqrt{b} \cdot a\sqrt{b^4} = a \cdot a \cdot \sqrt{b} \cdot \sqrt{b^4} = a^2\sqrt{b^5}$$

Following the examples, simplify these:

a) $3\sqrt{5x} \cdot x\sqrt{2} =$ ______________________

b) $2d\sqrt{p} \cdot d^3\sqrt{dp} =$ ______________________

Answers to 21: a) $3x\sqrt{10x}$ b) $2d^4\sqrt{dp^2}$

22. The multiplications below involve both non-radicals and radicals.

$$7\sqrt{x} \cdot \sqrt{x + 2} = 7\sqrt{(x)(x + 2)} = 7\sqrt{x^2 + 2x}$$
$$a\sqrt{y} \cdot b\sqrt{y - 1} = ab\sqrt{(y)(y - 1)} = ab\sqrt{y^2 - y}$$

Following the examples, complete these:

a) $5\sqrt{m} \cdot \sqrt{m - 8} =$ ______________________

b) $2\sqrt{y + 4} \cdot x\sqrt{y - 1} =$ ______________________

Answers to 22: a) $5\sqrt{m^2 - 8m}$ $\qquad$ b) $2x\sqrt{y^2 + 3y - 4}$

10-4 SIMPLIFYING RADICALS

In this section, we will show how some radicals can be simplified by factoring out perfect squares.

---

23. We can factor a radical by reversing the procedure for multiplying. That is:

$$\sqrt{6} = \sqrt{2} \cdot \sqrt{3} \qquad \sqrt{35} = \sqrt{5} \cdot \sqrt{7}$$

When we factor and get a radicand that is a perfect square, we can simplify. For example:

$$\sqrt{12} = \sqrt{4} \cdot \sqrt{3} = 2\sqrt{3}$$

$$\sqrt{50} = \sqrt{25} \cdot \sqrt{2} = 5\sqrt{2}$$

The process above is called <u>simplifying by factoring out a perfect square</u>. Simplify each of these by factoring out a perfect square.

a) $\sqrt{8}$ = ________ c) $\sqrt{45}$ = ________

b) $\sqrt{20}$ = ________ d) $\sqrt{75}$ = ________

---

24. Some numbers have more than one perfect-square factor. For example, 72 has 4, 9, and 36. When a number of that type is the radicand, we can simplify the radical completely <u>in one step</u> only by factoring out the largest perfect square. We get:

$$\sqrt{72} = \sqrt{36} \cdot \sqrt{2} = 6\sqrt{2}$$

If we <u>do not</u> factor out the largest perfect square, the remaining radical (like $\sqrt{18}$ below) will still contain a perfect-square factor.

$$\sqrt{72} = \sqrt{4} \cdot \sqrt{18} = 2\sqrt{18}$$

Therefore, to simplify completely, we have to factor again. We get:

$$\sqrt{72} = 2\sqrt{18} = 2\sqrt{9} \cdot \sqrt{2} = 2 \cdot 3\sqrt{2} = 6\sqrt{2}$$

Simplify completely. Try to do so in one step by factoring out the largest perfect square.

a) $\sqrt{48}$ = ________ c) $\sqrt{80}$ = ________

b) $\sqrt{200}$ = ________ d) $\sqrt{300}$ = ________

Answers to frame 23:

a) $2\sqrt{2}$ c) $3\sqrt{5}$

b) $2\sqrt{5}$ d) $5\sqrt{3}$

---

25. We simplified each radical below by factoring out the largest perfect square.

$$\sqrt{36x} = \sqrt{36} \cdot \sqrt{x} = 6\sqrt{x}$$

$$\sqrt{50xy} = \sqrt{25} \cdot \sqrt{2xy} = 5\sqrt{2xy}$$

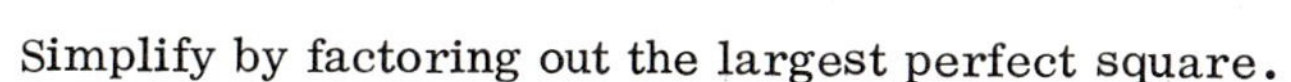

Simplify by factoring out the largest perfect square.

a) $\sqrt{81y}$ = ________ c) $\sqrt{32ab}$ = ________

b) $\sqrt{27x}$ = ________ d) $\sqrt{24xy}$ = ________

Answers to frame 24:

a) $4\sqrt{3}$ c) $4\sqrt{5}$

b) $10\sqrt{2}$ d) $10\sqrt{3}$

26. Since the radicands below do not contain a perfect-square factor, the radicals cannot be simplified.

$$\sqrt{14} \qquad \sqrt{33x} \qquad \sqrt{58ab}$$

Simplify each radical if possible.

a) $\sqrt{40}$ = ________ c) $\sqrt{21x}$ = ________

b) $\sqrt{55}$ = ________ d) $\sqrt{45xy}$ = ________

Answers: a) $9\sqrt{y}$ b) $3\sqrt{3x}$ c) $4\sqrt{2ab}$ d) $2\sqrt{6xy}$

---

27. When multiplying radicals, we simplify the product (if possible) by factoring out the largest perfect square. For example:

$$\sqrt{2} \cdot \sqrt{6} = \sqrt{12} = 2\sqrt{3}$$
$$\sqrt{5} \cdot \sqrt{40x} = \sqrt{200x} = 10\sqrt{2x}$$

Multiply and simplify each product if possible.

a) $\sqrt{2} \cdot \sqrt{32}$ = ________ c) $\sqrt{2x} \cdot \sqrt{14y}$ = ________

b) $\sqrt{5x} \cdot \sqrt{2}$ = ________ d) $\sqrt{8} \cdot \sqrt{6t}$ = ________

Answers: a) $2\sqrt{10}$ b) Not possible c) Not possible d) $3\sqrt{5xy}$

---

28. In $\sqrt{x^2}$, the radicand is a perfect square. Since squares are always positive, we can substitute either a positive or a negative value for x.

If $x = 4$, $\sqrt{x^2} = \sqrt{4^2} = \sqrt{16} = 4$

If $x = -4$, $\sqrt{x^2} = \sqrt{(-4)^2} = \sqrt{16} = 4$

Notice these points about $\sqrt{x^2}$:

If x is positive, $\sqrt{x^2} = x$. (If $x = 4$, $\sqrt{x^2} = 4$.)

If x is negative, $\sqrt{x^2} = -x$. (If $x = -4$, $\sqrt{x^2} = 4$.)

Note: In this text, we will assume that negative values are not substituted in $\sqrt{x^2}$. Therefore, in this text, $\sqrt{x^2} = x$.

Answers: a) 8 b) $\sqrt{10x}$ c) $2\sqrt{7xy}$ d) $4\sqrt{3t}$

---

29. To square a power, we double its exponent. Therefore, the exponent of its square is always an even number. For example:

$$(x)^2 = x^2 \qquad (x^3)^2 = x^6 \qquad (y^8)^2 = y^{16}$$

Any power whose exponent is even is a perfect square. To find the square root of a perfect square power, we divide its exponent by 2. That is:

$$\sqrt{x^2} = x^{\frac{2}{2}} = x^1 = x \qquad \sqrt{y^{10}} = y^{\frac{10}{2}} = y^5$$

Find the square root of each perfect square below.

a) $\sqrt{y^2}$ = ______ b) $\sqrt{x^6}$ = ______ c) $\sqrt{t^{20}}$ = ______

30. Radicals containing perfect-square powers can also be simplified by factoring out perfect squares. For example:

$$\sqrt{4x^2} = \sqrt{4}\cdot\sqrt{x^2} = 2x$$
$$\sqrt{5y^4} \quad \sqrt{5}\cdot\sqrt{y^4} = y^2\sqrt{5}$$
$$\sqrt{12t^6} \quad \sqrt{4}\cdot\sqrt{3}\cdot\sqrt{t^6} = 2t^3\sqrt{3}$$

Following the examples, simplify these:

a) $\sqrt{25x^6}$ = ________ c) $\sqrt{18t^4}$ = ________

b) $\sqrt{3y^8}$ = ________ d) $\sqrt{28d^2}$ = ________

a) $y^1 = y$
b) x^3
c) t^{10}

---

31. Three more simplifications are shown below.

$$\sqrt{x^2y^8} = \sqrt{x^2}\cdot\sqrt{y^8} = xy^4$$
$$\sqrt{3x^{10}y^4} = \sqrt{3}\cdot\sqrt{x^{10}}\cdot\sqrt{y^4} = x^5y^2\sqrt{3}$$
$$\sqrt{50ax^6} = \sqrt{25}\cdot\sqrt{2a}\cdot\sqrt{x^6} = 5x^3\sqrt{2a}$$

Following the examples, simplify these:

a) $\sqrt{x^4y^2}$ = ________ c) $\sqrt{49x^2y^6}$ = ________

b) $\sqrt{cd^{12}}$ = ________ d) $\sqrt{32at^8}$ = ________

a) $5x^3$ c) $3t^2\sqrt{2}$
b) $y^4\sqrt{3}$ d) $2d\sqrt{7}$

---

32. When the exponent of a power is odd, it is not a perfect square. But if the odd exponent is 3 or more, we can factor into a perfect-square power and a first power. That is:

$$x^3 = x^2\cdot x^1 \qquad y^7 = y^6\cdot y^1$$

Using the factoring process above, we simplified each radical below by factoring out the largest perfect square.

$$\sqrt{x^5} = \sqrt{x^4}\cdot\sqrt{x^1} = x^2\sqrt{x}$$
$$\sqrt{y^{11}} = \sqrt{y^{10}}\cdot\sqrt{y^1} = y^5\sqrt{y}$$

Simplify these by factoring out the largest perfect square.

a) $\sqrt{x^3}$ = ________ b) $\sqrt{y^9}$ = ________ c) $\sqrt{t^{13}}$ = ________

a) x^2y c) $7xy^3$
b) $d^6\sqrt{c}$ d) $4t^4\sqrt{2a}$

---

33. We simplified each radical below by factoring out the largest perfect squares.

$$\sqrt{3x^5} = \sqrt{x^4}\cdot\sqrt{3x} = x^2\sqrt{3x}$$
$$\sqrt{8y^3} = \sqrt{4}\cdot\sqrt{y^2}\cdot\sqrt{2y} = 2y\sqrt{2y}$$

Following the examples, simplify these:

a) $\sqrt{9x^7}$ = ________ c) $\sqrt{12t^5}$ = ________

b) $\sqrt{2y^9}$ = ________ d) $\sqrt{27p^3}$ = ________

a) $x\sqrt{x}$
b) $y^4\sqrt{y}$
c) $t^6\sqrt{t}$

34. We simplified each radical below by factoring out the largest perfect squares.

$$\sqrt{a^7b} = \sqrt{a^6} \cdot \sqrt{ab} = a^3\sqrt{ab}$$
$$\sqrt{3x^4y^3} = \sqrt{x^4} \cdot \sqrt{y^2} \cdot \sqrt{3y} = x^2y\sqrt{3y}$$

Following the examples, simplify these:

a) $\sqrt{a^4b^5}$ = ________ b) $\sqrt{20cd^9}$ = ________

Answers (previous frame): a) $3x^3\sqrt{x}$ b) $y^4\sqrt{2y}$ c) $2t^2\sqrt{3t}$ d) $3p\sqrt{3p}$

---

35. After multiplying below, we simplified each product.

$$\sqrt{2x} \cdot \sqrt{10x} = \sqrt{20x^2} = 2x\sqrt{5}$$
$$\sqrt{3a^3} \cdot \sqrt{12b^5} = \sqrt{36a^3b^5} = 6ab^2\sqrt{ab}$$

Multiply and simplify each product.

a) $\sqrt{3x} \cdot \sqrt{6x^2}$ = ________________

b) $\sqrt{20a^4} \cdot \sqrt{5b^7}$ = ________________

Answers: a) $a^2b^2\sqrt{b}$ b) $2d^4\sqrt{5cd}$

---

36. Notice how we used the distributive principle to factor out a perfect square from these:

$$\sqrt{9x + 9y} = \sqrt{9(x + y)} = 3\sqrt{x + y}$$
$$\sqrt{25a - 100b} = \sqrt{25(a - 4b)} = 5\sqrt{a - 4b}$$

Following the examples, simplify these:

a) $\sqrt{36x - 36y}$ = ________________

b) $\sqrt{8a + 4b}$ = ________________

Answers: a) $3x\sqrt{2x}$ b) $10a^2b^3\sqrt{b}$

---

37. We also used the distributive principle to simplify each of these.

$$\sqrt{a^2x - a^2y} = \sqrt{a^2(x - y)} = a\sqrt{x - y}$$
$$\sqrt{16x^2y + 16x^2z} = \sqrt{16x^2(y + z)} = 4x\sqrt{y + z}$$

Following the examples, simplify these:

a) $\sqrt{d^2x - d^2y}$ = ________________

b) $\sqrt{9ac^2 + 9bc^2}$ = ________________

Answers: a) $6\sqrt{x - y}$ b) $2\sqrt{2a + b}$

Answers: a) $d\sqrt{x - y}$ b) $3c\sqrt{a + b}$

10-5 DIVIDING RADICALS

In this section, we will discuss the procedure for dividing radicals. We will assume that radicands are non-negative.

---

38. Any division of radicals is equal to the square root of a fraction. That is:

$$\frac{\sqrt{36}}{\sqrt{9}} = \sqrt{\frac{36}{9}} \qquad \frac{\sqrt{64}}{\sqrt{4}} = \sqrt{\frac{64}{4}}$$

To show that the statements are true, we evaluated both sides below.

$$\frac{\sqrt{36}}{\sqrt{9}} = \sqrt{\frac{36}{9}} \qquad \frac{\sqrt{64}}{\sqrt{4}} = \sqrt{\frac{64}{4}}$$

$$\frac{6}{3} = \sqrt{4} \qquad \frac{8}{2} = \sqrt{16}$$

$$2 = 2 \qquad 4 = 4$$

Therefore, to divide radicals, we convert to the square root of a fraction and simplify. For example:

$$\frac{\sqrt{50}}{\sqrt{2}} = \sqrt{\frac{50}{2}} = \sqrt{25} = 5 \qquad \frac{\sqrt{27}}{\sqrt{3}} = \sqrt{\frac{27}{3}} = \sqrt{} = \underline{}$$

---

39. When the radicand of the quotient is not a perfect square, we simplify it as much as possible by factoring out perfect squares. For example:

$$\frac{\sqrt{25}}{\sqrt{5}} = \sqrt{\frac{25}{5}} = \sqrt{5} \qquad \frac{\sqrt{36}}{\sqrt{3}} = \sqrt{\frac{36}{3}} = \sqrt{12} = 2\sqrt{3}$$

Simplify each quotient as much as possible.

a) $\frac{\sqrt{20}}{\sqrt{2}} = \sqrt{\frac{20}{2}} =$ ______________

b) $\frac{\sqrt{150}}{\sqrt{3}} = \sqrt{\frac{150}{3}} =$ ______________

Answer: $\sqrt{9} = 3$

---

40. The same procedure was used below. Complete the other division.

$$\frac{\sqrt{8x}}{\sqrt{2x}} = \sqrt{\frac{8x}{2x}} = \sqrt{4} = 2 \qquad \frac{\sqrt{100y}}{\sqrt{2y}} = \underline{}$$

Answers: a) $\sqrt{10}$ b) $5\sqrt{2}$

Answer: $5\sqrt{2}$

41. The same procedure is used for divisions of radicals containing powers. For example:

$$\frac{\sqrt{18x^5}}{\sqrt{2x^3}} = \sqrt{\frac{18x^5}{2x^3}} = \sqrt{9x^2} = 3x$$

Following the example, complete these.

a) $\frac{\sqrt{c^7d^9}}{\sqrt{c^3d^3}} =$ ______________

b) $\frac{\sqrt{75y^{10}}}{\sqrt{3y^6}} =$ ______________

Answers: a) c^2d^3 b) $5y^2$

42. In the division below, we simplified the quotient as much as possible.

$$\frac{\sqrt{40y^4}}{\sqrt{2y}} = \sqrt{\frac{40y^4}{2y}} = \sqrt{20y^3} = 2y\sqrt{5y}$$

Simplify each quotient as much as possible.

a) $\frac{\sqrt{6x^6}}{\sqrt{2x^4}} =$ ______________

b) $\frac{\sqrt{p^6q^8}}{\sqrt{pq}} =$ ______________

Answers: a) $x\sqrt{3}$ b) $p^2q^3\sqrt{pq}$

43. After converting a division of two radicals to the square root of a fraction, we cannot always simplify. For example:

$$\frac{\sqrt{37}}{\sqrt{11}} = \sqrt{\frac{37}{11}} \qquad \frac{\sqrt{cd}}{\sqrt{V}} = \text{\_\_\_\_\_\_}$$

Answer: $\sqrt{\frac{cd}{V}}$

SELF-TEST 31 (pages 420-433)

1. Find the two square roots of 144.

2. Find the principal square root of 49.

3. When $x = 25$, $\sqrt{2x - 14} =$

4. When $y = 6$, $\sqrt{\frac{y^2 - 1}{7}} =$

Do these multiplications.

5. $\sqrt{3x} \cdot \sqrt{2y}$

6. $2\sqrt{x} \cdot \sqrt{5y}$

7. $\sqrt{m} \cdot 3\sqrt{m - 1}$

Simplify.

8. $\sqrt{80}$

9. $\sqrt{54x^3}$

10. $\sqrt{75a^2b^5}$

Multiply and simplify each product.

11. $\sqrt{6y} \cdot \sqrt{12}$

12. $\sqrt{2r^3} \cdot \sqrt{32s^4}$

Divide and simplify each quotient.

13. $\frac{\sqrt{60}}{\sqrt{3}}$

14. $\frac{\sqrt{8x^3y^2}}{\sqrt{2xy}}$

ANSWERS:

1. 12 and -12
2. 7
3. 6
4. 2.236
5. $\sqrt{6xy}$
6. $2\sqrt{5xy}$
7. $3\sqrt{m^2 - m}$
8. $4\sqrt{5}$
9. $3x\sqrt{6x}$
10. $5ab^2\sqrt{3b}$
11. $6\sqrt{2y}$
12. $8rs^2\sqrt{r}$
13. $2\sqrt{5}$
14. $2x\sqrt{y}$

10-6 FRACTIONAL RADICANDS

In this section, we will discuss the procedure for simplifying radicals when the radicand is a fraction. We will assume that all radicands are non-negative.

---

44. To simplify a radical containing a fraction, we can reverse the procedure for dividing radicals. That is, we can convert the radical to a division of two radicals and then simplify both terms. For example:

$$\sqrt{\frac{9}{25}} = \frac{\sqrt{9}}{\sqrt{25}} = \frac{3}{5}$$

$$\sqrt{\frac{4a^2}{b^6}} = \frac{\sqrt{4a^2}}{\sqrt{b^6}} = \frac{2a}{b^3}$$

Following the examples, simplify these:

a) $\sqrt{\frac{100}{49}} =$ ______ b) $\sqrt{\frac{x^4}{16y^2}} =$ ______

---

45. When simplifying a radical, remember that "1" is a perfect square For example:

$$\sqrt{\frac{1}{4}} = \frac{\sqrt{1}}{\sqrt{4}} = \frac{1}{2} \qquad \sqrt{\frac{1}{x^2}} = \frac{\sqrt{1}}{\sqrt{x^2}} = \text{\_\_\_\_}$$

Answers: a) $\frac{10}{7}$ b) $\frac{x^2}{4y}$

---

46. When simplifying a radical, we can get a radical in the numerator or denominator. For example:

$$\sqrt{\frac{5}{4}} = \frac{\sqrt{5}}{\sqrt{4}} = \frac{\sqrt{5}}{2}$$

$$\sqrt{\frac{1}{2x}} = \frac{\sqrt{1}}{\sqrt{2x}} = \frac{1}{\sqrt{2x}}$$

Following the examples, simplify these:

a) $\sqrt{\frac{9}{10}} =$ ______ b) $\sqrt{\frac{13}{x^2y^2}} =$ ______

Answer: $\frac{1}{x}$

---

47. Notice how we factored out perfect squares in the examples below.

$$\sqrt{\frac{8}{9}} = \frac{\sqrt{8}}{\sqrt{9}} = \frac{2\sqrt{2}}{3}$$

$$\sqrt{\frac{81}{t^5}} = \frac{\sqrt{81}}{\sqrt{t^5}} = \frac{9}{t^2\sqrt{t}}$$

Following the examples, simplify these:

a) $\sqrt{\frac{12}{25}} =$ ______ b) $\sqrt{\frac{49}{20m}} =$ ______

Answers: a) $\frac{3}{\sqrt{10}}$ b) $\frac{\sqrt{13}}{xy}$

48. Simplify each radical.

a) $\sqrt{\frac{25x^2}{16y^4}}$ = ________

b) $\sqrt{\frac{11}{25}}$ = ________

c) $\sqrt{\frac{18y}{49}}$ = ________

d) $\sqrt{\frac{1}{ab^6}}$ = ________

Answers: a) $\frac{2\sqrt{3}}{5}$ b) $\frac{7}{2\sqrt{5m}}$

---

49. Simplify each radical.

a) $\sqrt{\frac{7}{4t^6}}$ = ________

b) $\sqrt{\frac{16}{3ax}}$ = ________

c) $\sqrt{\frac{28m^4}{81}}$ = ________

d) $\sqrt{\frac{64x^2}{9y^3}}$ = ________

Answers: a) $\frac{5x}{4y^2}$ b) $\frac{\sqrt{11}}{5}$ c) $\frac{3\sqrt{2y}}{7}$ d) $\frac{1}{b^3\sqrt{a}}$

---

50. We simplified the radical below by factoring out perfect squares.

$$\sqrt{\frac{4x + 4y}{b^2}} = \frac{\sqrt{4(x + y)}}{\sqrt{b^2}} = \frac{2\sqrt{x + y}}{b}$$

Following the example, simplify these:

a) $\sqrt{\frac{9a - 9b}{16}}$ = ________

b) $\sqrt{\frac{1}{cy^2 + dy^2}}$ = ________

Answers: a) $\frac{\sqrt{7}}{2t^3}$ b) $\frac{4}{\sqrt{3ax}}$ c) $\frac{2m^2\sqrt{7}}{9}$ d) $\frac{8x}{3y\sqrt{y}}$

---

51. After multiplying radicals with fractional radicands, we simplify the product. For example:

$$\sqrt{\frac{5}{2}} \cdot \sqrt{\frac{4}{5}} = \sqrt{\frac{20}{10}} = \sqrt{2}$$

$\sqrt{18} \cdot \sqrt{\frac{1}{2}}$ = ________

Answers: a) $\frac{3\sqrt{a - b}}{4}$ b) $\frac{1}{y\sqrt{c + d}}$

---

52. Following the example, simplify the other product by factoring out perfect squares.

$$\sqrt{\frac{4}{3}} \cdot \sqrt{\frac{x}{3}} = \sqrt{\frac{4x}{9}} = \frac{2\sqrt{x}}{3}$$

$\sqrt{\frac{3}{y}} \cdot \sqrt{\frac{6}{y^3}}$ = ________

Answer: 3

Answer: $\frac{3\sqrt{2}}{y^2}$

10-7 INDICATED-SQUARE RADICANDS

In this section, we will discuss the procedure for simplifying radicals when the radicand is an indicated square. We will assume that negative values are not substituted for variables.

---

53. The square root of the indicated square of a monomial is the monomial itself. For example:

$$\sqrt{(7)^2} = 7 \,, \quad \text{since} \quad \sqrt{(7)^2} = \sqrt{49} = 7$$

$$\sqrt{(5x)^2} = 5x \,, \quad \text{since} \quad \sqrt{(5x)^2} = \sqrt{25x^2} = 5x$$

$$\sqrt{(pq)^2} = pq \,, \quad \text{since} \quad \sqrt{(pq)^2} = \sqrt{p^2q^2} = pq$$

Find each square root.

a) $\sqrt{(10)^2}$ = ______ b) $\sqrt{(2y)^2}$ = ______ c) $\sqrt{(7ab)^2}$ = ______

---

54. Notice the difference between the two radicands below. One is an indicated square; the other is a perfect square.

$$\sqrt{(4x)^2} = 4x \qquad \sqrt{4x^2} = 2x$$

Following the examples, complete these:

a) $\sqrt{9x^2}$ = ______ c) $\sqrt{(36t)^2}$ = ______

b) $\sqrt{(9x)^2}$ = ______ d) $\sqrt{36t^2}$ = ______

Answers to 53: a) 10 b) 2y c) 7ab

---

55. The square root of the indicated square of a fraction is the fraction itself. For example:

$$\sqrt{\left(\frac{3}{4}\right)^2} = \frac{3}{4} \,, \quad \text{since} \quad \sqrt{\left(\frac{3}{4}\right)^2} = \sqrt{\frac{9}{16}} = \frac{3}{4}$$

$$\sqrt{\left(\frac{2a}{b}\right)^2} = \frac{2a}{b} \,, \quad \text{since} \quad \sqrt{\left(\frac{2a}{b}\right)^2} = \sqrt{\frac{4a^2}{b^2}} = \frac{2a}{b}$$

Find each square root.

a) $\sqrt{\left(\frac{1}{5}\right)^2}$ = ______ b) $\sqrt{\left(\frac{5x}{6y}\right)^2}$ = ______

Answers to 54: a) 3x b) 9x c) 36t d) 6t

Answers to 55: a) $\frac{1}{5}$ b) $\frac{5x}{6y}$

56. Notice the difference between the two simplifications below.

$$\sqrt{\left(\frac{p}{q}\right)^2} = \frac{p}{q} \qquad \sqrt{\frac{p^2}{q}} = \frac{p}{\sqrt{q}}$$

Following the examples, simplify these:

a) $\sqrt{\left(\frac{x}{5}\right)^2} =$ ______ b) $\sqrt{\frac{x^2}{5}} =$ ______

Answers: a) $\frac{x}{5}$ b) $\frac{x}{\sqrt{5}}$

57. The square root of the indicated square of any quantity equals the quantity itself. For example:

$$\sqrt{(x+3)^2} = x + 3$$

$$\sqrt{\left(\frac{y-2}{5}\right)^2} = \frac{y-2}{5}$$

Following the examples, complete these:

a) $\sqrt{(p-q)^2} =$ ______ b) $\sqrt{\left(\frac{x+1}{x-1}\right)^2} =$ ______

Answers: a) $p - q$ b) $\frac{x+1}{x-1}$

10-8 SQUARING RADICALS

In this section, we will discuss the procedure for squaring radicals. We will assume that all radicands are non-negative.

---

58. To indicate the square of a radical, we put parentheses around the radical. For example:

$(\sqrt{7})^2$ means: square $\sqrt{7}$

To square a radical, we multiply the radical by itself. That is:

$$(\sqrt{7})^2 = \sqrt{7} \cdot \sqrt{7}$$

The square of a radical is the radicand. For example:

$$(\sqrt{7})^2 = 7, \quad \text{since} \quad (\sqrt{7})^2 = \sqrt{7} \cdot \sqrt{7} = \sqrt{(7)^2} = 7$$

$$(\sqrt{3y})^2 = 3y, \quad \text{since} \quad (\sqrt{3y})^2 = \sqrt{3y} \cdot \sqrt{3y} = \sqrt{(3y)^2} = 3y$$

Write each square.

a) $(\sqrt{10})^2 =$ ______ b) $(\sqrt{5x})^2 =$ ______ c) $(\sqrt{ab})^2 =$ ______

59. The square of a radical is also the radicand when the radicand is a fraction. For example:

$$\left(\sqrt{\frac{x}{3}}\right)^2 = \frac{x}{3}, \text{ since } \left(\sqrt{\frac{x}{3}}\right)^2 = \sqrt{\frac{x}{3}}\cdot\sqrt{\frac{x}{3}} = \sqrt{\left(\frac{x}{3}\right)^2} = \frac{x}{3}$$

Write each square.

a) $\left(\sqrt{\frac{1}{5}}\right)^2 =$ ____ b) $\left(\sqrt{\frac{7}{y}}\right)^2 =$ ____ c) $\left(\sqrt{\frac{ab}{c}}\right)^2 =$ ____

Answers: a) 10 b) 5x c) ab

---

60. The square of a radical always equals the radicand. For example:

$$(\sqrt{x+3})^2 = x + 3 \qquad \left(\sqrt{\frac{y-5}{7}}\right)^2 = \frac{y-5}{7}$$

Write each square.

a) $(\sqrt{a-b})^2 =$ ________ b) $\left(\sqrt{\frac{m+1}{2}}\right)^2 =$ ________

Answers: a) $\frac{1}{5}$ b) $\frac{7}{y}$ c) $\frac{ab}{c}$

---

61. Multiplying a radical by itself is the same as squaring the radical. Therefore, the product is the radicand. For example:

$$\sqrt{7}\cdot\sqrt{7} = 7 \qquad \sqrt{\frac{y}{4}}\cdot\sqrt{\frac{y}{4}} = \frac{y}{4}$$

Complete these:

a) $\sqrt{2x}\cdot\sqrt{2x} =$ ________ b) $\sqrt{y-4}\cdot\sqrt{y-4} =$ ________

Answers: a) a - b b) $\frac{m+1}{2}$

---

62. All three expressions below are equivalent because each equals 3x.

$$(\sqrt{3x})^2 \qquad \sqrt{3x}\cdot\sqrt{3x} \qquad \sqrt{(3x)^2}$$

Complete these:

a) $\sqrt{T}\cdot\sqrt{T} =$ ____ b) $\sqrt{(140)^2} =$ ____ c) $\sqrt{(4ab)^2} =$ ____

Answers: a) 2x b) y - 4

---

63. All three expressions below are equivalent because each equals x + 7 .

$$(\sqrt{x+7})^2 \qquad \sqrt{x+7}\cdot\sqrt{x+7} \qquad \sqrt{(x+7)^2}$$

Complete these:

a) $\sqrt{m-1}\cdot\sqrt{m-1} =$ ____ c) $(\sqrt{t+4})^2 =$ ____

b) $\sqrt{\frac{a+b}{c}}\cdot\sqrt{\frac{a+b}{c}} =$ ____ d) $\sqrt{\left(\frac{y-3}{5}\right)^2} =$ ____

Answers: a) T b) 140 c) 4ab

Answers: a) m - 1 b) $\frac{a+b}{c}$ c) t + 4 d) $\frac{y-3}{5}$

10-9 RATIONALIZING DENOMINATORS

When a fraction contains a radical in its denominator, we can convert it to an equivalent form without a radical in its denominator. The process of doing so is called rationalizing the denominator. We will discuss that process in this section. We will assume that all radicands are non-negative.

---

64. The denominator of a fraction is rationalized only if it does not contain a radical. That is:

$\frac{\sqrt{2}}{3}$ and $\frac{a\sqrt{b}}{c}$ are rationalized.

$\frac{2}{\sqrt{3}}$ and $\frac{1}{x\sqrt{y}}$ are not rationalized.

Which fractions below have denominators that are rationalized? ____________

a) $\frac{a}{\sqrt{b}}$ b) $\frac{\sqrt{c}}{d}$ c) $\frac{2\sqrt{x}}{5}$ d) $\frac{7}{9\sqrt{y}}$

---

65. The procedure for rationalizing a denominator uses these facts:

(b) and (c)

1) Any radical divided by itself equals "1".

$$\frac{\sqrt{5}}{\sqrt{5}} = 1 \qquad \frac{\sqrt{3x}}{\sqrt{3x}} = 1$$

2) To obtain equivalent forms of a fraction, we multiply by a fraction that equals "1".

$$\frac{2}{\sqrt{3}} = \frac{2}{\sqrt{3}}\left(\frac{\sqrt{2}}{\sqrt{2}}\right) = \frac{2\sqrt{2}}{\sqrt{6}}$$

To rationalize a denominator containing a square root, we multiply by a fraction whose terms are identical to the radical in the denominator. For example:

$$\frac{5}{\sqrt{7}} = \frac{5}{\sqrt{7}}\left(\frac{\sqrt{7}}{\sqrt{7}}\right) = \frac{5\sqrt{7}}{7}$$

$$\frac{1}{\sqrt{ab}} = \frac{1}{\sqrt{ab}}\left(\frac{\sqrt{ab}}{\sqrt{ab}}\right) = \frac{\sqrt{ab}}{ab}$$

Notice that we got rid of the radical in each denominator above by multiplying it by itself. Rationalize each denominator below.

a) $\frac{1}{\sqrt{15}}$ = ____________

b) $\frac{3}{\sqrt{2y}}$ = ____________

66. Following the example, rationalize the other denominators.

$$\frac{1}{5\sqrt{2}} = \frac{1}{5\sqrt{2}}\left(\frac{\sqrt{2}}{\sqrt{2}}\right) = \frac{\sqrt{2}}{5(2)} = \frac{\sqrt{2}}{10}$$

a) $\frac{3}{8\sqrt{5}}$ = ______________________

b) $\frac{x}{a\sqrt{y}}$ = ______________________

a) $\frac{\sqrt{15}}{15}$

b) $\frac{3\sqrt{2y}}{2y}$

---

67. Following the example, rationalize the other denominator.

$$\frac{3}{4\sqrt{5x}} = \frac{3}{4\sqrt{5x}}\left(\frac{\sqrt{5x}}{\sqrt{5x}}\right) = \frac{3\sqrt{5x}}{4(5x)} = \frac{3\sqrt{5x}}{20x}$$

$\frac{1}{2\sqrt{7x}}$ = ______________________

a) $\frac{3\sqrt{5}}{40}$

b) $\frac{x\sqrt{y}}{ay}$

---

68. Following the example, rationalize the other denominator.

$$\frac{2\sqrt{3}}{\sqrt{5}} = \frac{2\sqrt{3}}{\sqrt{5}}\left(\frac{\sqrt{5}}{\sqrt{5}}\right) = \frac{2\sqrt{15}}{5}$$

$\frac{7\sqrt{x}}{\sqrt{y}}$ = ______________________

$\frac{\sqrt{7x}}{14x}$

---

69. Following the example, rationalize the other denominator.

$$\frac{3}{\sqrt{x - 1}} = \frac{3}{\sqrt{x - 1}}\left(\frac{\sqrt{x - 1}}{\sqrt{x - 1}}\right) = \frac{3\sqrt{x - 1}}{x - 1}$$

$\frac{a}{\sqrt{b + c}}$ = ______________________

$\frac{7\sqrt{xy}}{y}$

---

70. When a denominator is rationalized, sometimes the new fraction can be reduced to lowest terms. For example:

$$\frac{7}{2\sqrt{7}} = \frac{7}{2\sqrt{7}}\left(\frac{\sqrt{7}}{\sqrt{7}}\right) = \frac{\cancel{7}\sqrt{7}}{2(\cancel{7})} = \frac{\sqrt{7}}{2}$$

$$\frac{x}{\sqrt{x}} = \frac{x}{\sqrt{x}}\left(\frac{\sqrt{x}}{\sqrt{x}}\right) = \frac{\cancel{x}\sqrt{x}}{\cancel{x}} = \sqrt{x}$$

Continued on following page.

$\frac{a\sqrt{b + c}}{b + c}$

70. Continued

Rationalize each denominator and then reduce to lowest terms.

a) $\frac{3}{\sqrt{3}}$ = ______________

b) $\frac{c}{d\sqrt{c}}$ = ______________

Answers to 70: a) $\sqrt{3}$ b) $\frac{\sqrt{c}}{d}$

---

71. After simplifying a radical containing a fraction, we can frequently rationalize the denominator of the new fraction. For example:

$$\sqrt{\frac{c^2x}{d^2y}} = \frac{c\sqrt{x}}{d\sqrt{y}} = \frac{c\sqrt{x}}{d\sqrt{y}}\left(\frac{\sqrt{y}}{\sqrt{y}}\right) = \frac{c\sqrt{xy}}{dy}$$

Simplify and then rationalize the denominator of the new fraction.

a) $\sqrt{\frac{8}{27}}$ = ______________

b) $\sqrt{\frac{m^2}{bt^4}}$ = ______________

Answers to 71: a) $\frac{2\sqrt{6}}{9}$ b) $\frac{m\sqrt{b}}{bt^2}$

---

72. Even when we cannot simplify a radical containing a fraction, we can still rationalize its denominator. For example:

$$\sqrt{\frac{5}{7}} = \frac{\sqrt{5}}{\sqrt{7}} = \frac{\sqrt{5}}{\sqrt{7}}\left(\frac{\sqrt{7}}{\sqrt{7}}\right) = \frac{\sqrt{35}}{7}$$

$$\sqrt{\frac{x}{ay}} = \frac{\sqrt{x}}{\sqrt{ay}} = \frac{\sqrt{x}}{\sqrt{ay}}\left(\frac{\sqrt{ay}}{\sqrt{ay}}\right) = \frac{\sqrt{axy}}{ay}$$

Following the examples, rationalize each denominator.

a) $\sqrt{\frac{3}{2x}}$ = ______________

b) $\sqrt{\frac{cp}{q}}$ = ______________

Answers to 72: a) $\frac{\sqrt{6x}}{2x}$ b) $\frac{\sqrt{cpq}}{q}$

10-10 ADDING AND SUBTRACTING RADICALS

In this section, we will discuss the procedure for adding and subtracting radicals. We will assume that all radicands are non-negative.

---

73. Two radical terms are called like terms if they contain the same radical. For example:

$2\sqrt{5}$ and $7\sqrt{5}$ are like terms because both contain $\sqrt{5}$.

$3\sqrt{x}$ and $4\sqrt{x}$ are like terms because both contain $\sqrt{x}$.

We use the distributive principle to add like radical terms. For example:

$$2\sqrt{5} + 7\sqrt{5} = (2 + 7)\sqrt{5} = 9\sqrt{5}$$

$$3\sqrt{x} + 4\sqrt{x} = (3 + 4)\sqrt{x} = 7\sqrt{x}$$

Notice that using the distributive principle is the same as adding the coefficients of the radicals. Using that method, do these:

a) $5\sqrt{y} + 6\sqrt{y}$ = ________ b) $8\sqrt{2} + 2\sqrt{2}$ = ________

---

74. If the coefficient of a radical is not explicitly shown, its coefficient is "1". For example:

$\sqrt{7} = 1\sqrt{7}$ $\sqrt{x} = 1\sqrt{x}$

Therefore: a) $\sqrt{7} + 3\sqrt{7} = 1\sqrt{7} + 3\sqrt{7}$ = ________

b) $5\sqrt{x} + \sqrt{x} = 5\sqrt{x} + 1\sqrt{x}$ = ________

c) $\sqrt{3} + \sqrt{3} = 1\sqrt{3} + 1\sqrt{3}$ = ________

Answers to 73: a) $11\sqrt{y}$ b) $10\sqrt{2}$

---

75. Complete each addition.

a) $9\sqrt{t} + \sqrt{t}$ = ________ c) $\sqrt{15} + 6\sqrt{15}$ = ________

b) $\sqrt{y} + \sqrt{y}$ = ________ d) $\sqrt{23} + \sqrt{23}$ = ________

Answers to 74: a) $4\sqrt{7}$ b) $6\sqrt{x}$ c) $2\sqrt{3}$

---

76. We use the distributive principle to subtract like radical terms. For example:

$$10\sqrt{x} - 7\sqrt{x} = (10 - 7)\sqrt{x} = 3\sqrt{x}$$

$$2\sqrt{5} - 6\sqrt{5} = (2 - 6)\sqrt{5} = -4\sqrt{5}$$

Notice that using the distributive principle is the same as subtracting the coefficients of the radicals. Using that method, do these:

a) $15\sqrt{y} - 5\sqrt{y}$ = ________ b) $4\sqrt{3} - 9\sqrt{3}$ = ________

Answers to 75: a) $10\sqrt{t}$ b) $2\sqrt{y}$ c) $7\sqrt{15}$ d) $2\sqrt{23}$

		Answers
77.	To perform the subtractions below, it is helpful to write the "1" coefficients explicitly. That is: a) $4\sqrt{7} - \sqrt{7} = 4\sqrt{7} - 1\sqrt{7} =$ ________ b) $\sqrt{x} - 3\sqrt{x} = 1\sqrt{x} - 3\sqrt{x} =$ ________ c) $\sqrt{2} - \sqrt{2} = 1\sqrt{2} - 1\sqrt{2} =$ ________	a) $10\sqrt{y}$ b) $-5\sqrt{3}$
78.	To do the addition and subtraction below, we added and subtracted the coefficients. $5\sqrt{x-1} + 4\sqrt{x-1} = 9\sqrt{x-1}$ $8\sqrt{a+b} - 3\sqrt{a+b} = 5\sqrt{a+b}$ Complete these: a) $10\sqrt{y-8} + 20\sqrt{y-8} =$ ________ b) $12\sqrt{p+q} - 6\sqrt{p+q} =$ ________	a) $3\sqrt{7}$ b) $-2\sqrt{x}$ c) 0
79.	If the coefficient of a radical is not shown, its coefficient is "1". That is: $\sqrt{x-5} = 1\sqrt{x-5}$ $\sqrt{R+S} = 1\sqrt{R+S}$ Using the fact above, complete these: a) $7\sqrt{y+3} - \sqrt{y+3} =$ ________ b) $\sqrt{a+b} + \sqrt{a+b} =$ ________ c) $\sqrt{c-d} - \sqrt{c-d} =$ ________	a) $30\sqrt{y-8}$ b) $6\sqrt{p+q}$
80.	Two radical terms cannot be combined into one term by addition **or** subtraction <u>if they do not contain the same radical</u>. For example: $4\sqrt{3} + 5\sqrt{7}$ cannot be combined into one term. $8\sqrt{x} + 3\sqrt{y}$ cannot be combined into one term. Perform these additions and subtractions if possible. a) $\sqrt{y} + 2\sqrt{y} =$ ________ c) $2\sqrt{7} - \sqrt{7} =$ ________ b) $\sqrt{3} + 5\sqrt{2} =$ ________ d) $4\sqrt{t} - \sqrt{5} =$ ________	a) $6\sqrt{y+3}$ b) $2\sqrt{a+b}$ c) 0
		a) $3\sqrt{y}$ b) Not possible c) $\sqrt{7}$ d) Not possible

81. The distributive principle is used to add or subtract radicals with variables as coefficients. For example:

$$c\sqrt{y} + d\sqrt{y} = (c + d)\sqrt{y}$$

$$a\sqrt{x} - b\sqrt{x} = (a - b)\sqrt{x}$$

Complete these:

a) $p\sqrt{t} + q\sqrt{t}$ = ________ b) $F\sqrt{V} - D\sqrt{V}$ = ________

---

82. Each expression below contains one radical term and one non-radical term.

$3 + \sqrt{2}$ $y - \sqrt{x}$

Since the terms are "unlike", they cannot be combined into one. That is:

$3 + \sqrt{2}$ does not equal $3\sqrt{2}$

$y - \sqrt{x}$ does not equal $-y\sqrt{x}$

Perform the following additions and subtractions if possible.

a) $2\sqrt{t} - \sqrt{t}$ = ________ c) $4 - \sqrt{m}$ = ________

b) $7 + \sqrt{R}$ = ________ d) $a\sqrt{x} + \sqrt{x}$ = ________

Answers to 81:
a) $(p + q)\sqrt{t}$
b) $(F - D)\sqrt{V}$

---

83. Sometimes we can add or subtract after simplifying a radical by factoring out perfect squares. For example:

$$\sqrt{3} + \sqrt{12} = \sqrt{3} + 2\sqrt{3} = 3\sqrt{3}$$

$$\sqrt{9x} - \sqrt{x} = 3\sqrt{x} - \sqrt{x} = 2\sqrt{x}$$

Following the examples, complete these:

a) $\sqrt{8} - \sqrt{2}$ = ____________________

b) $\sqrt{5x} + \sqrt{20x}$ = ____________________

c) $\sqrt{y} - \sqrt{4y}$ = ____________________

Answers to 82:
a) $\sqrt{t}$
b) Not possible
c) Not possible
d) $(a + 1)\sqrt{x}$

---

84. Following the example, do the other operations.

$$\sqrt{x^3} + 3\sqrt{x} = x\sqrt{x} + 3\sqrt{x} = (x + 3)\sqrt{x}$$

a) $2\sqrt{y} - \sqrt{y^3}$ = ____________________

b) $\sqrt{t^3} + \sqrt{16t}$ = ____________________

c) $\sqrt{4a} - \sqrt{9a^3}$ = ____________________

Answers to 83:
a) $\sqrt{2}$
b) $3\sqrt{5x}$
c) $-\sqrt{y}$

85. Following the example, do the other operation.

$$\sqrt{x^3 + x^2} + \sqrt{9x + 9} = \sqrt{x^2(x + 1)} + \sqrt{9(x + 1)}$$
$$= x\sqrt{x + 1} + 3\sqrt{x + 1}$$
$$= (x + 3)\sqrt{x + 1}$$

$\sqrt{12y - 12} - \sqrt{3y - 3} =$

a) $(2 - y)\sqrt{y}$
b) $(t + 4)\sqrt{t}$
c) $(2 - 3a)\sqrt{a}$

---

86. A radical containing a binomial cannot be broken up into an addition or subtraction of two radicals. For example:

$$\sqrt{36 + 64} \neq \sqrt{36} + \sqrt{64}$$

Since: $\sqrt{100} \neq 6 + 8$

$10 \neq 14$

Two more examples of the statement above are:

$$\sqrt{x + 9} \neq \sqrt{x} + \sqrt{9}$$
$$\sqrt{y - 25} \neq \sqrt{y} - \sqrt{25}$$

Which of the following statements are true? ______________

a) $\sqrt{9 + 16} = \sqrt{9} + \sqrt{16}$
b) $\sqrt{t - 100} = \sqrt{t} - \sqrt{100}$
c) $\sqrt{49 - 25} = \sqrt{24}$
d) $\sqrt{R + 81} = \sqrt{R} + \sqrt{81}$

$\sqrt{3y - 3}$

---

87. Even when both terms are perfect squares, a radical containing a binomial is in its simplest form. It cannot be broken up into an addition or subtraction of two radicals. For example:

$$\sqrt{c^2 - a^2} \neq \sqrt{c^2} - \sqrt{a^2}$$

To show that the statement above is true, we can substitute numbers for the variables. If $c = 10$ and $a = 8$:

a) $\sqrt{c^2 - a^2} = \sqrt{10^2 - 8^2} = \sqrt{100 - 64} = \sqrt{36} =$ ______

b) $\sqrt{c^2} - \sqrt{a^2} = \sqrt{10^2} - \sqrt{8^2} = \sqrt{100} - \sqrt{64} = 10 - 8 =$ ______

c) Does $\sqrt{c^2 - a^2} = \sqrt{c^2} - \sqrt{a^2}$? ______

Only (c)

---

88. If the radicand is a multiplication with two perfect-square factors, it can be simplified. For example:

$$\sqrt{9x^2} = \sqrt{9} \cdot \sqrt{x^2} = 3x$$
$$\sqrt{a^2b^2} = \sqrt{a^2} \cdot \sqrt{b^2} = ab$$

Continued on following page.

a) 6
b) 2
c) No

88. Continued

If the radicand is a binomial with two perfect-square terms, it cannot be simplified. For example:

$$\sqrt{x^2 + 9} \neq \sqrt{x^2} + \sqrt{9} \text{ or } x + 3$$
$$\sqrt{a^2 - b^2} \neq \sqrt{a^2} - \sqrt{b^2} \text{ or } a - b$$

Simplify if possible:

a) $\sqrt{y^2 - 25}$ = ______ c) $\sqrt{p^2 + q^2}$ = ______

b) $\sqrt{25y^2}$ = ______ d) $\sqrt{p^2q^2}$ = ______

---

a) Not possible b) 5y c) Not possible d) pq

SELF-TEST 32 (pages 434-446)

Simplify.

1. $\sqrt{\dfrac{7}{16}}$
2. $\sqrt{\dfrac{4d^4}{25b^2}}$
3. $\sqrt{\dfrac{x^3}{y^2}}$
4. $\sqrt{\dfrac{9a^2}{4b^5}}$

Simplify.

5. $\sqrt{(4m)^2}$
6. $\left(\sqrt{\dfrac{9a}{b}}\right)^2$
7. $\sqrt{2x} \cdot \sqrt{2x}$
8. $\sqrt{16t^2}$

Rationalize each denominator. Report each answer in lowest terms.

9. $\dfrac{1}{\sqrt{3x}}$
10. $\dfrac{2\sqrt{3}}{\sqrt{6}}$
11. $\dfrac{rt}{k\sqrt{r}}$

Add or subtract.

12. $3\sqrt{5} + 4\sqrt{5}$
13. $\sqrt{y} + 2\sqrt{y}$
14. $7\sqrt{3x} - \sqrt{3x}$
15. $a\sqrt{t} - p\sqrt{t}$

ANSWERS:

1. $\dfrac{\sqrt{7}}{4}$
2. $\dfrac{2d^2}{5b}$
3. $\dfrac{x\sqrt{x}}{y}$
4. $\dfrac{3a}{2b^2\sqrt{b}}$
5. 4m
6. $\dfrac{9a}{b}$
7. 2x
8. 4t
9. $\dfrac{\sqrt{3x}}{3x}$
10. $\sqrt{2}$
11. $\dfrac{t\sqrt{r}}{k}$
12. $7\sqrt{5}$
13. $3\sqrt{y}$
14. $6\sqrt{3x}$
15. $(a - p)\sqrt{t}$

10-11 RADICAL EQUATIONS

In this section, we will discuss a method for solving radical equations.

---

89. A radical equation is an equation in which the variable appears in a radicand. For example:

$$\sqrt{2x} = 4 \qquad\qquad \sqrt{x-1} = 9$$

To solve a radical equation, we must get rid of the radical. To do so, we use the SQUARING PRINCIPLE FOR EQUATIONS. That principle says:

> If we square both sides of a true equation, the new equation is true. That is:
>
> If $a = b$
>
> Then $a^2 = b^2$

By using the squaring principle, we can get rid of the radical in each equation above.

$$\sqrt{2x} = 4 \qquad\qquad \sqrt{x-1} = 9$$

$$(\sqrt{2x})^2 = (4)^2 \qquad\qquad (\sqrt{x-1})^2 = (9)^2$$

$$2x = 16 \qquad\qquad x - 1 = 81$$

Write the non-radical equation obtained by squaring both sides of these.

a) $\sqrt{x} = 5$ b) $6 = \sqrt{4a - 3}$ c) $\sqrt{\frac{4x}{7}} = 2$

______ ______ ______

---

90. When using the squaring principle, both sides of the equation must be squared. The common error is forgetting to square the non-radical side.

Write the non-radical equation obtained by squaring both sides of these:

a) $\sqrt{x} = \frac{4}{3}$ b) $3 = \sqrt{\frac{100}{y+1}}$ c) $\sqrt{\frac{5m}{7}} = \frac{1}{2}$

______ ______ ______

Answers to 89:

a) $x = 25$

b) $36 = 4a - 3$

c) $\frac{4x}{7} = 4$

91. When the radical is isolated on one side, only two steps are needed to solve a radical equation. They are:

1. Eliminate the radical by squaring both sides.
2. Then solve the resulting non-radical equation.

We used the two steps to solve the equation below and then checked the solution. Solve and check the other equation.

$$\sqrt{x-3} = 2$$
$$(\sqrt{x-3})^2 = (2)^2$$
$$x - 3 = 4$$
$$x = 7$$

$$\sqrt{3y} = 6$$

Check

$$\sqrt{x-3} = 2$$
$$\sqrt{7-3} = 2$$
$$\sqrt{4} = 2$$
$$2 = 2$$

Check

a) $x = \frac{16}{9}$

b) $9 = \frac{100}{y+1}$

c) $\frac{5m}{7} = \frac{1}{4}$

---

92. Some radical equations have no solution. For example, we get 25 as a solution below, but 25 does not check.

$$\sqrt{x} = -5$$
$$(\sqrt{x})^2 = (-5)^2$$
$$x = 25$$

Check

$$\sqrt{x} = -5$$
$$\sqrt{25} = -5$$
$$5 \neq -5$$

The equation above has no solution because a principal square root cannot be negative. The same applies to the equation below. Solve it and show that the obtained solution does not check.

$$\sqrt{x+1} = -2$$

Check

$y = 12$

Check:

$$\sqrt{3(12)} = 6$$
$$\sqrt{36} = 6$$
$$6 = 6$$

93. When solving a radical equation, it is important to check the solution because some radical equations do not have solutions. Following the example, solve the other equation below and check your solution.

$$\sqrt{\frac{3x}{4}} = 6 \qquad\qquad 1 = \sqrt{\frac{7}{2y}}$$

$$\frac{3x}{4} = 36$$

$$3x = 144$$

$$x = 48$$

Check

$$\sqrt{\frac{3x}{4}} = 6$$

$$\sqrt{\frac{3(\overset{12}{\cancel{48}})}{\cancel{4}}} = 6$$

$$\sqrt{36} = 6$$

$$6 = 6$$

x = 3 does not check, since:

$$\sqrt{3+1} \neq -2$$

$$\sqrt{4} \neq -2$$

$$2 \neq -2$$

---

94. Following the example, solve the other equation.

$$\sqrt{\frac{3x-1}{2}} = 4 \qquad\qquad \sqrt{\frac{200}{y+1}} = 5$$

$$\frac{3x-1}{2} = 16$$

$$3x - 1 = 32$$

$$3x = 33$$

$$x = 11$$

$$y = \frac{7}{2}$$

Check:

$$1 = \sqrt{\frac{7}{\cancel{2}\left(\frac{7}{\cancel{2}}\right)}}$$

$$1 = \sqrt{\frac{7}{7}}$$

$$1 = \sqrt{1}$$

$$1 = 1$$

---

95. Following the example, solve the other equation.

$$\sqrt{5x} = \frac{3}{4} \qquad\qquad \frac{1}{2} = \sqrt{7y}$$

$$5x = \frac{9}{16}$$

$$80x = 9$$

$$x = \frac{9}{80}$$

$y = 7$, since:

$$\frac{200}{y+1} = 25$$

$$200 = 25(y+1)$$

$$200 = 25y + 25$$

$$y = \frac{1}{28}$$

96. Following the example, solve the other equation.

$$\frac{1}{3} = \sqrt{\frac{1}{d}} \qquad\qquad \sqrt{\frac{p+1}{2}} = \frac{3}{2}$$

$$\frac{1}{9} = \frac{1}{d}$$

$$\not{9}d\left(\frac{1}{\not{9}}\right) = 9\not{d}\left(\frac{1}{\not{d}}\right)$$

$$d = 9$$

$p = \frac{7}{2}$

97. When the radical is not isolated, we isolate the radical before squaring both sides. An example is shown. Use the same steps to solve the other equation.

$$2 + \sqrt{3x} = 8 \qquad\qquad 5 + \sqrt{T + 10} = 10$$

$$\sqrt{3x} = 6$$

$$3x = 36$$

$$x = 12$$

T = 15, since:

$$\sqrt{T+10} = 5$$

$$T + 10 = 25$$

$$T = 15$$

98. Notice how we isolated the radical below. Solve the other equation.

$$4\sqrt{y} = 12 \qquad\qquad 3\sqrt{2F} = 18$$

$$\sqrt{y} = 3$$

$$y = 9$$

$F = 18$

99. Following the example, solve the other equation.

$$2\sqrt{\frac{m}{3}} = 8 \qquad\qquad 30 = 10\sqrt{\frac{3t}{5}}$$

$$\sqrt{\frac{m}{3}} = 4$$

$$\frac{m}{3} = 16$$

$$m = 48$$

$t = 15$

100. Following the example, solve the other equation.

$$\sqrt{\frac{5x}{x-1}} - 3 = 0 \qquad\qquad \sqrt{\frac{V}{2V-3}} + 7 = 8$$

$$\sqrt{\frac{5x}{x-1}} = 3$$

$$\frac{5x}{x-1} = 9$$

$$5x = 9(x - 1)$$

$$5x = 9x - 9$$

$$-4x = -9$$

$$x = \frac{9}{4}$$

---

101. Following the example, solve the other equation.

Answer to 100: $V = 3$

$$\frac{3\sqrt{2x}}{4} = 6 \qquad\qquad \frac{1}{2\sqrt{P}} = 4$$

$$3\sqrt{2x} = 24$$

$$\sqrt{2x} = 8$$

$$2x = 64$$

$$x = 32$$

---

102. The equation below contains a radical on each side. We got rid of both radicals in one step by squaring both sides. Solve and check the other equation.

Answer to 101: $P = \frac{1}{64}$

$$\sqrt{3x-7} = \sqrt{x+1} \qquad\qquad \sqrt{5y-9} = \sqrt{4y-2}$$

$$(\sqrt{3x-7})^2 = (\sqrt{x+1})^2$$

$$3x - 7 = x + 1$$

$$2x = 8$$

$$x = 4$$

Check

$$\sqrt{3x-7} = \sqrt{x+1}$$

$$\sqrt{3(4)-7} = \sqrt{4+1}$$

$$\sqrt{12-7} = \sqrt{5}$$

$$\sqrt{5} = \sqrt{5}$$

103. Notice how we squared y - 1 below. Notice also how we factored and used the principle of zero products. We got two solutions.

$$y - 1 = \sqrt{4y - 7}$$
$$(y - 1)^2 = (\sqrt{4y - 7})^2$$
$$y^2 - 2y + 1 = 4y - 7$$
$$y^2 - 6y + 8 = 0$$
$$(y - 2)(y - 4) = 0$$

$y - 2 = 0$	$y - 4 = 0$
$y = 2$	$y = 4$

The two solutions are 2 and 4. We checked 2 below. Check 4 in the space provided.

Checking 2

$$y - 1 = \sqrt{4y - 7}$$
$$2 - 1 = \sqrt{4(2) - 7}$$
$$1 = \sqrt{8 - 7}$$
$$1 = \sqrt{1}$$
$$1 = 1$$

Checking 4

$$y - 1 = \sqrt{4y - 7}$$

Answer: $y = 7$

Check:

$$\sqrt{5(7) - 9} = \sqrt{4(7) - 2}$$
$$\sqrt{35 - 9} = \sqrt{28 - 2}$$
$$\sqrt{26} = \sqrt{26}$$

---

104. Sometimes when we get two solutions, one of them does not satisfy the original radical equation. A solution of that type is called an extraneous solution. For example, we got two solutions below.

$$\sqrt{x + 3} = x + 1$$
$$(\sqrt{x + 3})^2 = (x + 1)^2$$
$$x + 3 = x^2 + 2x + 1$$
$$x^2 + x - 2 = 0$$
$$(x - 1)(x + 2) = 0$$

$x - 1 = 0$	$x + 2 = 0$
$x = 1$	$x = -2$

The two solutions for $x^2 + x - 2 = 0$ are "1" and -2. We checked both solutions in the original radical equation below.

Checking "1"

$$\sqrt{1 + 3} = 1 + 1$$
$$\sqrt{4} = 2$$
$$2 = 2$$

Checking -2

$$\sqrt{-2 + 3} \neq -2 + 1$$
$$\sqrt{1} \neq -1$$
$$1 \neq -1$$

Of the two solutions, only "1" satisfies the original equation. Therefore, -2 is an extraneous solution. -2 is not a solution of the original equation.

Answer:

$$4 - 1 = \sqrt{4(4) - 7}$$
$$3 = \sqrt{9}$$
$$3 = 3$$

105. Whenever you get two solutions for a radical equation, the only way to detect an extraneous solution is to check both solutions in the original equation. Solve the equation below and check both solutions.

$$m - 5 = \sqrt{3m - 5}$$

a) The two obtained solutions are ______ and ______.

b) Is either solution an extraneous solution? __________

a) 3 and 10 b) Yes, 3 is extraneous.

10-12 RADICAL FORMULAS

In this section, we will do some evaluations and rearrangements with formulas containing radicals.

106. To do the evaluation below, we substituted and found the square root.

In $s = \sqrt{A}$, find s when $A = 64$.

$$s = \sqrt{A} = \sqrt{64} = 8$$

When the radicand is a more complex expression, we simplify before finding the square root. For example:

In $v = \sqrt{2as}$, find v when $a = 3$ and $s = 6$.

$$v = \sqrt{2as} = \sqrt{2(3)(6)} = \sqrt{36} = 6$$

In $E = \sqrt{PR}$, find E when $P = 3$ and $R = 27$.

$E = \sqrt{PR} =$ ______________________________

$E = 9$

107. Notice how we simplified before finding the square root below. Complete the other evaluations.

In $a = \sqrt{\frac{b-c}{c}}$, find a when b = 100 and c = 10.

$$a = \sqrt{\frac{b-c}{c}} = \sqrt{\frac{100-10}{10}} = \sqrt{\frac{90}{10}} = \sqrt{9} = 3$$

a) In $t = \sqrt{\frac{2s}{g}}$, find t when s = 64 and g = 32.

$t = \sqrt{\frac{2s}{g}} =$ ______________________

b) In $R = \sqrt{Z^2 - X^2}$, find R when Z = 10 and X = 8.

$R = \sqrt{Z^2 - X^2} =$ ______________________

---

a) t = 2

b) R = 6

108. In the formula below, the radical is a factor in a multiplication. Notice that we found the square root before multiplying. Complete the other evaluations.

In $r = a\sqrt{w}$, find r when a = 3 and w = 16.

$$r = a\sqrt{w} = 3\sqrt{16} = 3(4) = 12$$

a) In $N = K\sqrt{\frac{V}{H}}$, find N when K = 7, V = 18, and H = 2.

$N = K\sqrt{\frac{V}{H}} =$ ______________________

b) In $p = m + r\sqrt{t}$, find p when m = 50, r = 6, and t = 25.

$p = m + r\sqrt{t} =$ ______________________

---

a) N = 21

b) p = 80

109. To perform the evaluation below, we had to solve a radical equation. Complete the other evaluation.

In the formula below, find A when s = 10.

$$s = \sqrt{A}$$

$$10 = \sqrt{A}$$

$$A = 100$$

In the formula below, find s when v = 8 and a = 2.

$$v = \sqrt{2as}$$

110. Following the example, complete the other evaluation.

In the formula below, find R when I = 6 and P = 72.

$$I = \sqrt{\frac{P}{R}}$$

$$6 = \sqrt{\frac{72}{R}}$$

$$36 = \frac{72}{R}$$

$$36R = 72$$

$$R = 2$$

In the formula below, find s when t = 10 and g = 3.

$$t = \sqrt{\frac{2s}{g}}$$

Answer: s = 16, from:

$$8 = \sqrt{4s}$$

$$64 = 4s$$

---

111. a) In the formula below, find I_c when d = 5, $I_a = 100$, and m = 2.

$$d = \sqrt{\frac{I_a - I_c}{m}}$$

b) In the formula below, find Q when R = 100 and P = 95.

$$R = P + \sqrt{Q}$$

Answer: s = 150, from:

$$100 = \frac{2s}{3}$$

---

112. a) In the formula below, find T when a = 100, R = 4, and V = 50.

$$\frac{a\sqrt{T}}{R} = V$$

b) In the formula below, find L_2 when K = 10, M = 100, and $L_1 = 5$.

$$K = \frac{M}{\sqrt{L_1 L_2}}$$

Answers:

a) $I_c = 50$, from:

$$25 = \frac{100 - I_c}{2}$$

b) Q = 25, from:

$$5 = \sqrt{Q}$$

113. Following the example, complete the other evaluation.

In the formula below, find g_p when $g_o = 4$, $r_o = 60$, and $r_p = 12$.

$$\sqrt{\frac{g_p}{g_o}} = \frac{r_o}{r_p}$$

$$\sqrt{\frac{g_p}{4}} = \frac{60}{12}$$

$$\sqrt{\frac{g_p}{4}} = 5$$

$$\frac{g_p}{4} = 25$$

$$g_p = 100$$

In the formula below, find S when T = 45, V = 30, and D = 10.

$$\sqrt{\frac{T}{S}} = \frac{V}{D}$$

a) T = 4, from:

$$\sqrt{T} = 2$$

b) $L_2 = 20$, from:

$$\sqrt{5L_2} = 10$$

---

114. If a radical is isolated on one side of a formula, we can use the squaring principle to eliminate the radical. For example:

$$a = \sqrt{b + c}$$

$$(a)^2 = (\sqrt{b + c})^2$$

$$a^2 = b + c$$

$$I = \sqrt{\frac{P}{R}}$$

$$(I)^2 = \left(\sqrt{\frac{P}{R}}\right)^2$$

$$I^2 = \frac{P}{R}$$

Write the non-radical formula obtained by squaring both sides of these:

a) $v = \sqrt{2as}$ b) $h = \sqrt{\frac{R - V}{t}}$ c) $\sqrt{ab} = d$

________   ________   ________

S = 5, from:

$$\sqrt{\frac{45}{S}} = 3$$

---

115. To solve for a variable under the radical, we use the squaring principle to eliminate the radical and then proceed in the usual way. As an example, we solved for P below. Solve for g in the other formula.

$$I = \sqrt{\frac{P}{R}}$$

$$I^2 = \frac{P}{R}$$

$$R(I^2) = R\left(\frac{P}{R}\right)$$

$$P = I^2R$$

$$t = \sqrt{\frac{2s}{g}}$$

a) $v^2 = 2as$

b) $h^2 = \frac{R - V}{t}$

c) $ab = d^2$

116. We solved for s below. Solve for b in the other formula.

$$v = \sqrt{2as} \qquad\qquad a = \sqrt{b + c}$$

$$v^2 = 2as$$

$$s = \frac{v^2}{2a}$$

$g = \frac{2s}{t^2}$

---

117. When the non-radical side is a monomial containing more than one factor, we square it by squaring each factor. For example:

$$(bc)^2 = b^2c^2$$

We used the fact above to solve for Q below. Solve for d in the other formula.

$$\sqrt{\frac{K}{Q}} = bt \qquad\qquad \sqrt{a + 2d} = cp$$

$$\frac{K}{Q} = b^2t^2$$

$$K = Qb^2t^2$$

$$Q = \frac{K}{b^2t^2}$$

$b = a^2 - c$

---

118. When the non-radical side is a fraction, we square it by squaring both terms. For example:

$$\left(\frac{a}{b}\right)^2 = \frac{a^2}{b^2}$$

Using the fact above, we solved for L_1 below. Solve for b_o in the other formula.

$$\sqrt{L_1L_2} = \frac{M}{K} \qquad\qquad \sqrt{\frac{b_p}{b_o}} = \frac{d_o}{d_p}$$

$$L_1L_2 = \frac{M^2}{K^2}$$

$$K^2L_1L_2 = M^2$$

$$L_1 = \frac{M^2}{K^2L_2}$$

$d = \frac{c^2p^2 - a}{2}$

---

119. To solve for r below, we isolated the radical before using the squaring principle. Solve for b in the other formula.

$$\frac{\sqrt{r}}{b} = t \qquad\qquad H = \frac{p}{\sqrt{b}}$$

$$\sqrt{r} = bt$$

$$r = b^2t^2$$

$b_o = \frac{b_pd_p^2}{d_o^2}$

$b = \frac{p^2}{H^2}$

120. When solving for a variable that is not part of the radicand, we do not have to use the squaring principle. As an example, we solved for m below. Solve for d in the other formula.

$$v = \frac{m}{\sqrt{t}} \qquad\qquad \frac{c\sqrt{y}}{d} = R$$

$$v\sqrt{t} = m$$

or $$m = v\sqrt{t}$$

$$d = \frac{c\sqrt{y}}{R}$$

SELF-TEST 33 (pages 447-458)

Solve each equation.

1. $\sqrt{5x} = 10$	2. $\sqrt{\frac{2x}{5}} = 1$	3. $\frac{2}{3} = \sqrt{\frac{y}{9}}$
4. $\frac{3\sqrt{2w}}{2} = 6$	5. $15 = 5\sqrt{\frac{18}{t}}$	6. $x + 1 = \sqrt{2x + 5}$

7. In $N = G - K\sqrt{P}$, find N when $G = 20$, $K = 3$, and $P = 25$.	8. In $h = \sqrt{\frac{2d}{a}}$, find d when $h = 6$ and $a = 3$.
9. In $R = \frac{B\sqrt{V}}{H}$, find V when $R = 20$, $B = 10$, and $H = 2$.	10. Solve for a. $\sqrt{h - a} = t$
11. Solve for B_2. $\frac{W}{P} = \sqrt{\frac{B_1}{B_2}}$	12. Solve for t. $m = \frac{a\sqrt{t}}{b}$

ANSWERS:

1. $x = 20$
2. $x = \frac{5}{2}$
3. $y = 4$
4. $w = 8$
5. $t = 2$
6. $x = 2$ (not $x = -2$)
7. $N = 5$
8. $d = 54$
9. $V = 16$
10. $a = h - t^2$
11. $B_2 = \frac{B_1P^2}{W^2}$
12. $t = \frac{b^2m^2}{a^2}$

SUPPLEMENTARY PROBLEMS - CHAPTER 10

Assignment 31

Find the two square roots of each of the following.

1. 9   2. 36   3. 81   4. 400

Using the square-root table, complete these:

5. $\sqrt{7}$ 6. $-\sqrt{19}$ 7. $\sqrt{51}$ 8. $-\sqrt{99}$

Do these evaluations. Use the square-root table when needed.

9. When $x = 9$, $\sqrt{x + 7} =$ 10. When $y = 10$, $\sqrt{69 - 2y} =$

11. When $t = 5$, $\sqrt{\frac{t^2 - 1}{2}} =$ 12. When $m = 2$, $\sqrt{\frac{m + 8}{m}} =$

Do the following multiplications.

13. $\sqrt{5x} \cdot \sqrt{3y}$ 14. $\sqrt{ax} \cdot \sqrt{ax^4}$ 15. $\sqrt{\frac{1}{2}} \cdot \sqrt{\frac{1}{3}}$

16. $\sqrt{3} \cdot \sqrt{x - 5}$ 17. $2\sqrt{2} \cdot x\sqrt{y}$ 18. $\sqrt{t - 4} \cdot 3\sqrt{t + 1}$

Simplify.

19. $\sqrt{27}$ 20. $\sqrt{300x}$ 21. $\sqrt{16t^4}$ 22. $\sqrt{cd^8}$

23. $\sqrt{t^3}$ 24. $\sqrt{28y^9}$ 25. $\sqrt{4h^5p^3}$ 26. $\sqrt{9a - 27b}$

Multiply and simplify each product.

27. $\sqrt{3} \cdot \sqrt{6}$ 28. $\sqrt{2} \cdot \sqrt{18y}$ 29. $\sqrt{3x} \cdot \sqrt{8x}$

30. $\sqrt{6a^3} \cdot \sqrt{8ab}$ 31. $\sqrt{10tw} \cdot \sqrt{20tw^2}$ 32. $\sqrt{h^5k} \cdot \sqrt{12k^3}$

Divide and simplify each quotient.

33. $\frac{\sqrt{40}}{\sqrt{5}}$ 34. $\frac{\sqrt{27x^2}}{\sqrt{3x}}$ 35. $\frac{\sqrt{d^2p^5}}{\sqrt{dp}}$ 36. $\frac{\sqrt{75a^5b^3}}{\sqrt{3a^2b}}$

Assignment 32

Simplify.

1. $\sqrt{\frac{4}{9}}$ 2. $\sqrt{\frac{36x^2}{25}}$ 3. $\sqrt{\frac{12y}{49}}$ 4. $\sqrt{\frac{1}{a^2b^4}}$

5. $\sqrt{\frac{t^3}{m^6}}$ 6. $\sqrt{\frac{16}{2a^2x}}$ 7. $\sqrt{\frac{4y^2}{9x^5}}$ 8. $\sqrt{\frac{4x - 8y}{81}}$

Multiply and simplify each product.

9. $\sqrt{48} \cdot \sqrt{\frac{1}{3}}$ 10. $\sqrt{\frac{8}{3}} \cdot \sqrt{\frac{3}{4}}$ 11. $\sqrt{\frac{1}{5}} \cdot \sqrt{\frac{x^3}{5}}$ 12. $\sqrt{\frac{4}{x^5}} \cdot \sqrt{\frac{5}{x}}$

Simplify.

13. $\sqrt{(16x)^2}$ 14. $\sqrt{16x^2}$ 15. $\sqrt{\left(\frac{y}{3}\right)^2}$ 16. $\sqrt{\frac{y^2}{3}}$

17. $\sqrt{(ab)^2}$ 18. $\sqrt{ab^2}$ 19. $\sqrt{(x - 7)^2}$ 20. $\sqrt{\left(\frac{a - b}{c}\right)^2}$

Simplify.

21. $(\sqrt{5})^2$ 22. $\sqrt{3} \cdot \sqrt{3}$ 23. $(\sqrt{xy})^2$ 24. $\sqrt{2t} \cdot \sqrt{2t}$

25. $(\sqrt{y - 4})^2$ 26. $\sqrt{x + 1} \cdot \sqrt{x + 1}$ 27. $\left(\sqrt{\frac{t - 2}{5}}\right)^2$ 28. $\sqrt{\frac{x - y}{7}} \cdot \sqrt{\frac{x - y}{7}}$

Rationalize each denominator. Report answer in lowest terms.

29. $\dfrac{1}{\sqrt{2x}}$ 30. $\dfrac{3}{2\sqrt{5}}$ 31. $\dfrac{2\sqrt{3}}{\sqrt{x}}$ 32. $\dfrac{d}{\sqrt{p-q}}$

33. $\dfrac{3}{2\sqrt{3}}$ 34. $\dfrac{y}{\sqrt{y}}$ 35. $\sqrt{\dfrac{4}{5}}$ 36. $\sqrt{\dfrac{8x^2}{12y^2}}$

Add or subtract.

37. $7\sqrt{3} + 2\sqrt{3}$ 38. $\sqrt{x} + 3\sqrt{x}$ 39. $a\sqrt{y} + b\sqrt{y}$

40. $2\sqrt{10} - \sqrt{10}$ 41. $\sqrt{y} - 5\sqrt{y}$ 42. $K\sqrt{A} - W\sqrt{A}$

Assignment 33

Solve each equation.

1. $\sqrt{x} = 4$ 2. $\sqrt{3y} = 9$ 3. $\sqrt{\dfrac{m}{2}} = 5$ 4. $1 = \sqrt{\dfrac{5}{7x}}$

5. $\sqrt{\dfrac{2P-8}{5}} = 2$ 6. $\sqrt{3x} = \dfrac{3}{2}$ 7. $\dfrac{1}{4} = \sqrt{\dfrac{1}{y}}$ 8. $4 + \sqrt{V-5} = 6$

9. $5\sqrt{x} = 40$ 10. $20 = 10\sqrt{\dfrac{2x}{5}}$ 11. $\sqrt{\dfrac{m-1}{m}} - 2 = 0$ 12. $\dfrac{3}{2\sqrt{y}} = 2$

13. $\sqrt{5t-4} = \sqrt{3t+2}$ 14. $w + 3 = \sqrt{9w+7}$ 15. $\sqrt{y-1} = y - 3$ 16. $\sqrt{m+3} = -4$

Do each evaluation.

17. In $t = \sqrt{\dfrac{2s}{g}}$, find $\underline{t}$ when $s = 144$ and $g = 32$.

18. In $a = \sqrt{c^2 - b^2}$, find $\underline{a}$ when $c = 5$ and $b = 3$.

19. In $N = K\sqrt{\dfrac{V}{H}}$, find N when $K = 5$, $V = 48$ and $H = 12$.

20. In $v = \sqrt{2as}$, find $\underline{a}$ when $v = 20$ and $s = 25$.

21. In $I = \sqrt{\dfrac{P}{R}}$, find R when $P = 100$ and $I = 5$.

22. In $d = \sqrt{\dfrac{I-R}{t}}$, find R when $I = 200$, $d = 4$ and $t = 10$.

23. In $\dfrac{a\sqrt{t}}{r} = v$, find $\underline{t}$ when $a = 8$, $r = 4$ and $v = 10$.

24. In $\sqrt{\dfrac{a}{b}} = \dfrac{c}{d}$, find $\underline{a}$ when $b = 3$, $c = 80$ and $d = 20$.

Do each rearrangement.

25. Solve for $\underline{s}$. $v = \sqrt{2as}$

26. Solve for R. $I = \sqrt{\dfrac{P}{R}}$

27. Solve for $\underline{q}$. $m = \sqrt{p+q}$

28. Solve for $\underline{c}$. $a = \sqrt{\dfrac{b-c}{d}}$

29. Solve for V. $\sqrt{\dfrac{T}{V}} = ab$

30. Solve for $\underline{t}$. $\sqrt{a+bt} = mp$

31. Solve for P_2. $\sqrt{P_1P_2} = \dfrac{c}{d}$

32. Solve for E. $\sqrt{\dfrac{E}{F}} = \dfrac{H}{P}$

33. Solve for R. $W = P\sqrt{R}$

34. Solve for $\underline{h}$. $s = \dfrac{b\sqrt{h}}{r}$

35. Solve for B. $P = \dfrac{B}{\sqrt{G}}$

36. Solve for A. $F = \dfrac{A\sqrt{R}}{D}$

Quadratic Equations

11

In this chapter, we will define quadratic equations and solve them by factoring, the square root method, completing the square, and the quadratic formula. Some evaluations and rearrangements with formulas are included, with special sections for the Pythagorean Theorem and the compound interest formula. Quadratic equations are also graphed.

11-1 SOLVING BY FACTORING

In this section, we will define quadratic equations and review the factoring method for solving quadratic equations.

1. A quadratic equation is a polynomial equation whose highest-degree term is second-degree. Quadratic equations are also called second-degree equations.

   THE STANDARD FORM OF A QUADRATIC EQUATION is:

   $$ax^2 + bx + c = 0$$

   where a, b, and c are real numbers and $a \neq 0$.

Continued on following page.

1. Continued

When both b and c do not equal 0, the left side is a trinomial. For example:

$$x^2 + 7x + 10 = 0 \qquad 2y^2 - 7y + 6 = 0$$

When c = 0, the left side is a binomial because there is no c-term. For example:

$$x^2 + 9x = 0 \qquad 3y^2 - 4y = 0$$

When b = 0, the left side is a binomial because there is no bx-term. For example:

$$x^2 - 25 = 0 \qquad 4y^2 - 9 = 0$$

---

2. In Chapter 6, we solved quadratic equations by the factoring method. An example is shown below. Notice that we factored the left side and then used the principle of zero products. Solve the other equation.

$$x^2 + 3x - 10 = 0 \qquad y^2 - 6y + 8 = 0$$

$$(x - 2)(x + 5) = 0$$

$x - 2 = 0$	$x + 5 = 0$
$x = 2$	$x = -5$

---

3. Following the example, solve the other equation.

Answer to 2: $y = 2$ and 4

$$5d^2 + 3d - 2 = 0 \qquad 2x^2 - x - 10 = 0$$

$$(5d - 2)(d + 1) = 0$$

$5d - 2 = 0$	$d + 1 = 0$
$5d = 2$	$d = -1$
$d = \frac{2}{5}$	

---

4. We got -3 as a root twice for the equation below. Sometimes that is called a double root. Solve the other equation.

Answer to 3: $x = \frac{5}{2}$ and -2

$$x^2 + 6x + 9 = 0 \qquad y^2 - 10y + 25 = 0$$

$$(x + 3)(x + 3) = 0$$

$x + 3 = 0$	$x + 3 = 0$
$x = -3$	$x = -3$

Answer to 4: $y = 5$ and 5

5. In the equations below, there is no <u>bx</u> term because b = 0. Following the example, solve the other equation.

$$9x^2 - 4 = 0 \qquad\qquad 36y^2 - 1 = 0$$

$$(3x + 2)(3x - 2) = 0$$

$3x + 2 = 0$	$3x - 2 = 0$
$3x = -2$	$3x = 2$
$x = -\frac{2}{3}$	$x = \frac{2}{3}$

$y = \frac{1}{6}$ and $-\frac{1}{6}$

6. In the equations below, there is no <u>c</u> term because c = 0. For equations of that type, one root is always 0. Solve the other equation.

$$x^2 + 3x = 0 \qquad\qquad 4y^2 - 5y = 0$$

$$x(x + 3) = 0$$

$x = 0$	$x + 3 = 0$
	$x = -3$

$y = 0$ and $\frac{5}{4}$

7. Before using the factoring method, we must get all terms on one side with 0 on the other side. An example is shown. Solve the other equation.

$$x^2 = 5x + 14 \qquad\qquad 3y^2 + 2 = 5y$$

$$x^2 - 5x - 14 = 0$$

$$(x + 2)(x - 7) = 0$$

$x + 2 = 0$	$x - 7 = 0$
$x = -2$	$x = 7$

$y = \frac{2}{3}$ and 1

8. Following the example, solve the other equation.

$$5x^2 = 4x \qquad\qquad 16y^2 = 25$$

$$5x^2 - 4x = 0$$

$$x(5x - 4) = 0$$

$x = 0$	$5x - 4 = 0$
	$5x = 4$
	$x = \frac{4}{5}$

$y = \frac{5}{4}$ and $-\frac{5}{4}$

11-2 SOLVING BY THE SQUARE ROOT METHOD

In this section, we will show how quadratic equations of the forms $x^2 = k$ and $(x + h)^2 = k$ can be solved by the square root method.

---

9. We used the factoring method to solve $x^2 = 49$ below.

$$x^2 = 49$$

$$x^2 - 49 = 0$$

$$(x - 7)(x + 7) = 0$$

$$x - 7 = 0 \quad \Big| \quad x + 7 = 0$$

$$x = 7 \quad \Big| \quad x = -7$$

We can also use the square root principle to solve equations of the form $x^2 = k$. The SQUARE ROOT PRINCIPLE is:

> If $x^2 = k$ and k is positive, then:
>
> $$x^2 = k$$
> $$\sqrt{x^2} = \pm\sqrt{k}$$
> $$x = \pm\sqrt{k}$$
>
> Note: $x = \pm\sqrt{k}$ means: $x = \sqrt{k}$ and $x = -\sqrt{k}$

We used the square root principle to solve $x^2 = 49$ below.

$$x^2 = 49$$

$$\sqrt{x^2} = \pm\sqrt{49}$$

$$x = \pm\sqrt{49}$$

$$x = \sqrt{49} \text{ and } -\sqrt{49}$$

$$x = 7 \text{ and } -7$$

Use the same method to solve this equation.

$$y^2 = 4$$

$$y = \pm\sqrt{4}$$

y = ______ and ______

---

2 and -2

10. To solve the equation below, we used the addition axiom to get t^2 on one side and 100 on the other side. Solve the other equation.

$$t^2 - 100 = 0 \qquad\qquad m^2 - 1 = 0$$
$$t^2 = 100$$
$$t = \pm\sqrt{100}$$
$$t = 10 \text{ and } -10$$

$m = 1$ and -1

11. In each equation below, the number is a perfect-square fraction. The two solutions are the two square roots of the fraction. That is:

$$x^2 = \frac{36}{25} \qquad\qquad y^2 = \frac{1}{4}$$
$$x = \pm\sqrt{\frac{36}{25}} \qquad\qquad y = \pm\sqrt{\frac{1}{4}}$$
$$x = \frac{6}{5} \text{ and } -\frac{6}{5} \qquad\qquad y = \_\_\_\_ \text{ and } \_\_\_\_$$

$\frac{1}{2}$ and $-\frac{1}{2}$

12. To solve the equation below, we used the multiplication axiom to isolate t^2 and then found both square roots of $\frac{9}{16}$. Solve the other equation.

$$16t^2 = 9 \qquad\qquad 49d^2 = 100$$
$$t^2 = \frac{9}{16}$$
$$t = \pm\sqrt{\frac{9}{16}}$$
$$t = \frac{3}{4} \text{ and } -\frac{3}{4}$$

$d = \frac{10}{7}$ and $-\frac{10}{7}$

13. To solve the equation below, we used both axioms to isolate x^2 and then found both square roots of $\frac{81}{64}$. Solve the other equation.

$$64x^2 - 81 = 0 \qquad\qquad 9b^2 - 1 = 0$$
$$64x^2 = 81$$
$$x^2 = \frac{81}{64}$$
$$x = \frac{9}{8} \text{ and } -\frac{9}{8}$$

$b = \frac{1}{3}$ and $-\frac{1}{3}$

14. We used the multiplication axiom to isolate x^2 below. Notice that we multiplied both sides by 2. Solve the other equation.

$$\frac{1}{2}x^2 = 8 \qquad\qquad 12 = \frac{1}{3}y^2$$

$$\not{2}\left(\frac{1}{\not{2}}x^2\right) = 2(8)$$

$$x^2 = 16$$

$$x = 4 \text{ and } -4$$

---

15. To clear the fraction below, we multiplied both sides by 3. Solve the other equation.

$y = 6$ and -6

$$\frac{x^2}{3} = 12 \qquad\qquad 16 = \frac{9y^2}{4}$$

$$\not{3}\left(\frac{x^2}{\not{3}}\right) = 3(12)$$

$$x^2 = 36$$

$$x = 6 \text{ and } -6$$

---

16. The solution of the equation below is 0. Solve the other equation.

$y = \frac{8}{3}$ and $-\frac{8}{3}$

$$x^2 = 0 \qquad\qquad \frac{1}{3}y^2 = 0$$

$$x = \sqrt{0}$$

$$x = 0$$

---

17. When the number is not a perfect square, we use a calculator or the square root table. Using the table, we get:

$y = 0$

$$x^2 = 69$$

$$x = \pm\sqrt{69}$$

$$x = \pm 8.307$$

Note: ± 8.307 means 8.307 and -8.307.

Use the square root table to solve these:

a) $t^2 - 94 = 0$ b) $\frac{1}{4}m^2 = 20$

18. Following the example, use the square root table to solve the other equation.

$$2p^2 = 30$$
$$p^2 = \frac{30}{2}$$
$$p^2 = 15$$
$$p = \pm 3.873$$

$$5d^2 - 475 = 0$$

a) $t = \pm\sqrt{94} = \pm 9.695$

b) $m = \pm\sqrt{80} = \pm 8.944$

---

19. Since there is no real number that is the square root of a negative number, the equation below has no real number solution.

$$y^2 = -100$$
$$y = \pm\sqrt{-100}$$

Solve each equation if possible.

a) $x^2 = 64$ b) $x^2 = -64$

$d = \pm 9.747$

---

20. The square root method also applies to equations of the form $(x + h)^2 = k$. That is:

> If $(x + h)^2 = k$ and <u>k</u> is positive, then:
>
> $$(x + h)^2 = k$$
> $$\sqrt{(x + h)^2} = \pm\sqrt{k}$$
> $$x + h = \pm\sqrt{k}$$
>
> <u>Note</u>: $x + h = \pm\sqrt{k}$ means: $x + h = \sqrt{k}$ <u>and</u> $x + h = -\sqrt{k}$.

We used the square root principle to solve $(x + 2)^2 = 25$ below.

$$(x + 2)^2 = 25$$
$$\sqrt{(x + 2)^2} = \pm\sqrt{25}$$
$$x + 2 = \pm\sqrt{25}$$

$x + 2 = 5$	$x + 2 = -5$
$x = 3$	$x = -7$

We checked 3 as a root below. Check -7 as a root.

$$(x + 2)^2 = 25$$
$$(3 + 2)^2 = 25$$
$$5^2 = 25$$
$$25 = 25$$

$$(x + 2)^2 = 25$$

a) $x = 8$ and -8

b) No real number solution.

21. Following the example, solve the other equation.

$$(x - 1)^2 = 16$$
$$\sqrt{(x-1)^2} = \pm\sqrt{16}$$
$$x - 1 = \pm\sqrt{16}$$

$x - 1 = 4$	$x - 1 = -4$
$x = 5$	$x = -3$

$$(y - 5)^2 = 49$$

Answers: $(-7 + 2)^2 = 25$; $(-5)^2 = 25$; $25 = 25$

---

22. Following the example, solve the other equation.

$$(2x - 1)^2 = 9$$
$$2x - 1 = \pm\sqrt{9}$$

$2x - 1 = 3$	$2x - 1 = -3$
$2x = 4$	$2x = -2$
$x = 2$	$x = -1$

$$(3y - 7)^2 = 4$$

Answer: $y = 12$ and -2

---

23. Notice how we left the solution below in radical form. Solve the other equation.

$$(x + 3)^2 = 5$$
$$x + 3 = \pm\sqrt{5}$$

$x + 3 = \sqrt{5}$	$x + 3 = -\sqrt{5}$
$x = -3 + \sqrt{5}$	$x = -3 - \sqrt{5}$

$$(y - 5)^2 = 11$$

Answer: $y = 3$ and $\frac{5}{3}$

---

24. Since there is no real number that is the square root of a negative number, the equation below has no real number solution.

$$(x + 1)^2 = -16$$
$$x + 1 = \pm\sqrt{-16}$$

Solve each equation if possible.

a) $(y - 2)^2 = 36$

b) $(y - 2)^2 = -36$

Answer: $y = 5 + \sqrt{11}$ and $5 - \sqrt{11}$

Answers:

a) $y = 8$ and -4

b) No real number solution.

11-3 SOLVING BY COMPLETING THE SQUARE

The factoring and square root methods are not general methods. That is, they cannot be used to solve all quadratic equations. In this section, we will solve quadratic equations by a general method called completing the square.

---

25. The perfect square trinomial below is the square of a binomial. The coefficient of the x^2 term is "1".

$$x^2 + 10x + 25 = (x + 5)^2$$

Notice the relationship between 25 and 10. 25 is the square of half of 10. That is:

$$\frac{1}{2}(10) = 5 \quad \text{and} \quad 5^2 = 25$$

We can use the relationship above to add a number to $x^2 + 12x$ to make it a perfect square trinomial. That is, we take half of 12 and square it.

Since $\frac{1}{2}(12) = 6$ and $6^2 = 36$, we add 36 to $x^2 + 12x$ and get:

$$x^2 + 12x + 36 = (x + 6)^2$$

The procedure above is called completing the square. By taking half the coefficient of x and squaring it, complete the square for each of these.

a) $x^2 + 4x +$ ______ $= (x +$ ______$)^2$

b) $x^2 + 14x +$ ______ $= (x +$ ______$)^2$

Answers to 25:

a) $x^2 + 4x + 4 = (x + 2)^2$

b) $x^2 + 14x + 49 = (x + 7)^2$

26. The perfect square trinomial below is the square of a binomial.

$$x^2 - 16x + 64 = (x - 8)^2$$

Notice the relationship between 64 and -16. 64 is the square of half of -16. That is:

$$\frac{1}{2}(-16) = -8 \quad \text{and} \quad (-8)^2 = 64$$

We can use the relationship above to add a number to $x^2 - 6x$ to make it a perfect square trinomial. That is, we take half of -6 and square it.

Since $\frac{1}{2}(-6) = -3$ and $(-3)^2 = 9$, we add 9 to $x^2 - 6x$ and get:

$$x^2 - 6x + 9 = (x - 3)^2$$

The procedure above is called completing the square. By taking half the coefficient of x and squaring it, complete the square for each of these.

a) $x^2 - 8x +$ ______ $= (x -$ ______$)^2$

b) $x^2 - 18x +$ ______ $= (x -$ ______$)^2$

Answers to 26:

a) $x^2 - 8x + 16 = (x - 4)^2$

b) $x^2 - 18x + 81 = (x - 9)^2$

27. To complete a square, we sometimes add a fraction. For example, to complete the square for $x^2 + 5x$ below, we took half of 5 and squared it.

Since $\frac{1}{2}(5) = \frac{5}{2}$ and $\left(\frac{5}{2}\right)^2 = \frac{25}{4}$, we add $\frac{25}{4}$ to $x^2 + 5x$ and get:

$$x^2 + 5x + \frac{25}{4} = \left(x + \frac{5}{2}\right)^2$$

By taking half the coefficient of x and squaring it, complete the square for each of these.

a) $x^2 + 3x +$ ______ $= (x +$ ______$)^2$

b) $x^2 - 7x +$ ______ $= (x -$ ______$)^2$

Answers: a) $x^2 + 3x + \frac{9}{4} = \left(x + \frac{3}{2}\right)^2$ b) $x^2 - 7x + \frac{49}{4} = \left(x - \frac{7}{2}\right)^2$

28. We could solve the equation below by the factoring method. However, we will solve it by a method called completing the square. In that method, we write the equation in the form $(x + h)^2 = k$ and then use the square root principle. To begin, we remove the 12 from the left side.

$$x^2 + 8x + 12 = 0$$
$$x^2 + 8x = -12$$

If we add 4^2 or 16 to both sides, we complete the square on the left side. We can then write the equation in the form $(x + h)^2 = k$ and solve it.

$$x^2 + 8x + 16 = -12 + 16$$
$$(x + 4)^2 = 4$$
$$x + 4 = \pm\sqrt{4}$$
$$x + 4 = \pm 2$$

$x + 4 = 2$	$x + 4 = -2$
$x = -2$	$x = -6$

We checked -2 as a root below. Check -6 as a root.

$$x^2 + 8x + 12 = 0$$
$$(-2)^2 + 8(-2) + 12 = 0$$
$$4 + (-16) + 12 = 0$$
$$0 = 0$$

$$x^2 + 8x + 12 = 0$$

Answer: $(-6)^2 + 8(-6) + 12 = 0$, $36 + (-48) + 12 = 0$, $0 = 0$

29. $x^2 - 6x + 4 = 0$ cannot be solved by the factoring method or square root method. However, it can be solved by completing the square. We solved it below. Notice that we left the solutions in radical form. Use the same method to solve the other equation.

$$x^2 - 6x + 4 = 0$$
$$x^2 - 6x = -4$$
$$x^2 - 6x + 9 = -4 + 9$$
$$(x - 3)^2 = 5$$
$$x - 3 = \pm\sqrt{5}$$

$x - 3 = \sqrt{5}$	$x - 3 = -\sqrt{5}$
$x = 3 + \sqrt{5}$	$x = 3 - \sqrt{5}$

$$y^2 - 4y - 9 = 0$$

30. To complete the square below, we added $\left(\frac{5}{2}\right)^2$ or $\frac{25}{4}$ to both sides. Use the same method to solve the other equation.

$$x^2 + 5x - 6 = 0$$
$$x^2 + 5x = 6$$
$$x^2 + 5x + \frac{25}{4} = 6 + \frac{25}{4}$$
$$\left(x + \frac{5}{2}\right)^2 = \frac{24}{4} + \frac{25}{4}$$
$$\left(x + \frac{5}{2}\right)^2 = \frac{49}{4}$$
$$x + \frac{5}{2} = \pm\sqrt{\frac{49}{4}}$$
$$x + \frac{5}{2} = \pm\frac{7}{2}$$

$x + \frac{5}{2} = \frac{7}{2}$	$x + \frac{5}{2} = -\frac{7}{2}$
$x = \frac{7}{2} - \frac{5}{2}$	$x = -\frac{7}{2} - \frac{5}{2}$
$x = \frac{2}{2}$	$x = \frac{-12}{2}$
$x = 1$	$x = -6$

$$y^2 - 3y - 10 = 0$$

Answer: $y = 2 + \sqrt{13}$ and $2 - \sqrt{13}$

---

31. To use completing the square, the coefficient of the x^2 term must be "1". If it is not "1", we can make it "1" by multiplying both sides by its reciprocal. For example, we multiplied both sides below by $\frac{1}{2}$. Then to complete the square, we added $\frac{9}{16}$ to both sides, since $\frac{1}{2}\left(-\frac{3}{2}\right) = \left(-\frac{3}{4}\right)$ and $\left(-\frac{3}{4}\right)^2 = \frac{9}{16}$. Use the same method to solve the other equation. Begin by multiplying both sides by $\frac{1}{3}$.

$$2x^2 - 3x - 3 = 0$$
$$\frac{1}{2}(2x^2 - 3x - 3) = \frac{1}{2}(0)$$
$$x^2 - \frac{3}{2}x - \frac{3}{2} = 0$$
$$x^2 - \frac{3}{2}x = \frac{3}{2}$$
$$x^2 - \frac{3}{2}x + \frac{9}{16} = \frac{3}{2} + \frac{9}{16}$$
$$\left(x - \frac{3}{4}\right)^2 = \frac{24}{16} + \frac{9}{16}$$
$$\left(x - \frac{3}{4}\right)^2 = \frac{33}{16}$$
$$x - \frac{3}{4} = \pm\sqrt{\frac{33}{16}}$$
$$x - \frac{3}{4} = \pm\frac{\sqrt{33}}{4}$$

$$3x^2 - 2x - 4 = 0$$

Answer (30): $y = 5$ and -2, from:

$$y - \frac{3}{2} = \sqrt{\frac{49}{4}}$$
$$y - \frac{3}{2} = -\sqrt{\frac{49}{4}}$$

Continued on following page.

31. Continued

$x - \frac{3}{4} = \frac{\sqrt{33}}{4}$

$x = \frac{3}{4} + \frac{\sqrt{33}}{4}$

$x = \frac{3 + \sqrt{33}}{4}$

$x - \frac{3}{4} = -\frac{\sqrt{33}}{4}$

$x = \frac{3}{4} - \frac{\sqrt{33}}{4}$

$x = \frac{3 - \sqrt{33}}{4}$

$x = \frac{1 + \sqrt{13}}{3}$ and $\frac{1 - \sqrt{13}}{3}$

SELF-TEST 34 (pages 461-472)

Solve by the factoring method.

1. $2x^2 + x - 15 = 0$

2. $y^2 = 5y$

Solve by the square root method.

3. $9x^2 - 4 = 0$

4. $y^2 - 11 = 0$

5. $(m + 2)^2 = 36$

6. $(y - 5)^2 = 7$

Complete the square.

7. $x^2 + 20x +$ ______ $= (x +$ ______$)^2$

8. $y^2 - 3y +$ ______ $= (y -$ ______$)^2$

9. Solve by completing the square.

$x^2 + 10x - 4 = 0$

ANSWERS:

1. $x = \frac{5}{2}$ and -3
2. $y = 0$ and 5
3. $x = \frac{2}{3}$ and $-\frac{2}{3}$
4. $y = \pm\sqrt{11} = \pm 3.317$
5. $m = 4$ and -8
6. $y = 5 \pm \sqrt{7}$
7. $x^2 + 20x + 100 = (x + 10)^2$
8. $y^2 - 3y + \frac{9}{4} = \left(y - \frac{3}{2}\right)^2$
9. $x = -5 \pm \sqrt{29}$

11-4 THE QUADRATIC FORMULA

Though completing the square is a general method that will solve any quadratic equation, it is not a very efficient method. In this section, we will derive and use a general formula called the quadratic formula that gives the solution for any quadratic equation.

---

32. The standard form of a quadratic equation is:

$$ax^2 + bx + c = 0 \text{ , where } a \neq 0$$

We can solve the standard equation by completing the square. To do so, we begin by multiplying both sides by $\frac{1}{a}$ and then getting rid of $\frac{c}{a}$ on the left side.

$$\frac{1}{a}(ax^2 + bx + c) = \frac{1}{a}(0)$$

$$x^2 + \frac{b}{a}x + \frac{c}{a} = 0$$

$$x^2 + \frac{b}{a}x = -\frac{c}{a}$$

Half of $\frac{b}{a} = \frac{1}{2}\left(\frac{b}{a}\right) = \frac{b}{2a}$, and $\left(\frac{b}{2a}\right)^2 = \frac{b^2}{4a^2}$. We add $\frac{b^2}{4a^2}$ to both sides.

$$x^2 + \frac{b}{a}x + \frac{b^2}{4a^2} = -\frac{c}{a} + \frac{b^2}{4a^2}$$

$$\left(x + \frac{b}{2a}\right)^2 = -\frac{4ac}{4a^2} + \frac{b^2}{4a^2}$$

$$\left(x + \frac{b}{2a}\right)^2 = \frac{b^2 - 4ac}{4a^2}$$

$$x + \frac{b}{2a} = \pm\sqrt{\frac{b^2 - 4ac}{4a^2}}$$

$$x + \frac{b}{2a} = \pm\frac{\sqrt{b^2 - 4ac}}{2a}$$

$$x = -\frac{b}{2a} \pm \frac{\sqrt{b^2 - 4ac}}{2a}$$

The two solutions of the standard equation are:

$$x = -\frac{b}{2a} + \frac{\sqrt{b^2 - 4ac}}{2a} \qquad x = -\frac{b}{2a} - \frac{\sqrt{b^2 - 4ac}}{2a}$$

$$x = \frac{-b + \sqrt{b^2 - 4ac}}{2a} \qquad x = \frac{-b - \sqrt{b^2 - 4ac}}{2a}$$

The two solutions are usually stated together as we have done below.

THE QUADRATIC FORMULA is: $x = \frac{-b \pm \sqrt{b^2 - 4ac}}{2a}$, $a \neq 0$

33. To use the quadratic formula, you must be able to identify a, b, and c in a quadratic equation. Two examples are shown.

$$x^2 + 5x + 3 = 0$$

a = 1
b = 5
c = 3

$$5y^2 - 4y - 1 = 0$$

a = 5
b = -4
c = -1

Identify a, b, and c in each equation below.

a) $t^2 - t + 30 = 0$ a = ______, b = ______, c = ______

b) $7d^2 - 2d - 8 = 0$ a = ______, b = ______, c = ______

---

a) a = 1
b = -1
c = 30

b) a = 7
b = -2
c = -8

34. Though we can solve the equation below by factoring, let's use the quadratic formula to solve it.

$$x^2 + 6x - 16 = 0$$

The quadratic formula is stated below.

$$x = \frac{-b \pm \sqrt{b^2 - 4ac}}{2a}$$

Since a = 1, b = 6, and c = -16, we can substitute and simplify.

$$x = \frac{-6 \pm \sqrt{(6)^2 - 4(1)(-16)}}{2(1)}$$

$$= \frac{-6 \pm \sqrt{36 - (-64)}}{2}$$

$$= \frac{-6 \pm \sqrt{36 + 64}}{2}$$

$$= \frac{-6 \pm \sqrt{100}}{2}$$

$$= \frac{-6 \pm 10}{2}$$

We can compute the two solutions below:

a) The first solution $= \frac{-6 + 10}{2} =$ ________

b) The second solution $= \frac{-6 - 10}{2} =$ ________

---

a) 2

b) -8

35. There are a few points you should notice about plugging values into the quadratic formula.

$$x = \frac{(-b) \pm \sqrt{b^2 - 4ac}}{2a}$$

1) -b means the opposite of b. If b = 4, -b = -4; If b = -5, -b = +5

2) b^2 (under the radical) is always a positive number.

If b = 4, $b^2 = 4^2 = 16$
If b = -5, $b^2 = (-5)^2 = 25$

3) Watch the signs with $b^2 - 4ac$ under the radical.

4ac is a set of three factors. Perform the multiplication before handling the subtraction. That is:

If a = 1, b = -10, c = -6:
$$b^2 - 4ac = (-10)^2 - [4(1)(-6)] = 100 - [-24] = 100 + 24 = 124$$

If a = -3, b = 12, c = -9:
$$b^2 - 4ac = 12^2 - [4(-3)(-9)] = 144 - [108] = 36$$

a) For the equation: $2x^2 - 5x - 3 = 0$

-b = ______ b^2 = ______ $b^2 - 4ac$ = ______

b) For the equation: $7m^2 - 6m - 1 = 0$

-b = ______ b^2 = ______ $b^2 - 4ac$ = ______

Answers to frame 35:
a) -b = 5; $b^2 = 25$; $b^2 - 4ac = 49$
b) -b = 6; $b^2 = 36$; $b^2 - 4ac = 64$

36. The quadratic formula should be memorized. Therefore, study the formula as stated at the bottom of frame 32 and then write it from memory below.

x =

Answer to frame 36:
$$x = \frac{-b \pm \sqrt{b^2 - 4ac}}{2a}$$

37. Though we could solve the equation below by factoring, let's use the quadratic formula to solve it.

$$3t^2 - 2t - 5 = 0$$

a) Substitute the values for a, b, and c into the quadratic formula.

x =

b) Complete the solution.

first solution = ______ second solution = ______

38. Though we cannot solve the equation below by factoring, we can use the quadratic formula to solve it.

$$y^2 + 9y + 7 = 0$$

$$a = 1, \quad b = 9, \quad c = 7$$

$$y = \frac{-9 \pm \sqrt{9^2 - 4(1)(7)}}{2(1)}$$

$$= \frac{-9 \pm \sqrt{81 - 28}}{2}$$

$$= \frac{-9 \pm \sqrt{53}}{2}$$

When the radicand is not a perfect square (like 53 above), we sometimes leave the solutions in radical form. However, we can also use the square root table to find $\sqrt{53}$ and then report the solutions as decimal numbers. We get:

$$y = \frac{-9 + 7.28}{2} = \frac{-1.72}{2} = -0.86$$

$$y = \frac{-9 - 7.28}{2} = \_\_\_\_ = \_\_\_\_$$

a) $x = \frac{2 \pm \sqrt{(-2)^2 - 4(3)(-5)}}{2(3)}$

b) $x = \frac{5}{3}$ and -1

---

39. Use the quadratic formula to solve this equation. Report each solution as a decimal number. Round to hundredths.

$$2x^2 - 6x - 1 = 0$$

$\frac{-16.28}{2} = -8.14$

---

40. Using the factoring method, we got -5 as a root twice for the equation below.

$$x^2 + 10x + 25 = 0$$

$$(x + 5)(x + 5) = 0$$

$x + 5 = 0$	$x + 5 = 0$
$x = -5$	$x = -5$

$x = 3.16$ and -0.16

Continued on following page.

40. Continued

We used the quadratic formula to solve the same equation below. Notice that $b^2 - 4ac = 0$ when a quadratic equation has a double root.

$$x^2 + 10x + 25 = 0$$

$$a = 1, \quad b = 10, \quad c = 25$$

$$x = \frac{-10 \pm \sqrt{10^2 - 4(1)(25)}}{2(1)}$$

$$x = \frac{-10 \pm \sqrt{100 - 100}}{2(1)}$$

$$x = \frac{-10 \pm \sqrt{0}}{2(1)}$$

The two solutions are:

$$x = \frac{-10 + 0}{2} = \frac{-10}{2} = -5 \qquad x = \frac{-10 - 0}{2} = \frac{-10}{2} = -5$$

---

41. When using the quadratic formula below, we get a negative number as the radicand. Therefore, the equation does not have real-number solutions.

$$x^2 - 2x + 7 = 0$$

$$x = \frac{2 \pm \sqrt{(-2)^2 - 4(1)(7)}}{2(1)}$$

$$= \frac{2 \pm \sqrt{4 - 28}}{2}$$

$$= \frac{2 \pm \sqrt{-24}}{2}$$

The expression $b^2 - 4ac$ is called the <u>discriminant</u>. When $b^2 - 4ac$ is positive or 0, the equation has real-number solutions. When $b^2 - 4ac$ is negative, the equation has no real-number solutions. We will avoid equations with a negative discriminant in this text.

---

42. The following suggestions can be used to pick a method for solving quadratic equations.

1. If the left side is a <u>binomial</u>:
   a) Use the square root method if the bx-term is missing.
   b) Use the factoring method if the c-term is missing.
2. If the left side is a <u>trinomial</u>:
   a) Try factoring.
   b) If factoring is impossible or difficult, use the quadratic formula.
   c) When in doubt, use the quadratic formula because it is a general method that solves all quadratic equations.

11-5 WRITING EQUATIONS IN STANDARD FORM

Before solving quadratic equations that are not in standard form, we write them in standard form. In this section, we will write equations in standard form. Only a few equations are solved.

---

43. The standard form of quadratic equations is $ax^2 + bx + c = 0$. In the standard form, a is always positive. Therefore, the equation below is not in standard form. To write it in standard form, we multiplied both sides by -1 to make a positive. Write the other equation in standard form.

$$-5x^2 + 7x - 3 = 0$$
$$-1(-5x^2 + 7x - 3) = -1(0)$$
$$5x^2 - 7x + 3 = 0$$

$$-y^2 - 4y + 9 = 0$$

Answer: $y^2 + 4y - 9 = 0$

44. We used the addition axiom to write the equation below in standard form.

$$2x^2 = x + 3$$
$$2x^2 - x - 3 = 0$$

Following the example, write these in standard form.

a) $y^2 = 7 - 5y$ b) $3m - 8 = m^2 + m$

Answer: a) $y^2 + 5y - 7 = 0$ b) $m^2 - 2m + 8 = 0$

45. Following the example, write the other equation in standard form.

$$(x + 2)(x - 5) = 9$$
$$x^2 - 3x - 10 = 9$$
$$x^2 - 3x - 19 = 0$$

$$(2y - 4)(y - 3) = 5$$

Answer: $2y^2 - 10y + 7 = 0$

46. Following the example, write the other equation in standard form.

$$x^2 + (x - 1)^2 = 5$$
$$x^2 + x^2 - 2x + 1 = 5$$
$$2x^2 - 2x - 4 = 0$$

$$y^2 + (y + 2)^2 = 9$$

Answer: $2y^2 + 4y - 5 = 0$

47. To write the equation below in standard form, we began by clearing the fraction. To do so, we multiplied both sides by $x + 3$. Write the other equation in standard form.

$$x + 1 = \frac{2}{x + 3} \qquad\qquad y + 4 = \frac{5}{y - 1}$$

$$(x + 3)(x + 1) = \left(\frac{2}{\cancel{x + 3}}\right)(\cancel{x + 3})$$

$$x^2 + 4x + 3 = 2$$

$$x^2 + 4x + 1 = 0$$

$y^2 + 3y - 9 = 0$

48. To write the equation below in standard form, we also began by clearing the fraction. To do so, we multiplied both sides by $x(x + 4)$. Write the other equation in standard form.

$$\frac{x + 4}{x} = \frac{1}{x + 4} \qquad\qquad \frac{1}{y - 5} = \frac{y - 5}{y}$$

$$\cancel{x}(x + 4)\left(\frac{x + 4}{\cancel{x}}\right) = \left(\frac{1}{\cancel{x + 4}}\right)(x)(\cancel{x + 4})$$

$$(x + 4)(x + 4) = (1)(x)$$

$$x^2 + 8x + 16 = x$$

$$x^2 + 7x + 16 = 0$$

$y^2 - 11y + 25 = 0$

49. To write the equation below in standard form, we began by multiplying both sides by $(x + 2)(x - 2)$.

$$\frac{5}{x + 2} + \frac{5}{x - 2} = 3$$

$$(x + 2)(x - 2)\left(\frac{5}{x + 2} + \frac{5}{x - 2}\right) = 3(x + 2)(x - 2)$$

$$(\cancel{x + 2})(x - 2)\left(\frac{5}{\cancel{x + 2}}\right) + (x + 2)(\cancel{x - 2})\left(\frac{5}{\cancel{x - 2}}\right) = 3(x + 2)(x - 2)$$

$$5(x - 2) + 5(x + 2) = 3(x + 2)(x - 2)$$

$$5x - 10 + 5x + 10 = 3(x^2 - 4)$$

$$10x = 3x^2 - 12$$

$$3x^2 - 10x - 12 = 0$$

Following the example, write this equation in standard form.

$$\frac{10}{x - 3} + \frac{10}{x + 3} = 1$$

50. Let's solve: $x + 2 = \frac{5}{x + 2}$.

a) Write the equation in standard form.

$$x + 2 = \frac{5}{x + 2}$$

b) Use the quadratic formula. Report the **solutions** as **decimal** numbers.

$x^2 - 20x - 9 = 0$

---

51. Let's solve: $\frac{2}{x + 1} + \frac{2}{x - 1} = 3$

a) Write the equation in standard form.

$$\frac{2}{x + 1} + \frac{2}{x - 1} = 3$$

b) Use the quadratic formula. Report the **solutions** as **decimal** numbers rounded to hundredths.

a) $x^2 + 4x - 1 = 0$

b) $x = 0.236$ and -4.236

---

a) $3x^2 - 4x - 3 = 0$

b) $x = 1.87$ and -0.54

11-6 WORD PROBLEMS

In this section, we will solve some word problems involving quadratic equations.

---

52. If an object is dropped, the distance d it falls in t seconds (disregarding air resistance) is given by the formula $d = 16t^2$. We used the formula to solve one problem below. Solve the other problem.

How long would it take an object to fall 80 ft?

$$d = 16t^2$$

$$80 = 16t^2$$

$$t^2 = \frac{80}{16}$$

$$t^2 = 5$$

$$t = \pm\sqrt{5}$$

$$t = \pm 2.236$$

Since a negative time interval does not make sense, it takes 2.236 seconds to drop 80 ft.

How long would it take an object to fall 320 ft?

---

53. To solve the problem below, we let x equal the width and x + 1 equal the length. We used the quadratic formula. Solve the other problem.

4.472 seconds

The length of a rectangle is 1 ft more than the width. The area of the rectangle is 3.75 ft^2. Find the length and width.

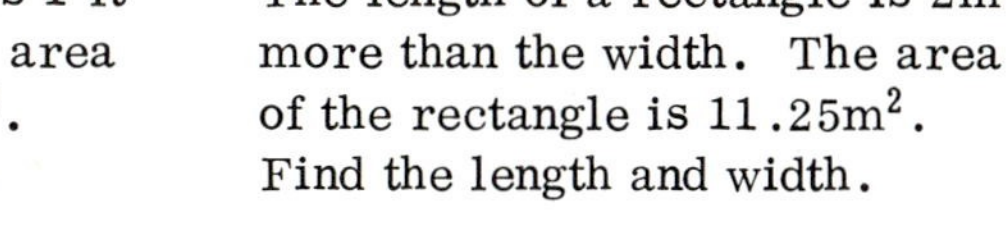

The length of a rectangle is 2m more than the width. The area of the rectangle is 11.25m^2. Find the length and width.

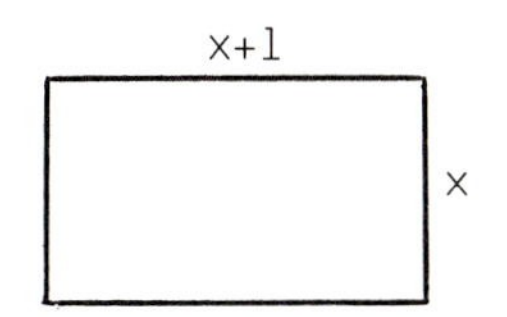

$$A = LW$$

$$3.75 = (x + 1)(x)$$

$$3.75 = x^2 + x$$

$$x^2 + x - 3.75 = 0$$

Continued on following page.

53. Continued

$$x = \frac{-1 \pm \sqrt{1^2 - 4(1)(-3.75)}}{2(1)}$$

$$= \frac{-1 \pm \sqrt{1 - (-15)}}{2}$$

$$= \frac{-1 \pm \sqrt{16}}{2}$$

$$= \frac{-1 \pm 4}{2}$$

$$x = \frac{-1 + 4}{2} = \frac{3}{2} = 1.5$$

and

$$x = \frac{-1 - 4}{2} = \frac{-5}{2} = -2.5$$

Since a negative length does not make sense, W = 1.5 ft and L = 2.5 ft.

---

54. The solution below is based on the formula $d = rt$. We used the factoring method.

A river has a 2 mph current. A boat travels 16 miles upstream and 16 miles downstream in a total of 6 hours. What is the speed of the boat in still water?

The information in the problem is given in the diagram below.

Upstream
16 miles (x-2) mph →

Downstream
← 16 miles (x+2) mph

We summarized the information in the problem in the chart below.

	Distance	Rate	Time
Upstream	16 miles	$x - 2$	t_1
Downstream	16 miles	$x + 2$	t_2

Continued on following page.

W = 2.5 ft

L = 4.5 ft

54. Continued

Since the total time is 6 hours, we know this fact:

$$t_1 + t_2 = 6$$

Rearranging $d = rt$ to solve for t, we get: $t = \frac{d}{r}$. Therefore:

$$t_1 = \frac{16}{x - 2} \quad \text{and} \quad t_2 = \frac{16}{x + 2}$$

Substituting those values in $t_1 + t_2 = 6$, we get:

$$\frac{16}{x - 2} + \frac{16}{x + 2} = 6$$

Clearing the fractions and solving, we get:

$$(x - 2)(x + 2)\left(\frac{16}{x - 2} + \frac{16}{x + 2}\right) = 6(x - 2)(x + 2)$$

$$\cancel{(x - 2)}(x + 2)\left(\frac{16}{\cancel{x - 2}}\right) + (x - 2)\cancel{(x + 2)}\left(\frac{16}{\cancel{x + 2}}\right) = 6(x^2 - 4)$$

$$16x + 32 + 16x - 32 = 6x^2 - 24$$

$$32x = 6x^2 - 24$$

$$6x^2 - 32x - 24 = 0$$

$$3x^2 - 16x - 12 = 0 \quad \text{(Multiplying by } \tfrac{1}{2} \text{ to simplify)}$$

$$(3x + 2)(x - 6) = 0$$

$3x + 2 = 0$	$x - 6 = 0$
$3x = -2$	$x = 6$
$x = -\frac{2}{3}$	

Since a negative time interval does not make sense, the speed of the boat in still water is __________.

---

55. Following the example in the last frame, solve this problem.

> A river has a 1 mph current. A boat travels 24 miles upstream and 24 miles downstream in a total of 10 hours. What is the speed of the boat in still water?

Answer (54): 6 mph

Answer (55): 5 mph

SELF-TEST 35 (pages 473-484)

Write in standard form.

1. $(x + 5)(x + 3) = 25$

2. $\frac{7}{x - 4} + \frac{7}{x + 4} = 1$

Solve with the quadratic formula.

3. $8y^2 - 2y - 3 = 0$

4. $\frac{1}{R} - R = 1$

5. Using the formula $d = 16t^2$, determine how many seconds it would take an object to fall 400 ft?

6. The width of a rectangle is 3 ft less than the length. If the area is 6.75 ft, find the length and width.

ANSWERS:

1. $x^2 + 8x - 10 = 0$

2. $x^2 - 14x - 16 = 0$

3. $y = \frac{3}{4}$ and $-\frac{1}{2}$

4. $R = \frac{-1 \pm \sqrt{5}}{2}$

or

$R = 0.618$ and -1.618

5. 5 seconds

6. L = 4.5 ft, W = 1.5 ft

11-7 FORMULAS

In this section, we will do some evaluations and rearrangements with formulas containing a squared variable.

---

56. In the evaluation below, we squared before multiplying. Complete the other evaluation.

In $P = I^2R$, find P when $I = 3$ and $R = 10$.

$$P = I^2R = (3^2)(10) = 9(10) = 90$$

In $A = 3.14r^2$, find A when $r = 10$.

$$A = 3.14r^2 = \_\_\_\_\_\_\_\_\_\_$$

---

57. Following the example, complete the other evaluation. | Answer: $A = 314$

In $s = \frac{1}{2}at^2$, find s when $a = 12$ and $t = 4$.

$$s = \frac{1}{2}at^2 = \frac{1}{2}(12)(4^2) = \frac{1}{2}(12)(16) = 6(16) = 96$$

In $E = \frac{1}{2}mv^2$, find E when $m = 8$ and $v = 10$.

$$E = \frac{1}{2}mv^2 = \_\_\_\_\_\_\_\_\_\_$$

---

58. In the evaluation below, we squared before dividing. Complete the other evaluation. | Answer: $E = 400$

In $P = \frac{E^2}{R}$, find P when $E = 9$ and $R = 10$.

$$P = \frac{E^2}{R} = \frac{9^2}{10} = \frac{81}{10} = 8.1$$

In $R = \frac{pL}{d^2}$, find R when $p = 2$, $L = 50$, and $d = 5$.

$$R = \frac{pL}{d^2} = \_\_\_\_\_\_\_\_\_\_$$

---

59. To do the evaluation below, we must solve a quadratic equation. | Answer: $R = 4$

Find the side of a square if its area is $144\ cm^2$.

$$A = s^2$$

$$144 = s^2$$

$$s = \pm\sqrt{144} = \pm 12\ cm$$

Of the two answers (12 cm and -12 cm), only one makes sense. Which one? ______

Answer: 12 cm

60. When solving for a variable that is squared in a formula, we report only the positive root because the negative root does not ordinarily make sense. An example is shown. Complete the other evaluation.

In the formula below, find I when P = 72 and R = 2.

$$P = I^2R$$
$$72 = I^2(2)$$
$$I^2 = \frac{72}{2} = 36$$
$$I = \sqrt{36} = 6$$

In the formula below, find V when h = 1.

$$h = \frac{V^2}{64}$$

Answer: V = 8

61. Following the example, complete the other evaluation. Use the square-root table.

In the formula below, find t when s = 100 and a = 10.

$$s = \frac{1}{2}at^2$$
$$100 = \frac{1}{2}(10)(t^2)$$
$$100 = 5t^2$$
$$t^2 = \frac{100}{5} = 20$$
$$t = \sqrt{20} = 4.472$$

In the formula below, find v when E = 600 and m = 20.

$$E = \frac{1}{2}mv^2$$

Answer: v = 7.746

62. In the formula below, find t when s = 100 and g = 5.

$$s = \frac{gt^2}{2}$$
$$100 = \frac{5t^2}{2}$$
$$2(100) = \not{2}\left(\frac{5t^2}{\not{2}}\right)$$
$$200 = 5t^2$$
$$t^2 = \frac{200}{5} = 40$$
$$t = \sqrt{40} = 6.325$$

In the formula below, find d when F = 3, $m_1 = 6$, $m_2 = 10$, and r = 2.

$$F = \frac{m_1m_2}{rd^2}$$

Answer: d = 3.162

63. The formula below shows the relationship between distance traveled (s), initial velocity (v_o), and time traveled (t) for an object that is thrown vertically downward.

$$s = v_o t + 16t^2$$

If the initial velocity is 48 feet per second, we can find the amount of time needed to travel 64 feet by solving the quadratic equation below.

$$64 = 48t + 16t^2$$

Multiplying both sides by $\frac{1}{16}$ to simplify, we get:

$$\frac{1}{16}(64) = \frac{1}{16}(48t + 16t^2)$$

$$4 = 3t + t^2$$

a) Write the equation in standard form. ______________

b) Using the factoring method, find the two solutions.

c) Only one of the two solutions makes sense. Which one? ________

---

64. When rearranging formulas to solve for a variable that is squared, negative values do not ordinarily make sense. Therefore, when using the square root principle to solve for v below, we only used the principle square root of 2as. Solve for s and t in the other two formulas.

$$v^2 = 2as$$

$$\sqrt{v^2} = \sqrt{2as}$$

$$v = \sqrt{2as}$$

a) $s^2 = A$

b) $t^2 = \frac{a}{h}$

Answers to 63:

a) $t^2 + 3t - 4 = 0$

b) $t = 1$ and -4

c) $t = 1$ second

---

65. When solving for a variable that is squared, <u>be sure to take the square root of all of the other side</u>. An example is shown. Solve for c in the other formula.

$$a^2 = \frac{b + c}{d}$$

$$a = \sqrt{\frac{b + c}{d}}$$

$$c^2 = a^2 + b^2$$

Answers to 64:

a) $s = \sqrt{A}$

b) $t = \sqrt{\frac{a}{h}}$

---

66. To solve for I below, we isolated I^2 first and then used the square root principle. Solve for d in the other formula.

$$P = I^2R$$

$$I^2 = \frac{P}{R}$$

$$I = \sqrt{\frac{P}{R}}$$

$$A = .7854d^2$$

Answer to 65:

$c = \sqrt{a^2 + b^2}$

67. To solve for D, we isolated D^2 and then used the square root principle. Solve for $\underline{d}$ in the other formula.

$$H = \frac{D^2N}{2.5} \qquad F = \frac{m_1m_2}{rd^2}$$

$$2.5H = D^2N$$

$$D^2 = \frac{2.5H}{N}$$

$$D = \sqrt{\frac{2.5H}{N}}$$

Answer: $d = \sqrt{\frac{A}{.7854}}$

---

68. To solve for $\underline{t}$ below, we began by writing $\frac{1}{2}at^2$ as $\frac{at^2}{2}$. Solve for $\underline{v}$ in the other formula.

$$s = \frac{1}{2}at^2 \qquad E = \frac{1}{2}mv^2$$

$$s = \frac{at^2}{2}$$

$$2s = at^2$$

$$t^2 = \frac{2s}{a}$$

$$t = \sqrt{\frac{2s}{a}}$$

Answer: $d = \sqrt{\frac{m_1m_2}{Fr}}$

---

69. To solve for v_f below, we used the addition axiom to isolate v_f^2 first. Solve for $\underline{c}$ in the other formula.

$$v_o^2 = v_f^2 - 2gs \qquad a^2 = c^2 - b^2$$

$$v_o^2 + 2gs = v_f^2 - 2gs + 2gs$$

$$v_o^2 + 2gs = v_f^2$$

$$v_f = \sqrt{v_o^2 + 2gs}$$

Answer: $v = \sqrt{\frac{2E}{m}}$

---

70. To solve for X below, we isolated X^2 first. Solve for $\underline{b}$ in the other formula.

$$Z^2 = X^2 + R^2 \qquad c^2 = a^2 + b^2$$

$$Z^2 + (-R^2) = X^2 + R^2 + (-R^2)$$

$$Z^2 - R^2 = X^2$$

$$X = \sqrt{Z^2 - R^2}$$

Answer: $c = \sqrt{a^2 + b^2}$

Answer: $b = \sqrt{c^2 - a^2}$

11-8 THE PYTHAGOREAN THEOREM

The Pythagorean Theorem is a formula relating the hypotenuse and legs of a right triangle. We will discuss that formula in this section and use it to solve some problems.

---

71. A right triangle is a **triangle** that contains a right angle (90°). To label the sides of a right triangle, we frequently use the letters a, b, and c.

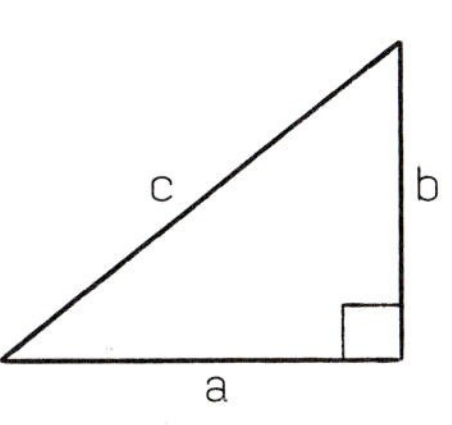

c is the hypotenuse, the side opposite the right angle.

a and b are the legs.

The Pythagorean Theorem says this: In any right triangle, the square of the hypotenuse is equal to the sum of the squares of the two legs. That is:

$$c^2 = a^2 + b^2$$

We used the formula to find the length of the hypotenuse of one right triangle below. Use it to find the length of the hypotenuse of the other triangle.

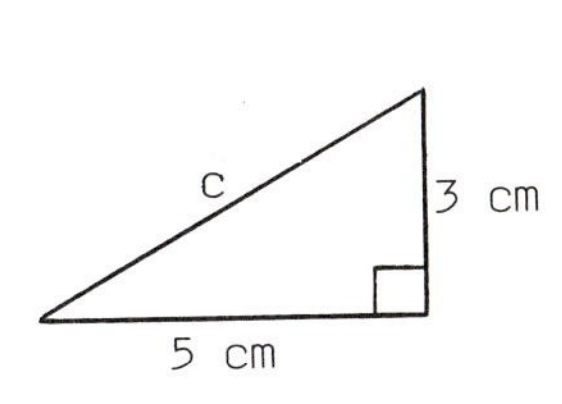

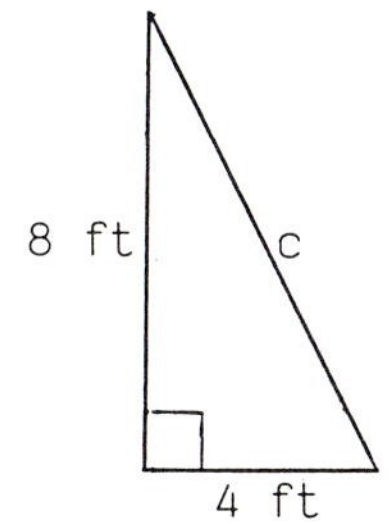

$c^2 = a^2 + b^2$

$c^2 = 5^2 + 3^2$

$c^2 = 25 + 9$

$c^2 = 34$

$c = \sqrt{34} = 5.831$ cm

$c = \sqrt{80} = 8.944$ ft

72. The Pythagorean Theorem can be used to find leg a in the right triangle at the right. Either of the two methods below can be used.

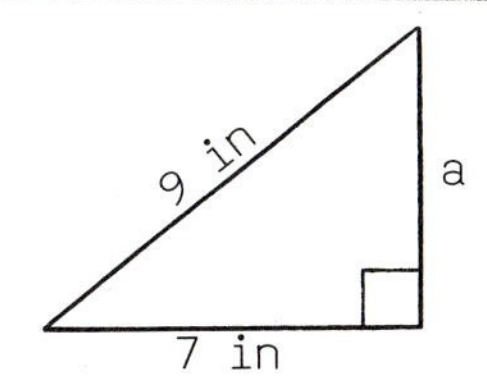

Equation-Solving Method

$c^2 = a^2 + b^2$

$9^2 = a^2 + 7^2$

$81 = a^2 + 49$

$a^2 = 32$

$a = \sqrt{32} = 5.657$ in

Rearrangement Method

$c^2 = a^2 + b^2$

$a^2 = c^2 - b^2$

$a = \sqrt{c^2 - b^2}$

$a = \sqrt{9^2 - 7^2}$

$a = \sqrt{81 - 49} = \sqrt{32} = 5.657$ in

Continued on following page.

72. Continued

Using either method, solve each problem below.

a) Find leg $\underline{a}$.

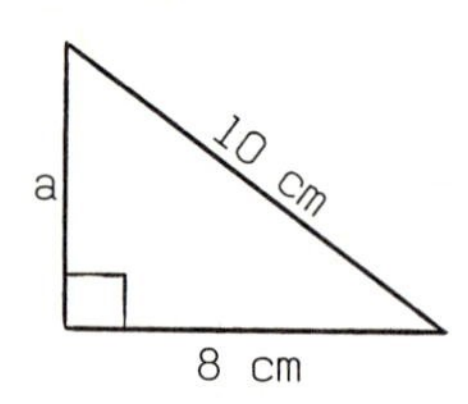

b) Find leg $\underline{b}$.

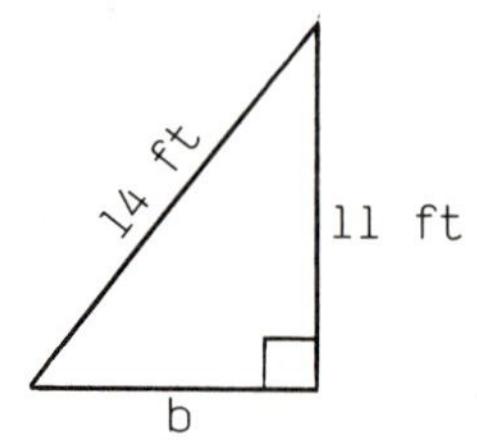

---

a) $a = \sqrt{36} = 6$ cm

b) $b = \sqrt{75} = 8.66$ ft

73. The distance between the opposite corners of a rectangle or square is called the diagonal. In each figure below, the diagonal (d) is the hypotenuse of a right triangle. We used the Pythagorean Theorem to find the diagonal of the rectangle. Find the diagonal of the square.

$d^2 = 7^2 + 5^2$

$d^2 = 49 + 25$

$d^2 = 74$

$d = \sqrt{74} = 8.602$ cm

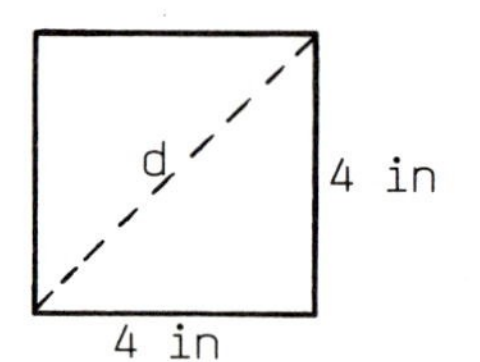

$d = \sqrt{32} = 5.657$ in

74. Use the Pythagorean Theorem to solve each problem below. Make a sketch and label the parts.

a) Find the length of a rectangle if its diagonal is 10 cm and its width is 4 cm.

b) Find the width of a rectangle if its diagonal is 9 ft and its length is 8 ft.

Answers to Frame 74:

a) $L = \sqrt{84} = 9.165$ cm

b) $W = \sqrt{17} = 4.123$ ft

11-9 COMPOUND INTEREST

In this section, we will use the compound interest formula to solve some problems.

---

75. Interest is paid on money put into a savings account. After interest is credited to an account, interest is paid on both the original amount invested and the interest. When interest is credited to an account only at the end of each full year, we say that the interest is compounded annually.

When an amount of money P is invested at an interest rate r and compounded annually, in t years it will grow to the amount A given by the formula:

$$A = P(1 + r)^t$$

We used the formula to solve one problem below. Use it to solve the other.

If $1,000 is invested at 12% for 2 years compounded annually, it will grow to what amount?

$A = P(1 + r)^t$

$A = 1,000(1 + 0.12)^2$

$A = 1,000(1.12)^2$

$A = 1,000(1.2544)$

$A = \$1,254.40$

If $1,000 is invested at 10% for 2 years compounded annually, it will grow to what amount?

$A = P(1 + r)^t$

$1,210.00

---

76. We used the same formula to solve a problem below. Notice how we used the square root principle with $\frac{256}{225} = (1 + r)^2$. Following the example, solve the other problem.

$2,250 is invested at interest rate r compounded annually. If it grows to $2,560 in 2 years, what is the interest rate?

$A = P(1 + r)^t$

$2560 = 2250(1 + r)^2$

$\frac{2560}{2250} = (1 + r)^2$

$\frac{256}{225} = (1 + r)^2$

$\sqrt{\frac{256}{225}} = 1 + r$

$\frac{16}{15} = 1 + r$

$r = \frac{16}{15} - 1$

$r = \frac{16}{15} - \frac{15}{15}$

$r = \frac{1}{15}$

$r = 6.67\%$

$1,000 is invested at interest rate r compounded annually. If it grows to $1,440 in 2 years, what is the interest rate?

$A = P(1 + r)^t$

Answer to Frame 76: $r = 20\%$

11-10 GRAPHING QUADRATIC EQUATIONS

In this section, we will discuss the graphs of quadratic equations in two variables.

---

77. The general form of a quadratic equation in two variables is:

$$y = ax^2 + bx + c \qquad (\text{where } a \neq 0)$$

Though a cannot be 0, a can be either positive or negative. For example:

$y = x^2$ $\qquad$ $y = -2x^2 + 3$

$y = 3x^2 - 4x$ $\qquad$ $y = -x^2 + 2x - 3$

The graph of a quadratic equation is always a cup-shaped curve called a parabola. Two parabolas are shown below.

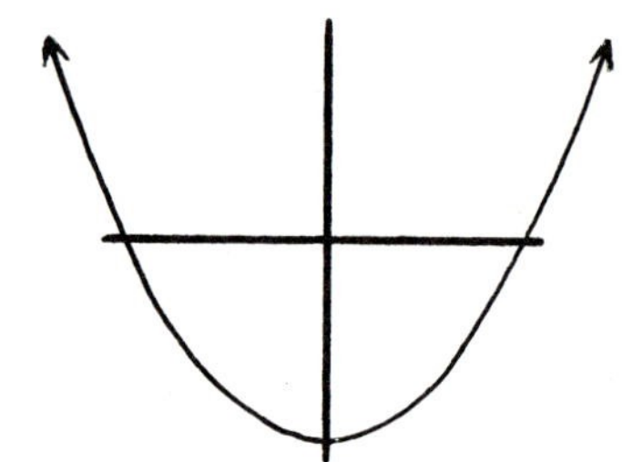

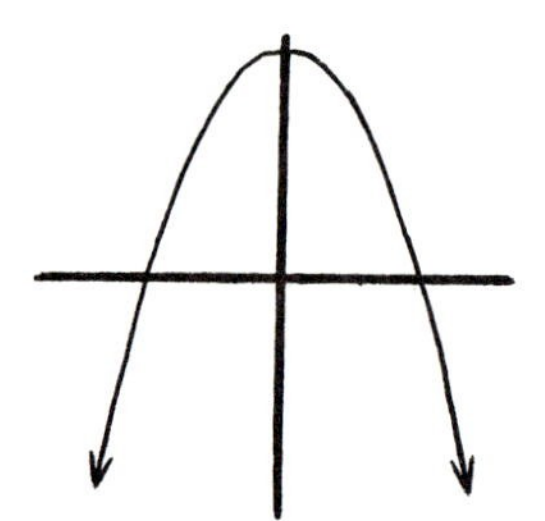

---

78. The two simplest quadratic equations are $y = x^2$ and $y = -x^2$. We graphed each below. To do so, we made up a solution-table, plotted the points, and drew a smooth curve through the plotted points.

$y = x^2$

x	y
3	9
2	4
1	1
0	0
-1	1
-2	4
-3	9

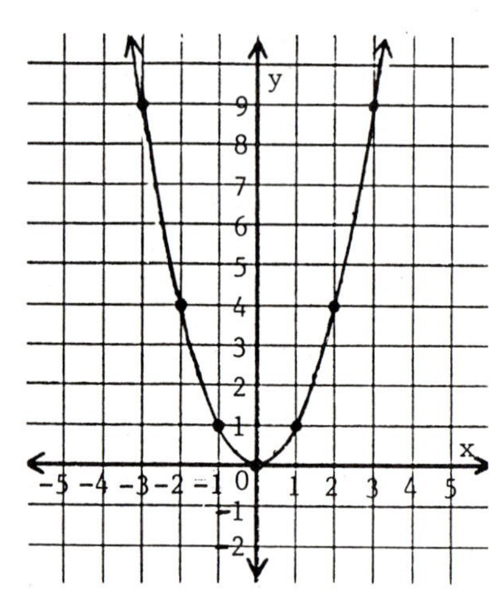

$y = -x^2$

x	y
3	-9
2	-4
1	-1
0	0
-1	-1
-2	-4
-3	-9

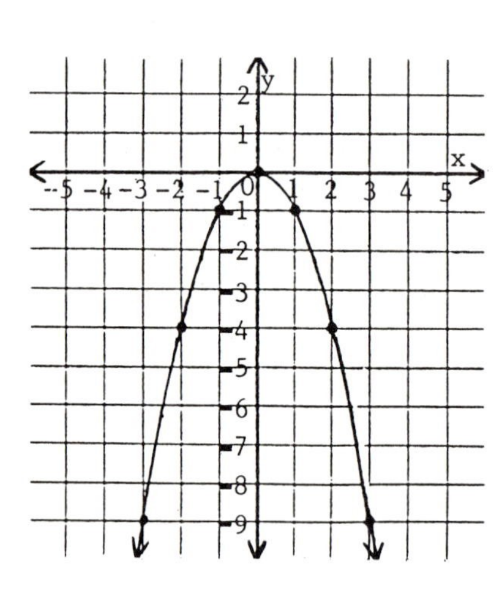

In $y = x^2$, the coefficient of x^2 is "1", a positive number. The parabola opens upward. In $y = -x^2$, the coefficient of x^2 is -1, a negative number. The parabola opens downward.

79. When graphing equations of the form $y = ax^2 + bx + c$:

the parabola opens upward if a is positive.

the parabola opens downward if a is negative.

Answer either upward or downward for these:

a) The graph of $y = 2x^2 - 3x + 1$ opens ______________.

b) The graph of $y = 4x^2 - x$ opens ______________.

c) The graph of $y = -3x^2 + x$ opens ______________.

---

80. When graphing a quadratic equation, plot enough points so that the outline of the parabola is clear. Then draw a smooth curve through the plotted points. Use the solution-table below to graph $y = 2x^2$.

$y = 2x^2$

x	y
5	50
3	18
2	8
0	0
-2	8
-3	18
-5	50

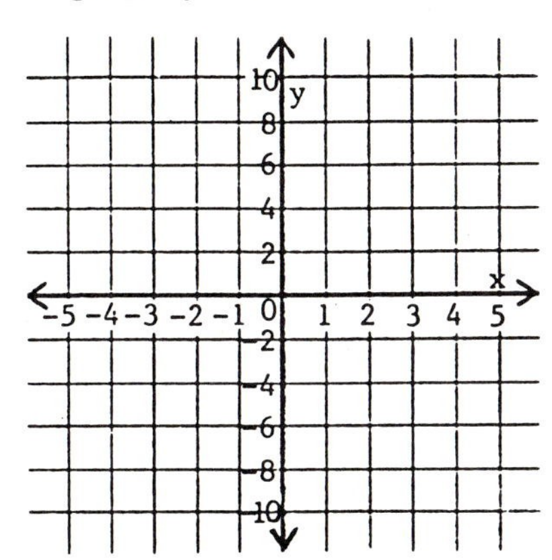

a) upward

b) upward

c) downward

---

81. Use the solution-table below to graph $y = x^2 - 2x$.

$y = x^2 - 2x$

x	y
4	8
3	3
2	0
1	-1
0	0
-1	3
-2	8

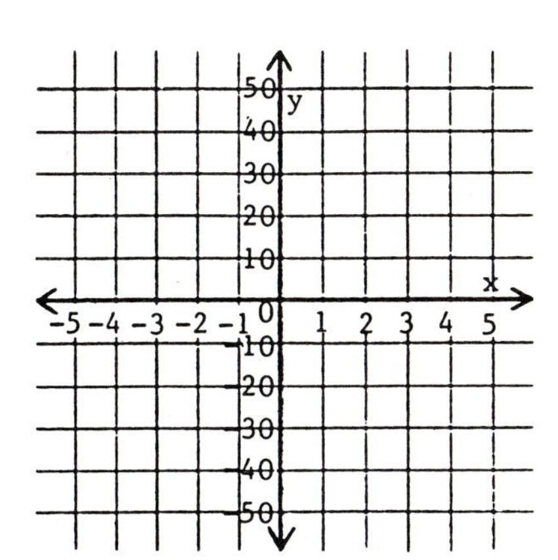

Answer to Frame 80:

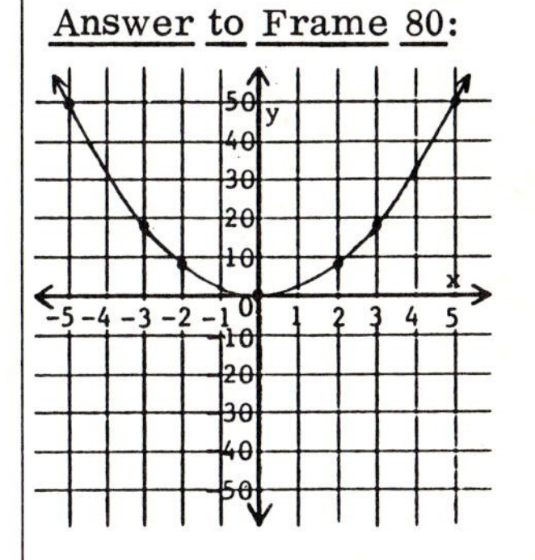

82. Use the solution-table below to graph $y = -2x^2 + 4x + 1$.

$$y = -2x^2 + 4x + 1$$

x	y
-1	-5
0	1
1	3
2	1
3	-5

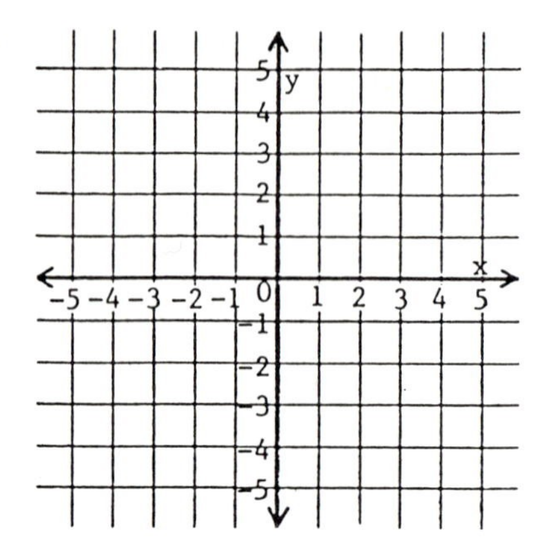

Answer to Frame 81:

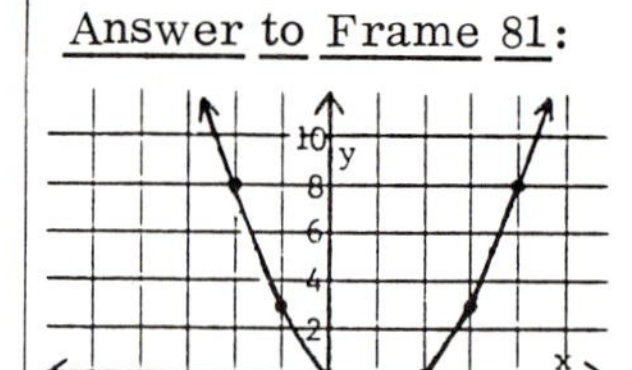

---

83. Graph $y = x^2 + 1$ below. Make up your own solution-table.

$$y = x^2 + 1$$

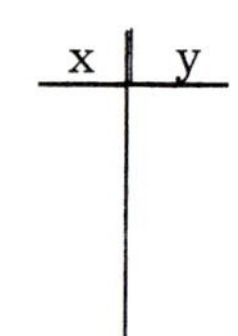

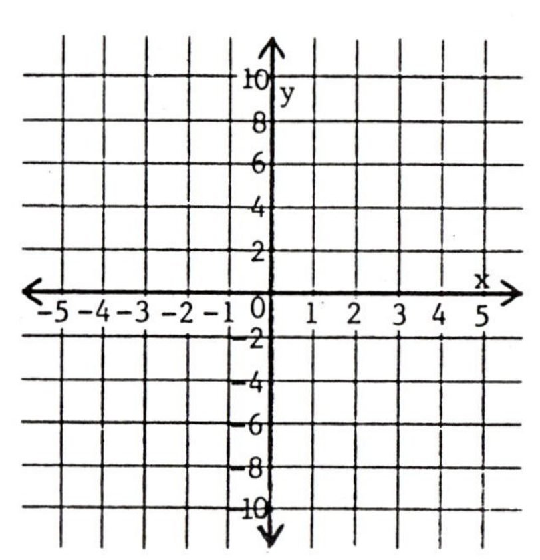

Answer to Frame 82:

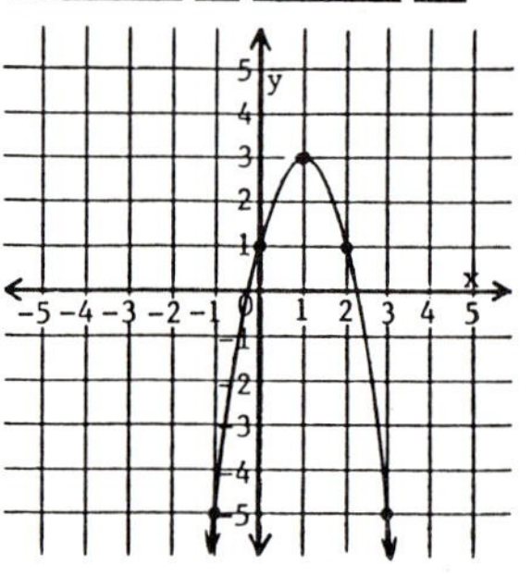

---

84. Graph $y = 4x - x^2$ below. Make up your own solution-table.

$$y = 4x - x^2$$

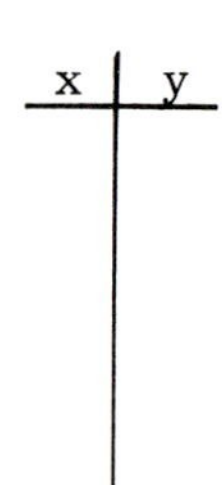

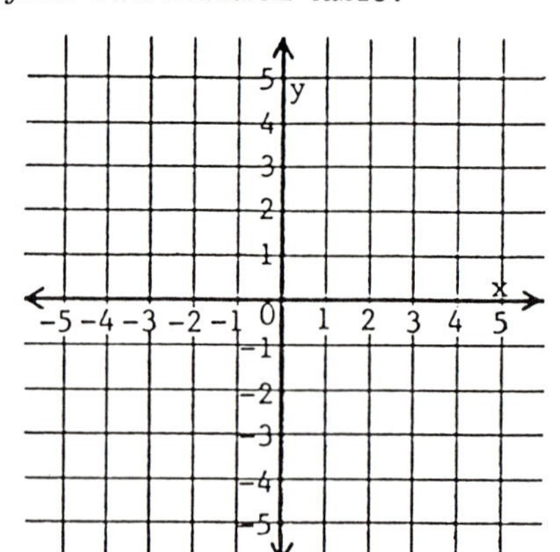

Answer to Frame 83:

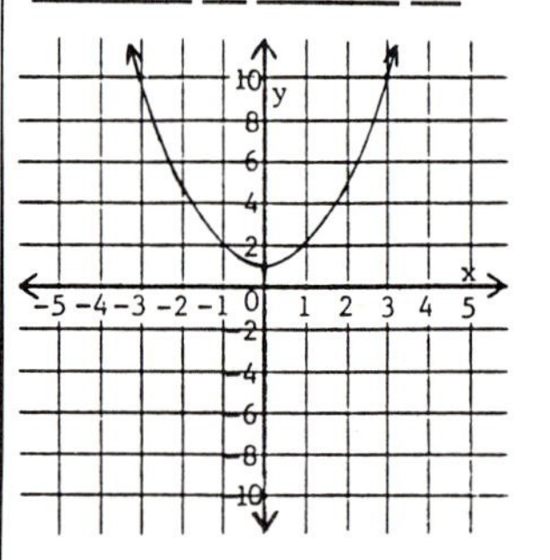

85. The graph below shows the distance (in meters) a dropped object will fall from rest in a given period of time (in seconds). Since negative time-intervals do not make sense, only quadrant 1 is used. Therefore, the graph is only half a parabola.

Using the graph, complete these:

a) How far would an object fall in 4 seconds?
_______ meters

b) How long would it take an object to fall 20 meters?
_______ seconds

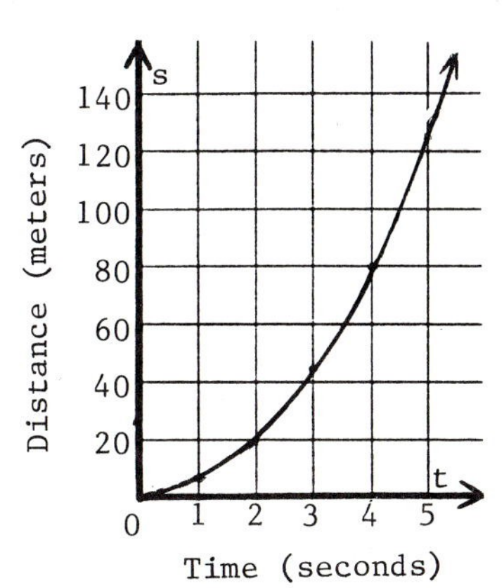

Answer to Frame 84:

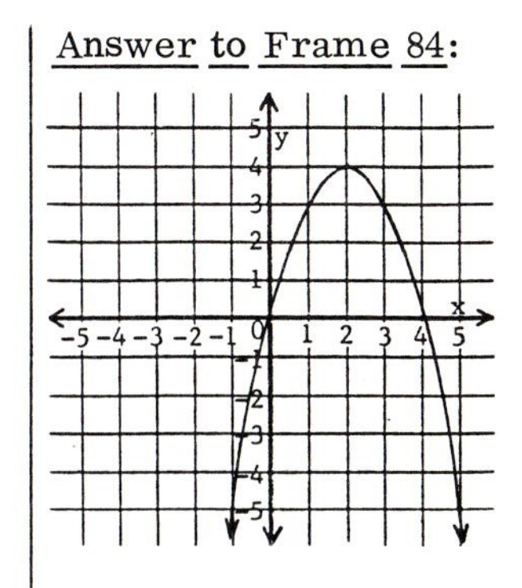

a) 80 meters

b) 2 seconds

SELF-TEST 36 (pages 485-496)

1. In $s = \frac{1}{2}at^2$, find t, when $s = 50$ and $a = 4$.

2. In $F = \frac{m_1 m_2}{rd^2}$, find d when $F = 8$, $m_1 = 8$, $m_2 = 10$, and $r = 2$.

3. Solve for p.

$$K = \frac{p^2}{12}$$

4. Solve for d.

$$V = \frac{1}{2}bd^2$$

5. Solve for Q.

$$P^2 = Q^2 + R^2$$

6. Find the diagonal of a rectangle if its length is 6 ft and its width is 5 ft.

7. Find the second leg of a right triangle if the first leg is 8 cm and the hypotenuse is 12 cm.

8. Graph $y = x^2 - 3$. Make up your own solution-table.

$$y = x^2 - 3$$

x	y

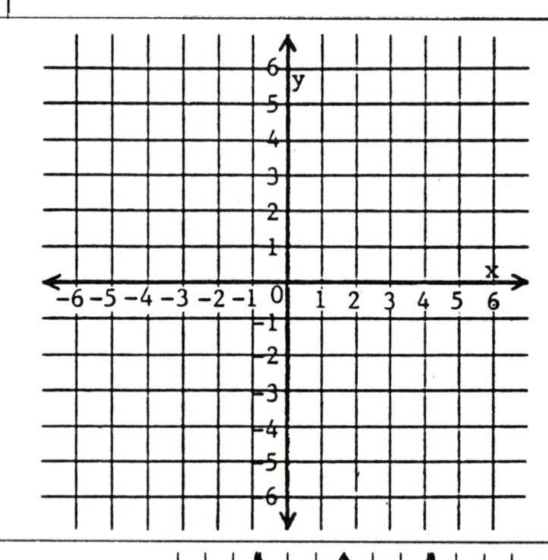

ANSWERS:

1. $t = 5$
2. $d = 2.236$
3. $p = \sqrt{12K}$
4. $d = \sqrt{\frac{2V}{b}}$
5. $Q = \sqrt{P^2 - R^2}$
6. $d = 7.81$ ft
7. leg = 8.944 cm
8.

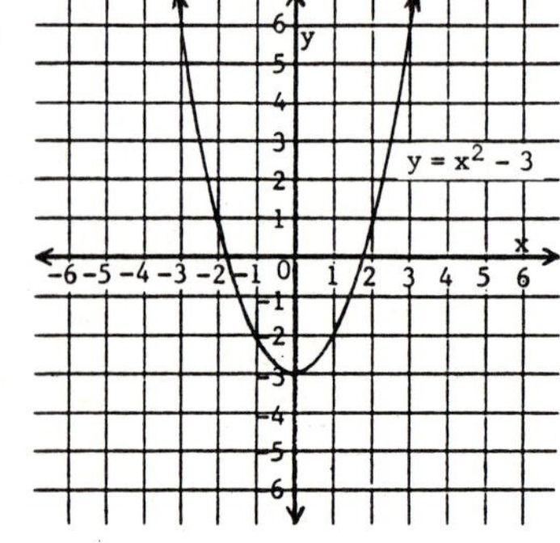

SUPPLEMENTARY PROBLEMS - CHAPTER 11

Assignment 34

Solve by the factoring method.

1. $x^2 - 2x - 35 = 0$
2. $y^2 - 10y + 24 = 0$
3. $3x^2 + 2x - 5 = 0$
4. $5t^2 + 13t + 6 = 0$
5. $m^2 - 4m + 4 = 0$
6. $d^2 - 81 = 0$
7. $16x^2 - 9 = 0$
8. $y^2 - 7y = 0$
9. $7p^2 + 8p = 0$
10. $2d^2 = 3 - d$
11. $4v^2 = 9$
12. $h^2 = 3h$

Solve by the square root method.

13. $x^2 = 100$
14. $y^2 - 9 = 0$
15. $t^2 = \frac{25}{81}$
16. $16x^2 = 1$
17. $36y^2 - 49 = 0$
18. $\frac{1}{4}m^2 = 9$
19. $1 = \frac{4a^2}{9}$
20. $\frac{1}{2}b^2 = 0$
21. $y^2 - 17 = 0$
22. $\frac{1}{3}x^2 = 10$
23. $2d^2 - 84 = 0$
24. $x^2 = -9$
25. $(x + 3)^2 = 25$
26. $(y - 1)^2 = 9$
27. $(t - 2)^2 = 3$
28. $(b + 5)^2 = -4$

Complete the square.

29. $x^2 + 20x + \_\_\_\_ = (x + \_\_\_\_)^2$
30. $y^2 - 2y + \_\_\_\_ = (y - \_\_\_\_)^2$
31. $t^2 + 24t + \_\_\_\_ = (t + \_\_\_\_)^2$
32. $m^2 - 40m + \_\_\_\_ = (m - \_\_\_\_)^2$
33. $d^2 + d + \_\_\_\_ = (d + \_\_\_\_)^2$
34. $b^2 - 5b + \_\_\_\_ = (b - \_\_\_\_)^2$

Solve by completing the square.

35. $x^2 - 6x + 5 = 0$
36. $x^2 - 4x - 11 = 0$
37. $y^2 + 10y + 20 = 0$
38. $t^2 - 3t + 1 = 0$
39. $3y^2 + 4y - 4 = 0$
40. $2x^2 - 5x - 1 = 0$

Assignment 35

Solve with the quadratic formula.

1. $x^2 - 2x - 8 = 0$
2. $6t^2 + t - 2 = 0$
3. $4w^2 - w - 3 = 0$
4. $2R^2 + 3R - 20 = 0$
5. $y^2 + 4y + 1 = 0$
6. $3d^2 - 5d - 1 = 0$
7. $5p^2 - 8p + 2 = 0$
8. $2F^2 + F - 7 = 0$
9. $h^2 + 2h - 1 = 0$

Write in standard form.

10. $-x^2 + 3x - 5 = 0$
11. $y^2 = 10 - 5y$
12. $5t - 4 = t^2 + 2t$
13. $7a + 5(a^2 - 1) = 0$
14. $(3x - 1)(x + 2) = 7$
15. $d^2 + (d + 1)^2 = 10$
16. $\frac{1}{N} = N - 1$
17. $\frac{1}{x - 3} = \frac{x + 3}{x}$
18. $\frac{5}{x - 2} + \frac{5}{x + 2} = 1$

Write in standard form and then solve with the quadratic formula.

19. $5x - 2 = 3x^2$
20. $t(2t - 5) = 3$
21. $11y + 3(y^2 + 2) = 0$
22. $\frac{3r}{2} + r^2 = 1$
23. $x - 2 = \frac{7}{x - 2}$
24. $\frac{3}{x + 2} + \frac{3}{x - 2} = 1$

If an object is dropped, the distance d it falls in t seconds is given by the formula $d = 16t^2$. How long would it take an object to fall:

25. 96 ft?
26. 144 ft?
27. 240 ft?
28. 480 ft?

29. The length of a rectangle is 1 ft more than the width. The area of the rectangle is 15.75 ft^2. Find the length and width.

30. The width of a rectangle is 2m less than the length. If the area is $1.25m^2$, find the length and width.

31. A river has a current of 2 mph. A boat travels 30 miles upstream and 30 miles downstream in a total of 8 hours. What is the speed of the boat in still water?

32. A river has a current of 3 mph. A boat travels 20 miles upstream and 20 miles downstream in a total of 7 hours. What is the speed of the boat in still water?

Assignment 36

Do these evaluations.

1. In $A = s^2$, find A when $s = 5$.
2. In $s = \frac{1}{2}at^2$, find s when $a = 40$ and $t = 2$.
3. In $h = \frac{V^2}{64}$, find h when $V = 16$.
4. In $F = \frac{m_1m_2}{rd^2}$, find F when $m_1 = 10$, $m_2 = 20$, $r = 2$, and $d = 5$.
5. If $P = I^2R$, find I when $P = 20$ and $R = 5$.
6. In $A = 3.14r^2$, find r when $A = 6.28$.
7. In $E = \frac{1}{2}mv^2$, find v when $E = 27$ and $m = 6$.
8. In $F = \frac{md^2}{r}$, find d when $F = 72$, $m = 8$, and $r = 16$.
9. In $Z^2 = R^2 + X^2$, find R when $Z = 8$ and $X = 6$.
10. In $P = ht^2 - dt$, find t when $P = 10$, $h = 2$, and $d = 1$.

Do these rearrangements.

11. Solve for K. $K^2 = ab$
12. Solve for H. $H^2 = D + T$
13. Solve for c. $E = mc^2$
14. Solve for v. $a = \frac{v^2}{2s}$
15. Solve for t. $2k = \frac{d}{t^2}$
16. Solve for h. $M = \frac{1}{2}bh^2$
17. Solve for R. $R^2 = P^2 + Q^2$
18. Solve for Z. $X^2 = Z^2 - R^2$

Use the Pythagorean Theorem to find the unknown dimensions.

19.

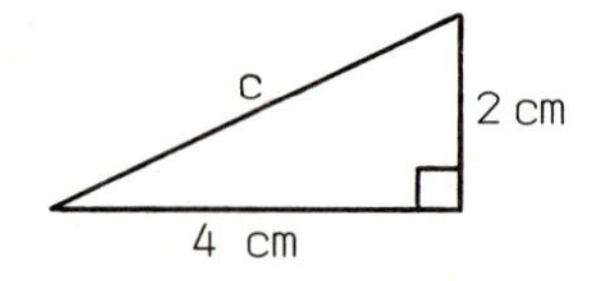

20.

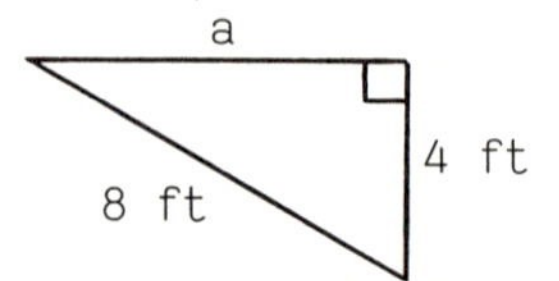

21.

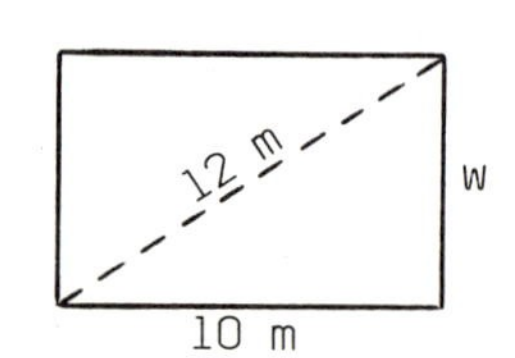

22. Find the second leg of a right triangle if the first leg is 6 ft and the hypotenuse is 10 ft.
23. Find the diagonal of a square whose side is 3 cm.
24. Find the length of a rectangle if the diagonal is 7m and its width is 4m.

Use the formula $A = P(1 + r)^t$ for these:

25. If \$1,000 is invested at 8% for two years compounded annually, it will grow to what amount?

26. If \$1,000 is invested at 11% for two years compounded annually, it will grow to what amount?

27. \$1,000 is invested at interest rate r compounded annually. If it grows to \$1,210 in 2 years, what is the interest rate?

28. If \$1,440 is invested at interest rate r compounded annually and it grows to \$1,690 in 2 years, what is the interest rate?

Graph each of the following equations.

29. $y = x^2 - 4$

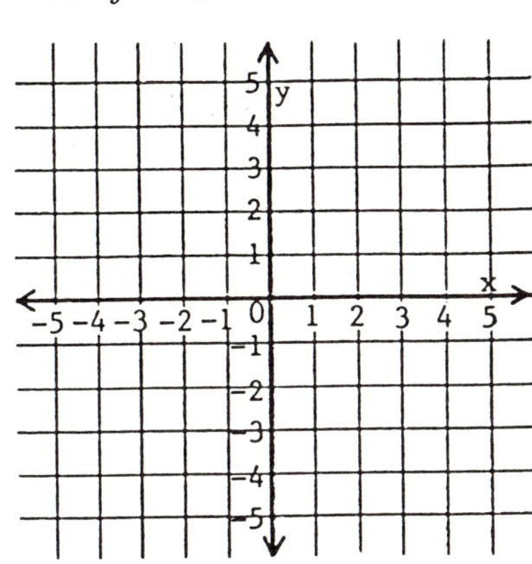

30. $y = -2x^2 + 3$

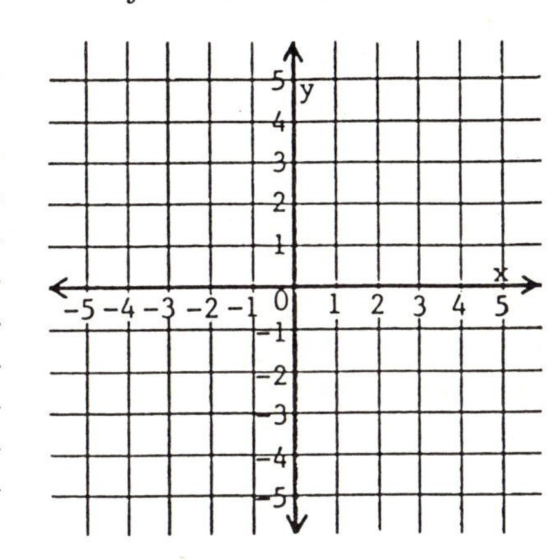

31. $y = x^2 - 2x - 3$

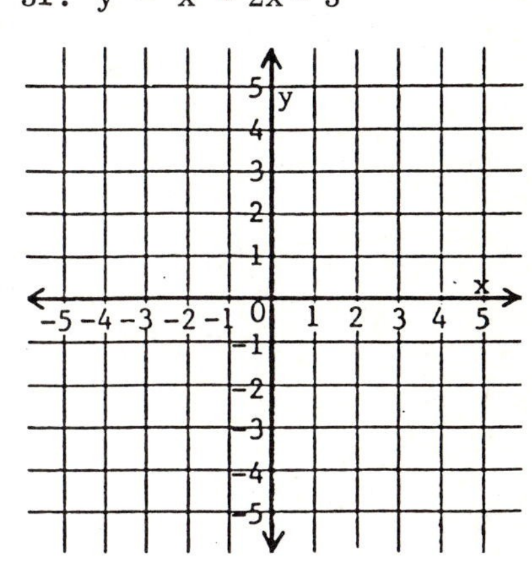

12 Inequalities

In this chapter, we will define inequalities and use the addition axiom and multiplication axiom to solve inequalities. A method for graphing linear inequalities is shown. Some word problems are included.

12-1 INEQUALITIES

In this section, we will define inequalities and discuss their solutions.

1. Part of the number line is shown below.

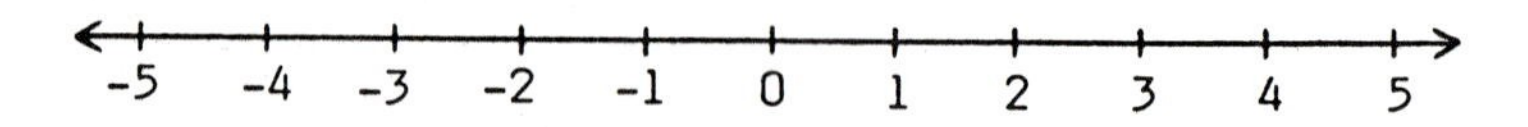

In an earlier chapter, we defined the symbols $>$ and $<$.

$>$ means "is greater than".

$<$ means "is less than".

Continued on following page.

1. Continued

A number is greater than a second number if it is to the right of the second on the number line. That is:

$4 > 2$ is true, since 4 is to the right of 2.
$4 > 5$ is false, since 4 is to the left of 5.

$-3 > -5$ is true, since -3 is to the right of -5.
$-3 > -1$ is false, since -3 is to the left of -1.

A number is less than a second number if it is to the left of the second on the number line. That is:

$3 < 5$ is true, since 3 is to the left of 5.
$3 < 2$ is false, since 3 is to the right of 2.

$-2 < -1$ is true, since -2 is to the left of -1.
$-2 < -4$ is false, since -2 is to the right of -4.

Answer true or false for these.

a) $5 > -2$ b) $0 < -1$ c) $-5 > -3$ d) $-4 < 0$

Answers: a) True b) False c) False d) True

---

2. The symbols $\geq$ and $\leq$ are defined below.

$\geq$ means "is greater than or equal to".
$\leq$ means "is less than or equal to".

Statements containing $\geq$ or $\leq$ can be either true or false. For example:

$4 \geq 2$ is true, since $4 > 2$ is true.
$-3 \leq -3$ is true, since $-3 = -3$ is true.

$5 \geq 7$ is false, since both $5 > 7$ and $5 = 7$ are false.
$0 \leq -4$ is false, since both $0 < -4$ and $0 = -4$ are false.

Answer either true or false for these.

a) $2 \geq 2$ b) $3 \leq 0$ c) $-4 \leq 0$ d) $-5 \geq -4$

Answers: a) True b) False c) True d) False

---

3. Statements containing $>$, $<$, $\geq$, or $\leq$ are called inequalities. An inequality can contain a variable. For example:

$x > 3$ $y < -1$ $x \leq 4$ $-2 \geq y$

If we substitute a number for the variable, the inequality becomes either true or false. For example:

If $x = 5$, $x > 3$ becomes $5 > 3$ which is true.
If $y = 0$, $2 \leq y$ becomes $2 \leq 0$ which is false.

Answer true or false for these.

a) If $x = 7$, $4 > x$ is ____________. c) If $x = 4$, $x \leq 4$ is __________.

b) If $y = -4$, $y < -1$ is __________. d) If $y = -1$, $-3 \geq y$ is ________.

4. Inequalities like $x < 3$ and $x \leq 3$ have an infinite number of solutions. We can graph their solutions on a number line.

For $x < 3$, the solutions are all real numbers less than 3. They are shown by the arrow to the left of 3 below. The open circle at 3 means that 3 is not included.

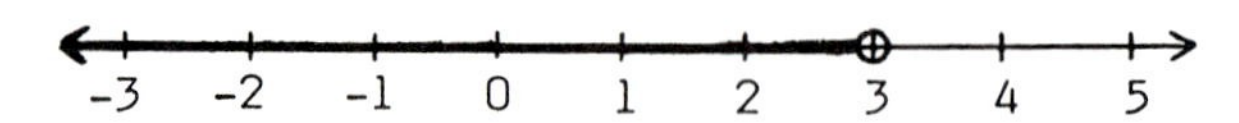

For $x \leq 3$, the solutions are 3 and all real numbers less than 3. They are shown by the arrow to the left of 3 below. The closed circle at 3 means that 3 is included.

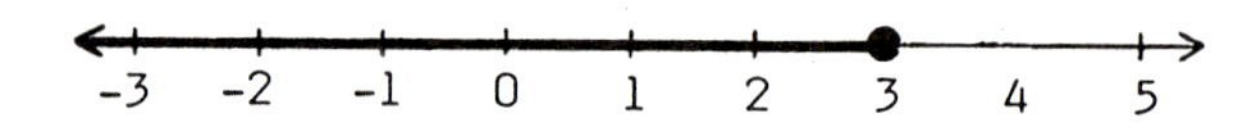

Following the examples, graph each inequality below.

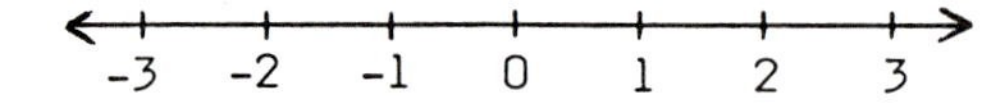

b) $x \leq -2$

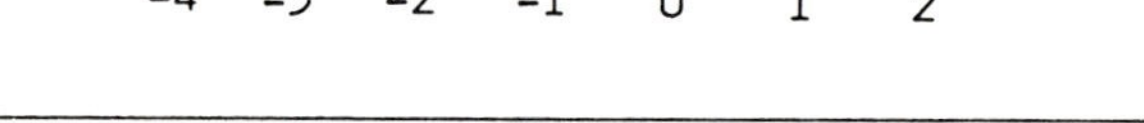

a) False

b) True

c) True

d) False

---

5. We graphed $x > -2$ and $x \geq -2$ below. The only difference is the open or closed circle at -2.

$x > -2$

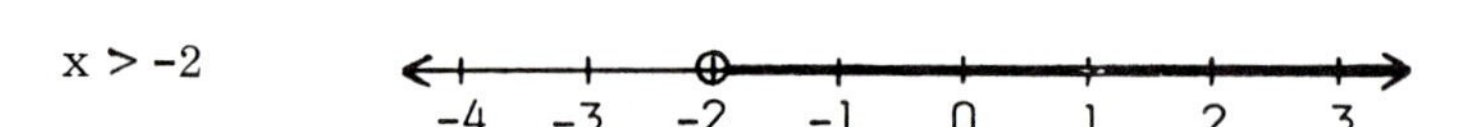

$x \geq -2$

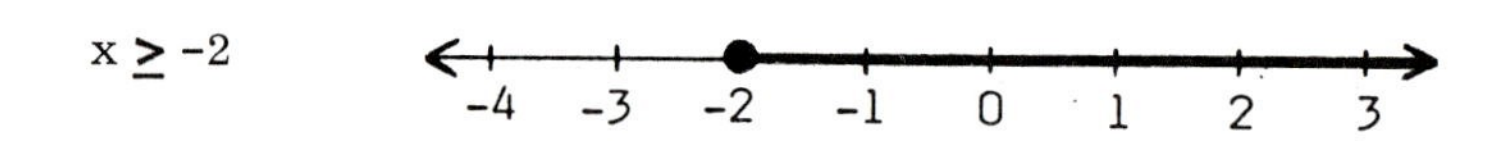

Following the examples, graph each inequality below.

a) $x > 0$

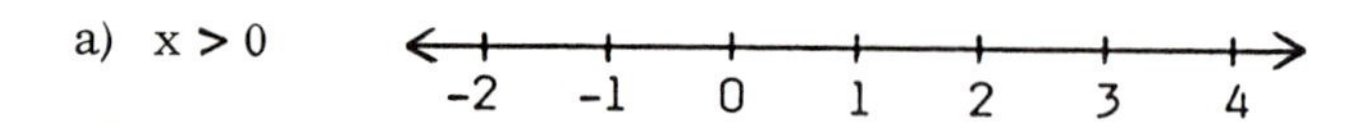

b) $x \geq -3$

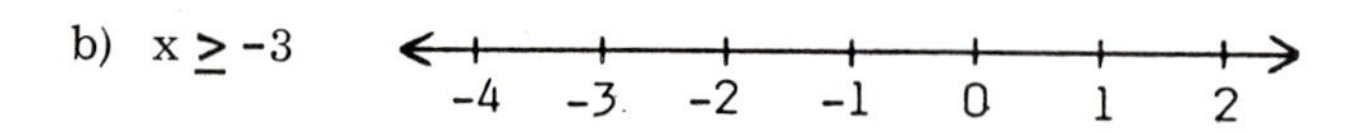

a) 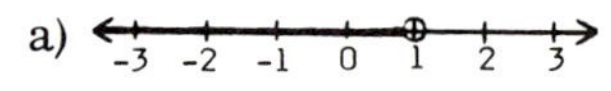

b)

---

a)

b) 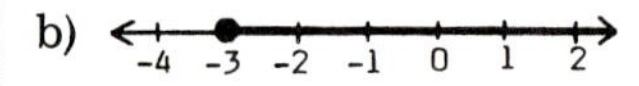

6. The inequality $5 \geq x$ says that 5 is greater than or equal to x. We can say the same thing with the inequality $x \leq 5$ which has the variable on the left side. $x \leq 5$ says that x is less than or equal to 5. Therefore:

$$5 \geq x \text{ is the same as } x \leq 5$$

Sometimes it is easier to understand an inequality if the variable is on the left side. Write each of these with the variable on the left side.

a) $3 > x$ ________ b) $-4 < y$ ________ c) $-2 \geq x$ ________ d) $7 \leq y$ ________

---

a) $x < 3$

b) $y > -4$

c) $x \leq -2$

d) $y \geq 7$

7. It is easier to determine the solution of an inequality if the variable is on the left side. Therefore, we can rewrite inequalities like $-1 < x$ and $4 \geq y$.

Since $-1 < x$ is the same as $x > -1$,
the solutions are all real numbers greater than -1.

Since $4 \geq x$ is the same as $x \leq 4$,
the solutions are 4 and all real numbers less than 4.

Graph each inequality below. Use the method above to determine the solutions.

a) $2 < x$

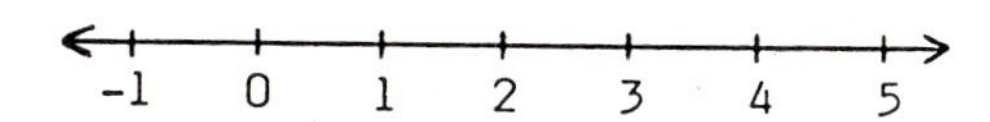

b) $-1 \geq x$

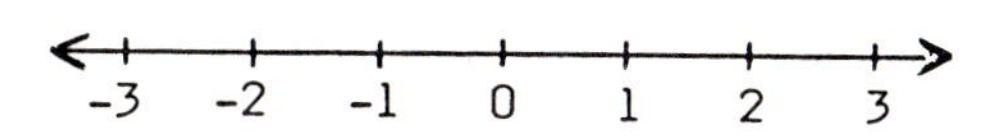

---

a) -1 0 1 2 3 4 5

b) -3 -2 -1 0 1 2 3

8. We translated each phrase below to an inequality.

All numbers greater than -1.	$x > -1$
5 and all numbers less than 5.	$x \leq 5$
All positive numbers.	$x > 0$

Using x as the variable, translate these to inequalities.

a) All numbers less than 7. ________

b) -1 and all numbers greater than -1. ________

c) All negative numbers. ________

---

a) $x < 7$

b) $x \geq -1$

c) $x < 0$

9. One side of each inequality below is an algebraic expression.

$x + 2 < 7$ $\quad$ $3x > 18$ $\quad$ $2y - 5 \geq 4$

If we substitute a number for the variable, the inequality becomes either true or false.

If $x = 3$, $x + 2 < 7$ becomes $3 + 2 < 7$ or $5 < 7$, which is true.

If $y = 3$, $2y - 5 \geq 4$ becomes $2(3) - 5 \geq 4$ or $1 \geq 4$, which is false.

Answer true or false for these.

a) If $x = 9$, $x - 5 > 2$ is ____________.

b) If $y = -4$, $2y < -10$ is ____________.

c) If $x = 3$, $3x + 2 \leq 9$ is ____________.

d) If $y = 4$, $14 > 3y$ is ____________.

---

a) True

b) False

c) False

d) True

10. We translated each sentence below to an inequality.

The sum of a number and 3 is less than 10. $\quad x + 3 < 10$

Twice a number, minus 1 is greater than or equal to -4. $\quad 2x - 1 \geq -4$

Using x as the variable, translate each sentence to an inequality.

a) The difference between a number and 4 is greater than 8. ____________

b) Four times a number is less than or equal to 20. ____________

c) 10 is greater than or equal to three times a number, plus 5. ____________

---

a) $x - 4 > 8$ $\quad$ b) $4x \leq 20$ $\quad$ c) $10 \geq 3x + 5$

12-2 THE ADDITION AXIOM

In this section, we will use the addition axiom to solve inequalities.

---

11. The inequality $7 > 5$ is true. If we add 2 or -3 to both sides, the new inequality is also true.

Adding 2 to both sides, we get:	Adding -3 to both sides, we get:
$7 > 5$	$7 > 5$
$7 + 2 > 5 + 2$	$7 - 3 > 5 - 3$
$9 > 7$	$4 > 2$

Are both $9 > 7$ and $4 > 2$ true? ____________

12. To <u>solve an inequality</u> means to find its solutions. One principle used to solve inequalities is <u>THE ADDITION AXIOM FOR INEQUALITIES</u>. It says:

> IF WE ADD THE SAME QUANTITY TO BOTH SIDES OF AN INEQUALITY THAT IS TRUE, THE NEW INEQUALITY IS ALSO TRUE. That is:
>
> If: $A > B$
>
> Then: $A + C > B + C$
>
> <u>Note</u>: The addition axiom also applies to $<$, $\geq$, and $\leq$.

We used the addition axiom to solve $x + 2 > 7$ below. To do so, we added -2 to both sides.

$$x + 2 > 7$$
$$x + 2 - 2 > 7 - 2$$
$$x > 5$$

The solution of $x + 2 > 7$ is $x > 5$. That is, any number greater than 5 satisfies $x + 2 > 7$. We checked 10 below. Check 6 in the space provided.

$x + 2 > 7$	$x + 2 > 7$
$10 + 2 > 7$	
$12 > 7$	

Answer: Yes

---

13. We used the addition axiom to solve two inequalities below.

To solve $x - 3 < 4$, we added 3 to both sides.	To solve $y + 5 \geq 10$, we added -5 to both sides.
$x - 3 < 4$	$y + 5 \geq 10$
$x - 3 + 3 < 4 + 3$	$y + 5 - 5 \geq 10 - 5$
$x < 7$	$y \geq 5$

Following the examples, solve each inequality.

a) $x + 5 > -1$ b) $y - 7 \leq 3$

Answer (12): $6 + 2 > 7$
$8 > 7$

Answer (13):
a) $x > -6$
b) $y \leq 10$

14. Following the example, solve the other inequality.

$$x - 9 > 0$$
$$x - 9 + 9 > 0 + 9$$
$$x > 9$$

$$y + 8 < 0$$

---

15. Following the example, solve the other inequality.

$y < -8$

$$y + 1.4 \leq 6.7$$
$$y + 1.4 - 1.4 \leq 6.7 - 1.4$$
$$y \leq 5.3$$

$$x - 2.6 \geq 3.3$$

---

16. Following the example, solve the other inequality.

$x \geq 5.9$

$$x - \frac{1}{4} < \frac{1}{2}$$
$$x - \frac{1}{4} + \frac{1}{4} < \frac{1}{2} + \frac{1}{4}$$
$$x < \frac{3}{4}$$

$$y + \frac{2}{3} \geq \frac{1}{6}$$

---

17. We solved an inequality below. Since we like to state solutions with the variable on the left side, we rewrote $2 > x$ and got $x < 2$. Solve the other inequality. Write the solution with the variable on the left side.

$y \geq -\frac{1}{2}$

$$5 > x + 3$$
$$5 - 3 > x + 3 - 3$$
$$2 > x$$
$$x < 2$$

$$-7 \leq y - 2$$

---

18. Solve these. Write each solution with the variable on the left side.

$y \geq -5$

a) $0 < x - 1.5$

b) $\frac{1}{2} \geq y + \frac{1}{3}$

---

a) $x > 1.5$

b) $y \leq \frac{1}{6}$

12-3 THE MULTIPLICATION AXIOM

In this section, we will use the multiplication axiom to solve inequalities.

---

19. The inequality $3 < 5$ is true. Let's multiply both sides by 2 and -4 and see whether we get true inequalities.

Multiplying both sides by 2, we get:	Multiplying both sides by -4, we get:
$3 < 5$	$3 < 5$
$2(3) < 2(5)$	$-4(3) < -4(5)$
$6 < 10$ True	$-12 < -20$ False

Though $6 < 10$ is true, $-12 < -20$ is false. To get a true inequality when multiplying by -4, we have to reverse the direction of the inequality symbol. That is:

$$3 < 5$$

$$-4(3) > -4(5) \quad \text{Reversed } < \text{ to } >$$

$$-12 > -20 \quad \text{True}$$

To get a true inequality when multiplying by a negative number, we have to reverse the inequality symbol. As another example, we multiplied both sides of $7 \geq -6$ by -3 below.

$$7 \geq -6$$

$$-3(7) \leq -3(-6) \quad \text{Reversed } \geq \text{ to } \leq$$

$$-21 \leq 18$$

Is $-21 \leq 18$ a true inequality? ________

---

20. A second principle used to solve inequalities is THE MULTIPLICATION AXIOM FOR INEQUALITIES. It says:

Answer to 19: Yes

> IF WE MULTIPLY BOTH SIDES OF AN INEQUALITY THAT IS TRUE BY A POSITIVE NUMBER, THE NEW INEQUALITY IS ALSO TRUE. That is:
>
> If: $B > C$ and $A > 0$
>
> Then: $AB > AC$
>
> IF WE MULTIPLY BOTH SIDES OF AN INEQUALITY THAT IS TRUE BY A NEGATIVE NUMBER AND REVERSE THE INEQUALITY SYMBOL, THE NEW INEQUALITY IS ALSO TRUE. That is:
>
> If: $B > C$ and $A < 0$
>
> Then: $AB < AC$
>
> Note: The multiplication axiom also applies to $<$, $\geq$, and $\leq$.

Continued on following page.

20. Continued

To solve the inequality below, we multiplied both sides by the positive number $\frac{1}{4}$.

$$4x > 12$$

$$\frac{1}{4}(4x) > \frac{1}{4}(12)$$

$$x > 3$$

The solution of $4x > 12$ is $x > 3$. That is, any number greater than 3 satisfies $4x > 12$. We checked 6 below. Check 4 in the space provided.

$$4x > 12$$

$$4(6) > 12$$

$$24 > 12$$

$$4x > 12$$

---

21. Following the example, solve the other inequality.

$$5x < -20$$

$$\frac{1}{5}(5x) < \frac{1}{5}(-20)$$

$$x < -4$$

$$3y \geq 18$$

Answer (20): $4(4) > 12$; $16 > 12$

---

22. To solve the inequality below, we multiplied both sides by the negative number $-\frac{1}{2}$. Notice that we reversed the inequality symbol.

$$-2x \leq 18$$

$$-\frac{1}{2}(-2x) \geq -\frac{1}{2}(18) \qquad \text{Reversed } \leq \text{ to } \geq$$

$$x \geq -9$$

The solution of $-2x \leq 18$ is $x \geq -9$. That is, any number greater than or equal to -9 satisfies $-2x \leq 18$. We checked -5. Check -9 is the space provided.

$$-2x \leq 18$$

$$-2(-5) \leq 18$$

$$10 \leq 18$$

$$-2x \leq 18$$

Answer (21): $y \geq 6$

---

23. Since we multiplied by a negative number below, we reversed the inequality symbol. Solve the other inequality.

$$-7x > -14$$

$$-\frac{1}{7}(-7x) < -\frac{1}{7}(-14)$$

$$x < 2$$

$$-6y \leq 48$$

Answer (22): $-2(-9) \leq 18$; $18 \leq 18$

24. Solve these. Reverse the inequality symbol when multiplying by a negative number.

a) $3x < 2$ b) $-5y \geq 0$ c) $-7t \leq -8$

Answer: $y \geq -8$

---

25. Following the example, solve the other inequality.

$$3x < \frac{2}{5}$$
$$\frac{1}{3}(3x) < \frac{1}{3}\left(\frac{2}{5}\right)$$
$$x < \frac{2}{15}$$

$$2y \geq -\frac{1}{2}$$

Answers: a) $x < \frac{2}{3}$ b) $y \leq 0$ c) $t \geq \frac{8}{7}$

---

26. Following the example, solve the other inequality.

$$-\frac{1}{2}x > 3$$
$$-2\left(-\frac{1}{2}x\right) < -2(3)$$
$$x < -6$$

$$-\frac{3}{5}y \leq -1$$

Answer: $y \geq -\frac{1}{4}$

---

27. To solve for x below, we multiplied both sides by -1. Notice that $-1(-x) = x$. Solve the other inequalities.

$$-x < 6$$
$$-1(-x) > -1(6)$$
$$x > -6$$

a) $-y > -3$ b) $-t \leq \frac{3}{5}$

Answer: $y \geq \frac{5}{3}$

---

28. After solving the inequality below, we wrote the solution with the variable on the left side. Solve the other inequality. Write the solution with the variable on the left side.

$$4 \geq 3x$$
$$\frac{1}{3}(4) \geq \frac{1}{3}(3x)$$
$$\frac{4}{3} \geq x$$
$$x \leq \frac{4}{3}$$

$$-1 < 5y$$

Answers: a) $y < 3$ b) $t \geq -\frac{3}{5}$

---

29. Following the example, write the other solution with the variable on the left side.

$$8 \leq -2x$$
$$-\frac{1}{2}(8) \geq -\frac{1}{2}(-2x)$$
$$-4 \geq x$$
$$x \leq -4$$

$$-6 > -7y$$

Answer: $y > -\frac{1}{5}$

Answer: $y > \frac{6}{7}$

12-4 USING BOTH AXIOMS

Both axioms are needed to solve many inequalities. We will discuss solutions of that type in this section.

---

30. We used both axioms to solve the inequality below. Solve the other inequality.

$$3x - 2 < 5$$
$$3x - 2 + 2 < 5 + 2$$
$$3x < 7$$
$$\frac{1}{3}(3x) < \frac{1}{3}(7)$$
$$x < \frac{7}{3}$$

$2x + 6 \geq 7$

---

31. We used both axioms below. Solve the other inequality.

Answer to 30: $x \geq \frac{1}{2}$

$$4 - 3x > 5$$
$$-4 + 4 - 3x > 5 - 4$$
$$-3x > 1$$
$$-\frac{1}{3}(-3x) < -\frac{1}{3}(1)$$
$$x < -\frac{1}{3}$$

$2 - 5y \leq 6$

---

32. Notice that we multiplied both sides by -1 below. Solve the other inequality.

Answer to 31: $y \geq -\frac{4}{5}$

$$3 - x \geq 9$$
$$-3 + 3 - x \geq 9 - 3$$
$$-x \geq 6$$
$$-1(-x) \leq -1(6)$$
$$x \leq -6$$

$7 - y < 4$

---

33. To solve the inequality below, we used the addition axiom twice to get 2x on the left side and 6 on the right side. Solve the other inequality.

Answer to 32: $y > 3$

$$7x + 3 \leq 5x + 9$$
$$7x + 3 - 3 \leq 5x + 9 - 3$$
$$7x \leq 5x + 6$$
$$-5x + 7x \leq -5x + 5x + 6$$
$$2x \leq 6$$
$$\frac{1}{2}(2x) \leq \frac{1}{2}(6)$$
$$x \leq 3$$

$4x - 1 > x - 3$

34. To solve the inequality below, we used the addition axiom twice to get -7x on the left side and -9 on the right side. Solve the other inequality.

$$8 - 3x < 4x - 1$$

$$-8 + 8 - 3x < 4x - 1 - 8$$

$$-3x < 4x - 9$$

$$-4x - 3x < -4x + 4x - 9$$

$$-7x < -9$$

$$-\frac{1}{7}(-7x) > -\frac{1}{7}(-9)$$

$$x > \frac{9}{7}$$

$$2y - 5 \geq 7y + 5$$

$x > -\frac{2}{3}$

---

35. In solving the inequality below, we got the variable on the right side. We rewrote the solution to get the variable on the left side. Solve the other inequality.

$$3 - 5x > x - 5$$

$$3 - 5x + 5x > 5x + x - 5$$

$$3 > 6x - 5$$

$$5 + 3 > 6x - 5 + 5$$

$$8 > 6x$$

$$\frac{1}{6}(8) > \frac{1}{6}(6x)$$

$$\frac{4}{3} > x$$

$$x < \frac{4}{3}$$

$$9 - 6x \leq 7 - 4x$$

$y \leq -2$

$x \geq 1$

SELF-TEST 37 (pages 500-512)

Answer true or false.

1. If $x = -3$, $x > -1$ is ________.
2. If $y = 4$, $2y - 5 \geq 3$ is ________.

Graph on the number line.

5. $x < 2$

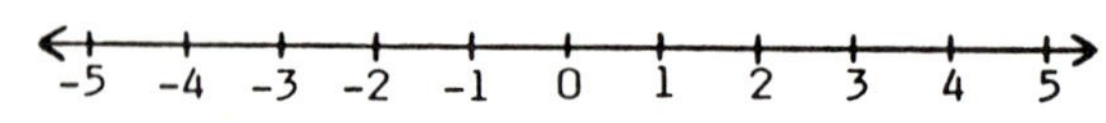

6. $x \geq 0$

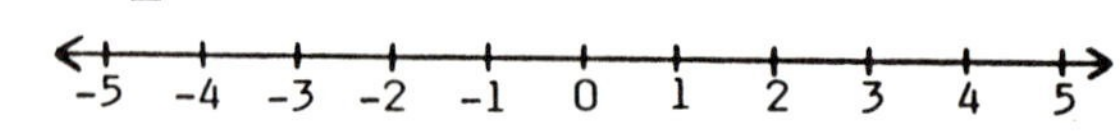

Write with the variable on the left side.

3. $4 < x$ ________
4. $-2 \geq y$ ________

Translate to an inequality. Use $\underline{x}$ as the variable.

7. All numbers greater than or equal to -1. ________
8. The difference between a number 2 is less than -3. ________

Solve each inequality.

9. $x + 5 < 1$

10. $y - \frac{1}{6} \geq \frac{1}{2}$

11. $0 > x - 2.6$

12. $-4y \leq 12$

13. $2x < -\frac{1}{4}$

14. $-1 \geq 5t$

15. $5 - x > 9$

16. $3y - 4 \leq y + 1$

ANSWERS:

1. false
2. true
3. $x > 4$
4. $y \leq -2$
5. -5 -4 -3 -2 -1 0 1 2 3 4 5
6. -5 -4 -3 -2 -1 0 1 2 3 4 5
7. $x \geq -1$
8. $x - 2 < -3$
9. $x < -4$
10. $y \geq \frac{2}{3}$
11. $x < 2.6$
12. $y \geq -3$
13. $x < -\frac{1}{8}$
14. $t \leq -\frac{1}{5}$
15. $x < -4$
16. $y \leq \frac{5}{2}$

12-5 WORD PROBLEMS

In this section, we will solve some word problems involving inequalities.

---

36. We used an inequality to solve the problem below. Solve the other problem.

If 4 is subtracted from twice a number, the result is less than 8. Find the numbers satisfying this condition.

$$2x - 4 < 8$$

$$2x < 12$$

$$x < 6$$

Any number less than 6 satisfies this condition.

The sum of a number and 10 is equal to or greater than 25. Find the numbers satisfying this condition.

---

37. Following the example, solve the other problem.

Answer (36): 15 or any number greater than 15 satisfies this condition.

The perimeter of a rectangle must be no greater than 100 ft. The width of the rectangle must be 18 ft. Find the longest possible length of the rectangle.

$$2L + 2W \le 100$$

$$2L + 2(18) \le 100$$

$$2L + 36 \le 100$$

$$2L \le 64$$

$$L \le 32$$

The length can be no longer than 32 ft.

The perimeter of a rectangle must be no less than 120 meters. The length of the rectangle must be 35 meters. Find the shortest possible width of the rectangle.

---

38. Following the example, solve the other problem.

Answer (37): The width can be no shorter than 25 meters.

The length of a rectangle is 20 meters. The area of the rectangle can be no more than 240 square meters. Find the longest possible width of the rectangle.

$$LW \le 240$$

$$20W \le 240$$

$$W \le 12$$

The longest possible width is 12 meters.

The width of a rectangle is 8 ft. The area of the rectangle must be at least 160 ft^2. Find the shortest possible length of the rectangle.

39. Following the example, solve the other problem.

The base of a triangle is 10 cm. The area of the triangle can be no less than 75 cm². Find the shortest possible height of the triangle.

$$\frac{1}{2}bh \geq 75$$

$$\frac{1}{2}(10)h \geq 75$$

$$5h \geq 75$$

$$h \geq 15$$

The shortest possible height is 15 cm.

The height of a triangle is 12 ft. The area of the triangle can be no more than 180 ft². Find the longest possible base of the triangle.

The shortest possible length is 20 ft.

---

40. Following the example, solve the other problem.

Three tests are given in an algebra course. To get a B, a student needs a total of 240 points. If a student gets scores of 82 and 75 on the first two tests, what score can he get on the third test to get a B?

$$T_1 + T_2 + T_3 \geq 240$$

$$82 + 75 + T_3 \geq 240$$

$$157 + T_3 \geq 240$$

$$T_3 \geq 83$$

A score of 83 or more on the third test will get him a B.

Four tests are given in a biology course. To get an A, a student needs a total of 360 points. If a student gets scores of 85, 93, and 89 on the first three tests, what score can she get on the fourth test to get an A?

The longest possible base is 30 ft.

A score of 93 or more on the fourth test will get her an A.

41. We solved one problem below. Notice how we multiplied both sides of $1.00 < .02x$ by 100 to get rid of the decimals. Solve the other problem.

Company A rents compact cars for \$22.95 plus 20¢ per mile. Company B rents compact cars for \$21.95 plus 22¢ per mile. For what mileages would the cost of renting from Company A be cheaper?

$$22.95 + .20x < 21.95 + .22x$$
$$1.00 < .02x$$
$$100 < 2x$$
$$50 < x$$
$$x > 50$$

If you drive more than 50 miles, it is cheaper to rent from Company A.

Company A rents intermediate-size cars at \$26.95 plus 20¢ per mile. Company B rents intermediate-size cars at \$24.95 plus 22¢ per mile. For what mileages would the cost of renting from Company A be cheaper?

If you drive more than 100 miles, it is cheaper to rent from Company A.

42. Following the example, solve the other problem.

A woman on a business trip cannot exceed a daily car rental allowance of \$76. She rents a car for \$25.95 plus 20¢ per mile. What mileage can she drive and stay within her allowance?

$$25.95 + .20x \le 76$$
$$.20x \le 50.05$$
$$20x \le 5005$$
$$x \le 250.25$$

To stay within her allowance, she must drive 250.25 miles per day or less.

A man on a business trip cannot exceed a daily car rental allowance of \$68. He rents a car for \$22.95 plus 22¢ per mile. What mileage can he drive and stay within his allowance?

To stay within his allowance, he must drive 204.77 miles per day or less.

12-6 GRAPHING LINEAR INEQUALITIES

The graph of any linear inequality is a half-plane. We will discuss the method for graphing linear inequalities in this section.

---

43. Three linear inequalities are shown below. They are the same as linear equations except that they contain a $>$, $<$, $\geq$, or $\leq$ instead of an $=$ sign.

$y \geq x + 1$ $\qquad$ $y < -2x$ $\qquad$ $2x - y > 7$

A solution of a linear inequality is a pair of values that satisfies the inequality. For example:

(3, 7) <u>is</u> a solution of $y \geq x + 1$, since:

$y \geq x + 1$

$7 \geq 3 + 1$

$7 \geq 4$ $\qquad$ True

(5, 3) <u>is not</u> a solution of $y \geq x + 1$, since:

$y \geq x + 1$

$3 \geq 5 + 1$

$3 \geq 6$ $\qquad$ False

a) Is (-2, -5) a solution of $y < -2x$? ________

b) Is (4, 3) a solution of $2x - y > 7$? ________

---

44. Any linear inequality has an infinite number of solutions. For example, all of the following are solutions of $y \geq x + 1$.

(3, 5) $\qquad$ (0, 2) $\qquad$ (-2, 0) $\qquad$ (-5, -3)

The inequality $y \geq x + 1$ means both $y > x + 1$ and $y = x + 1$. We know that the graph of $y = x + 1$ is a straight line. We graphed the line below. It is a boundary that divides the coordinate system into two half-planes. One of the half-planes contains the solutions for $y > x + 1$. To determine the correct half-plane, we plotted the four solutions above. All of them lie in the half-plane <u>above</u> the line.

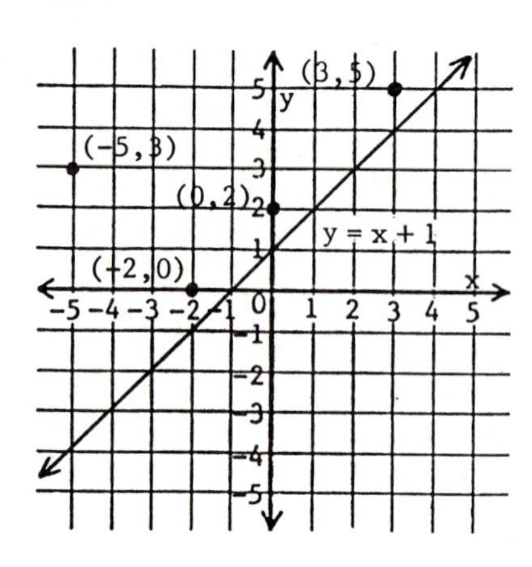

Continued on following page.

a) Yes, since

$-5 < 4$ is true.

b) No, since

$5 > 7$ is false.

44. Continued

The graph of $y \geq x + 1$ includes the line for $y = x + 1$ and the half-plane above the line for $y > x + 1$. To show that on the graph, we shade the half-plane above the line as we have done below.

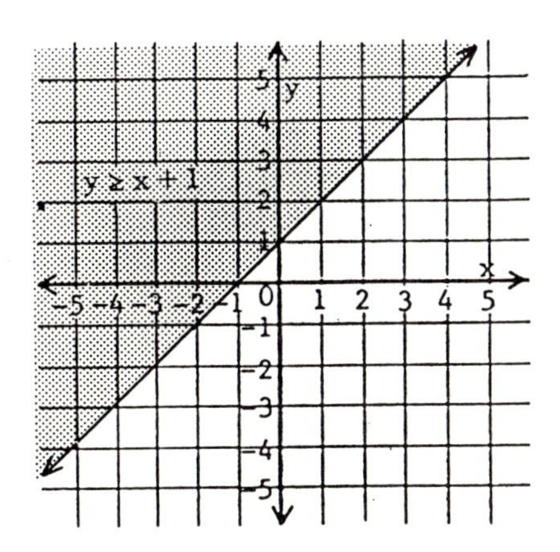

a) (1, 5) is a solution of $y \geq x + 1$. Does it lie in the shaded half-plane? ________

b) (4, 2) is not a solution of $y \geq x + 1$. Does it lie in the shaded half-plane? ________

---

a) Yes

b) No

45. All of the following are solutions of $y < -2x$.

(-2, -5)   (-3, 3)   (-4, -2)   (-5, 1)

Though the inequality does not include $y = -2x$, we still use that line as a boundary between two half-planes. To show that the points on the line are not part of the graph, we used a dashed line below. Then to determine which half-plane contains the solutions for $y < -2x$, we plotted the four solutions above. All of the points lie in the half-plane below the line.

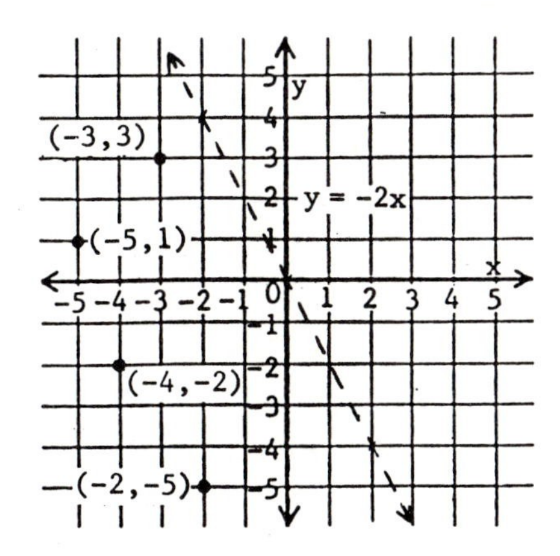

Continued on following page.

45. Continued

To show that the half-plane below the line is the graph of $y < -2x$, we shade the half-plane below the line as we have done below. The points on the dashed line are not part of the graph.

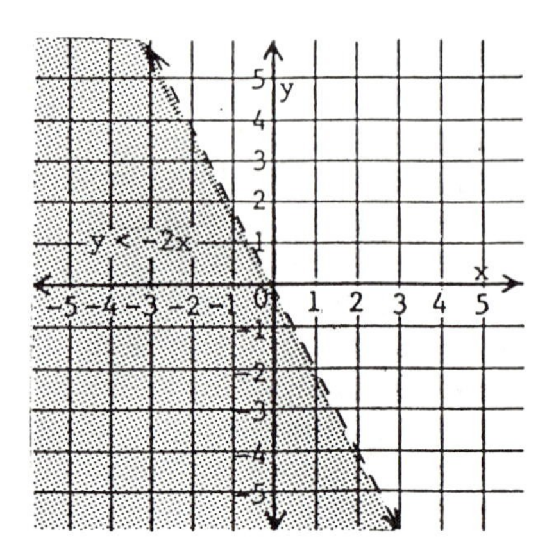

a) $(1,-5)$ is a solution of $y < -2x$. Does it lie in the shaded half-plane? ____________

b) $(3,-2)$ is not a solution of $y < -2x$. Does it lie in the shaded half-plane? ____________

---

a) Yes

b) No

46. The following can be used to determine whether to use a solid boundary line or a dashed boundary line.

Use a solid line if equality is included. The line is part of the graph. We use a solid line for these:

$$y \geq 3x + 2 \qquad y \leq x$$

Use a dashed line if equality is not included. The line is not part of the graph. We use a dashed line for these:

$$y > -3x \qquad y < x - 4$$

Answer either solid or dashed for these.

a) $y < -x$ ____________ c) $y \leq 2x$ ____________

b) $y \geq 2x - 3$ ____________ d) $y > x + 5$ ____________

---

a) dashed

b) solid

c) solid

d) dashed

47. We can use the following method to graph $3x + 2y < 6$.

1. Use the intercept method to graph the boundary $3x + 2y = 6$. The intercepts are (0, 3) and (2, 0). The boundary is dashed.

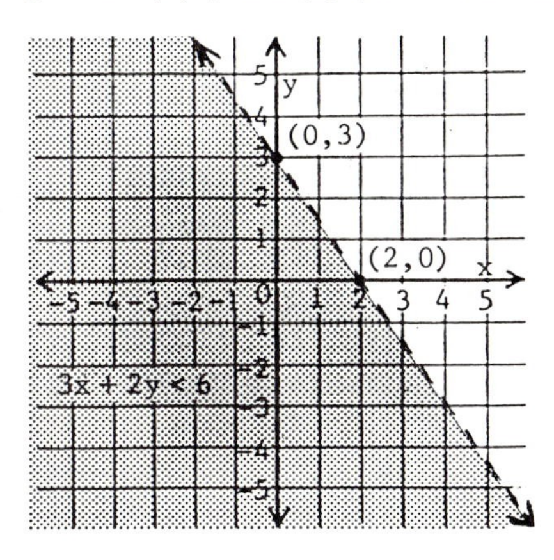

2. Then use a test point to determine the correct half-plane. Pick a point not on the line. We ordinarily use the origin (0, 0).

$$3(0) + 2(0) < 6 \text{ is true.}$$

Since the origin is a solution, the half-plane below the boundary is correct.

---

48. We can use the same method to graph $x - y \geq 4$.

1. Use the intercept method to graph the boundary $x - y = 4$. The intercepts are (4, 0) and (0, -4). The boundary is solid.

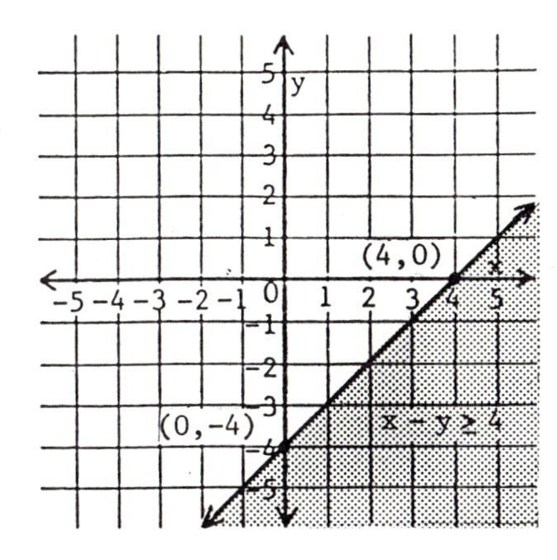

2. Then use the origin (0, 0) as a test point to determine the correct half-plane.

$$0 - 0 \geq 4 \text{ is false.}$$

Since the origin is not a solution, the half-plane below the boundary is correct.

49. We can use the same method when the boundary is a line through the origin. As an example, we graphed $y - 3x > 0$ below. $y = 3x$ is the boundary. We used a dashed line.

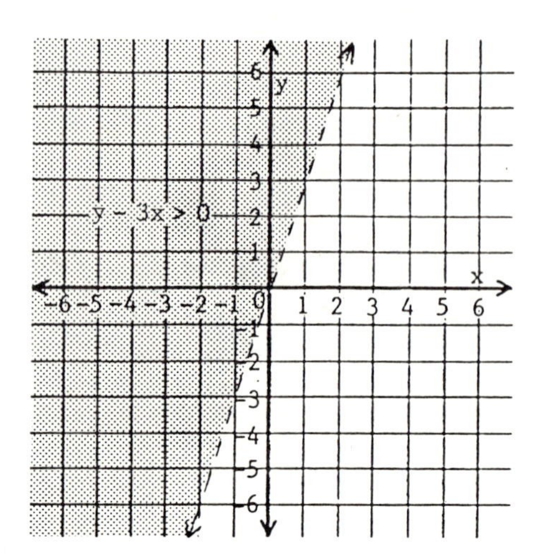

To determine the correct half-plane, we cannot use the origin because it is on the boundary line. We use some other point. If (1, 0) is not on the line, it is easy to use.

$$0 - 3(1) > 0 \text{ is false.}$$

Since (1, 0) is not a solution, the half-plane above the boundary is correct.

---

50. The following general method can be used to graph a linear inequality:

1. Graph the boundary. Use a solid line if the inequality contains $\geq$ or $\leq$. Use a dashed line if the inequality contains $>$ or $<$.
2. If the boundary line <u>does not</u> go through the origin, use the origin (0, 0) as a test point. If (0, 0) satisfies the inequality, shade the half-plane containing (0, 0). If (0, 0) does not satisfy the inequality, shade the other half-plane.
3. If the boundary line <u>does</u> go through the origin, use any other point on the test point. (1, 0) is easy to use if it is not on the boundary line.

Continued on following page.

50. Continued

Using the method on the preceding page, graph each inequality.

a) $3x + 4y < 12$

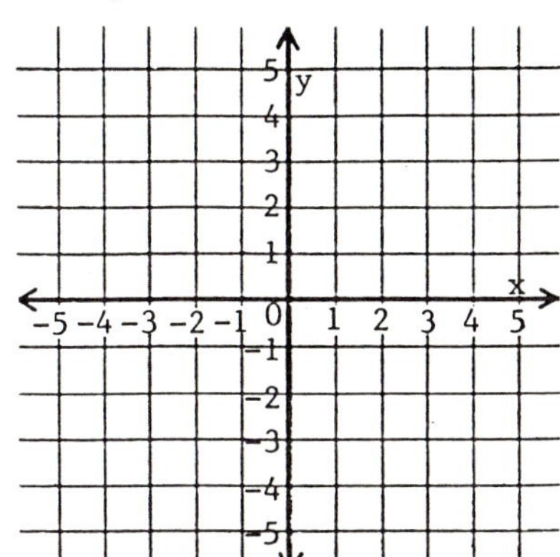

b) $y \leq 2x$

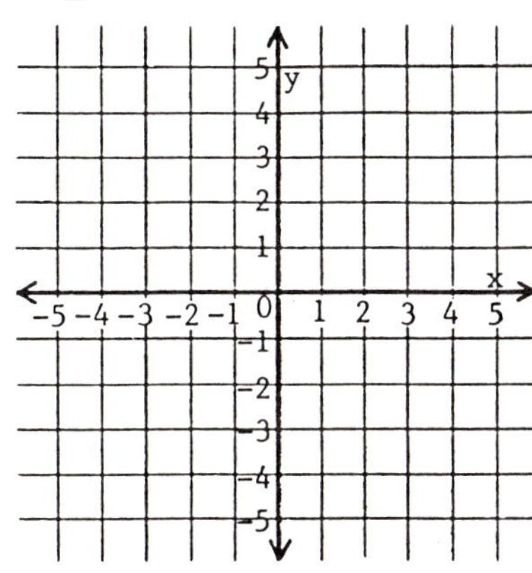

Answers to Frame 50:

a)

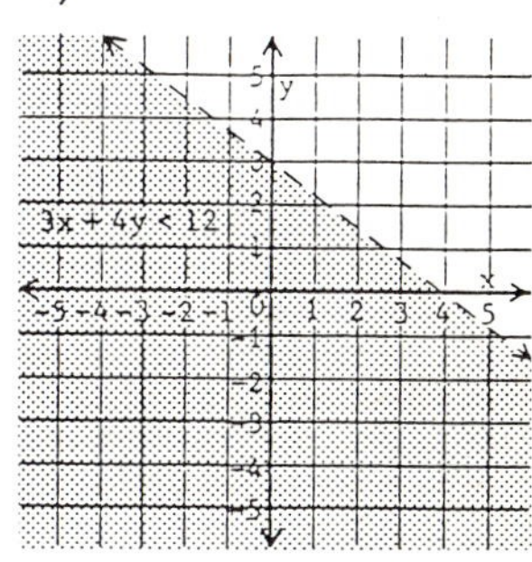

b)

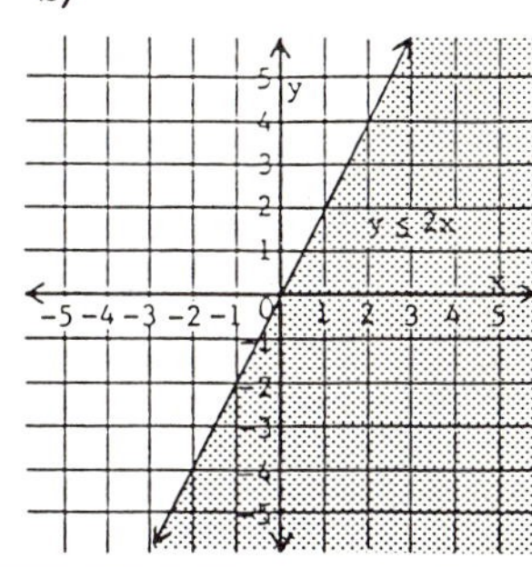

51. Using the same method, graph each inequality.

a) $2x - 5y \leq 10$

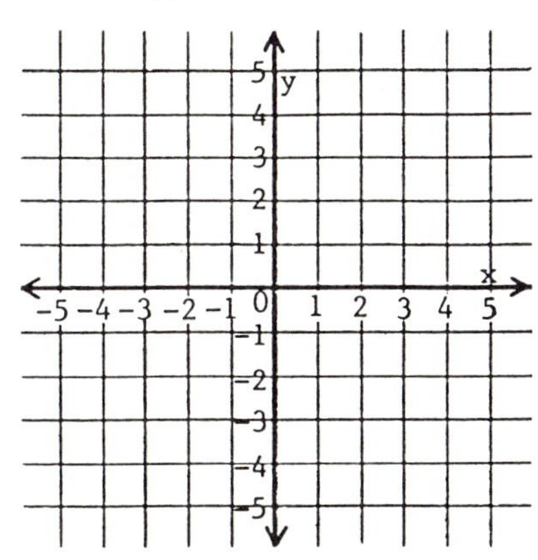

b) $y > -x + 2$

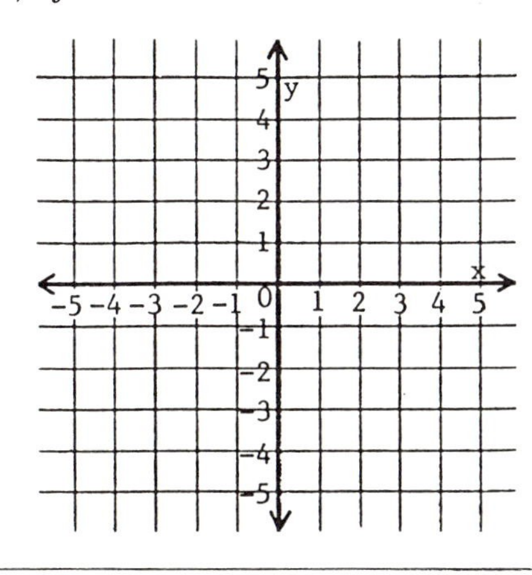

Answers to Frame 51:

a)

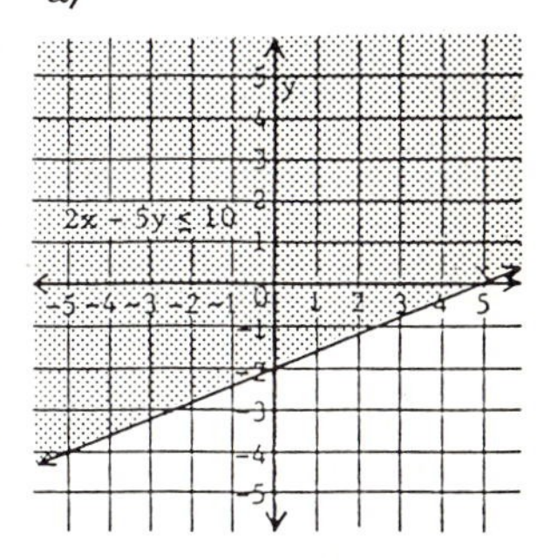

b)

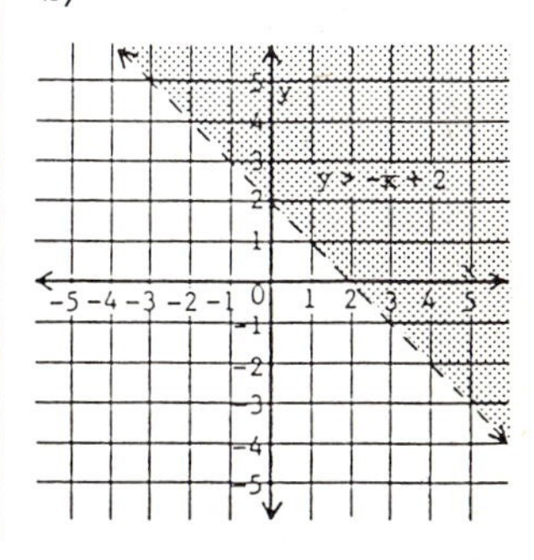

52. Using the same method, graph each inequality.

a) $4x - 5y \geq 20$

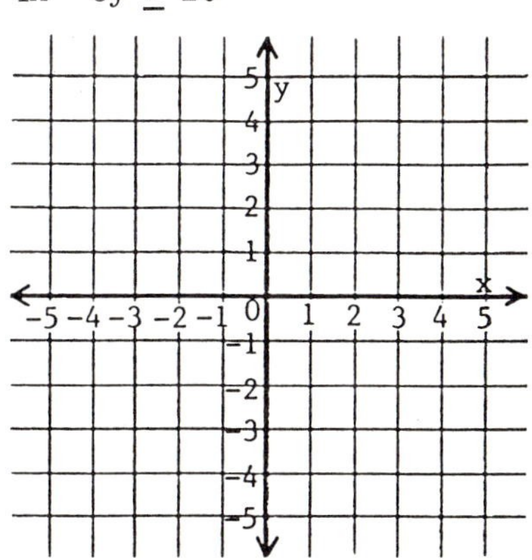

b) $x + y > 0$

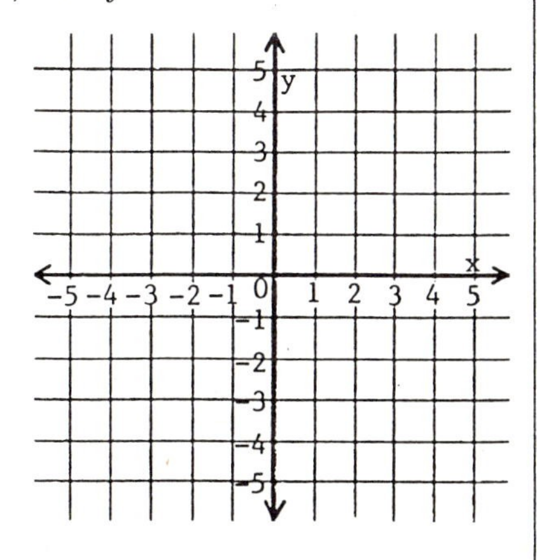

53. We graphed $y > 2$ below. The graph includes all points above the line $y = 2$. The line is dashed because $y = 2$ is not included. Graph $y \leq -1$ on the other graph.

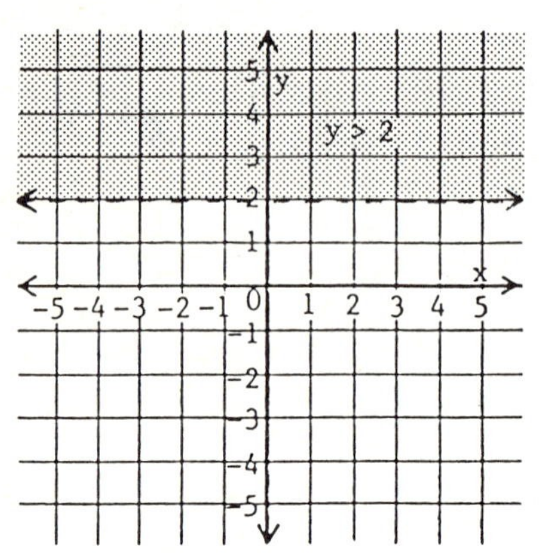

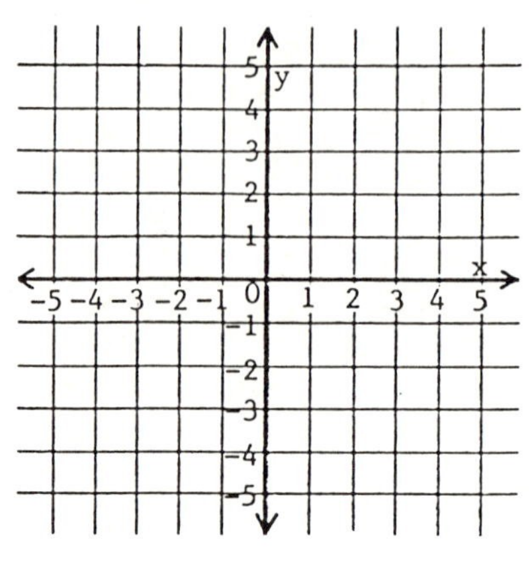

Answers to Frame 52:

a)
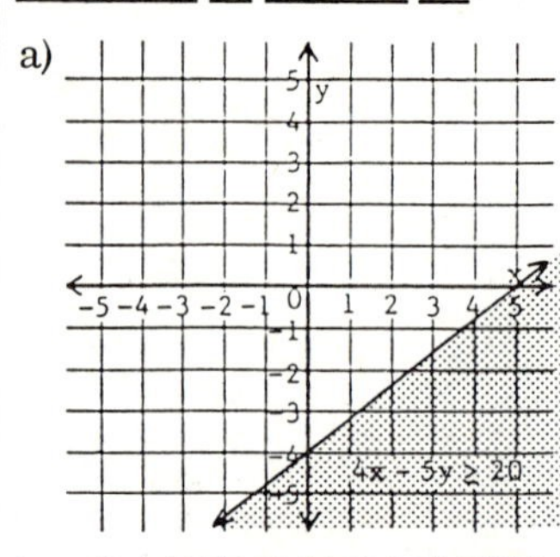

b)
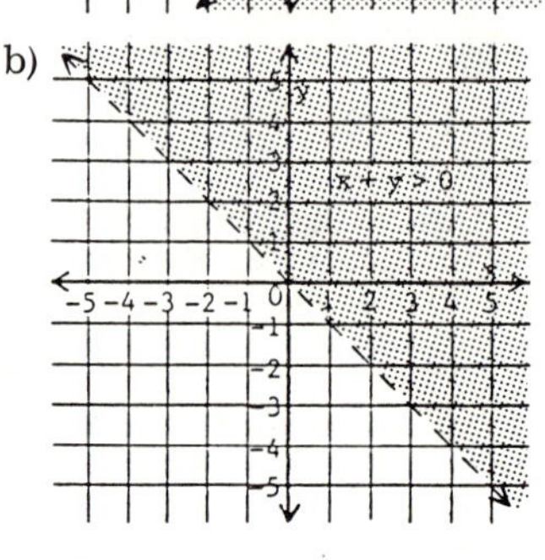

---

54. As we saw in an earlier section, $x \geq -2$ can be graphed on the number line. We get:

$x \geq -2$ can also be graphed on the coordinate system. We did so below. Its graph includes the line $x = -2$ and all points to the right of that line. Graph $x < 3$ on the other graph.

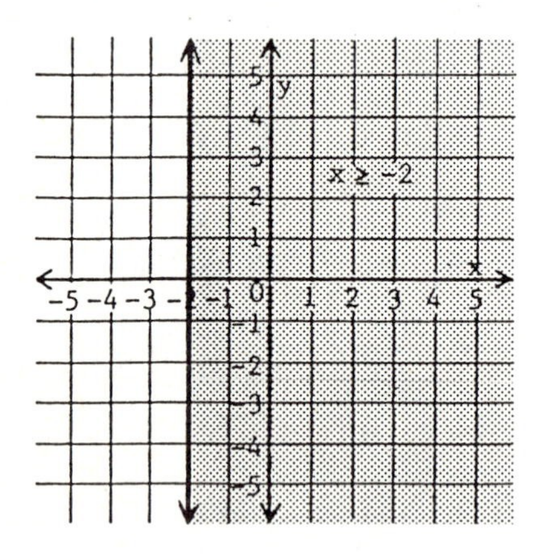

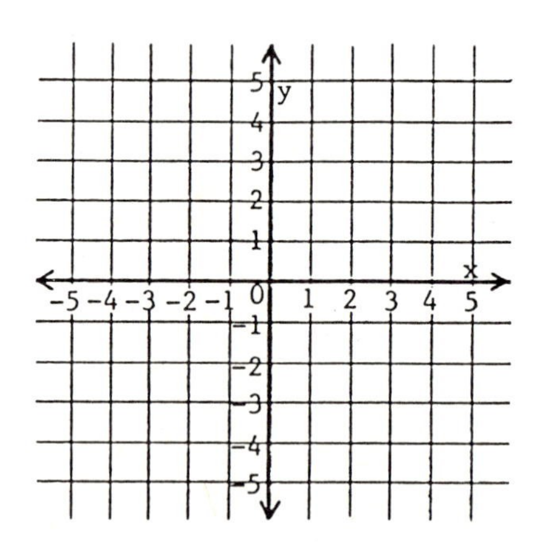

Answer to Frame 53:

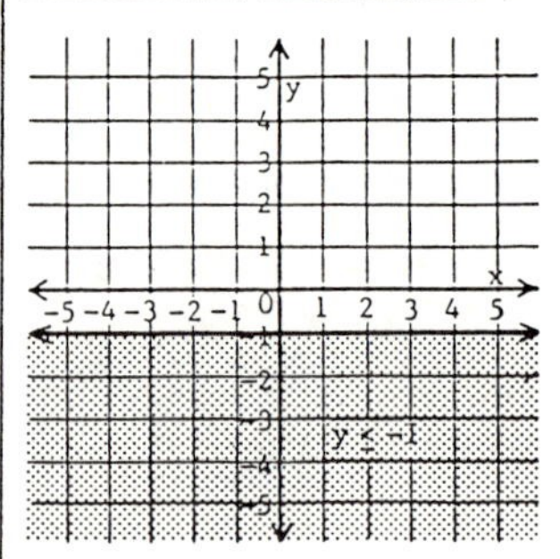

Answer to Frame 54:

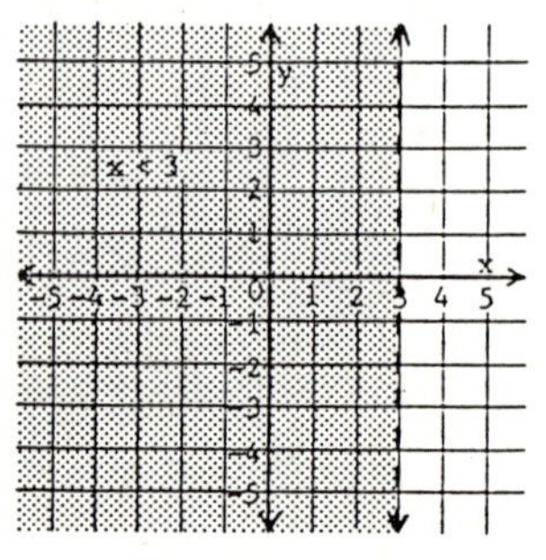

SELF-TEST 38 (pages 513-523)

Solve each problem.

1. The length of a rectangle is 15 ft. The area of the rectangle can be no more than 135 ft^2. Find the longest possible width of the rectangle.

2. A woman on a business trip cannot exceed a daily car rental allowance of $80. She rents a car for $25.95 plus 20¢ per mile. What mileage can she drive and stay within her allowance?

Graph each inequality.

3. $y < x - 2$

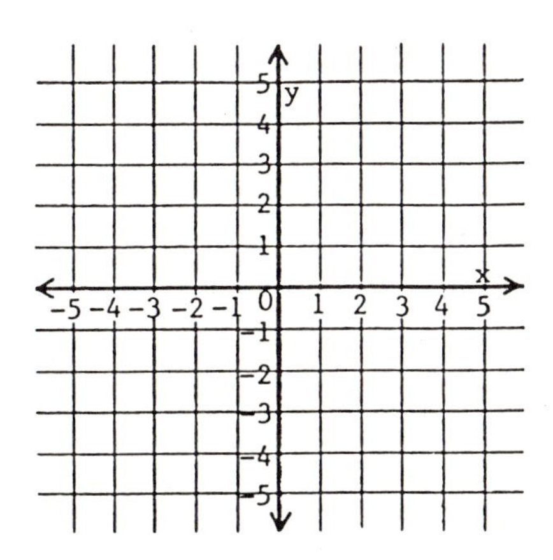

4. $2x + 3y \geq 6$

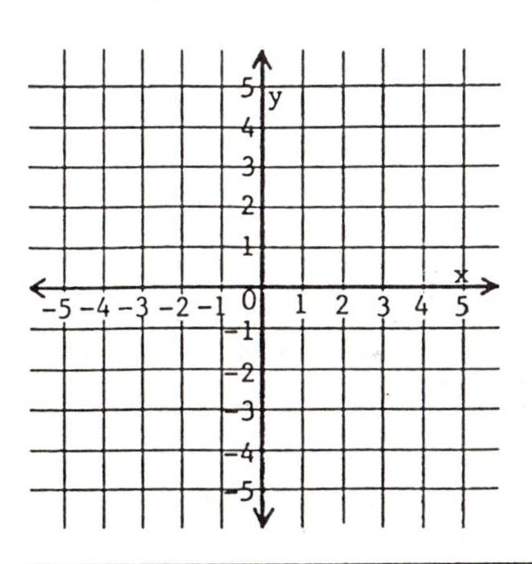

ANSWERS:

1. 9 ft

2. 270.25 miles per day or less

3. 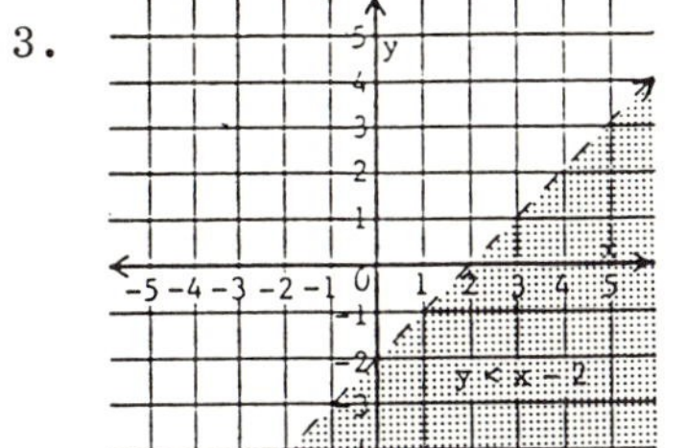

4. 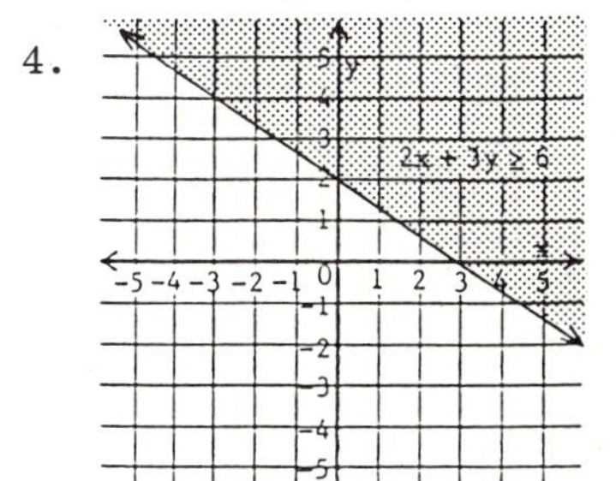

SUPPLEMENTARY PROBLEMS - CHAPTER 12

Assignment 37

Answer true or false for these.

1. If $x = 0$, $x \geq 1$ is ____________.
2. If $y = -4$, $y < -2$ is ____________.
3. If $x = 5$, $2x - 7 \leq 3$ is ____________.
4. If $y = -3$, $4y > -10$ is ____________.

Write with the variable on the left side.

5. $7 \geq x$
6. $-2 < y$
7. $-5 > x$
8. $3 \leq y$

Graph on the number line.

9. $x < -2$

10. $x \geq 1$

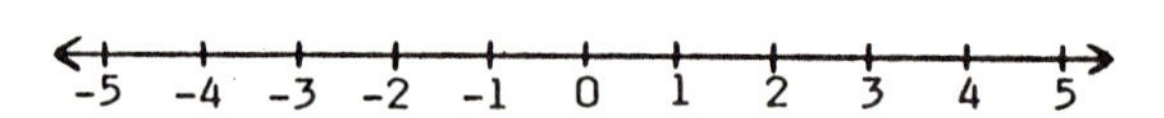

11. $x \leq 2$

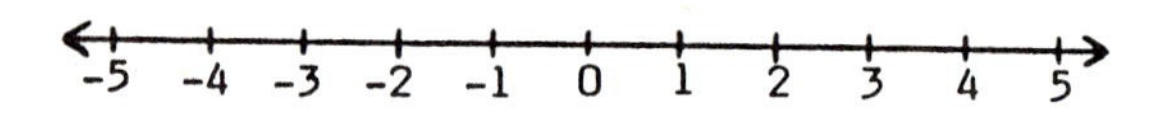

12. $-1 < x$

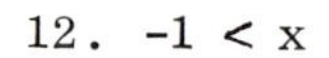

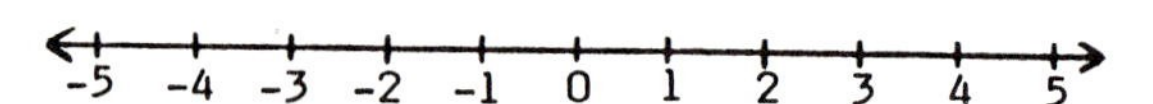

Translate to an inequality. Use x as the variable.

13. -2 and all numbers less than -2.
14. All positive numbers.
15. All numbers less than 4.
16. -3 and all numbers greater than -3.
17. The sum of a number and 5 is less than 7.
18. The difference between a number and 4 is greater than -1.
19. Three times a number is greater than or equal to -6.
20. 5 is less than or equal to twice a number, minus 3.

Solve each inequality.

21. $x + 3 > 2$
22. $y - 1 \leq 4$
23. $x + 5 < 0$
24. $y - 1.2 \geq 2.4$
25. $x - \frac{1}{3} < \frac{1}{2}$
26. $y + \frac{1}{2} \geq \frac{1}{4}$
27. $-7 > x - 5$
28. $0 \leq y - \frac{4}{5}$

Solve each inequality.

29. $4x \geq 20$
30. $-2y < 10$
31. $5x \leq 0$
32. $-3y > -2$
33. $3x \leq -\frac{1}{4}$
34. $-y > \frac{2}{5}$
35. $4 \geq \frac{2}{3}x$
36. $-5 < -4y$

Solve each inequality.

37. $2x - 3 < -2$
38. $4 - 3y \geq 1$
39. $10 - x > 6$
40. $3x - 2 \leq x - 5$
41. $7 - 4x < 5x - 1$
42. $5 - 3y \geq y - 7$

Assignment 38

Solve each problem.

1. Four times a number is less than -12. Find the number satisfying this condition.

2. The difference between a number and 10 is equal to or greater than 50. Find the number satisfying this condition.

3. The perimeter of a rectangle must be no greater than 80 meters. The length of the rectangle must be 28 meters. Find the longest possible width of the rectangle.

4. The perimeter of a rectangle must be no less than 110 ft. The width of the rectangle must be 22 ft. Find the shortest possible length of the rectangle.

5. The length of a rectangle is 12 cm. The area of the rectangle must be at least 108 cm^2. Find the shortest possible width of the rectangle.

6. The width of a rectangle is 15 meters. The area of the rectangle can be no more than 300m^2. Find the longest possible length of the rectangle.

7. The base of a triangle is 20 ft. The area of the triangle can be no more than 120 ft^2. Find the longest possible height of the triangle.

8. The height of a triangle is 14 cm. The area of the triangle must be at least 70 cm^2. Find the shortest possible base of the triangle.

9. Four tests are given in a math course. A total of 280 points is needed to get a C. If a student gets scores of 72, 65, and 68 on the first three tests, what score can she get on the third test to get a C?

10. Three tests are given in a history course. A total of 240 points is needed to get a B. If a student gets scores of 83 and 81 on the first two tests, what score can he get on the third test to get a B?

11. Company A rents compact cars for $21.95 plus 20¢ per mile. Company B rents compacts for $20.95 plus 21¢ per mile. For what mileages would the cost of renting from Company A be cheaper?

12. Company A rents intermediate-size cars for $26.95 plus 21¢ per mile. Company B rents intermediate-size cars for $24.95 plus 23¢ per mile. For what mileages would the cost of renting from Company B be cheaper?

13. A man on a business trip cannot exceed a daily car rental allowance of $72. He rents a car for $21.95 plus 20¢ per mile. What mileage can he drive and stay within his allowance?

14. A woman on a business trip cannot exceed a daily car rental allowance of $78. She rents a car for $24.95 plus 22¢ per mile. What mileage can she drive and stay within her allowance?

Graph each inequality.

15. $y < x$

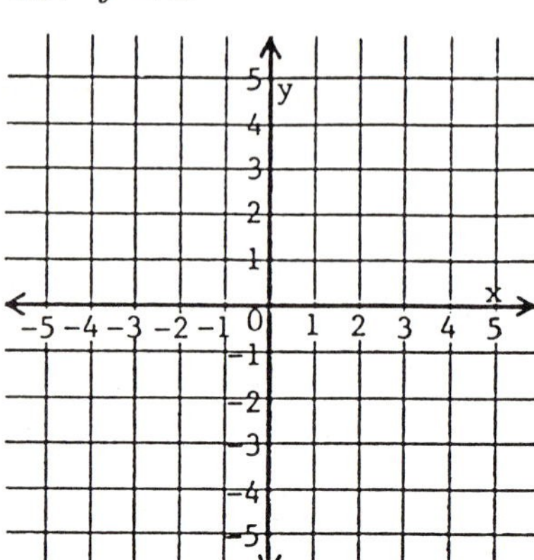

16. $y \geq x + 3$

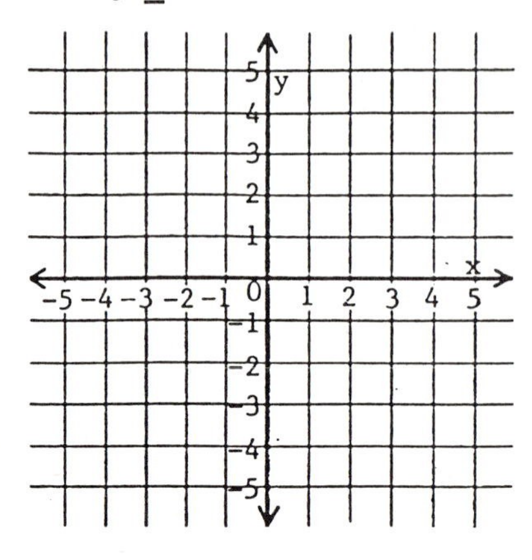

17. $y > 1 - x$

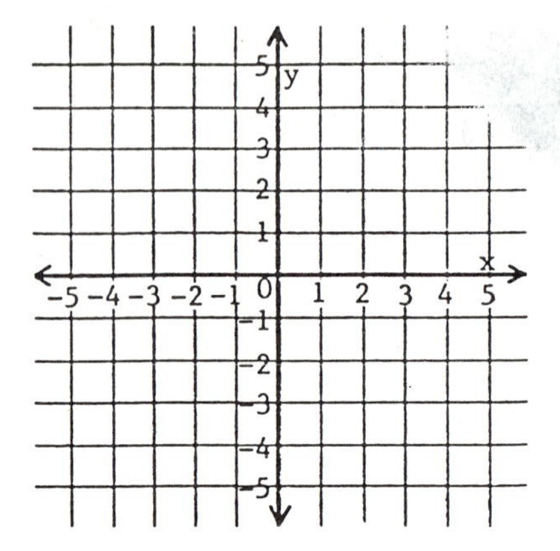

18. $x - 2y \leq 2$

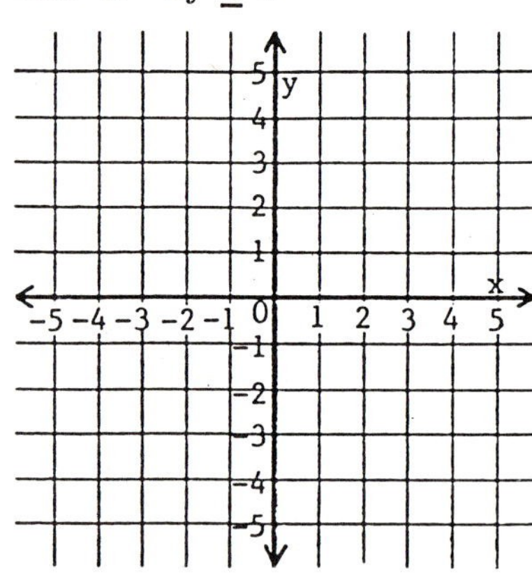

19. $x + 2y > 4$

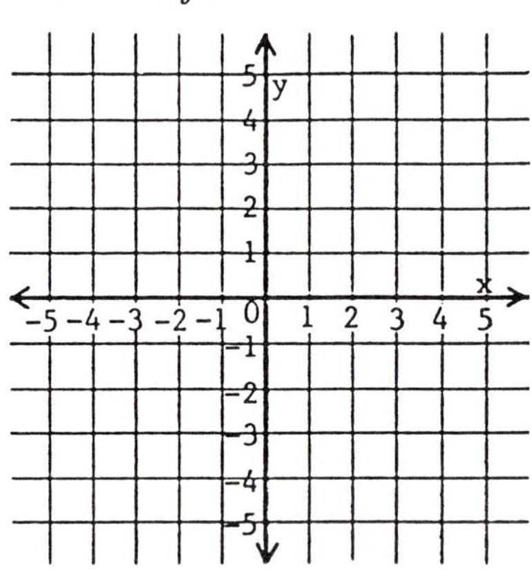

20. $3x - 5y \geq 15$

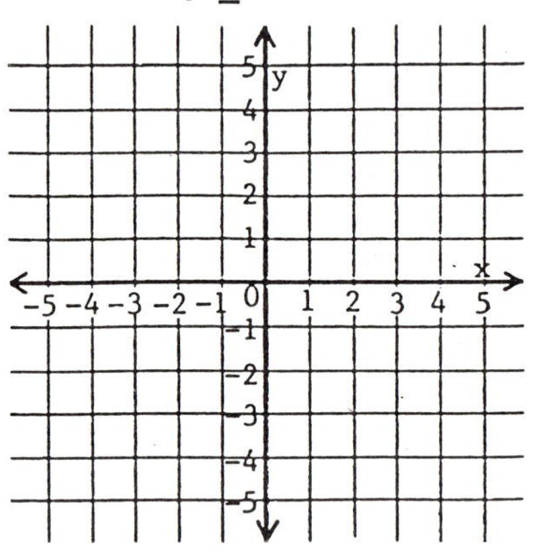

21. $5x + 2y < 10$

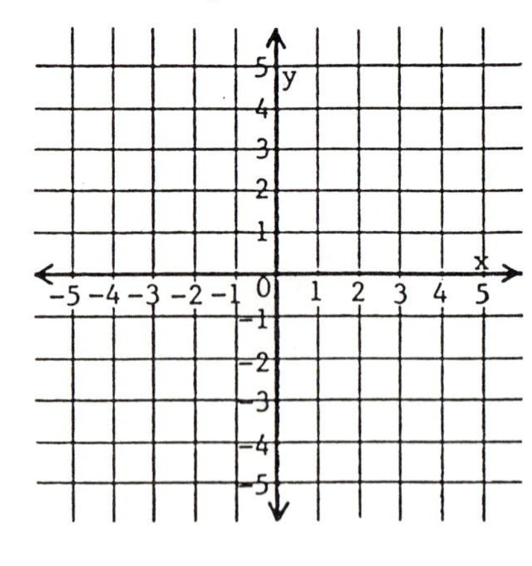

22. $y < -2$

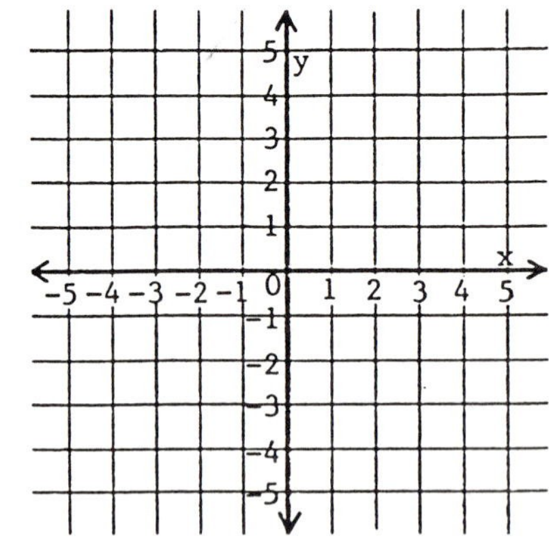

23. $x \geq 1$

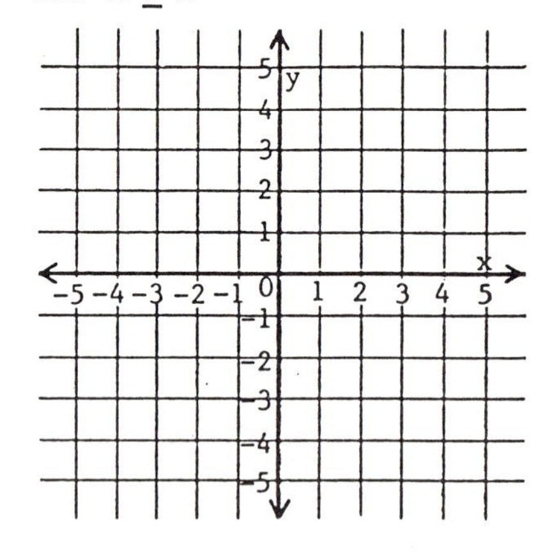

CHAPTER 1 - INTEGERS

Assignment 1

1. $>$ 2. $>$ 3. $<$ 4. $<$ 5. $<$ 6. $<$ 7. $<$ 8. $>$ 9. 6 10. 1 11. 0 12. 18 13. -6
14. -3 15. -4 16. 7 17. -7 18. -8 19. 0 20. 5 21. -30 22. 5 23. -30 24. 4
25. -4 26. 3 27. -3 28. -9 29. 6 30. -5 31. 1 32. -8 33. 10 34. -8 35. 1
36. 8 37. 0 38. -3 39. -2 40. -11 41. 32 42. -6 43. -10

Assignment 2

1. -56 2. -20 3. 0 4. 10 5. -12 6. 0 7. 24 8. -50 9. 28 10. -30 11. 0
12. 126 13. -6 14. -5 15. 0 16. -15 17. -9 18. 8 19. -32 20. 4 21. 5 22. 0
23. 64 24. 8 25. 256 26. 1 27. 9 28. 0 29. 81 30. -81 31. -1 32. -1

Assignment 3

1. 2 2. 22 3. -5 4. -16 5. 0 6. -10 7. -8 8. -8 9. -10 10. -1 11. 1 12. 5
13. -4 14. -2 15. 30 16. 16 17. -1 18. -7 19. 50 20. 3 21. -8 22. -3 23. 8
24. -8 25. 6 26. 10 27. 3 28. $20 - x$ 29. $6x^2$ 30. $3x + 8$ 31. $2x - 3$ 32. $6x^3 + 9$
33. $\frac{5}{3x}$ 34. $x^2 - 2x$ 35. $7(x + 2)$ 36. $4xy$ 37. $\frac{x}{y} + 2$ 38. $x^2 - y^2$ 39. x^3y^3 40. $20 - (x + y)$
41. $5(x - y)$

CHAPTER 2 - RATIONAL NUMBERS

Assignment 4

1. $-\frac{1}{15}$ 2. $-\frac{3}{4}$ 3. $\frac{10}{3}$ 4. $\frac{15}{8}$ 5. $-\frac{7}{4}$ 6. $-\frac{3}{4}$ 7. $\frac{9}{5}$ 8. 2 9. 0 10. $\frac{1}{4}$ 11. $-\frac{7}{6}$ 12. -27

13. $-\frac{12}{27}$ 14. $-\frac{70}{40}$ 15. $\frac{100}{20}$ 16. $-\frac{18}{6}$ 17. $\frac{3}{10}$ 18. $-\frac{4}{3}$ 19. $-\frac{1}{3}$ 20. -5 21. $-\frac{1}{6}$ 22. $\frac{50}{9}$

23. $-\frac{7}{4}$ 24. 6 25. $-\frac{4}{7}$ 26. 25 27. $\frac{2}{3}$ 28. -1

Assignment 5

1. 1 2. 0 3. $-\frac{7}{2}$ 4. -3 5. $-\frac{13}{20}$ 6. $\frac{8}{3}$ 7. $-\frac{5}{24}$ 8. $-\frac{7}{15}$ 9. $-\frac{7}{4}$ 10. $\frac{8}{3}$ 11. $-\frac{14}{5}$

12. $-\frac{37}{8}$ 13. $\frac{11}{9}$ 14. -1 15. -5 16. 0 17. $-\frac{39}{8}$ 18. $\frac{19}{5}$ 19. $-\frac{1}{30}$ 20. $-\frac{3}{20}$ 21. $\frac{18}{7}$

22. $-\frac{1}{3}$ 23. $-\frac{43}{8}$ 24. $-\frac{7}{16}$ 25. $\frac{1}{32}$ 26. $\frac{16}{81}$ 27. $\frac{64}{27}$ 28. $\frac{81}{100}$ 29. $\frac{5}{2}$ 30. $-\frac{1}{64}$ 31. $\frac{1}{16}$

32. $\frac{49}{64}$ 33. $-\frac{6}{5}$ 34. $-\frac{81}{100}$ 35. 2 36. $\frac{17}{8}$ 37. $-\frac{5}{4}$ 38. $\frac{20}{9}$ 39. $\frac{3}{2}$ 40. $\frac{4}{7}$ 41. -2

42. $\frac{9}{2}$ 43. $\frac{1}{2}$ 44. $-\frac{6}{5}$ 45. $-\frac{1}{4}$ 46. $\frac{2}{7}$ 47. $2\frac{2}{5}$ 48. $-3\frac{1}{5}$ 49. $1\frac{1}{2}$ 50. $-3\frac{3}{4}$

Assignment 6

1. 4.58 2. 5.757 3. 1.7 4. .1 5. -.57 6. 2.44 7. -2.86 8. -.59 9. 2.7 10. .021 11. -.8 12. .61 13. -1.1 14. -.6 15. -1.65 16. -1.7 17. 6.6 18. .032 19. -.414 20. .18 21. -.42 22. -3.52 23. 14.52 24. -.54 25. .08 26. .4 27. -.4 28. 1 29. -.2 30. -3.9 31. 8 32. -76 33. -9.3 34. -92 35. .442 36. .5 37. .25 38. .875 39. .8 40. .86 41. 1.11 42. .08 43. 2.77 44. .1 45. .29 46. .067 47. .03 48. $\frac{9}{100}$ 49. $\frac{769}{1000}$ 50. $\frac{56}{10}$ 51. $\frac{187}{100}$ 52. $x + \frac{3}{4}$ 53. $17.5 - x^2$ 54. $\frac{1}{2}x^3$ 55. 3.7x 56. $5x - \frac{2}{3}$ 57. 8.8 - 2.3x 58. $\frac{44.6}{x^2}$ 59. $\frac{2}{5}x + x^3$ 60. $x - y - \frac{5}{6}$ 61. 2.5 + xy 62. $\frac{x + y}{4.4}$ 63. $\frac{1}{4}(x - y)$

Assignment 7

1. 1 2. $\frac{1}{8}$ 3. $\frac{1}{49}$ 4. $\frac{1}{216}$ 5. $\frac{1}{81}$ 6. $\frac{1}{64}$ 7. 4^4 8. 2^{-1} 9. 6^8 10. 7^{-9} 11. 10^{-6} 12. 5^3 13. 9^{-1} 14. 1 15. 8^{-4} 16. 3^7 17. 3^8 18. 10^{20} 19. 4^{-8} 20. 7^{-12} 21. 2^{21} 22. 100 23. .001 24. 10 25. .1 26. 1 27. .00001 28. 1,000,000 29. .00000001 30. 1,000,000,000 31. .0000000001 32. 10^1 33. 10^4 34. 10^{-1} 35. 10^{-3} 36. 10^0 37. 10^6 38. 10^{-5} 39. 10^3 40. 10^{-8} 41. 10^{10} 42. 28 43. 65,600 44. 70,000,000 45. 9,400,000,000 46. .41 47. .00114 48. .000002 49. .000000088 50. 2.77×10^1 51. 5.19×10^2 52. 4.96×10^4 53. 6×10^6 54. 4.25×10^8 55. 3.14×10^{-1} 56. 7.5×10^{-2} 57. 4×10^{-6} 58. 6.39×10^{-5} 59. 1.1×10^{-8} 60. 8×10^9 61. 3.57×10^7 62. 4×10^{-5} 63. 4.1×10^{-6} 64. 6.02×10^{23} 65. .000000782 centimeter 66. 1.1574×10^{-5} day 67. 45,000,000,000 years

CHAPTER 3 - SOLVING EQUATIONS

Assignment 8

1. y = -4 2. t = 11 3. x = 41 4. N = 53 5. G = 1 6. d = -5 7. x = 6.6 8. p = 1.8 9. $d = \frac{2}{3}$ 10. $x = \frac{7}{8}$ 11. $V = -\frac{3}{7}$ 12. $m = -\frac{2}{3}$ 13. $p = \frac{4}{9}$ 14. $R = \frac{15}{8}$ 15. F = -1 16. $y = \frac{1}{7}$ 17. $w = -\frac{5}{3}$ 18. h = 0 19. x = -8 20. y = 3 21. $p = \frac{4}{3}$ 22. $x = -\frac{2}{5}$ 23. $y = \frac{5}{4}$ 24. H = 0 25. x = 4 26. r = -3 27. $m = -\frac{2}{3}$ 28. $s = \frac{1}{4}$ 29. $H = -\frac{3}{2}$ 30. A = -1 31. $y = -\frac{7}{4}$ 32. x = 1 33. $c = -\frac{1}{3}$ 34. p = 0 35. t = -4 36. m = 4.5 37. A = 10 38. $y = \frac{9}{4}$ 39. $x = \frac{3}{16}$

Assignment 9

1. 3x + 21 2. 18 + 27y 3. 50R + 10 4. 7x - 14 5. 15 - 20y 6. 8 - 40d 7. -4m - 12 8. -7 + 6P 9. 6x 10. 6y 11. -7m 12. 5d 13. -6r 14. -1 + 5p 15. 5V - 7 16. -3a - 9 17. R = -4 18. $y = -\frac{13}{5}$ 19. $P = \frac{1}{3}$ 20. $x = -\frac{5}{2}$ 21. w = 0 22. $h = \frac{10}{3}$ 23. $F = -\frac{3}{4}$ 24. x = 5 25. $y = \frac{4}{3}$ 26. $r = \frac{5}{4}$ 27. b = -2 28. $x = -\frac{2}{3}$ 29. $a = \frac{1}{2}$ 30. V = 1 31. t = 3 32. $x = \frac{1}{3}$ 33. $N = \frac{1}{3}$ 34. c = 0 35. E = 1 36. h = -12 37. y = 2 38. $w = -\frac{5}{2}$ 39. k = -1 40. $d = \frac{1}{3}$

Assignment 10

1. $a = -\frac{1}{4}$ 2. $R = \frac{4}{3}$ 3. $s = -\frac{1}{2}$ 4. $x = -\frac{1}{6}$ 5. $w = 0$ 6. $b = \frac{5}{2}$ 7. $G = -2$ 8. $h = -\frac{7}{2}$

9. $r = \frac{2}{5}$ 10. $t = \frac{1}{5}$ 11. $x = -\frac{7}{4}$ 12. $y = 10$ 13. $P = \frac{3}{2}$ 14. $F = \frac{5}{2}$ 15. $d = 2$ 16. $E = -\frac{3}{5}$

17. $x = -2$ 18. $V = -5$ 19. $w = 5$ 20. $d = -\frac{1}{4}$ 21. $x = -\frac{15}{4}$ 22. $y = 15$ 23. $E = -\frac{1}{2}$

24. $r = 3$ 25. $w = 54$ 26. $x = 3$ 27. $y = 25$ 28. $G = \frac{12}{5}$ 29. $v = \frac{1}{10}$ 30. $d = \frac{7}{2}$ 31. $p = \frac{8}{3}$

32. $t = 0$ 33. $h = -3$ 34. $x = -1$ 35. $y = \frac{1}{2}$ 36. $w = -\frac{3}{4}$ 37. $F = 1$ 38. $a = 6$ 39. $t = -5$

Assignment 11

1. $v = 96$ 2. $E = 72$ 3. $B = 70$ 4. $V = 400$ 5. $F = 68$ 6. $a = 5$ 7. $W = 12$ 8. $I = 4$
9. $I = 5$ 10. $h = 8$ 11. $N = 11$ 12. $W = 8$ 13. $t = 4$ 14. $A = 20$ 15. $W = 6$ 16. $v = 4$

17. $i = 3$ 18. $t_2 = 8$ 19. $s = \frac{P}{4}$ 20. $L = \frac{A}{W}$ 21. $r = \frac{C}{2\pi}$ 22. $R = \frac{E^2}{P}$ 23. $h = \frac{V}{\pi r^2}$

24. $v = \frac{d}{t_2 - t_1}$ 25. $M = \frac{F(p - a)}{2m}$ 26. $V_2 = \frac{T_2 V_1}{T_1}$ 27. $F_t = F_1 + F_2$ 28. $P_2 = P_1 + EI$

29. $Q = P - R$ 30. $W = \frac{P - 2L}{2}$ 31. $R = AB - Q$ 32. $F_1 = F_t - F_2$ 33. $D = C - B$ 34. $k = \frac{ah - w}{s}$

35. $W = dF$ 36. $f = \frac{1}{t}$ 37. $c = \frac{Q}{mT}$ 38. $G = \frac{WR^2}{KN}$ 39. $h = \frac{3V}{B}$ 40. $r = \frac{E - e}{i}$ 41. $G_1 = WP - G_2$

42. $d_2 = d_1 - vt$

CHAPTER 4 - WORD PROBLEMS

Assignment 12

1. $x - 15 = 32$ 2. $4x = x + 45$ 3. $\frac{1}{5}x = x - 16$ 4. $2x + 17 = 91$ 5. $10(x + 7) = 120$ 6. $x - \frac{1}{2}x = 24$

7. $\frac{3x}{5} = 30$ 8. $\frac{x - 12}{x} = 3$ 9. 26 10. 16 11. 12 12. 116 and 118 13. 17, 18, and 19
14. 26, 27, and 28 15. 45 cm and 80 cm 16. 799 adults 17. 38 amperes, 52 amperes and 28 amperes

Assignment 13

1. $\frac{3}{4}$ 2. $\frac{2}{5}$ 3. $\frac{1}{3}$ 4. $\frac{9}{10}$ 5. $\frac{1}{4}$ 6. .19 7. .039 8. .0875 9. 1.47 10. 3 11. 81%
12. 10.5% 13. 4.25% 14. 130% 15. 500% 16. 50% 17. 80% 18. 30% 19. $66\frac{2}{3}$% 20. 59%

21. 35% 22. 131% 23. 4.2% 24. 42.5% 25. 40% 26. 300 27. 50 28. \$21.25 29. 700
30. 4% 31. \$473.50 32. 76% 33. \$12,500 34. 60.7% 35. 667 grams 36. 27.4 kilograms

Assignment 14

1. $P = 16$ cm, $A = 15$ cm^2 2. $P = 32$m, $A = 64$ m^2 3. $C = 31.4$ cm, $A = 78.5$ cm^2
4. $C = 18.84$ ft, $A = 28.26$ ft^2 5. $A = 320$ m^2 6. $A = 20$ cm^2 7. $A = 30$ ft^2 8. $\angle A = 50°$
9. $\angle B = 50°$ 10. $\angle A = 30°$ 11. $V = 125$ cm^3 12. $V = 160$ in^3 13. $V = 62,800$ cm^3 14. $W = 7$ cm
15. $L = 20$ ft 16. $s = 12$ m 17. $h = 5$ cm 18. $b = 10$ yd 19. $A = 64$ in^2 20. $l_2 = 20$ m
21. $r = 7$ cm 22. $d = 8$ in 23. $W = 4$ ft 24. $W = 40$ m, $L = 65$ m 25. $L = 70$ ft, $W = 50$ ft
26. 24°, 72°, and 84° 27. 32°, 64°, and 84°

CHAPTER 5 - POLYNOMIALS

Assignment 15

1. x^{-1} 2. y^3 3. a^3b^4 4. p^2r^{-4} 5. $c^{-1}d^{-4}$ 6. $s^4v^3w^5$ 7. $m^{-1}p^2q^2$ 8. ac^2d^{-3} 9. p^4 10. q^{-4} 11. $c^{-1}d$ 12. $x^{-4}y^3$ 13. $a^{-8}b^{-3}$ 14. mp^7q^{-1} 15. $x^{-1}yz^{-4}$ 16. $a^3s^{-1}t^7$ 17. x^3y^6 18. $t^{-8}v^{-6}$ 19. $a^{-5}b^5$ 20. c^2d^{-8} 21. $3x^3 - 2x^2 + 5x - 1$ 22. $b^2x^3 + b^3x^2 + bx + 5b$ 23. $2x^5 - dx^3 + cx - 4y$ 24. 1 25. 2 26. 4 27. 5 28. 6 29. 2 30. 7 31. 9 32. binomial 33. monomial 34. trinomial 35. binomial 36. $9t^3$ 37. not possible 38. $-4cd^2$ 39. not possible 40. $-2y^3 + 7y$ 41. $ax^4 - bx^2$ 42. $4t^2 - 4$ 43. $7x^2 - x - 1$ 44. $-3x^3y^2 + 4xy + 5y^2$ 45. $3x^2y + xy^2 - 6$

Assignment 16

1. $2x^2 + 1$ 2. $3x^3 - x$ 3. $4x^4 + x^2 - 1$ 4. $6x^2 - 5$ 5. $kx^4 + ax^2 - b$ 6. $2y^2 + 4$ 7. $-y^3 + 3y + 2$ 8. $y^2 - y + 3$ 9. $xy^4 - 3$ 10. $2by^2 + dy$ 11. $ty + 2w$ 12. $4x^2$ 13. $-7V^2$ 14. $48y^2$ 15. $-63b^2$ 16. $-12r^4t^3$ 17. abx^3y^2 18. $10d^4k^6$ 19. $-18a^3p^4s^2$ 20. $9V^2$ 21. $49h^2$ 22. $144d^2$ 23. $400T^2$ 24. $16c^2d^2$ 25. $4x^6$ 26. $s^4t^2w^8$ 27. $25h^6p^2$ 28. $x^2 + x$ 29. $3y^2 - 7y$ 30. $5d^2 - 5d$ 31. $30a^2 + 70a$ 32. $40h^2 - 48h$ 33. $4x^3 - 12x$ 34. $2a^3 + a^2b$ 35. $3my^4 - 2m^2y$ 36. $2t^3 - 2t^2 + 10t$ 37. $x^3y + x^2y^2 + xy^3$ 38. $3m^3v^3 - 6mv^4 + 3m^2v^2$ 39. $15h^3s^3 - 5h^4s^2 - 10h^3s^2$ 40. $x^2 - x - 20$ 41. $8y^2 - 10y - 3$ 42. $2t^3 + t^2 - 10t - 8$ 43. $2d^2 + 3d - 5$ 44. $3x^2 - 8x - 35$ 45. $y^4 - 2y^3 + 2y^2 + 2y - 3$ 46. $2p^4 + p^3 - 7p^2 + 19p - 15$

Assignment 17

1. $x^2 + 8x + 15$ 2. $b^2 - 3b - 4$ 3. $t^2 + t - 6$ 4. $y^2 - 15y + 56$ 5. $6a^2 + 7a + 2$ 6. $5h^2 - 8h - 4$ 7. $4m^2 - 4m - 35$ 8. $12p^2 - 19p + 5$ 9. $cp + cq + dp + dq$ 10. $mt - 2m + 4t - 8$ 11. $2bs + bw - 6ds - 3dw$ 12. $a^2m - 2a^2t - b^2m + 2b^2t$ 13. $6m^2 - 7mp - 3p^2$ 14. $20d^2 + 11dk - 3k^2$ 15. $x^4 - 3x^2y + 2y^2$ 16. $12r^4 + 19r^2s^2 + 5s^4$ 17. $a^2 - 64$ 18. $R^2 - 1$ 19. $16x^2 - 9$ 20. $49y^2 - 100$ 21. $m^2 - n^2$ 22. $49a^2 - 4b^2$ 23. $c^4 - d^4$ 24. $p^2q^2 - 25$ 25. $x^2 + 6x + 9$ 26. $y^2 - 18y + 81$ 27. $9a^2 + 24a + 16$ 28. $100 - 60d + 9d^2$ 29. $4a^2 + 4ab + b^2$ 30. $9m^2 - 24mt + 16t^2$ 31. $x^4 - 12x^2y^2 + 36y^4$ 32. $25d^4 + 70d^2h + 49h^2$ 33. $8y^3$ 34. $-a$ 35. $-3cf$ 36. $4x^4 - 3$ 37. $4mp + 3p$ 38. $5t^2 - 3t + 4$ 39. $3x^2y + xz - 2$ 40. $m^4s^2 - m^3s - m$ 41. $x + 6$ 42. $d - 5$ 43. $a - 6$ 44. $3y - 4$ 45. $2r + 3$ 46. $5t - 2$

CHAPTER 6 - FACTORING

Assignment 18

1. 4 2. $6x$ 3. $8y$ 4. $6t$ 5. $7d^2$ 6. x^2y 7. $4b^3$ 8. 1 9. $4(2x + 5)$ 10. $6(3y - 2)$ 11. $4(t + 1)$ 12. $5(5d - 1)$ 13. $3(3a^2 + 5)$ 14. $12(5V^2 - 3)$ 15. $x(3x + 5)$ 16. $y^4(4y^2 - 1)$ 17. $12d(d^2 + 2)$ 18. $5p^3(8p^2 - 5)$ 19. $xy(x^2 + 2)$ 20. $5p^2q^2(2p^3q^3 - 1)$ 21. $6(2x^2 - x + 5)$ 22. $y(y^3 + 7y - 1)$ 23. $2t(t^2 - 3t + 2)$ 24. $3ab^2(3a^3b + 2a^2 - 1)$ 25. $(x + 2)(x + 8)$ 26. $(y + 3)(y - 7)$ 27. $(4t + 5)(t + 2)$ 28. $(3d + 7)(2d - 1)$ 29. $(4a + 3t)(a + 3t)$ 30. $(4p + 5q)(3p - 2q)$ 31. $(x - 3)(x + 2)$ 32. $(y - 2)(y + 6)$ 33. $(b - 4)(b - 3)$ 34. $(x + 1)(x + 2)$ 35. $(a + 2)(a + 5)$ 36. $(y + 3)(y + 3)$ 37. $(t - 1)(t - 3)$ 38. $(R - 2)(R - 4)$ 39. $(w - 4)(w + 3)$ 40. $(h + 5)(h - 1)$ 41. $(d - 5)(d + 2)$ 42. $(m + 7)(m - 3)$ 43. $(p + 3q)(p + 5q)$ 44. $(a - 3b)(a + 2b)$ 45. $(cd - 5)(cd - 1)$ 46. $(xy + 7)(xy - 2)$ 47. $(t^2 - 3)(t^2 - 4)$ 48. $(p^3q^3 - 6)(p^3q^3 + 4)$

Assignment 19

1. $(2y + 1)(y + 1)$ 2. $(3x + 1)(x + 2)$ 3. $(3p + 2)(p + 3)$ 4. $(2r - 1)(r - 2)$ 5. $(5t - 3)(t - 1)$ 6. $(3w - 4)(2w - 1)$ 7. $(2E + 3)(2E - 1)$ 8. $(d + 2)(3d - 4)$ 9. $(7x + 3)(x - 1)$ 10. $(4m + 3)(2m - 3)$ 11. $(2b + t)(b + 3t)$ 12. $(2r - s)(r + 3s)$ 13. $(5a + b)(a + 2b)$ 14. $(3tw + 2)(tw - 3)$ 15. $(4x - 3y)(x + 2y)$ 16. $(x + 3)^2$ 17. $(y - 6)^2$ 18. $(2b - 1)^2$ 19. $(3w + 2)^2$ 20. $(2a + 5b)^2$ 21. $(4dh - 1)^2$ 22. $(x + 7)(x - 7)$ 23. $(y + 10)(y - 10)$ 24. $(5m + 1)(5m - 1)$ 25. $(2t + 9)(2t - 9)$ 26. $(a^2 + 6)(a^2 - 6)$ 27. $(3b^3 + 2)(3b^3 - 2)$ 28. $(x + y)(x - y)$ 29. $(c + 4d)(c - 4d)$ 30. $(8p + 3q)(8p - 3q)$ 31. $(5x^2 + y)(5x^2 - y)$ 32. $(1 + bc)(1 - bc)$ 33. $(6ad^3 + 7)(6ad^3 - 7)$ 34. $(x - 1)(x^2 + x + 1)$ 35. $(y + 5)(y^2 - 5y + 25)$ 36. $(2b - 3)(4b^2 + 6b + 9)$ 37. $(5p + 1)(25p^2 - 5p + 1)$ 38. $(c - d)(c^2 + cd + d^2)$ 39. $(x + y^2)(x^2 - xy^2 + y^4)$ 40. $(m - 4t)(m^2 + 4mt + 16t^2)$ 41. $(3x + y)(9x^2 - 3xy + y^2)$

Assignment 20

1. $5(x+2)(x-2)$ 2. $2a(5b^2-2d^2)$ 3. $c(p-1)(p-7)$ 4. $2(y+3)(y-4)$ 5. $b(p+q)^2$
6. $(9t^2+1)(3t+1)(3t-1)$ 7. $(y+2)(y-2)(y^2+1)$ 8. $a(2x+1)(4x^2-2x+1)$ 9. $x=0$ and $\frac{3}{5}$
10. $d=0$ and $\frac{1}{3}$ 11. $y=0$ and $-\frac{1}{2}$ 12. $t=0$ and 1 13. $w=0$ and $\frac{1}{4}$ 14. $R=0$ and $\frac{3}{2}$
15. $x=5$ and -2 16. $k=-7$ and 1 17. $r=1$ and 1 18. $t=5$ and -3 19. $F=4$ and 1
20. $x=-3$ and 1 21. $t=\frac{1}{2}$ and -2 22. $d=-\frac{1}{3}$ and 1 23. $p=\frac{1}{5}$ and 2 24. $x=\frac{1}{3}$ and $-\frac{3}{2}$
25. $h=-\frac{4}{3}$ and 2 26. $y=\frac{5}{2}$ and 3 27. -2 and 5 28. $-\frac{5}{2}$ and 3 29. 6 and -6
30. 8 and 9, or -9 and -8 31. 7 and 9, or -9 and -7 32. 10 and 12, or -12 and -10
33. L = 10 ft, W = 4 ft 34. h = 7m, b = 10m 35. 3 seconds 36. 5 seconds

CHAPTER 7 - RATIONAL EXPRESSIONS

Assignment 21

1. $\frac{7}{8}$ 2. 3 3. $\frac{1}{5m}$ 4. $\frac{2}{5}$ 5. $\frac{3d}{2}$ 6. $\frac{1}{2}$ 7. $\frac{2(x-1)}{2x+3}$ 8. $\frac{t-2}{t+3}$ 9. $\frac{y+5}{y-2}$ 10. $\frac{2a}{a-3}$ 11. -1
12. $-(x+y)$ 13. $\frac{2x}{3y}$ 14. $\frac{4(x+1)}{5(x-1)}$ 15. $\frac{6}{5}$ 16. 2 17. $\frac{5x}{4}$ 18. $\frac{5}{6y}$ 19. $\frac{y-1}{y-2}$ 20. $\frac{x(x+5)}{2(2x+3)}$
21. $\frac{t-3}{t+1}$ 22. $\frac{p-5}{3}$ 23. $x-3$ 24. $\frac{1}{a}$ 25. $\frac{2}{7}$ 26. $\frac{y-1}{3y}$ 27. $\frac{a}{b}$ 28. $\frac{2(x-6)}{x+3}$ 29. $\frac{3y}{4(x+y)}$
30. $\frac{x+3}{x-1}$ 31. $x(x-y)$ 32. $\frac{a+5}{7}$

Assignment 22

1. $\frac{2x+1}{3}$ 2. $\frac{4y}{y+2}$ 3. $\frac{4t-5}{t-1}$ 4. $\frac{3x}{5}$ 5. $\frac{2y-5}{y-1}$ 6. $\frac{2p-7}{p+3}$ 7. $\frac{2}{3x}$ 8. $\frac{2y-5}{7}$ 9. $\frac{d+2}{d-3}$
10. $\frac{3+2x}{x}$ 11. $\frac{1}{y+2}$ 12. $\frac{1}{t-3}$ 13. $(2)(2)(2)(3)(3) = 72$ 14. $(2)(2)(3)(5) = 60$
15. $(2)(2)(2)(3)(x) = 24x$ 16. $(2)(2)(x)(y) = 4xy$ 17. $(2)(3)(5)(t^2) = 30t^2$ 18. $3y(y-3)$
19. $2x(2x+1)$ 20. $(y+1)(y-1)(y+4)$ 21. $(m+2)^2(m-5)$ 22. $\frac{17}{20x}$ 23. $\frac{2m^2+3m-3}{2m(m-1)}$
24. $\frac{2y^2-5y+17}{(y-2)(y+3)}$ 25. $\frac{x^2+15}{3x(x-2)}$ 26. $\frac{2y^2+7y+2}{(y+3)(y-3)(y+2)}$ 27. $\frac{5}{6x}$ 28. $\frac{p-7}{(p-2)(p-3)}$
29. $\frac{3x-8}{(x+4)(x-4)}$ 30. $\frac{x^2+y^2}{(x-y)(x-y)(x+y)}$ 31. $\frac{5(x+3)}{2x}$ 32. $\frac{2(y+12)}{7}$ 33. $\frac{15}{14}$ 34. $\frac{t-1}{t+1}$
35. $a-b$ 36. $\frac{2(3x-1)}{2x+1}$

Assignment 23

1. $x = \frac{7}{12}$ 2. $y = -1$ 3. $x = \frac{9}{4}$ 4. $y = 9$ 5. $x = \frac{10}{7}$ 6. $t = 9$ 7. $m = 11$ 8. $y = -\frac{8}{3}$
9. $x = -6$ 10. $p = \frac{1}{2}$ 11. $d = 8$ 12. $y = 2$ 13. $x = \frac{24}{7}$ 14. $y = 3$ 15. $m = 5$ and -4

16. $x = -8$ 17. $x = \frac{3}{2}$ 18. 2 and 12 19. 16 and 20 20. $\frac{8}{11}$ 21. $\frac{10}{30}$ 22. 40 23. $\frac{10}{7}$

24. 10 25. 15 and 16 26. 5 or -1 27. $2\frac{2}{5}$ hours 28. $2\frac{2}{9}$ hours 29. $3\frac{3}{7}$ hours

30. 9 hours for Joe, 18 hours for Mike

Assignment 24

1. $2\frac{1}{4}$ hours 2. 322 miles in 7 hours on first day; 260 miles in 5 hours on second day 3. 8 hours

4. $\frac{2}{3}$ hours 5. 300 km 6. 21 mph 7. 46 mph and 54 mph 8. 140 mph 9. $1\frac{1}{2}$ mph 10. $\frac{5}{3}$

11. $\frac{24}{1}$ 12. 5 hours 13. 375 miles 14. \$16.50 15. 30 minutes 16. 99 inches

17. 4.5 ounces 18. 72 lbs 19. 28 men 20. 64 votes 21. $F_1 = 36$ 22. $V_1 = 36$

23. $R_2 = 30$ 24. $D = 180$ 25. $P_1 = \frac{P_2V_2T_1}{V_1T_2}$ 26. $F = \frac{Df}{d}$ 27. $P_2 = \frac{P_1V_1}{V_2}$ 28. $T_2 = \frac{V_2T_1}{V_1}$

29. $A = \frac{B}{C + D}$ 30. $t = \frac{mp}{m + p}$ 31. $R = \frac{FS}{F - S}$ 32. $D = \frac{C}{1 - C}$ 33. $T = \frac{HP}{P - H}$

CHAPTER 8 - GRAPHING LINEAR EQUATIONS

Assignment 25

1. Yes 2. Yes 3. No 4. Yes 5. No 6. No 7. (-1, 3) 8. (-4, -6) 9. (0, 6) 10. (-2, 0)
11. (3, -4) 12. (1, 4) 13. (0, 8) 14. (2, 0) 15. A (3, 3), B (3, -2), C (-2, -1), D (-4, 2), E (2, 0),
F (0, -3) 16. G (1, 25), H $(-2\frac{1}{2}, 40)$, I (-4, -23), J $(1\frac{1}{2}, -32)$, K $(-3\frac{1}{2}, 0)$, L (0, 15) 17. a) 2 b) 4 c) 1
d) 3 e) 2 18. b, c, d 19. a, d, e 20. the origin 21. 9 22. -10 23. (17, -9)

24-25.

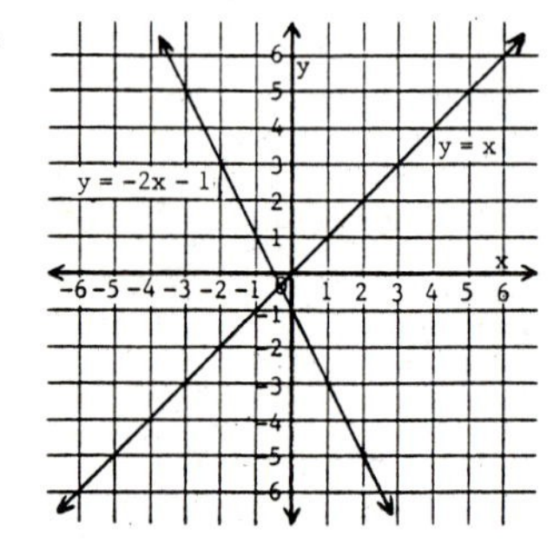

26-27.

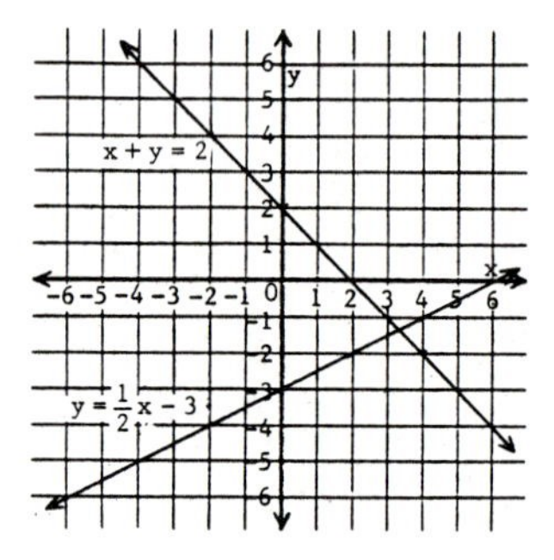

Assignment 26

1. (-3, 0) 2. (5, 0) 3. $(\frac{1}{2}, 0)$ 4. $(-\frac{8}{5}, 0)$ 5. (0, -9) 6. (0, 2) 7. $(0, \frac{1}{3})$ 8. $(0, -\frac{5}{2})$

9.

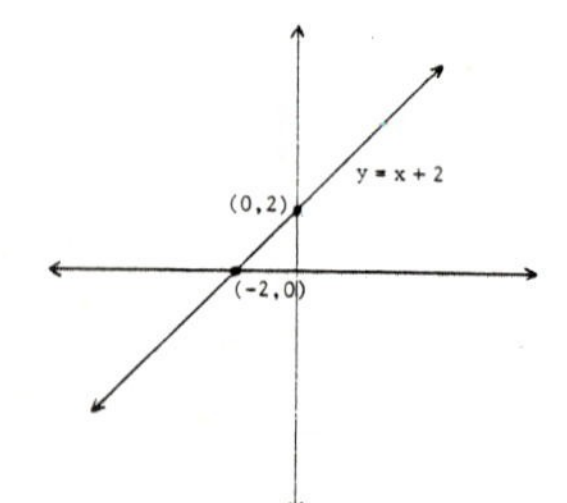

10.

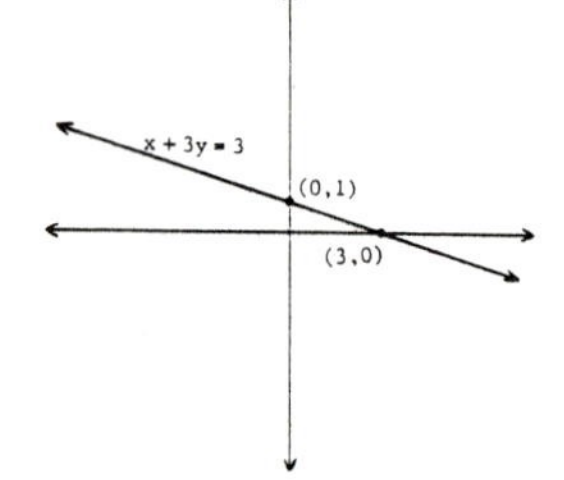

11.

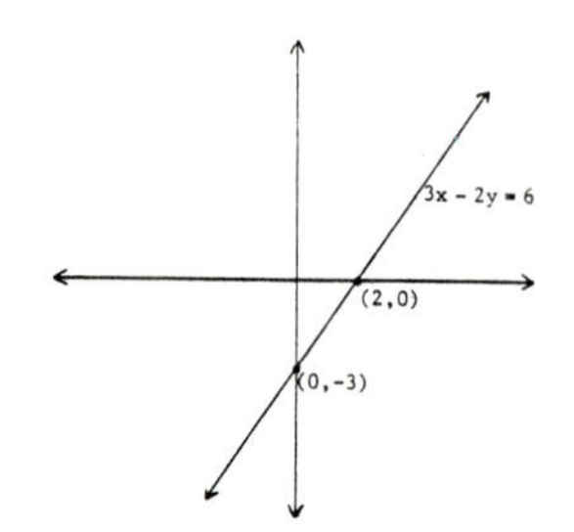

12. $y = 3$ 13. $x = -2$ 14. $x = 1$ 15. $y = -4$ 16. $y = 0$ 17. $x = 0$ 18. 1 19. $-\frac{1}{2}$ 20. -4
21. $\frac{1}{6}$ 22. A, D 23. A 24. B, C, E 25. C 26. C 27. 0 28. undefined 29. 2 30. $-\frac{3}{2}$
31. 1 32. -1 33. $\frac{1}{4}$ 34. $-\frac{3}{5}$ 35. $\frac{1}{2}$ 36. 2 37. -4

Assignment 27

1. $m = 5$, (0,2) 2. $m = 1$, (0,-3) 3. $m = -\frac{2}{5}$, $(0,\frac{3}{2})$ 4. $m = \frac{1}{2}$, (0,0) 5. $y = -x + 2$
6. $y = \frac{1}{2}x - 5$ 7. $y = \frac{1}{3}x - 1$ 8. $y = \frac{8}{3}x$ 9. $y = -4x + 6$ 10. $y = 3x + 2$ 11. $y = x - 5$
12. $y = \frac{1}{4}x - 2$ 13. $y = -\frac{1}{2}x + \frac{5}{3}$ 14. $y = \frac{1}{5}x$ 15. $y = 3x - 6$ 16. $y = -x + 2$ 17. $y = \frac{1}{3}x + 5$
18. $y = -4x + 6$ 19. $y = -\frac{1}{2}x + 3$ 20. $y = 5x$ 21. 4 amperes 22. 11 amperes 23. 60 volts
24. 35 volts 25. a) $k = 3$ b) $y = 3x$ 26. (a) and (e) 27. $G = 100$ 28. $w = 40$ 29. 375 kilometers
30. 35 centimeters

CHAPTER 9 - SYSTEMS OF EQUATIONS

Assignment 28

1. (3,1)

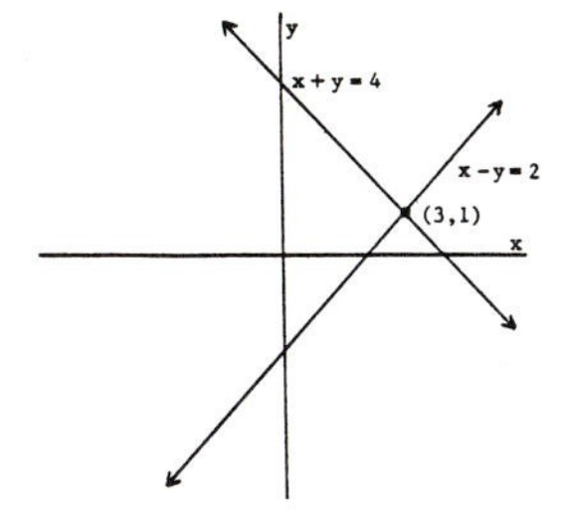

2. (-1,2)

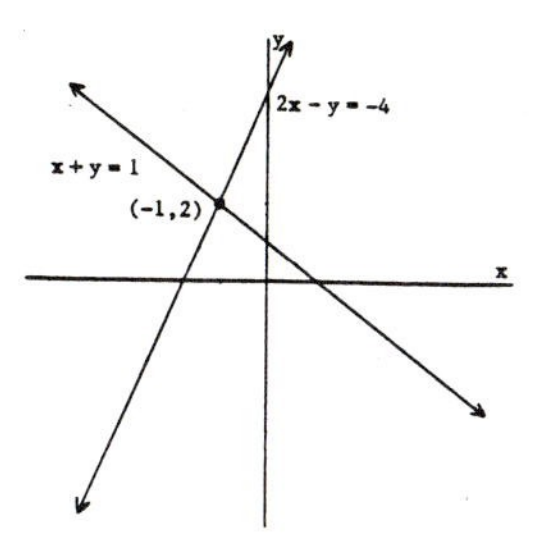

3. $x = 4$, $y = -1$ 4. $h = 6$, $m = 3$ 5. $x = 3$, $y = -2$ 6. $r = 10$, $v = 3$ 7. $h = 7$, $k = 0$
8. $b = 4$, $m = -8$ 9. $p = 20$, $r = 5$ 10. $x = 4$, $y = 1.2$ 11. $x = 5$, $y = 1$ 12. $x = 1$, $y = -1$
13. $h = 6$, $k = 5$ 14. $r = -1$, $s = 4$ 15. $c = 1$, $d = -2$ 16. $v = 3$, $w = -1$ 17. $x = 2$, $y = 4$
18. $p = 1$, $r = -1$ 19. $h = 5$, $k = 1$ 20. $x = -1$, $y = -6$ 21. $m = -2$, $n = -5$ 22. $b = 3$, $d = -5$

Assignment 29

1. $x = 5$, $y = 3$ 2. $a = 3$, $b = 9$ 3. $t = 6$, $w = 1$ 4. $r = 4$, $v = -1$ 5. $h = 20$, $m = 5$
6. $p = 1$, $s = 4$ 7. $d = 8$, $h = -2$ 8. $k = -3$, $m = 1$ 9. $r = 4$, $t = 1$ 10. $p = 2$, $w = 1$
11. $x = 2$, $y = 2$ 12. $d = 3$, $p = 4$ 13. $d = 6$, $t = 2$ 14. $E = 12$, $R = 3$ 15. $x = 1$, $y = -3$
16. $p = -1$, $v = -5$ 17. $b = 10$, $m = 4$ 18. $x = 6$, $y = 8$ 19. $F_1 = 20$, $F_2 = 10$ 20. $h = 2$, $k = -\frac{2}{3}$
21. $x = 4$, $y = -2$ 22. $p = -3$, $r = 2$ 23. $s = 10$, $w = \frac{5}{2}$ 24. $a = 1$, $b = -3$ 25. $b = 4$, $d = 6$
26. $x = 7$, $y = 30$ 27. $b = 20$, $d = 5$ 28. $r = \frac{3}{4}$, $w = 5$

Assignment 30

1. 150 and 68 2. 48 and 16 3. \$220 for the camera, \$55 for the calculator 4. 155 females 130 males 5. 55 cm and 45 cm 6. 16.4m and 8.2m 7. L = 18 cm, W = 12 cm 8. L = 22 ft, W = 11 ft 9. speed of wind is 13 mph, speed in still air is 143 mph 10. speed of current is $1\frac{1}{2}$ mph, speed in still water is $5\frac{1}{2}$ water 11. 40 liters of solution A, 20 liters of solution B 12. 75 milliliters of 10% solution, 125 milliliters of 50% solution 13. 100 pounds of alloy A, 200 pounds of alloy B 14. 5 kilograms of pure zinc, 10 kilograms of 40% zinc 15. 160 gallons of 6% milk, 40 gallons of skimmed milk

CHAPTER 10 - RADICAL EXPRESSIONS

Assignment 31

1. 3 and -3 2. 6 and -6 3. 9 and -9 4. 20 and -20 5. 2.646 6. -4.359 7. 7.141 8. -9.95 9. 4 10. 7 11. 3.464 12. 2.236 13. $\sqrt{15xy}$ 14. $\sqrt{a^2x^5}$ 15. $\sqrt{\frac{1}{6}}$ 16. $\sqrt{3x - 15}$ 17. $2x\sqrt{2y}$ 18. $3\sqrt{t^2 - 3t - 4}$ 19. $3\sqrt{3}$ 20. $10\sqrt{3x}$ 21. $4t^2$ 22. $d^4\sqrt{c}$ 23. $t\sqrt{t}$ 24. $2y^4\sqrt{7y}$ 25. $2h^2p\sqrt{hp}$ 26. $3\sqrt{a - 3b}$ 27. $3\sqrt{2}$ 28. $6\sqrt{y}$ 29. $2x\sqrt{6}$ 30. $4a^2\sqrt{3b}$ 31. $10tw\sqrt{2w}$ 32. $2h^2k^2\sqrt{3h}$ 33. $2\sqrt{2}$ 34. $3\sqrt{x}$ 35. $p^2\sqrt{d}$ 36. $5ab\sqrt{a}$

Assignment 32

1. $\frac{2}{3}$ 2. $\frac{6x}{5}$ 3. $\frac{2\sqrt{3y}}{7}$ 4. $\frac{1}{ab^2}$ 5. $\frac{t\sqrt{t}}{m^3}$ 6. $\frac{4}{a\sqrt{2x}}$ 7. $\frac{2y}{3x^2\sqrt{x}}$ 8. $\frac{2\sqrt{x - 2y}}{9}$ 9. 4 10. $\sqrt{2}$ 11. $\frac{x\sqrt{x}}{5}$ 12. $\frac{2\sqrt{5}}{x^3}$ 13. 16x 14. 4x 15. $\frac{y}{3}$ 16. $\frac{y}{\sqrt{3}}$ 17. ab 18. $b\sqrt{a}$ 19. $x - 7$ 20. $\frac{a - b}{c}$ 21. 5 22. 3 23. xy 24. 2t 25. $y - 4$ 26. $x + 1$ 27. $\frac{t - 2}{5}$ 28. $\frac{x - y}{7}$ 29. $\frac{\sqrt{2x}}{2x}$ 30. $\frac{3\sqrt{5}}{10}$ 31. $\frac{2\sqrt{3x}}{x}$ 32. $\frac{d\sqrt{p - q}}{p - q}$ 33. $\frac{\sqrt{3}}{2}$ 34. $\sqrt{y}$ 35. $\frac{2\sqrt{5}}{5}$ 36. $\frac{x\sqrt{6}}{3y}$ 37. $9\sqrt{3}$ 38. $4\sqrt{x}$ 39. $(a + b)\sqrt{y}$ 40. $\sqrt{10}$ 41. $-4\sqrt{y}$ 42. $(K - W)\sqrt{A}$

Assignment 33

1. $x = 16$ 2. $y = 27$ 3. $m = 50$ 4. $x = \frac{5}{7}$ 5. $P = 14$ 6. $x = \frac{3}{4}$ 7. $y = 16$ 8. $V = 9$ 9. $x = 64$ 10. $x = 10$ 11. $m = -\frac{1}{3}$ 12. $y = \frac{9}{16}$ 13. $t = 3$ 14. $w = 1$ and 2 15. $y = 5$ (not $y = 2$) 16. no solution 17. $t = 3$ 18. $a = 4$ 19. $N = 10$ 20. $a = 8$ 21. $R = 4$ 22. $R = 40$ 23. $t = 25$ 24. $a = 48$ 25. $s = \frac{v^2}{2a}$ 26. $R = \frac{P}{I^2}$ 27. $q = m^2 - p$ 28. $c = b - a^2d$ 29. $V = \frac{T}{a^2b^2}$ 30. $t = \frac{m^2p^2 - a}{b}$ 31. $P_2 = \frac{c^2}{d^2P_1}$ 32. $E = \frac{FH^2}{P^2}$ 33. $R = \frac{W^2}{P^2}$ 34. $h = \frac{s^2r^2}{b^2}$ 35. $B = P\sqrt{G}$ 36. $A = \frac{DF}{\sqrt{R}}$

CHAPTER 11 - QUADRATIC EQUATIONS

Assignment 34

1. $x = 7$ and -5 2. $y = 4$ and 6 3. $x = -\frac{5}{3}$ and 1 4. $t = -\frac{3}{5}$ and -2 5. $m = 2$ and 2 6. $d = 9$ and -9 7. $x = \frac{3}{4}$ and $-\frac{3}{4}$ 8. $y = 0$ and 7 9. $p = 0$ and $-\frac{8}{7}$ 10. $d = -\frac{3}{2}$ and 1 11. $v = \frac{3}{2}$ and $-\frac{3}{2}$ 12. $h = 0$ and 3 13. $x = 10$ and -10 14. $y = 3$ and -3 15. $t = \frac{5}{9}$ and $-\frac{5}{9}$ 16. $x = \frac{1}{4}$ and $-\frac{1}{4}$ 17. $y = \frac{7}{6}$ and $-\frac{7}{6}$ 18. $m = 6$ and -6 19. $a = \frac{3}{2}$ and $-\frac{3}{2}$ 20. $b = 0$

21. $y = \pm 4.123$ 22. $x = \pm 5.477$ 23. $d = \pm 6.481$ 24. No real number solution 25. $x = 2$ and -8 26. $y = 4$ and -2 27. $t = 2 + \sqrt{3}$ and $2 - \sqrt{3}$ 28. No real number solution 29. $x^2 + 20x + 100 = (x + 10)^2$ 30. $y^2 - 2y + 1 = (y - 1)^2$ 31. $t^2 + 24t + 144 = (t + 12)^2$ 32. $m^2 - 40m + 400 = (m - 20)^2$ 33. $d^2 + d + \frac{1}{4} = \left(d + \frac{1}{2}\right)^2$ 34. $b^2 - 5b + \frac{25}{4} = \left(b - \frac{5}{2}\right)^2$ 35. $x = 1$ and 5 36. $x = 2 \pm \sqrt{15}$ 37. $y = -5 \pm \sqrt{5}$ 38. $t = \frac{3 \pm \sqrt{5}}{2}$ 39. $y = \frac{2}{3}$ and -2 40. $x = \frac{5 \pm \sqrt{33}}{4}$

Assignment 35

1. $x = 4$ and -2 2. $t = \frac{1}{2}$ and $-\frac{2}{3}$ 3. $w = 1$ and $-\frac{3}{4}$ 4. $R = -4$ and $\frac{5}{2}$ 5. $y = -2 \pm \sqrt{3}$ or $y = -0.268$ and -3.732 6. $d = \frac{5 \pm \sqrt{37}}{6}$ or $d = 1.847$ and -0.181 7. $p = \frac{4 \pm \sqrt{6}}{5}$ or $p = 1.29$ and 0.31

8. $F = \frac{-1 \pm \sqrt{57}}{4}$ or $F = 1.638$ and -2.138 9. $h = -1 \pm \sqrt{2}$ or $h = 0.414$ and -2.414

10. $x^2 - 3x + 5 = 0$ 11. $y^2 + 5y - 10 = 0$ 12. $t^2 - 3t + 4 = 0$ 13. $5a^2 + 7a - 5 = 0$

14. $3x^2 + 5x - 9 = 0$ 15. $2d^2 + 2d - 9 = 0$ 16. $N^2 - N - 1 = 0$ 17. $x^2 - x - 9 = 0$

18. $x^2 - 10x - 4 = 0$ 19. $x = 1$ and $\frac{2}{3}$ 20. $t = 3$ and $-\frac{1}{2}$ 21. $y = -3$ and $-\frac{2}{3}$ 22. $r = \frac{1}{2}$ and -2

23. $x = 2 \pm \sqrt{7}$ or $x = 4.646$ and -0.646 24. $x = 3 \pm \sqrt{13}$ or $x = 6.606$ and -0.606 25. 2.449 sec

26. 3 sec 27. 3.873 sec 28. 5.477 sec 29. L = 4.5 ft, W = 3.5 ft 30. L = 2.5m, W = 0.5m

31. 8 mph 32. 7 mph

Assignment 36

1. $A = 25$ 2. $s = 80$ 3. $h = 4$ 4. $F = 4$ 5. $I = 2$ 6. $r = \sqrt{2} = 1.414$ 7. $v = 3$ 8. $d = 12$

9. $R = \sqrt{28} = 5.292$ 10. $t = \frac{5}{2}$ 11. $K = \sqrt{ab}$ 12. $H = \sqrt{D + T}$ 13. $c = \sqrt{\frac{E}{m}}$ 14. $v = \sqrt{2as}$

15. $t = \sqrt{\frac{d}{2k}}$ 16. $h = \sqrt{\frac{2M}{b}}$ 17. $R = \sqrt{P^2 + Q^2}$ 18. $Z = \sqrt{X^2 + R^2}$ 19. c = 4.472 cm

20. a = 6.928 ft 21. W = 6.633m 22. leg = 8 ft 23. d = 4.243 cm 24. L = 5.745m

25. $1,166.40 26. $1,232.10 27. r = 10% 28. r = 8.33%

29.

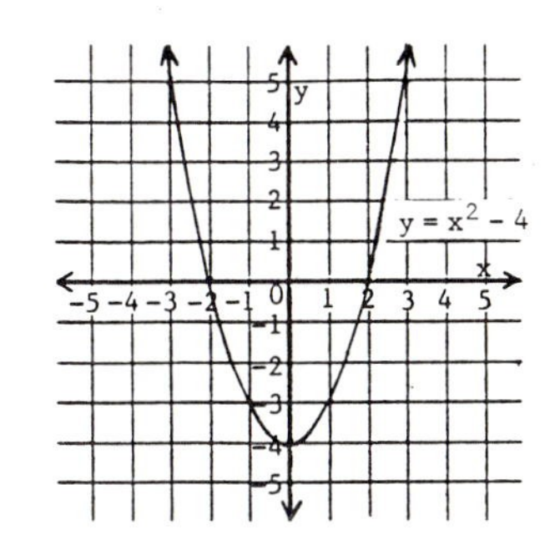

30.

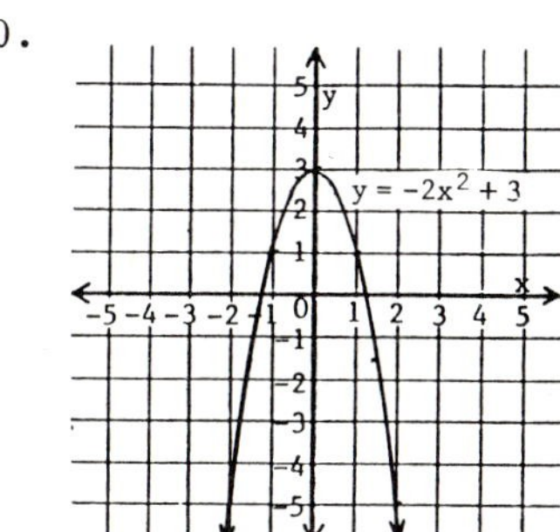

31.

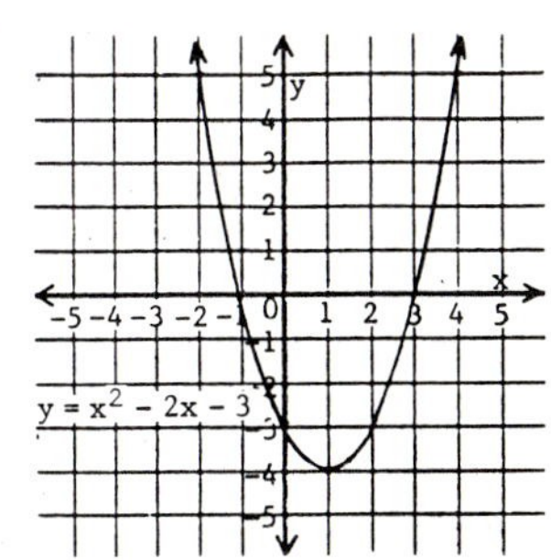

CHAPTER 12 - INEQUALITIES

Assignment 37

1. false 2. true 3. true 4. false 5. $x \leq 7$ 6. $y > -2$ 7. $x < -5$ 8. $y \geq 3$

9. 

10. 

11. 

12. 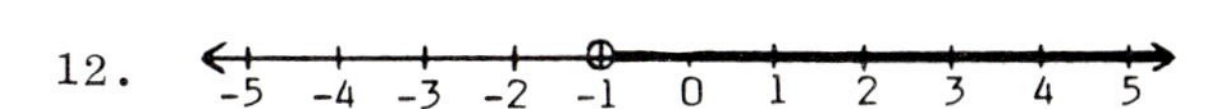

13. $x \leq -2$ 14. $x > 0$ 15. $x < 4$ 16. $x \geq -3$ 17. $x + 5 < 7$ 18. $x - 4 > -1$ 19. $3x \geq -6$

20. $5 \leq 2x - 3$ 21. $x > -1$ 22. $y \leq 5$ 23. $x < -5$ 24. $y \geq 3.6$ 25. $x < \frac{5}{6}$ 26. $y \geq -\frac{1}{4}$

27. $x < -2$ 28. $y \geq \frac{4}{5}$ 29. $x \geq 5$ 30. $y > -5$ 31. $x \leq 0$ 32. $y < \frac{2}{3}$ 33. $x \leq -\frac{1}{12}$ 34. $y < -\frac{2}{5}$

35. $x \leq 6$ 36. $y < \frac{5}{4}$ 37. $x < \frac{1}{2}$ 38. $y \leq 1$ 39. $x < 4$ 40. $x \leq -\frac{3}{2}$ 41. $x > \frac{8}{9}$ 42. $y \leq 3$

Assignment 38

1. $x < -3$ 2. $x \geq 60$ 3. 12 meters 4. 33 ft 5. 9 cm 6. 20 meters 7. 12 ft 8. 10 cm
9. 75 or more 10. 76 or more 11. more than 100 miles 12. less than 100 miles
13. 250.25 miles or less 14. 241.14 miles or less

15.

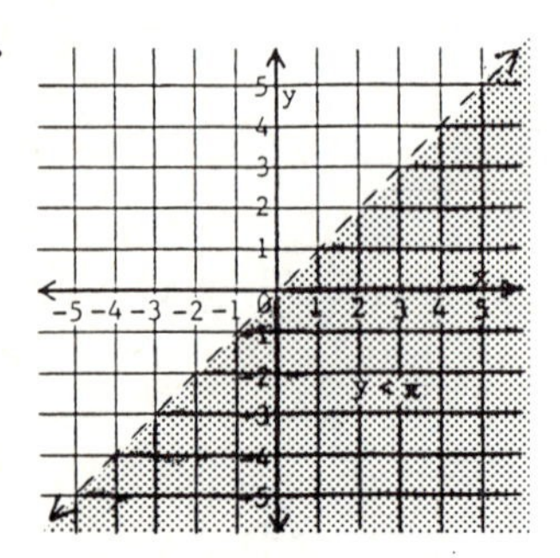

16.

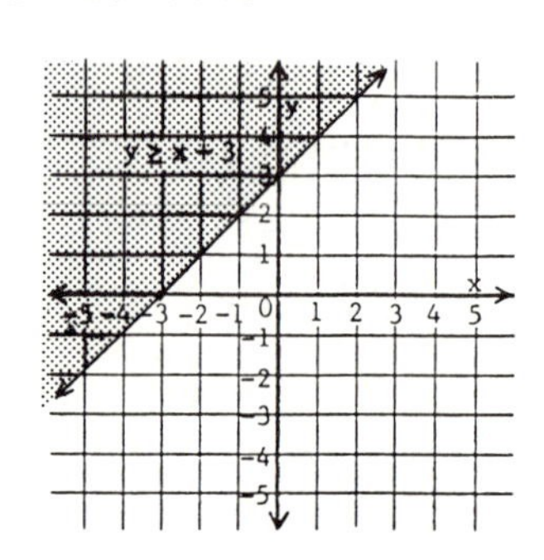

17.

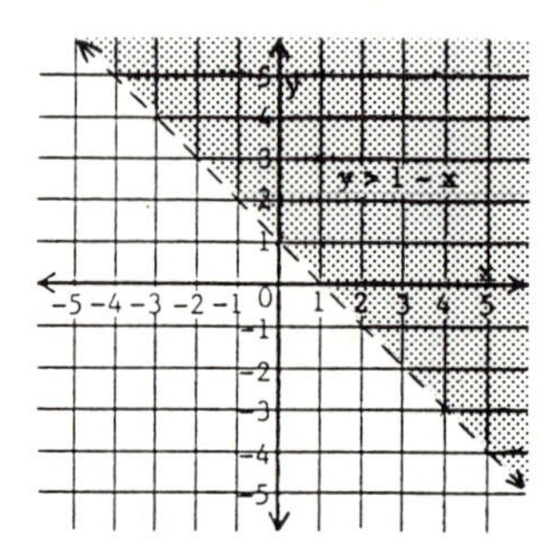

18.

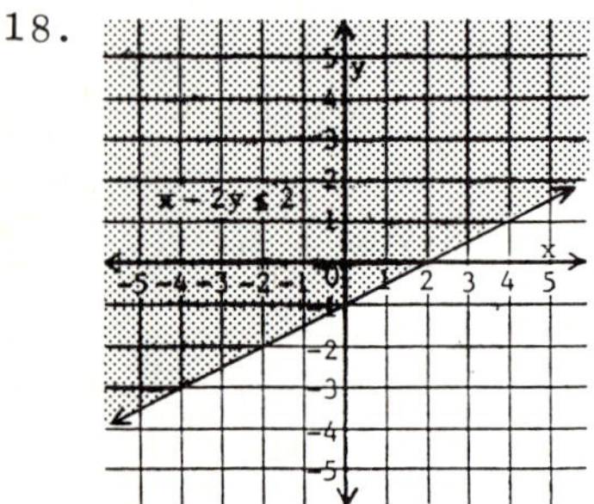

19.

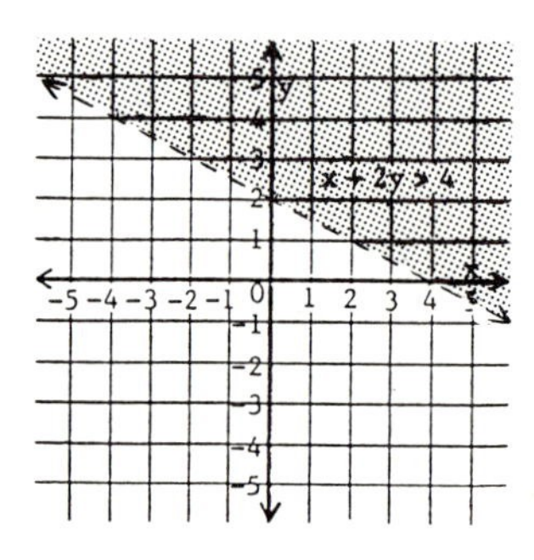

20.

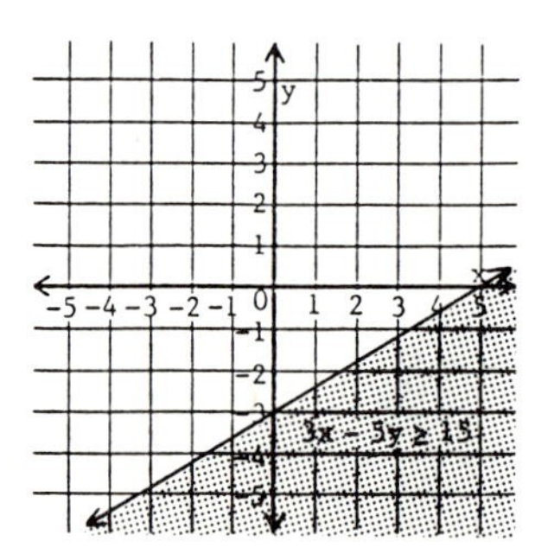

21.

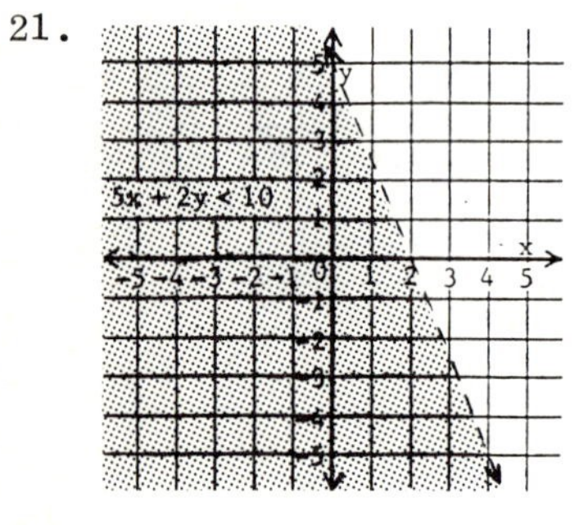

22.

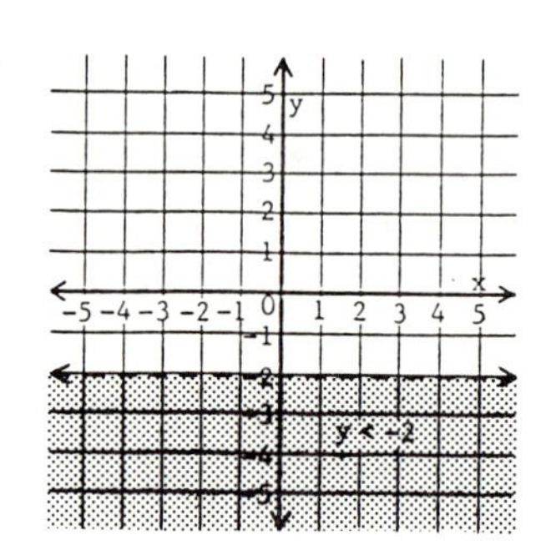

23.

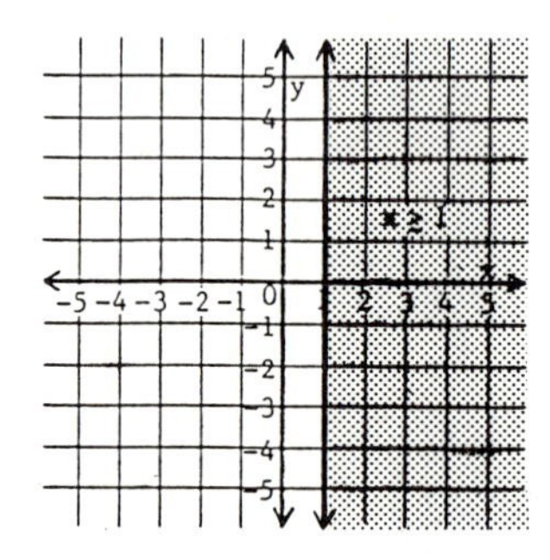

INDEX